Geotechnical, Geological and Earthquake Engineering

Volume 38

More information about this series at http://www.springer.com/series/6011

Ioannis Avramidis • Asimina Athanatopoulou
Konstantinos Morfidis • Anastasios Sextos
Agathoklis Giaralis

Eurocode-Compliant Seismic Analysis and Design of R/C Buildings

Concepts, Commentary and Worked Examples with Flowcharts

Springer

Ioannis Avramidis
Department of Civil Engineering
Aristotle University of Thessaloniki
Thessaloniki, Greece

Asimina Athanatopoulou
Department of Civil Engineering
Aristotle University of Thessaloniki
Thessaloniki, Greece

Konstantinos Morfidis
Institute of Engineering Seismology and Earthquake Engineering
Thessaloniki, Greece

Anastasios Sextos
Department of Civil Engineering
University of Bristol & Aristotle University of Thessaloniki
Bristol, UK & Thessaloniki, Greece

Agathoklis Giaralis
Department of Civil Engineering
City University London
London, UK

ISSN 1573-6059 ISSN 1872-4671 (electronic)
Geotechnical, Geological and Earthquake Engineering
ISBN 978-3-319-25269-8 ISBN 978-3-319-25270-4 (eBook)
DOI 10.1007/978-3-319-25270-4

Library of Congress Control Number: 2015940405

This book is an extended translation based on the original edition in Greek: "Αντισεισμικός Σχεδιασμός κτιρίων Ο/Σ και αριθμητικά παραδείγματα ανάλυσης και διαστασιολόγησης σύμφωνα με τους Ευρωκώδικες". Publisher: The authors (Ioannis Avramidis, Asimina Athanatopoulou, Konstantinos Morfidis, Anastasios Sextos), 2011; Distributor: Techartbooks (http://www.techartbooks.gr)

Springer Cham Heidelberg New York Dordrecht London

Printed on acid-free paper

Springer International Publishing AG Switzerland is part of Springer Science+Business Media (www.springer.com)

Preface

This book aims to serve as an essential reference to facilitate civil engineers involved in the design of new conventional (ordinary) reinforced concrete (r/c) buildings regulated by the current European Eurocode 8 or EC8 (EN 1998-1:2004) and EC2 (EN 1992-1-1:2004) codes of practice. It is addressed to practitioners working in consulting and designing engineering companies and to advanced undergraduate and postgraduate level civil engineering students attending modules and curricula in the earthquake-resistant design of structures and/or undertaking pertinent design projects. The book constitutes an updated and significantly extended version of a textbook co-authored by the first four authors published in 2011 in the Greek language. The changes and amendments incorporated into the current book discuss the recent trends in performance-based seismic design of structures and provide additional practical guidance on finite element modelling of r/c building structures for code-compliant seismic analysis methods.

It is emphasized that this book is neither a comprehensive text on the design of earthquake-resistant structures nor does it offer a complete commentary on the EC8 provisions. To this end, it presumes that the "user":

- Has sufficient knowledge of the fundamental concepts, principles, and methods of structural analysis for both static and dynamic loads pertinent to the earthquake-resistant design of structures and of r/c design
- Has access to and appreciation of the EC8 (EN 1998-1:2004) and EC2 (EN 1992-1-1:2004) codes of practice

The book is split notionally into two parts. The first part comprises the first three chapters in which:

- The fundamental principles for earthquake-resistant design are introduced and discussion and comments are included on how these principles reflect on the

current EC8 (EN 1998-1:2004) code and on several international guidelines for performance-based seismic design of structures (Chap. 1).
- Important practical aspects on the conceptual design, finite element modelling, analysis, and detailing of code-compliant earthquake-resistant r/c buildings are discussed and the relevant requirements prescribed by the EC8 (EN 1998-1:2004) provisions are critically commented upon (Chap. 2).
- All the required logic steps, computations, and verification checks for the design (seismic analysis and structural member detailing) of ordinary r/c buildings according to the EC8 (EN 1998-1:2004) and EC2 (EN 1992-1-1:2004) are presented in a sequential and methodological manner by means of self-contained flowcharts and additional explanatory comments (Chap. 3).

The second part of the book (Chap. 4) includes three numerical example problems, solved in detail, to illustrate the implementation of various clauses of the EC8 for the seismic analysis and design of three different multistorey buildings. The properties and structural layouts of the considered buildings are judicially chosen to achieve the necessary simplicity to serve as general benchmark structures while maintaining important features commonly encountered in real-life design scenarios. In this regard, these benchmark example problems provide for:

- A comprehensive illustration of complete and detailed numerical applications to gain a better appreciation of the flow and the sequence of the required logic and computational steps involved in the earthquake-resistant design of structures regulated by the EC8
- Verification tutorials to check the reliability of custom-made computer programs and of commercial finite element software developed/used for the design of earthquake-resistant r/c buildings complying with the EC8

The book is complemented by an Appendix discussing the inelastic static (pushover) analysis of the EC8 which is allowed to be used as an alternative method to the standard equivalent linear types of analysis for the design of EC8 compliant r/c building structures. In a second Appendix, the concepts of torsional sensitivity are delineated using analytical formulae and numerical examples. Lastly, to further facilitate practitioners, all requirements posed by both the EC2 (EN 1992-1-1:2004) and the EC8 (EN 1998-1:2004) codes regarding the detailing of r/c structural members are collected in a concise tabular/graphical format in a third Appendix. Notably, pertinent selected bibliography is included at the end of each chapter to direct the reader to appropriate sources discussing some of the herein introduced material in greater detail.

As a final note, we are thankful to Thanasis Spiliopoulos (School Buildings Organization SA, Athens, Greece) and to Prof. Ploutarchos Giannopoulos (National Technical University of Athens, Greece) who provided us with various photos from their personal archive and allowed us to include them in the book. We further acknowledge all colleagues of the academic community and practising engineers who contributed to the enhancement of the original manuscript through discussions, reviewing, and independent numerical verification checks.

Thessaloniki, Greece — Ioannis Avramidis
Thessaloniki, Greece — Asimina Athanatopoulou
Thessaloniki, Greece — Konstantinos Morfidis
Bristol, UK & Thessaloniki, Greece — Anastasios Sextos
London, UK — Agathoklis Giaralis
January 2015

Contents

Chapter 1
Fundamental Principles for the Design of Earthquake-Resistant Structures

Abstract This chapter provides a concise qualitative overview of the philosophy for earthquake resistant design of ordinary structures adopted by relevant international codes of practice, including Eurocode 8. The aim is to facilitate practicing engineers with the interpretation of the code-prescribed design objectives and requirements for the seismic design of ordinary reinforced concrete (r/c) building structures which allow for structural damage to occur for a nominal design seismic action specified in a probabilistic manner. In this regard, the structural properties of stiffness, strength, and ductility are introduced along with the standard capacity design rules and requirements. Further, the role of these structural properties in the seismic design of r/c building structures following a force-based approach in conjunction with equivalent linear analysis methods is explained. Emphasis is placed on delineating the concept of the behaviour factor, or force reduction factor, which regulates the intensity of the seismic design loads and ductility demands. Moreover, the development and current trends in the emerging performance-based design approach for earthquake resistance are briefly reviewed. Lastly, practical recommendations to achieve higher-than-the-minimum-required by current codes of practice structural performance within the force-based design approach are provided.

Keywords Seismic design objectives • Stiffness • Strength • Ductility • Capacity design • Force-based design • Performance-based design • Behaviour factor

Despite minor differences, current codes of practice and guidelines regulating the earthquake resistant design of structures share a common rationale in setting and achieving the requirements for structural performance under strong earthquake shaking. Developing a sufficient level of familiarity with this rationale, sometimes called the "philosophy of earthquake resistant design", is essential before embarking on conceptual design for earthquake resistance followed by the required structural analysis and detailing calculations prescribed by seismic codes of practice.

In this regard, this first chapter aims to provide the reader with a concise qualitative overview of the philosophy for earthquake resistant design as is currently implemented by codes of practice including Eurocode 8, hereafter EC8 (CEN

I. Avramidis et al., *Eurocode-Compliant Seismic Analysis and Design of R/C Buildings*, Geotechnical, Geological and Earthquake Engineering 38,
DOI 10.1007/978-3-319-25270-4_1

2004a). It further provides some recommendations as to how the current prescriptive regulations and requirements can be used to address the more recent trends in earthquake resistant design towards a performance-based approach. In this respect, this chapter forms a basis upon which subsequent chapters focusing exclusively on EC8 builds.

In particular, Sect. 1.1 provides a brief introduction on the rationale of the fundamental design objectives and requirements set by current codes of practice. Section 1.2 introduces the common force-based seismic design approach adopted by codes of practice to achieve the sought requirements for earthquake resistance. Next, Sect. 1.3 provides a brief overview on the development and current trends in the emerging performance-based design approach for earthquake resistance. Lastly, Sect. 1.4 lists practical recommendations to achieve higher-than-the-minimum-required structural performance levels as prescribed by current codes of practice within the traditional force-based design framework.

1.1 Partial Protection Against Structural Damage as the Underlying Design Philosophy for Earthquake Resistance

1.1.1 The Uncertain Nature of the Seismic Action

The uniqueness of the earthquake induced (seismic) action compared to other actions, such as gravitational live loads, which building structures must resist during their lifetime, stems from the following facts:

1. Many important parameters of the seismic action affecting the structural response are *inherently strongly uncertain* such as (Elnashai and Di Sarno 2008):
   - the earthquake magnitude (related to the energy released along the seismic fault),
   - the focal (hypo-central) depth of the earthquake,
   - the distance of the structure site from the source or the epicenter,
   - the directivity, the frequency content, and the duration of the earthquake induced ground motion at the foundation of the structure.
2. The probability of a certain structure being exposed to "extreme" seismic ground motion accelerations (and to consequent "extreme" lateral inertial forces) during its conventional lifetime (typically estimated to be 50 years) is relatively low due to the considerable geographic (spatial) dispersion of high seismic intensities. In other words, a severe (destructive) earthquake is considered to be a "rare" event which will, most likely, affect a relatively small percentage of the structural stock of a region or of a country. Still, its potential consequences to the built environment may be too high to be neglected.

1.1.2 *Can an "Absolute" Level of Protection Against the Seismic Hazard Be Achieved?*

An important implication of the *uncertain* nature of the seismic action is that the design and construction of structures that are "*seismically invulnerable*" under *any* future earthquake is not practically feasible since the location, the time instant, and the level of the earthquake-induced demand on the structures cannot be deterministically defined.

However, the expected level of earthquake ground motion can be quantified by the *seismic hazard* in a given geographical area, which naturally, is expressed in a *statistical/probabilistic* sense (McGuire 1995). To this end, the seismic intensity that the structure is designed for according to current seismic codes of practice and design guidelines (most commonly denoted as the "design earthquake") is typically defined in terms of the probability to be exceeded within a specific time interval (e.g., 10 % probability to be exceeded in 50 years).

Further, the uncertain nature of the earthquake induced action on structures necessitates making a critical *decision* regarding the *desired* level of protection against structural damage. This is achieved by defining the *minimum* level of protection or else, a minimum *performance*, in relation to prescribed specific levels of seismic action, an issue that is further discussed in the following section.

1.1.3 *Full and Partial Protection Against Structural Damage for a Given Design Seismic Action*

Let us assume that, based on appropriate seismological studies, the peak seismic hazard of a certain region is accurately mapped and that the earthquake scenarios corresponding to the "design earthquake" are determined for all structures within this region in a statistical/probabilistic context. In deciding the sought level of protection against structural damage for the above design earthquake, the perspective of different stakeholders need to be examined separately as explained below.

1.1.3.1 Regulatory Agencies and State Governments

From the authorities' viewpoint, a requirement to design *all* structures to *remain elastic (i.e., undamaged)* under the design ("rare") earthquake, and therefore to ensure full protection against structural damage for the design seismic action, is considered to be economically prohibitive. Such a decision would involve channeling excessive financial resources to address a relatively low risk in terms of casualties compared to risks associated with other critical public functions (e.g., traffic safety). Further, it is believed that in many cases, mandating such a stringent level of structural safety may lead to "bulky" and aesthetically displeasing structures of reduced architectural functionality.

For these reasons, State authorities and regulatory agencies set forth *minimum requirements* for the seismic design of structures described in the relevant building codes of practice and guidelines which aim to compromise social welfare (i.e., life and property safety) given reasonably limited financial resources. This is achieved by adopting an acceptable level of protection against the seismic hazard, both in social/psychological terms and in cost-effectiveness terms. To this aim, structures are commonly classified into two categories: (a) "special" structures, and (b) "ordinary" structures.

For "special" structures (e.g., nuclear reactor complexes and petrochemical facilities housing poisonous gas and liquid material), that is, for structures whose potential damage, downtime, or even global instability and collapse, would negatively affect large populations and the environment of large geographical regions (Fig. 1.1), a *zero structural damage* requirement is prescribed. In other words, it is required that these structures subjected to the design earthquake behave linearly and hence achieve full protection against structural damage for the design seismic action.

However, the requirement for full protection against the seismic hazard is relaxed for the case of "ordinary" structures (e.g., residential, retail, and office buildings, schools, hospitals, etc.), that is, for structures whose potential damage, downtime, or even global instability/collapse, would affect inhabitants and the environment within their immediate vicinity only, despite their -maybe- different importance (Fig. 1.2). Such ordinary structures are allowed to exhibit a certain level of inelastic behaviour and even to sustain irreversible plastic deformations (structural damage) under the design earthquake, which should not, however, lead to partial or global structural instability/collapse ("*partial*" protection against structural damage for the design seismic action).

Therefore, structural damage under the design earthquake is considered acceptable by State authorities for ordinary structures as long as the life of the occupants/users of structures is not endangered (life safety requirement). In this manner, a *minimum allowed* level of protection from the seismic hazard is set which, presumably, constitutes a socially and economically acceptable compromise in prioritizing limited available resources to meet public needs. In this regard, structural engineers and infrastructure owners should be aware that the required level of protection against the seismic hazard prescribed by seismic codes of practice constitutes solely the lowest permissible limit (lower bound) of structural safety and that there is no restriction in choosing to design a particular structure to achieve a higher level of protection or of seismic performance, if so desired.

1.1.3.2 Building Owners

From the infrastructure owner's viewpoint, a requirement to achieve higher than the minimum level of structural safety against a nominal "design earthquake" set by

Fig. 1.1 Special structures for which zero structural damage requirement is specified

Fig. 1.2 Global instability (collapse) of an ordinary multistorey residential building under a strong earthquake affecting the neighboring structures

codes of practice is, by all means, legitimate. Of course, as the choice of a higher level of safety against the minimum required by seismic codes is at the discretion of the owner, the additional required construction cost is covered by himself and not by the State. It has to be noted at this point that, as will be discussed in more detail in Sect. 1.4, the additional cost of full earthquake protection is (i) not as high as is often deemed, and (ii) does not necessarily lead to undesirably large structural member sections and dysfunctional buildings.

1.1.4 Design Objectives and Requirements for Partial Protection Against Structural Damage in Current Seismic Codes of Practice

Allowing for structural damage to occur for a certain level of "design seismic action", lies at the core of all modern seismic design codes of practice for ordinary structures. In this regard, code-compliant seismic design for ordinary structures, as defined in Sect. 1.1.3.1, requires a partial protection against structural damage. The practical implementation of this seismic design philosophy can be qualitatively framed via the three *fundamental structural design objectives for earthquake resistance,* as prescribed in early seismic codes (see e.g., the commentary of the Structural Engineers Association of California Blue Book (SEAOC 1967) which introduced the general philosophy of the earthquake resistant design of buildings that is still conceptually valid today). Each design objective can be implicitly associated with a specific limit state as follows:

1. Structures should withstand minor levels of earthquake induced ground motion without any damage to structural and to non-structural members. This design objective sets a *no damage requirement for frequently occurring earthquakes during the lifetime of structures and* corresponds to the "serviceability" limit state.
2. Structures should withstand moderate levels of earthquake induced ground motion with negligible (insignificant and readily repairable if deemed essential for aesthetical reasons) damage to structural members. Damage to non-structural members may occur (e.g., Fig. 1.3). This design objective sets a *damage limitation requirement for occasionally occurring earthquakes during the lifetime of structures.* It defines a "quasi" limit state lying in between the serviceability and the ultimate limit states which may be viewed as a "cost" limit state.
3. Structures should withstand major levels of earthquake induced ground motion without collapsing. Damage to structural and non-structural members is acceptable as long as it is not life threatening and the probability of partial or global collapse is sufficiently low (e.g., Fig. 1.4). This design objective sets a *no collapse requirement for rare earthquakes with relatively low probability of occurrence during the lifetime of structures.* In reference to modern seismic codes (i.e., ASCE 2010), it can be further approximately mapped onto a dual requirement of "life safety" and collapse prevention, and corresponds to the "ultimate" limit state.

A fourth design objective can be further added for the case of "important" structures, whose unobstructed operation is essential in the aftermath of a major seismic event (e.g., hospitals, conventional power plants, etc.) or whose collapse entails significant social, economic, or cultural loss (e.g., schools, museums, etc.), as follows:

4. Essential and large occupancy structures should withstand major levels of earthquake induced ground motion with minor or insignificant damage to structural members. In other words, more stringent requirements from the "no-collapse" requirement should be observed.

Note that the above (1)–(3) design objectives and related requirements are only qualitatively defined. That is, neither the limiting values of the "minor", "moderate", and "major" earthquake shaking, nor the permissible levels of damage corresponding to each of the above levels of the input seismic action are defined in an explicit quantitative manner. In fact, conventional design of ordinary

Fig. 1.3 Requirement for moderate level earthquakes: damage to non-structural in-fill walls is acceptable, yet only minor, repairable damage to structural elements is permitted

Fig. 1.4 Requirement for major level earthquakes: damage to structural elements is accepted, yet collapse probability must remain sufficiently low

structures for earthquake resistance involves the consideration of only one seismic-action-level ("design seismic action") corresponding to a rare "design earthquake", typically defined as the one having a 10 % probability of being exceeded in 50 years, that is, having a return period of approximately 475 years (see also Sect. 2.3.1). Next, a series of qualitative conceptual design rules, prescriptive verification checks, and empirical local detailing requirements are considered to ensure that the first three (1)–(3) performance objectives are met. In this respect, *current (conventional) seismic design practices for ordinary structures focus on the no collapse requirement to ensure life protection for a nominal "design seismic action"*. Consequently, it is natural to expect that the structural properties of the thus designed structures and, hence, their achieved level of structural safety exhibit significant variability.

Despite their qualitative nature, the above design objectives illustrate the current consensus of what is considered to be the acceptable (by Governments and Regulatory Agencies, but not necessarily by individual infrastructure owners) minimum requirements of structural performance in seismically prone areas and are commonly adopted, with minor variations, by most modern seismic codes of practice.

In this respect, EC8, in particular, prescribes the following two fundamental requirements to be satisfied by ordinary structures constructed in seismic regions with an *adequate degree of reliability* (clause 2.1 of EC8)

(a) **No collapse requirement**
"The structure shall be designed and constructed to withstand the design seismic action defined in Section 3 without local or global collapse, thus retaining its structural integrity and a residual load bearing capacity after the seismic events."
(b) **Damage limitation requirement**
"The structure shall be designed and constructed to withstand a seismic action having a larger probability of occurrence than the design seismic action, without the occurrence of damage and the associated limitations of use, the costs of which would be disproportionately high in comparison with the costs of the structure itself."

The above two requirements are assumed to cover the three first seismic design objectives (1)–(3). Further, in clause 2.1 of the EC8, the design objective (4) is also covered, by requiring that an enhanced "level of reliability" in meeting the above requirements is achieved depending on a classification of structures according to their "importance" to communities. This increased level of reliability is accomplished by increasing the return period of the design seismic action, that is, by increasing the considered design seismic loads (see also Sect. 2.3.1.1).

1.1.5 Stiffness, Strength, and Ductility: The Key Structural Properties in Earthquake Resistant Design

In general, the fundamental seismic design objectives discussed in the previous section are met by *judicially equipping structures with adequate and appropriately distributed (in plan and in elevation) stiffness, strength, and ductility* (Villaverde 2009). Specifically:

- *An adequate level of stiffness* for the lateral load-resisting structural system is required such that (i) under moderate earthquake shaking, the structure exhibits small deformations and remains elastic (no damage to structural members occurs), and (ii) under severe earthquake shaking (design seismic action), lateral deflections are sufficiently small rendering the contribution of second order phenomena negligible (see further Sect. 2.4.3.3). Furthermore, *the distribution of stiffness should be sufficiently uniform* to avoid significant localized stress accumulation at critical regions of structural members.
- *An adequate level of strength* for structural members of the lateral load-resisting system is required, so that only insignificant, if any, damage is observed under moderate earthquake shaking. Furthermore, the *distribution of strength should*

be sufficiently uniform to avoid large differences between the observed demand-capacity ratios across structural members (see further Sect. 2.4.4.1).

- *An adequate level of ductility capacity is required (herein, "ductility capacity" refers to the ability of a structure to undergo large inelastic deformations without significant reduction of its stiffness and strength during repetitive dynamic loading-unloading-reloading cycles)*. In such a case, local damage at structural members induced by severe earthquake shaking (design seismic action) exhibit a non-brittle (ductile) behaviour, and, thus, they do not lead to a premature global structural instability/collapse. Furthermore, a *proper distribution of ductility* within the lateral load-resisting structural system is required to avoid the formation of kinematic collapse mechanisms under severe earthquake shaking (see further Sect. 2.2.4).

The association of the (minimum) seismic design objectives with the above key structural properties is summarized in Table 1.1.

Focusing on reinforced concrete (r/c) structures, it is noted that certain geometric and/or material parameters may affect the value of more than one of the above structural properties. Thus, it may not always be possible to modify these properties independently at will. For example, increasing the longitudinal reinforcement provided to an r/c beam without changing its cross-sectional dimensions increases its flexural strength (yielding moment) as well as its (cracked and uncracked) flexural stiffness. However, despite this fact, seismic codes typically allow for using reinforcement-independent flexural and shear stiffness values in the (force-based) design and analysis of buildings (see Sect. 2.3.2.1).

For seismic design purposes, it is instructive to represent lateral load resisting structural systems of buildings in terms of total static lateral load (base shear) versus top storey building lateral sway graphs (e.g., Fig. 1.5). This relationship is commonly referred to as *capacity curve* and is an important proxy of the nonlinear response of the structure. It is noted that the storey shears are applied according to the corresponding profile of the modal forces along the direction of interest.

It is also recalled that the slope of the initial (pre-yielding) branch of these plots is proportional to the (lateral) stiffness of the structural system and the peak attained value of the base shear can be interpreted as the strength of the structure. Further, for static loads, ductility is commonly quantified as the ratio u_{tot}/u_y of a nominal peak elastoplastic displacement u_{tot} (e.g., translation, deflection, rotation, curvature, etc.) signifying the initiation of collapse/instability over a nominal yield displacement u_y signifying the initiation of nonlinear material behaviour (yielding point). It is noted in passing, that the above ratio might be in some cases a quite misleading proxy of the seismic energy absorption. To illustrate this issue, let a structure be characterized by ultimate and yield displacements u_{tot-1} and u_{y-1}, and another one by $u_{tot-2} = 2u_{tot-1}$ and $u_{y-2} = 2u_{y-1}$, respectively; both structures present the same nominal ductility despite the fact that the latter structure may obviously absorb more seismic energy in absolute terms (see further discussion in Sect. 1.2.1).

The capacity curve, as defined above, facilitates the interpretation and quantification of stiffness, strength, and ductility properties of structures at the system level

Table 1.1 Mapping of minimum design objectives for key structural properties for different levels of earthquake shaking severity following the partial protection to structural damage philosophy (see also Sect. 1.3.2)

Level of earthquake shaking (return period)	Desirable structural properties	Minimum required design objectives
Frequent/minor (~45 years)	Adequate and uniformly distributed **stiffness**	No damage (serviceability limit state)
Occasional/moderate (~75 years)	. . . plus adequate and properly distributed **strength**	Fairly limited readily repairable damage, if required. (cost limit state)
Rare/major "Design earthquake" (~475 years)	. . . plus adequate and properly distributed **ductility**	Extensive damage are acceptable, without local or global collapse (ultimate limit state-life protection)

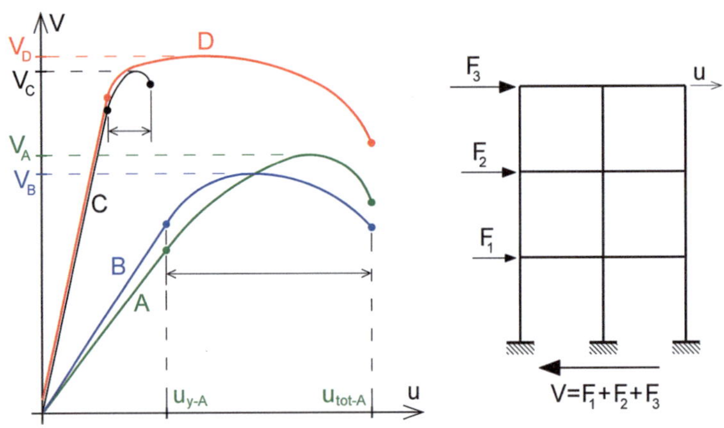

Fig. 1.5 Capacity curves of structures with different stiffness, strength, and ductility properties

(reflecting the effect of multiple damage mechanisms at the local level). It also renders possible the comparative assessment of different structural systems at initial seismic design stages. To further illustrate this point, four base shear-top storey displacement graphs A, B, C, and D corresponding to four different structural systems are considered in Fig 1.5.

Systems A and B have different stiffness (i.e., slope inclination of the elastic branch) and also different strength ($V_A \neq V_B$), yet the same ductility ratio u_{tot-A}/u_{y-A}. System C is stiffer and has greater strength compared to systems A and B ($F_C > F_A$, F_B), but is significantly less ductile (i.e., more brittle). Finally, system D has the same stiffness as system C, but is significantly more ductile and, thus, more capable of resisting seismic input loads exceeding the strength of the structure as will be explained in detail in subsequent sections.

It has to be noted herein that horizontal earthquake ground motions induce lateral inertial loads to structures that are essentially time-variant (*dynamic*) and their attributes (e.g., intensity, distribution, etc.) depend on the dynamic characteristics of the structure itself (e.g., natural frequencies, mode shapes, etc.). Since ordinary structures are mostly designed (and, thus, are expected) to exhibit *nonlinear inelastic* behaviour under the nominal "design seismic action", the seismic response of (yielding) structures is, ultimately, an inherently nonlinear dynamic problem.

In the remainder of this section, certain practical implications due to the nonlinear dynamic nature of the problem at hand are discussed from a design viewpoint. The interested reader is referred to standard texts in the field of earthquake engineering and structural dynamics for a more elaborate treatment of the following topics (Chopra 2007). It is further noted in passing that, apart from the aforementioned structural properties, the choice of the lateral load-resisting structural system along with compliance with certain conceptual design rules are of paramount importance in achieving the design objectives of Table 1.1 (see also Sect. 2.1).

1.1.5.1 Dependency of Input Seismic Loads on Structural Properties

Important differences between static gravitational loads vis-à-vis seismic loads considered in the design of r/c buildings are discussed below.

Within the range of linear elastic structural behaviour (prior to yielding):

- The distribution and intensity of *static (gravitational) design live loads* are prescribed by standard building codes based on the purpose/usage of structures (e.g., standard occupancy residential buildings, heavy occupancy public buildings, etc.). These loads are constant from the outset of the design process and do not depend on the properties of the structural load-resisting system. *Thus, increasing the stiffness of a building would not alter the design static live loading.* It would only increase the self-weight (dead load) of the structure, as stiffening is commonly accomplished by increasing the dimensions of structural members or by adding new ones. It is further reminded that, for a given level of live loads, a stiffer structural design generally achieves reduced displacements and strains in structural members.
- The *design seismic loading* depends not only on the site-specific characteristics of the design seismic action as defined by codes of practice (e.g., site seismic hazard, local soil conditions, etc.), but also on the dynamic (modal) characteristics of the lateral load-resisting structural system of the building under design (e.g., natural frequencies, mode shapes, etc.). These characteristics depend, in turn, on the stiffness, mass (inertia), and damping properties of the structure and their distribution within the lateral load-resisting system. *Thus, increasing the lateral stiffness of a building would generally (though not always) alter (increase or decrease) the design seismic load that the structure must be able to resist.* For example, increasing the stiffness of a rather flexible structure subjected to a given design seismic load leads, most probably, to higher levels of stresses at structural

members while it has a mixed effect on strains. In this regard, two adjacent buildings of the same geometry and dead load but with different lateral load-resisting structural systems (e.g., one with a relatively flexible moment resisting frame (MRF) system vis-à-vis one with a stiffer coupled MRF-shear wall system) subjected to the same strong ground motion will develop significantly different base shear and, therefore, internal forces at structural members due to their different dynamic characteristics with respect to those of the earthquake ground motion.

Within the range of nonlinear inelastic structural behaviour (post-yielding):

– Static (design) gravitational loads applied to a building remain constant up to local and/or global structural instability (building collapse). Although second order effects may develop at large displacements and various force distributions may take place, static loads themselves do not vary due to inelastic behaviour.
– The intensity and distribution of seismic loads applied to structures change continuously beyond yielding. Stiffness is gradually reduced both at the member and the system level; however, the intensity of the seismic loading in light of structural nonlinear response may vary depending on the interplay between the (modified) structural dynamic characteristics and the frequency content of seismic motion.

  Further details on the influence of inelastic structural behaviour on the input seismic loads from the viewpoint of earthquake resistance design are provided in Sects. 1.2.1 and 1.2.2.

1.1.5.2 Structural Properties Influencing the Design for Earthquake Resistance

The total mass (inertia) of ordinary r/c buildings depends mainly on the outer dimensions (envelop) of the structure in plan and in elevation and on its intended usage (e.g., standard occupancy, heavy occupancy, etc.). Therefore, although ensuring a favourable (i.e., uniform) mass distribution within the lateral load-resisting structural system of a building is an important consideration in designing for earthquake resistance (see also Sect. 1.2.5), the total mass of ordinary r/c buildings is not a property that can be significantly influenced at the seismic design stage.

Further, the intensity of the damping forces, commonly assumed to be velocity proportional (viscous damping model), depends on the structural material of choice for the lateral load-resisting system. In code-compliant seismic design of r/c structures, a 5 % of critical viscous damping ratio for all modes of vibration is the usual assumption. Therefore, similar to the case of the total structural mass, the intensity of damping forces resisting the seismically induced vibratory motion of structures is not a parameter that can be controlled at the seismic design stage, unless supplemental damping is introduced by means of special energy dissipation devices. However, the use of such devices falls within the non-conventional design approaches for earthquake resistance, and as such, it is not addressed in this book.

In this respect, stiffness, strength, and ductility of structural members are the main structural properties leveraging the earthquake resistant design of ordinary r/c building structures (Lindeburg and McMullin 2014). To summarize:

- *stiffness*, mass, and (viscous) *damping* properties determine the natural frequencies of vibration of linear structures (prior to yielding), which, in turn, determine the intensity of the design seismic input loads.
- for design purposes, *strength* is assumed to coincide with the elastic limit of structural behaviour above which the structure suffers *permanent* plastic strains and, thus, structural damage. Such damage occurs locally at specific critical regions of structural members where the flexural capacity of the section is exceeded, thus forming a series of "*plastic hinges*". Similarly, the shear capacity of a section (or a beam-column joint) may also be exceeded leading to unfavourable modes of brittle failure. However, special capacity design rules are enforced in design to minimize the probability of the latter failure modes which will be further discussed in Sect. 1.2.6. Large parts of the input seismic (kinetic) energy can be absorbed at plastic hinges via hysteretic (inelastic) mechanisms provided that sufficient *ductile behaviour* is exhibited. That is, no premature failure takes place during strong ground shaking either locally, at plastic hinges, or globally leading to structural collapse.
- *ductility* or ductile behaviour can be viewed as a means to dissipate the input seismic energy through inelastic/hysteretic mechanisms of structural behaviour. These mechanisms are activated by allowing the structure to yield in a controlled manner (i.e., without leading to global instability/ collapse) and, thus, by allowing the occurrence of structural damage under the design seismic action. Conveniently, the latter consideration is in alignment with the adoption of partial protection against structural damage philosophy of earthquake resistant design. In fact, it is the ductility capacity property of structures that renders the partial protection against structural damage philosophy practically possible and historically acceptable.

1.1.5.3 The Role of Ductility in Seismic Design

The role of ductility, that is, the ability of the lateral load-resisting structural system to exhibit large plastic strains with no significant stiffness and strength degradation during a large number of repetitive dynamic loading cycles, both at the system and the component level, is crucial in designing for earthquake resistance (Fardis et al. 2005; Elghazouli 2009; Bisch et al. 2012).

Specifically, in case a structure is designed to resist the design seismic action by developing inelastic/hysteretic energy dissipation mechanisms (i.e., by allowing for structural damage to occur – partial protection against structural damage for the design earthquake), ensuring adequate and reliable ductile behaviour (ductility capacity) via appropriate local detailing and conceptual design considerations is of *paramount importance* (Booth 2012).

On the antipode, in the case of a structure designed to resist the design seismic action through linear behaviour on a strength-based design (full protection against structural damage for the design earthquake), no special measures for ductile behaviour are needed to resist the design earthquake. This may even be the case for earthquake shaking levels beyond the design seismic action since a certain level of *inherent* ductility capacity is always existent in r/c structures due to internal section mechanisms able to absorb seismic energy. Apparently, for significantly higher levels of seismic force, both the inherent and the additional ductility, provided through capacity design and appropriate detailing, shall be mobilized, hence the latter is deemed necessary to minimize the probability of collapse (see also Sect. 1.2.6).

1.1.5.4 The Use of Reduced "Effective" Stiffness Properties for R/C Structures

In the case of r/c structures, partial protection against structural damage entails that significant concrete cracking occurs at certain regions of structural members (local structural damage). Thus, the mechanical properties of structural members, and especially their stiffness (in flexure, shear, tension/compression and torsion), deteriorate compared to the case of an "intact" structure with "uncracked" members. The determination of this level of deterioration and its influence on the structural member stiffness values to be used in seismic analysis in order to achieve realistic results is an active area of research. Current codes of practice are taking into account this influence in undertaking "equivalent linear" types of analyses by considering reduced ("effective") flexural and shear stiffness properties at structural members compared to the uncracked values (see also Sect. 2.3.2.1)

1.2 Implementation of the Partial Protection Against Structural Damage Seismic Design Philosophy in Current Codes of Practice

The three fundamental seismic design objectives (1)–(3) of Sect. 1.1.4 correspond to three different levels of seismic action associated with "frequent/minor", "occasional/moderate", and "rare/major" earthquake events. Accordingly, one would expect the design of ordinary structures to involve verification checks using three separate sets of structural analysis results corresponding to the above seismic action levels. Such a rigorous design practice would require explicit quantification of three distinct levels of the seismic action at each site along with undertaking three separate structural analyses of, perhaps, different type (linear elastic, nonlinear).

Further, it would also require the prescription of appropriate design verification checks and corresponding permissible criteria for each of these seismic action levels.

However, the current consensus is that such an explicit three-seismic-action-level design procedure would be too cumbersome and not readily applicable in practice. Thus, with the exception of very few current codes of practice (e.g., the Japanese code, (Midorikawa et al. 2003)), code-compliant seismic design of ordinary structures involves one analysis for a single-seismic-action-level ("design seismic action") corresponding to the "no collapse" requirement. The bulk of design verification checks, typically concerned with the ultimate strength of structural members, is performed for this single structural analysis run to ensure that the life safety requirement is met.

Further, special "capacity design" rules are prescribed to achieve a sufficient level of global ductile behaviour (see Sects. 1.2.5 and 1.2.6), along with local detailing provisions to ensure local ductility capacity at critical regions of structural members. It is implicitly assumed that capacity design provisions ensure the no-collapse requirement for higher-than-the-nominal-"design seismic action"-level earthquake ground motions without performing any additional quantitative assessment.

Finally, certain verification checks corresponding to the "damage limitation" requirement are also undertaken involving structural deflections and relative deformations. These structural response quantities are determined without performing structural analysis for an additional input seismic action level lower-than-the-nominal-"design seismic action". Instead, empirical reduction factors are applied to numerical results obtained from the analysis for the design seismic action level to implicitly account for the fact that damage limitation requirements correspond to a "moderate" earthquake event. In this respect, the above code-compliant seismic design framework, which is closely followed by EC8 among other international seismic codes, may not be characterized as a "full-fledged" two-seismic-action-level design procedure. However, it does constitute a "quasi" *two-tier seismic design procedure*, as it includes verification checks for two different levels of the seismic action (Fardis et al. 2005; Fardis 2009).

It is further noted that *a force-based design approach* is commonly adopted in conjunction with the above framework. Consequently, the verification checks for the no collapse requirement do not involve an explicit quantitative assessment for structural damage assumed to occur under the design seismic action (Fardis 2009). For the case of reinforced concrete (r/c) buildings, structural damage entails the formation of localised "plastic hinges" at certain "critical regions" of r/c structural members. Conveniently, the design verification checks specified by the herein described design framework do not require determining the total number of plastic hinges, their sequence of occurrence, and their overall distribution within the structure. Further, no assessment of the severity of inelastic deformations at each plastic hinge is mandated.

In summary, *the force-based design approach for earthquake resistance commonly adopted by most of the current seismic codes relies on performing linear types of analysis for a single-seismic-action-level allowing for structural damage to occur implicitly without any special provision to quantify the actual severity of this damage*. At the core of this approach lies the practice of defining the level of design seismic action by means of a "design response spectrum" of reduced coordinates. This issue is further discussed in Sect. 1.2.4 upon reviewing, in some detail, the concepts of *ductility capacity*, *ductility demand*, *force reduction factor (behaviour factor)*, and *overstrength*.

1.2.1 Ductility Demand and Ductility Capacity

Allowing for a structure to sustain damage under the "design seismic action" implies that, under this action, the structure exhibits significant inelastic deformations beyond its yielding deformation u_y without collapsing (ductile behaviour). This "demand" for inelastic behaviour posed by the design seismic action is illustrated in Fig. 1.6 by the red line which plots the seismic input load versus the exhibited structural deflection assuming a structure with elastic-perfectly plastic behaviour under dynamic cyclic loading. On the same figure, a second plot is included (green line) corresponding to a structure of equal stiffness which is designed to remain linear (no structural damage occurs) under the same design seismic action, assuming that damping is negligible.

Compared to the undamped structure exhibiting linear-elastic behavior, the inelastic structure absorbs a significant amount of the kinetic seismic input energy at each dynamic response cycle (loading-unloading-reloading) represented by the area of the observed hysteretic loops (*note:* actual dynamically excited linear structures possess inherent damping properties and, thus, they do dissipate a portion of the seismic input energy during each dynamic response cycle without yielding).

There are *two key considerations* to achieve a design that utilizes the capacity of a ductile structure to dissipate the seismic input energy by exhibiting hysteretic/inelastic behaviour for a set design seismic action. *The first one is that the structure is designed for a seismic load* F_d *significantly lower than the load* $\max F_{el}$ *that the structure would have to be designed for to remain linear*. For example, in the particular case of Fig. 1.6, the inelastic structure is assumed to be designed for $F_d = F_y$, where F_y is the yielding strength of the structure. In general, the reduction of the seismic design loads in the context of earthquake resistant design of yielding structures is commonly quantified by the so-called *force reduction factor*:

$$R = \max F_{el}/F_d. \tag{1.1}$$

Depending on the assumed reduced design seismic loads ($R > 1$) compared to a linear design ($R = 1$), a more or less substantial reduction to the initial construction cost of the lateral load-resisting structural system might be achieved. For instance, in the case of r/c buildings, assuming a force reduction factor significantly larger than unity would result in considerable savings in longitudinal reinforcement for

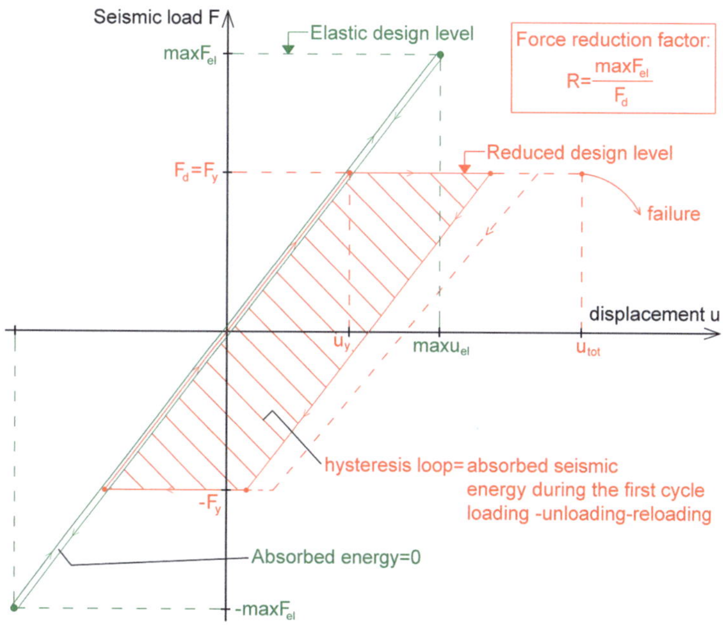

Fig. 1.6 Elastic-perfectly plastic system (*red line*) and corresponding elastic system (*green line*) under the design earthquake (load-unload-reload diagram)

the same lateral stiffness, i.e., fixed dimensions of structural r/c members comprising the lateral load-resisting system of choice.

However, as already noted above, structural designs for $R > 1$ rely on the available *capacity* of structures to behave in a ductile manner, that is, on their *ductility capacity*. The latter constitutes *the second key consideration in the earthquake resistant design of yielding structures: structures need to be conceptually designed and detailed such that their ductility capacity is larger than the ductility demand posed by the design seismic action.* In this regard, it is important to distinguish between *ductility demand* μ_{dem} and *ductility capacity* μ_{cap} of a structure (or, similarly, of a cross-section, or of a structural member):

– *Ductility demand* μ_{dem} *is the peak ductility (peak deformation/yielding deformation) that a yielding structure will exhibit under a specific earthquake induced strong ground motion without any partial or global collapse*. In other words, this is the ductility "demanded" by the particular strong ground motion (seismic action) to avoid failure. In this regard, ductility demand depends not only on the properties of the structure, but also on the characteristics of the considered strong ground motion (e.g., peak ground acceleration, duration, frequency content, etc.).
– In case a structure exhibits inelastic behaviour without failing under a specific earthquake ground motion, it can be stated that the ductility demand μ_{dem} posed

by this particular ground motion to the structure is smaller than the ductility capacity μ_{cap} of the structure. However, it is possible that the same structure fails under a ground motion of different characteristics (even if the peak ground acceleration remains the same). In the latter case, $\mu_{dem} > \mu_{cap}$, that is, the ductility capacity of the structure is not sufficient to meet the earthquake ductility demand. The latter implies that the actual performance of the structure, that is, the ratio between the *ductility capacity and demand, depends on both the properties of the yielding structure exhibiting strongly non-linear behaviour and on the characteristics of the input seismic load* (e.g., frequency content, duration, etc.). To further elucidate this point, consider a particular framed lateral load-resisting structural system designed to develop ductile plastic hinges at the ends of beams (primarily) and columns (well beyond the onset of structural yielding) driven to a state of collapse (mechanism) under a strong ground motion. It can be intuitively argued that the activation of its reserved (capacity) ductility in terms of the number of plastic hinges developed to form the particular collapse mechanism, their location, and the energy dissipation at each of these plastic hinges will depend on the characteristics of the induced strong ground motion. It is noted, however, that the conventional approach in defining the ductility capacity does not account for the dynamic nature of the seismic input action. It usually considers the lateral force-deformation capacity curves of yielding structures under statically applied incrementally increasing external loads as shown in Fig. 1.5. Further details on the structural properties influencing the ductility capacity of a structure are provided in Sect. 2.2.2.

An important conclusion from the above discussion on ductility demand and ductility capacity is that a ductile structure designed to yield under the design seismic action ($R > 1$) should have a ductility capacity μ_{cap} larger or at least equal to the ductility demand μ_{dem} imposed by the design seismic action to prevent collapse.

1.2.2 *The "Interplay" Between Ductility Capacity and Force Reduction or Behaviour Factor*

To gain further insight into the relation between the concept of ductility capacity and the use of reduced design seismic loads adopting a force reduction factor $R > 1$ (or behaviour factor $q > 1$), consider a yielding structure exhibiting an ideal elastic-perfectly plastic force-deformation relationship under a lateral statically applied and incrementally increasing external "seismic" load (red line in Fig. 1.7). Conveniently, for the considered structure, the displacement ductility capacity can be readily defined as the ratio of the peak displacement u_{tot} at which the structure fails/collapses over the yielding displacement u_y. That is,

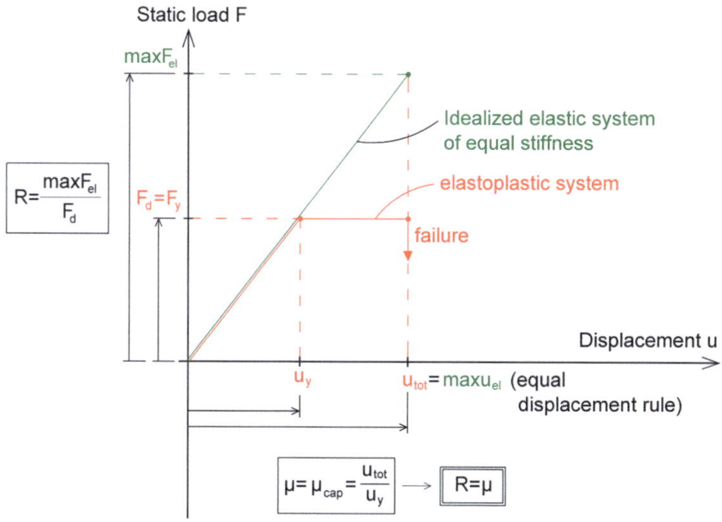

Fig. 1.7 Idealized elastic-perfectly plastic system with $T \geq T_c$ (assumption of equal displacement rule) – Available ductility $\mu = \mu_{avail}$ and force reduction factor R

$$\mu_{cap}=u_{tot}/u_y. \tag{1.2}$$

The peak lateral "seismic" design load (base shear) $maxF_d$ that can be undertaken by the assumed elastic-perfectly plastic structure is equal to its (yielding) strength F_y. This base shear remains constant and equal to F_y until the elastoplastic deformation reaches the maximum value u_{tot} at which the structure fails/collapses.

Further, consider an idealized linear elastic structure with stiffness equal to the initial (pre-yield) stiffness of the considered non-linear structure (green line in Fig. 1.7). The peak lateral load undertaken by this linear structure when it exhibits an (elastic) deformation $maxu_{el}$ equal to the maximum elastoplastic deformation u_{tot} (equal displacements rule $maxu_{el} = u_{tot}$) is equal to $maxF_{el}$. Under this assumption, the underlying force reduction factor R of the design lateral "seismic" load for the non-linear structure is given by the expression

$$R = \max F_{el}/F_d = \max F_{el}/F_y. \tag{1.3}$$

Moreover, geometric considerations (similarity of triangles) suggest that

$$max\,F_{el}/F_y = max\,u_{el}/u_y = u_{tot}/u_y. \tag{1.4}$$

Thus, by using Eqs. (1.2), (1.3) and (1.4), the following relationship between the force reduction factor and the ductility capacity is reached

$$R = \mu_{cap}. \tag{1.5}$$

In terms of terminology, the EC8 (§1.5.2(1)) uses the term *"behaviour factor"* denoted by the symbol q for the force reduction factor R (Eq. (1.1)). It is further noted that Eq. (1.5) holds only under the assumption of equal peak inelastic and corresponding peak elastic displacements ($\max u_{el} = u_{tot}$). Veletsos and Newmark (1964) provided numerical data involving response history analyses for a number of recorded accelerograms, suggesting that Eq. (1.5) is a valid approximation for relatively flexible elastic-perfectly plastic structures with fundamental periods of oscillation T_1 (before yielding occurs) equal or greater than 0.5 s. In the same pioneering work, it has been reported that, for stiffer structures (0.1 s $\leq T_1 \leq$ 0.5 s), the "equal energy assumption" holds approximately, while for "almost rigid" structures, it holds that $\max F_{el} = F_y$. The above empirical observations lead to the following relationships between the force reduction factor or q factor and ductility (Chopra 2007):

$$q = \begin{cases} \mu, T_1 \geq 0.5\,\mathrm{s} & (equal\ \ displacement\ \ rule) \\ \sqrt{2\mu - 1}, 1\,\mathrm{s} \leq T_1 \leq 0.5\,\mathrm{s} & (equal\ \ energy\ \ rule)\ . \\ 1, T_1 \leq 0.1\,\mathrm{s} & (equal\ \ force\ \ rule) \end{cases} \tag{1.6}$$

The third of the above relations suggests that very stiff (almost rigid) structures should be designed to remain linear under the "design seismic action". Further, according to the second relation, there exists an intermediate range between almost rigid structures and flexible structures for which the behaviour factor q is smaller than the ductility μ. Thus, allowing for structures whose fundamental natural period (before yielding) lies within this range, to yield will result in smaller reductions to the design seismic loads than for more flexible structures for which the "equal displacement rule" holds.

It is noted in passing that more elaborate force reduction factor or behaviour factor versus ductility capacity relationships than the one in Eq. (1.6) suggests have been proposed in the literature based on extensive numerical results for structures tracing various hysteretic force-deformation relationships and for earthquake induced strong ground motions corresponding to different seismogenetic environments (Miranda 2000; Ruiz-García and Miranda 2006). However, most of the contemporary codes of practice, including EC8, adopt Eq. (1.6) to define inelastic "design" spectra of reduced spectral ordinates. These are derived by dividing elastic response spectra with an assumed behaviour factor q. The thus obtained inelastic spectra are used to represent the seismic input action for the design of yielding structures as discussed in Sect. 2.3.1.2. Focusing on the elastic response spectrum of EC8 (§3.2.2.2) and on the corresponding inelastic design spectrum of EC8 (§3.2.2.5), it is noted that the range of natural periods for which the "equal energy rule" is assumed corresponds to a horizontal segment (flat plateau in Fig. 3.1 of EC8; see also Fig. 2.20) which is limited to the right by the corner period T_c. For $T \geq T_c$, the "equal displacement rule" holds. For "Type 1" elastic response spectra,

T_c ranges within 0.4–0.8 s (Table 3.2 of EC8) depending on the soil conditions ("ground type").

In the remainder of this section, three different design scenarios are considered in detail to further elaborate on the idea of "utilizing" (taking advantage of) the available ductility (ductility capacity) of structures to reduce the design seismic loads and, thus, to presumably achieve more economical designs. The three considered structures A, B, C follow an elastic-perfectly plastic force (base shear)-deformation relationship under static lateral loads. They all attain the same initial stiffness (before yielding) corresponding to $T_1 \geq T_c$ (equal displacement rule holds) as shown in Fig. 1.8. However, they are characterized by different yielding strengths $F_{y\text{-}P}$ and ductility capacities μ_{cap-P} (P = A,B,C). It is worth noticing that, in the case of r/c structural members, codes of practice, including EN1992-Part 1 (CEN 2004b), hereafter EC2, and EC8, assume that their stiffness depends on the dimensions of their (uncracked) cross-sections but not on the longitudinal steel reinforcement. Thus, under this assumption, it is possible to vary the strength of a structure assuming no change in its stiffness by varying the rebar without changing the cross-sectional dimensions. Of course, in reality, an increase of the longitudinal reinforcement does increase the flexural stiffness of r/c members, apart from their strength. However, this influence is deemed minor for practical (force-based) design and, therefore, is widely neglected. Similarly, the available ductility of r/c structural members may be increased without noticeably affecting their stiffness, e.g., through a denser arrangement of stirrups.

All three considered structures A, B, C are designed for the same level of "design seismic action", which is represented by the seismic load $maxF_{el}$ undertaken by an idealized elastic system of equal initial stiffness.

Structure A: No utilization of the ductility capacity (green line in Fig. 1.8)

Let the ductility capacity of structure A be $\mu_{cap-A} = u_{tot-A}/u_{y-A} > 1$ (Fig. 1.8). Suppose that the design engineer decides not to take advantage of the ductility capacity of the structure in designing it for earthquake resistance. In this case, the structure has to be designed to elastically resist the total design seismic action $maxF_{el}$, without yielding, i.e., $maxu_{el} \leq u_{y-A}$, and, thus, it is "demanded" that it undertake a design load F_{d-A} equal to $maxF_{el}$ (see also Fig. 1.7 for the case $maxu_{el} = u_{y-A}$). That is, $F_{d-A} = maxF_{el} \leq F_{y-A}$, where F_{y-A} is the yielding strength of the structure. Thus, no reduction in the "elastic" seismic demand load takes place and, consequently, the force reduction factor R_A or the behaviour factor q_A is taken to be equal to the unity. Clearly, in this case no ductility demand is posed to the structure ($\mu_{dem-A} = 1$) and, thus, $\mu_{dem-A} \leq \mu_{cap-A}$. It is emphasized that the fact that no ductility is required from the structure does not mean that the structure is brittle (i.e., does not possess any ductility capacity at all). It only means that, under the design seismic action, the existing ductility capacity is not utilized to resist the seismic loads. All the available ductility capacity is kept as a "reserve" for seismic loads exceeding the considered design ones. This issue is further discussed in Sect. 1.2.6.4.

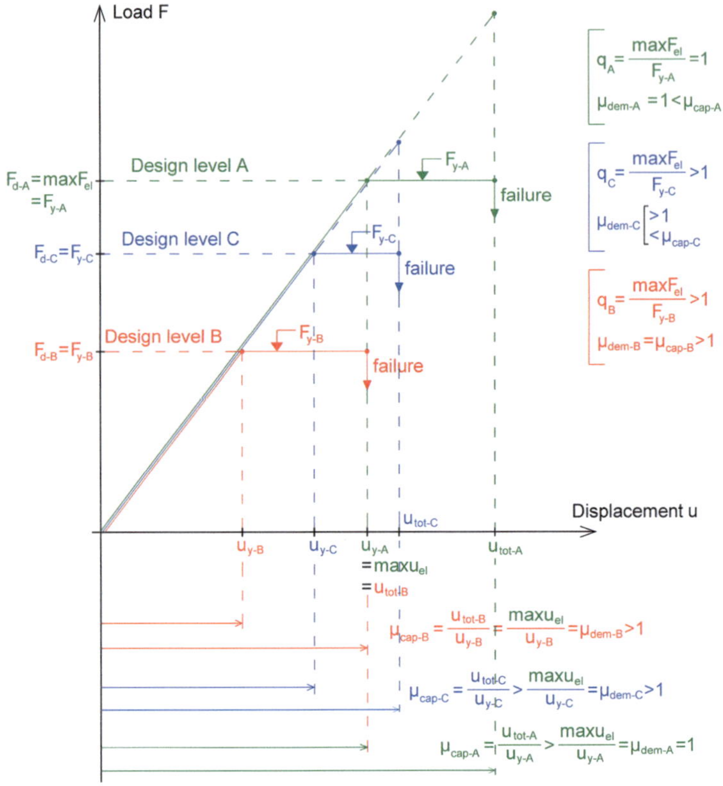

Fig. 1.8 Choice of different earthquake design levels A, B, C ('earthquake protection levels') for a given design earthquake

Overall, the following conditions hold in the case of linear elastic design for which no utilization of ductility capacity is foreseen for the design earthquake:

$$1 = q_A = \mu_{dem-A} \leq \mu_{cap-A}. \tag{1.7}$$

Table 1.2 collects all important expressions listed above for the considered design scenario of structure A.

Structure B: Full utilization of the ductility capacity (red line in Fig. 1.8)

Suppose that for structure B (Fig. 1.8), the design engineer decides to utilize its ductility capacity $\mu_{cap-B} = u_{tot-B}/u_{y-B}$ in full to undertake the considered design seismic action corresponding to the lateral seismic load $maxF_{el}$. In this case, the design seismic load F_{d-B} (i.e., the base shear for which the analysis is to be performed) will be smaller than the $maxF_{el}$ and equal to the yielding strength of

Table 1.2 Ductility capacity, behaviour factor, and design seismic load for the three design scenarios of elastic-perfectly plastic structures of Fig. 1.8 for the same design seismic action ($maxF_{el}$)

	Structure A	Structure B	Structure C
	No utilization of ductility capacity	Full utilization of ductility capacity	Partial utilization of ductility capacity
Ductility capacity μ_{cap-P} (P = A,B,C)	$\mu_{cap-A} = u_{tot-A}/u_{y-A}$	$\mu_{cap-B} = u_{tot-B}/u_{y-B}$	$\mu_{cap-C} = u_{tot-C}/u_{y-C}$
Ductility demand μ_{dem-P} (P = A,B,C)	$\mu_{dem-A} = 1$	$\mu_{dem-B} = u_{tot-B}/u_{y-B}$	$\mu_{dem-C} = u_{pl-C}/u_{y-C}$
Behaviour factor q_P (P = A,B,C)	$1 = q_A = \mu_{dem-A} \leq \mu_{cap-A}$	$1 < q_B = \mu_{dem-B} = \mu_{cap-B}$	$1 < q_C = \mu_{dem-C} < \mu_{cap-C}$
Design seismic load F_P (P = A,B,C)	$F_{d-A} = maxF_{el} \leq F_{y-A}$	$F_{d-B} = F_{y-B} = maxF_{el}/q_B$	$F_{d-C} = F_{yC} = maxF_{el}/q_C$

the structure F_{y-B}. Using Eqs. (1.5) and (1.6) under the equal displacement rule, the following equalities can be written

$$F_{d-B} = F_{y-B} = \max F_{el}/q_B = \max F_{el}/\mu_{cap-B}. \tag{1.8}$$

Clearly, the utilization of the ductility capacity of structure B in full maximizes the achieved reduction to the design seismic load (base shear) for a given design seismic action. Thus, $q_B = \mu_{cap-B} > 1$. Further, in this case, the required ductility demand is also maximized. That is, $\mu_{dem-B} = q_B$ and, consequently, $\mu_{dem-B} = \mu_{cap-B}$. Overall, a similar set of conditions as in Eq. (1.7) can be written for the case considered as

$$1 < q_B = \mu_{dem-B} = \mu_{cap-B}. \tag{1.9}$$

Table 1.2 collects all important expressions listed above for the considered design scenario of structure B.

Structure C: Partial utilization of the ductility capacity (blue line in Fig. 1.8)

In this third scenario, the design engineer decides not to utilize the ductility capacity of the elastic-perfectly plastic structure C, $\mu_{cap-C} = u_{tot-C}/u_{y-C}$, in full to design for the considered design seismic action corresponding to the $maxF_{el}$ load (Fig. 1.8). Instead, only a part of the ductility capacity is utilized expressed in terms of ductility demand as $\mu_{dem-C} = u_{pl-C}/u_{y-C}$, where $u_{pl-C} < u_{tot-C}$. In this case, it holds that $\mu_{dem-C} < \mu_{cap-C}$ (see also Fig. 1.8). Further, under the equal displacement rule, the corresponding idealized elastic structure undertaking a load $maxF_{el}$ under the considered design seismic action will exhibit a peak displacement $maxu_{el}$ which should be equal to u_{pl-C}. Thus, using Eqs. (1.5) and (1.6) under the equal displacement rule, the behaviour factor q_C can be expressed as (Table 1.2)

$$q_C = \max F_{el}/F_{y-C} = \max u_{el}/u_{y-C} = u_{pl-C}/u_{y-C} = \mu_{dem-C} > 1. \quad (1.10)$$

Finally, the design seismic load F_{d-C} (i.e., the base shear for which the analysis is to be performed) is equal to the yielding strength F_{y-C} of the elastic-perfectly plastic structure C, as in the previous design scenarios considered, and is determined by the ratio $\max F_{el}/q_C$ or $\max F_{el}/\mu_{dem-C}$ (see also Table 1.2).

In view of the previously considered design scenarios and the derived relations shown in tabular form (Table 1.2), several important observations can be made:

- *Behaviour factor q and ductility demand μ_{dem}*
  In all three cases considered spanning all possible design choices, the behaviour factor q coincides with the ductility demand ($q = \mu_{dem}$).
- *Minimum and maximum values of the behaviour factor q*
  Conceptually, the minimum value of the behaviour factor equals to unity ($\min q = 1$), in which case it is "demanded" that the structure remains linear (elastic) under the design seismic action. Further, the maximum possible value of the behaviour factor, maxq, equals the ductility capacity ($\max q = \mu_{cap}$).
- *Maximum allowed value of the behaviour factor q*
  In the context of code regulated design of structures for earthquake resistance, a maximum allowed value $\max q_{allow}$ of the behaviour factor q can be prescribed to set an upper limit for the portion of the ductility capacity permitted to be utilized to undertake the prescribed design seismic action. This upper limit ensures the existence of sufficient ductility capacity reserves to resist seismic actions beyond the "nominal" design one. That is, $\max q_{allow} \leq \mu_{cap}$.
- *Behaviour factor demand (q_{dem}) and behaviour factor capacity (q_{cap})*
  In analogy to the notions of "demand" and "capacity" to characterize ductility, that is μ_{dem} and μ_{cap}, one may assign the same notions to the behaviour factor q. The "behaviour factor demand" q_{dem} coincides with the behaviour factor q adopted in analysis to reduce the seismic design load (force reduction factor). In this regard, q_{dem} is the required behaviour factor chosen by the designer, or rather, "demanded" by the owner. The "behaviour factor capacity" q_{cap} is the maximum value (maxq) that the behaviour factor can attain (e.g., in case of full utilization of the ductility capacity to design for the design seismic action).

A collective consideration of the above observations yields the following relations

$$\min q = 1 \leq \mu_{dem} = q_{dem} = q \leq \max q_{allow} \leq q_{cap} = \max q = \mu_{cap}. \quad (1.11)$$

In Eq. (1.11), the relation $\max q_{allow} \leq q_{cap}$ is valid only if $q = \max q_{allow}$, that is, only if the q is selected to the maximum allowable value $\max q_{allow}$. If a smaller value is chosen for q, the actually available q_{cap} may well be smaller than $\max q_{allow}$.

Focusing on the specifics of EC8, it is emphasized that no particular distinction is made between ductility demand and ductility capacity. Rather, EC8 refers to the concepts of the behaviour factor q and of ductility μ as defined in Table 1.3.

Table 1.3 Definitions of behaviour factor and ductility according to EC8

Behaviour factor q: reduction factor of the spectral ordinates of the (linear) response spectrum and, thus, of the lateral seismic loads corresponding to the design seismic action.
In this respect, as in EC8, the q is the "behaviour factor demand" ($q = q_{dem}$) for which a maximum allowed value $maxq_{allow}$ is defined depending on certain criteria (§3.2.5 of EC8)
Ductility μ: ductility capacity ($\mu = \mu_{cap}$)
EC8 ensures that minimum levels of ductility capacity are achieved ***indirectly*** depending on the classification of structures in three different Ductility Classes, Low (DCL), Medium (DCM), and High (DCH). This is accomplished by prescribing a series of verification checks and requirements that any lateral load-resisting structural system must satisfy depending on the class it belongs to.

From the definitions of Table 1.3, it can be concluded that the following relationship should always hold within the context of earthquake resistant design according to EC8

$$q \leq \mu = \mu_{cap}. \tag{1.12}$$

The equality between behaviour factor and ductility in the above equation holds in case full utilization of the ductility capacity of the structure is decided to undertake the design seismic action (design scenario for structure B in Fig. 1.8). Then, q is set equal to $maxq_{allow}$ and, thus, no actual distinction needs to be made between ductility demand and ductility capacity since $\mu_{dem} = q = maxq_{allow} = \mu_{cap}$.

However, it should be emphasized that *choosing a behaviour factor equal to the maximum allowed value prescribed by EC8 is not mandatory.* In fact, a smaller value of q can be chosen down to the smallest nominal $q = 1.5$, that is prescribed for structures designed to dissipate seismic energy. This corresponds to an almost linear structural behaviour under the design seismic action, given the desired level of utilization of the ductility capacity (design scenario for structure C in Fig. 1.8) and the "overstrength" that actual structures possess, as is discussed in the following section. It is noted herein however, that the relevant provision 5.2.2.2 (1)P of EC8 is quite ambiguous, as it prescribes $q = 1.5$ as the lower bound value for buildings designed to dissipate seismic energy via yielding, hence not explicitly excluding the adoption of a behaviour factor $q = 1$ towards linear elastic response under the design earthquake without utilizing the overstrength resources. After all, the designer is required to adopt a class of low, medium, or high ductility (DCL, DCM, DCH), and, therefore, to specify measures for providing a certain level of ductility capacity prior to the choice of the behaviour factor q. Thus, in any case, a minimum ductility supply is always ensured (see also Sects. 3.1.3 and 3.1.4). It is also noted that in certain European countries exposed to high seismic risk (such as Greece and Cyprus for instance), the use of the Low Ductility Class is restricted by the building importance and seismic zone as per the respective National Annex provisions.

1.2.3 The Relationship Among the Behaviour Factor, the Ductility Capacity and the Overstrength of R/C Buildings

The relationship between the total lateral seismic load (base shear) versus top-storey lateral sway attained by actual r/c buildings traces a smooth curve ("pushover" or capacity curve), as shown in Fig. 1.5 (see also Fig. 2.13). Consequently, the lateral stiffness of r/c buildings, defined by the slope of the base shear-lateral sway pushover curve, varies continuously with the observed sway as the applied base shear increases. However, for design purposes, pushover curves are commonly replaced by a simplistic elastic-perfectly plastic force-deformation law, discussed in the previous section, following certain fit criteria (e.g., Luca et al. 2013). In this respect, a constant pre-yield "effective stiffness" equal to the ratio F_d/u_d, where F_d is the design seismic load (or design base shear) and u_d is the displacement corresponding to the load F_d, can be defined. An associated idealized linear structure with stiffness F_d/u_d can be also defined as shown in Fig. 1.9.

It is further noted that, in practical code-compliant design scenarios of r/c buildings, several cross-sections are usually overdesigned for various reasons, a common one being that the required reinforcement corresponding to the design seismic load F_d is smaller than the minimum reinforcement required by the code. In this regard, the maximum "nominal" base shear F_y that a structure can resist under the assumption of an elastic-perfectly plastic force-deformation law is usually higher than the design seismic load. That is, $F_y > F_d$ as depicted in Fig. 1.9. In such cases, the behaviour factor q (or force reduction factor R) considering full utilization of the available structural ductility capacity and under the equal displacement rule assumption is written as (see also Fig. 1.9)

$$q = \max F_{el}/F_d = \left(\max F_{el}/F_y\right)\left(F_y/F_d\right) = \left(\max u_{el}/u_y\right)\left(u_y/u_d\right) \\ = \mu_{cap}\, f, \tag{1.13}$$

where f is the overstrength ratio defined as

$$f = u_y/u_d > 1. \tag{1.14}$$

Equation (1.13) delineates that, *in the case of full utilization of the available structural ductility capacity, the behaviour factor is equal to the product of the ductility capacity times the overstrength*. Clearly, for ideally brittle structures with zero plastic deformation capacity ($\mu_{cap} = 1$), the behaviour factor becomes equal to the overstrength. This observation implies that the minimum behaviour factor minq attained by actual structures is equal to the overstrength which, in turn, is always greater than 1 (provided material strengths are not below specified values). In this respect, Eq. (1.11) is revisited and amended as follows:

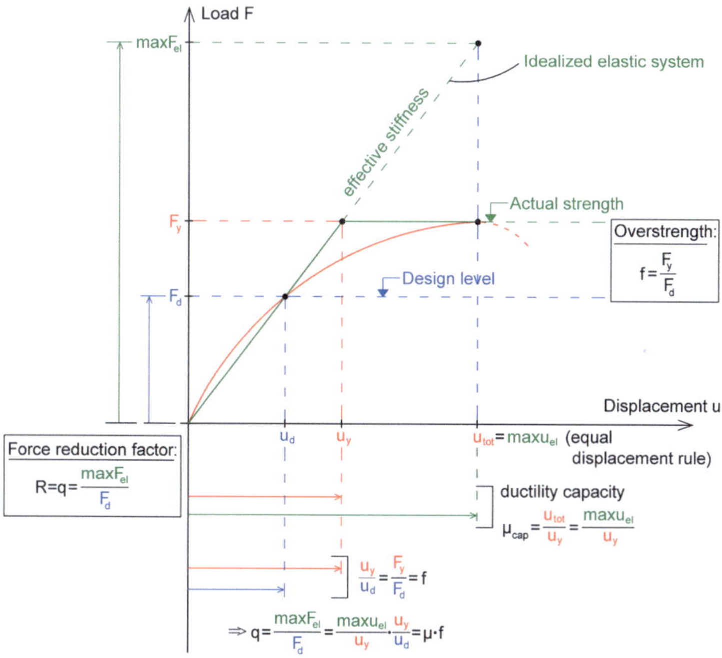

Fig. 1.9 Actual and idealized system with $T \geq T_c$ and overstrength $R_{d,o} = \mu{\cdot}f$

$$1 \leq f = \min q \leq \mu_{dem} = q_{dem} = q \leq \max q_{allow} \leq q_{cap} = \max q = \mu_{cap}. \quad (1.15)$$

The levels of the *available* ductility capacity and the overstrength of actual structures depend on many factors, and their relationship to the behaviour factor q may be more complex than the one previously discussed which holds only for the elastic-perfectly plastic force-deformation law. To this end, the most important factors influencing the overstrength of r/c buildings are listed below:

1. The difference between the actual yielding strength of the materials (concrete and steel) used in construction vis-à-vis their nominal characteristic strength assumed in design.
2. The difference between the actual ("as-built") dimensions of structural members vis-à-vis those assumed in the analysis.
3. The difference between the actual reinforcement placed in structural members vis-à-vis the required reinforcement area obtained from design calculations.
4. The achieved level of concrete confinement at critical regions of structural members using transverse reinforcement (stirrups).
5. The contribution of non-structural elements (e.g., brittle infill walls) in resisting lateral seismic forces which is commonly ignored in analysis.

6. The conservative assumptions made in structural modeling (e.g., the adoption of an approximate "flange width" for slab-supporting T-section beams vis-à-vis the actual slab contribution to the flexural resistance of beams).
7. The use of conservative analysis methods (e.g., the use of the simplified lateral load method as opposed to the response spectrum method).

It is noted that the observed overstrength in a typical r/c beam member may reach the order of $f = 1.5$ due to the factors (1)–(4). That is, the actual resisting capacity of a beam under flexure is 50 % larger than the one it was designed for. Further, the overstrength of the total lateral load resisting structural system can be higher than the overstrength of its individual structural members due to the factors (5)–(7). In this regard, it is emphasized that, in the context of seismic design of yielding ductile structures, higher than anticipated levels of overstrength may not be favourable, as is further explained in Sect. 1.2.6.2. To this end, the inherent overstrength of structures is taken into account at design by specifying appropriate values for the behaviour factor q. However, an accurate prediction of the overstrength of real structures is hard to achieve and, thus, the contribution of the overstrength to the behaviour factor is empirically quantified based on field observations in the aftermath of major earthquake events.

The values of the behaviour factor q specified by EC8 depend on numerous factors and are discussed in detail in Sect. 3.1.4.

1.2.4 Force-Based Seismic Design Using a Linear Single-Seismic-Action-Level Analysis

Most current codes of practice, including the EC8, adopt a force-based framework for the seismic design of ordinary structures which relies on linear structural analysis results for a single (design) seismic action level. In brief, the design seismic action is represented via a response (design) spectrum with reduced spectral ordinates according to the chosen force reduction factor R or behaviour factor q as shown in Fig. 1.10 (Penelis and Kappos 1997; Chopra 2007; Meskouris et al. 2011; Bisch et al. 2012; Fardis et al. 2014; Penelis and Penelis 2014). The thus defined spectral values are proportional to the seismic design base shear for which the structure is assumed to behave elastically even for the case of $q > 1$. For r/c structures, reduced ("effective") cross-sectional mechanical properties (namely flexural and shear stiffness) are assigned to account for the expected loss of stiffness of structural members due to concrete cracking under the design seismic action.

In this respect, the above design framework assumes that only a portion of the total design seismic input energy will be accommodated by means of linear elastic structural response mechanisms. That is, by conversion to (i) kinetic energy through elastic vibrations, (ii) elastic strain energy at structural members, and (iii) radiating heat through friction-related phenomena captured by adopting the viscous damping model. This portion corresponds to a $(1/q) \times 100\%$ of the considered design seismic

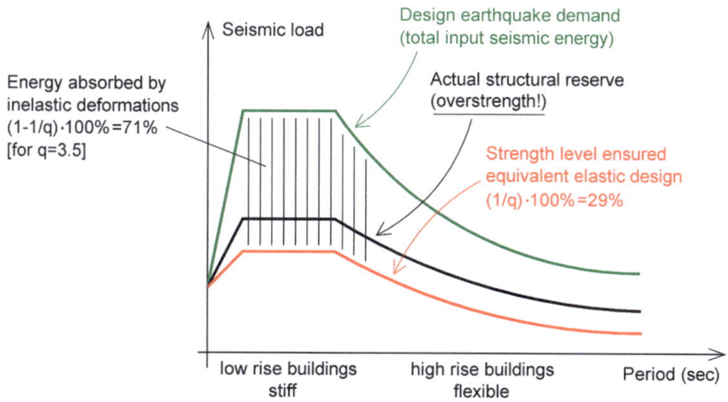

Fig. 1.10 Graphical representation (in design spectrum form) of the seismic energy dissipated through yielding

action in terms of base shear (red line in Fig. 1.10). Consequently, the rest of the design seismic input energy, corresponding to $(1-1/q) \times 100\%$ of the design base shear, *must* be dissipated by means of inelastic structural response mechanisms, as illustrated in Fig. 1.10. Ideally, this should be achieved through hysteretic energy dissipation at specific "critical" regions where "plastic hinges" are allowed to form. For r/c buildings, proper local detailing rules are enforced at these critical regions of structural members to avoid premature brittle types of local failures and to ensure that large inelastic deformations take place without significant loss of strength and stiffness (local ductile behaviour at plastic hinges). Further, certain global conceptual design considerations and capacity design rules are also imposed to ensure relatively even distribution of plastic hinges in plan and in elevation at structural members which are easier to repair and achieve higher levels of local ductile behaviour. The general qualitative requirements for accomplishing ductile seismic design for r/c buildings are further presented in Sect. 1.2.5. At this point, it is important to highlight that no additional structural analysis steps are prescribed for the quantitative assessment of the expected non-linear behaviour of the structure for the case of $q > 1$ within the herein discussed force-based design framework.

In view of the above, it can be argued that selecting the value of the behaviour factor q is the most critical consideration in code-compliant seismic design of ordinary r/c buildings. Evidently, the prescribed by codes of practice behaviour factor should be treated as the maximum allowed value that the designer can choose. In this respect, it is instructive to discuss further the following two extreme cases.

Selection of the minimum possible value for the behaviour factor (q = 1)

This case corresponds to the design scenario A presented in Sect. 1.2.2. In this case, the total design seismic load *must* be accommodated exclusively via elastic mechanisms and, thus, the design response spectrum (red line in Fig. 1.10) coincides with the elastic response spectrum (green line in Fig. 1.10). Consequently, structural members should not yield for the "design earthquake scenario", commonly taken as the one having a probability of 10 % to be exceeded in 50 years. Further, the overstrength and the (always existent though not explicitly quantified) inherent ductility capacity (see Sect. 1.1.5.3) are reserved to partially ensure collapse prevention for the non-negligible probability that a seismic event larger than the design earthquake occurs. In this respect, it is *advisable* that, even in the case of $q = 1$, the designer ensures that some appropriate, readily achievable local detailing measures are taken (e.g., denser stirrups at the critical regions of beams and columns) or even capacity design at the structural joints is performed in order to further increase -if only empirically- the inherent ductility so that sufficient ductility capacity exists to prevent collapse for seismic loads significantly larger than the design seismic action that might occur.

Selection of the maximum allowed behaviour factor ($q = maxq_{allow}$)

This case corresponds to the design scenario C of Sect. 1.2.2. The value of $q = 3.5$ is herein assumed as an *indicative* maximum allowed value for r/c buildings (see also Sect. 3.1.4). As shown in Fig. 1.10, for $q = 3.5$, more than 70 % of the total seismic load *must* be undertaken via inelastic/hysteretic behaviour of the structure, though in reality the existence of overstrength entails that the structure will accommodate somewhat higher seismic loads via elastic behaviour than what is assumed in design (black curve in Fig. 1.10). In this case, structural members are allowed to be designed for significantly less strength compared to the $q = 1$ case, however, sufficiently large ductility *must* be achieved under the design earthquake. It is important to note that if the required level of global and/or local ductility capacity is not exhibited (e.g., plastic hinges form within a single storey of the building and/or premature failure occurs at plastic hinges formed due to poor local detailing), the structure will collapse under the design earthquake. Thus, ensuring local ductile behaviour at critical cross-sections along with proper global conceptual design of the lateral load-resisting structural system are major concerns in designing for large values of the q factor.

At this point, it is important to emphasize that *plastic hinges may entail significant local damage* as shown in Fig. 1.11 *in need of repair in the aftermath of a seismic event corresponding to the "design seismic action"*. Depending on the damage severity and on the location of plastic hinges, such retrofitting steps may not always be cost-efficient and it may be the case that demolition of the damaged structure is deemed preferable or even necessary. Clearly, such a non-negligible likelihood should be taken into account in selecting the behaviour factor q at the initial stages of the design process, as will be discussed in detail in Sects. 1.3 and 1.4.

Fig. 1.11 Flexural damage observed at the end of a beam and at the base of a column

As a final remark, it is reiterated that, even for large values of the q factor for which significant inelastic behaviour is expected for the design earthquake, "equivalent" elastic types of analysis are allowed to be used within the herein discussed force-based design framework (see also Sect. 2.4.3). Specifically, structures are assumed to behave elastically considering "effective" reduced values for the pertinent mechanical properties of structural members to account for local loss of stiffness due to concrete cracking and spalling. For instance, clause 4.3.1(7) of EC8 states that, in the absence of a more accurate analysis, the flexural and shear stiffness of all r/c members can be taken to be equal to 50 % of the stiffness corresponding to the intact (uncracked) members. Clearly, this is a quite crude assumption, as the use of a cracked stiffness equal to 50 % of the gross stiffness does not account for whatever different level of inelastic demand that the member may be subjected to depending on the value of the behaviour factor used. Nevertheless it is an assumption inextricably incorporated into the *code-specified "equivalent" linear analyses for design seismic actions reduced by the behaviour factor q. It is quite evident that the reliability of such an "equivalent" linear analysis heavily depends* on the "regularity" of the lateral load resisting structural system. In turn, structural regularity entails that damage in the form of plastic hinges formed under the design seismic action is distributed in a relatively uniform manner within the lateral load resisting system in plan and in elevation. In fact, structures are categorized as regular or irregular in plan and in elevation based on certain conceptual design considerations (see Sects. 2.1.1 and 3.1.1). As a general rule of thumb, the following empirical limits apply regarding the perceived degree of reliability of "equivalent" linear analysis methods:

- They are reliable for small values of the q factor (e.g., $q < 2$), for regular as well as for irregular structures.
- They are reliable for large values of the q factor (e.g., $q > 3$), only for regular structures.

In the case of highly irregular structures undergoing large inelastic deformations under the design seismic action, non-linear types of analysis should preferably be performed, as further discussed in Sect. 2.4.1. These types of analysis account for the local inelastic behaviour exhibited at individual structural members explicitly instead of the consideration of a single scalar quantity (the force reduction factor or behaviour factor q) to capture the inelastic behaviour at a global/structural level. However, the capability of these non-linear methods to predict reliably the inelastic response of heavily yielding structures depends on the proper modeling of local non-linear material behaviour and the adequate representation of the input seismic action. Therefore, as the implementation of such methods in practical design presents certain limitations and assumptions (see Appendix A), the area is still open for further research. *It is the authors' opinion that, at first instance, structural regularity along with relatively small values of force reduction factor or behaviour factor should be sought in the design of ordinary structures for earthquake resistance.*

1.2.5 Additional Qualitative Requirements for Ductile Earthquake Resistant Design

The adopted by current codes of practice force-based design framework presented in the previous section does not involve quantitative assessment steps to verify whether the fundamental seismic design objectives of Table 1.1 and the associated requirements of structural performance are achieved. This is an important consideration, especially in the case of adopting large values for the behaviour factor q. Such a choice requires the designed structure to exhibit sufficient ductile behaviour, that is, to develop localized damage at plastic hinges exhibiting significant inelastic deformations without local or global collapse, to resist the design seismic action. In this regard, seismic codes of practice prescribe *additional local detailing and global conceptual design rules to ensure ductile behaviour by equipping structures with adequate and appropriately distributed stiffness, strength, and ductility properties*, as discussed in Sect. 1.1.5. It is assumed that these rules ensure that the following three fundamental sets of requirements for ductile structural behaviour under the design seismic action are met.

(a) *Maximization of the dispersion of the (kinetic) seismic input energy within the lateral load-resisting structural system*

The above requirement suggests that inelastic strain demands induced by the design seismic action are evenly distributed across the entire structure. In this manner, the severity of localized damage at each individual plastic hinge is minimized. It is assumed that requirement (a) for ductile behaviour is met by designing the lateral load-resisting structural system such that:

(i) structural *simplicity* (simple, clear, continuous and direct stress paths),
(ii) structural *uniformity* and *symmetry* (regularity in plan and in elevation),

(iii) *diaphragmatic action* of floors (in-plane perfectly rigid floors), and
(iv) *strong foundation* (elimination of differential displacements)

are achieved. A set of conceptual design rules and criteria discussed in Sect. 2.1.1 are prescribed to accomplish structural layouts observing attributes (i)–(iv).

(b) *Prevention of global structural instability/collapse*

The development of collapse mechanisms due to non-linear structural behaviour under the design seismic action can be avoided by controlling the type and location of local modes of failure/damage (see also Sect. 1.2.6 on "capacity design"). This is mainly accomplished by ensuring that

(v) *ductile modes of local failure* (e.g., failure in flexure) precede brittle modes of local failure (e.g., failure in shear and local buckling), and that
(vi) the *relative strength of all "neighboring" structural members* (i.e., structural members framing at the same joint) is such that the sequence of plastic hinges (local ductile failures) occur in a predetermined manner activating ductile types of global mechanisms (see also Sect. 2.2.4).

Further, the probability of developing collapse mechanisms is also reduced by

(vii) supplying the lateral load-resisting system with a large degree of *redundancy* allowing for the redistribution of stress demands upon each consecutive formation of a new plastic hinge and, thus, for maximum utilization of available strength in a large number of structural members.

(c) *Maximization of seismic input energy dissipation*

Energy dissipation through ductile (inelastic/hysteretic) structural behaviour is maximized by ensuring that

(viii) structural damage occurs (i.e., plastic hinges are formed) at designated zones of specific structural members which can potentially exhibit high levels of ductility capacity, and that
(ix) the above designated zones are equipped, by means of proper local detailing rules and practices, with the maximum possible level of ductility capacity to eliminate the probability of a premature failure or of a brittle failure.

Focusing on r/c building structures, *local brittle types of failure* to be avoided include failures due to shearing stresses, due to premature buckling of longitudinal reinforcing bars and due to premature pull-out (loss of steel-concrete bond) of longitudinal reinforcing bars because of insufficient anchorage length or lap splice. *The main desirable type of ductile mode of flexural failure* involves yielding of the longitudinal reinforcement under tensile stresses prior to concrete failure at compression zones. Local detailing rules ensuring sufficient ductility capacity at critical locations of structural members are discussed in Sect. 2.2.3 As a final remark, it is emphasized that attributes (v), (vi), and (viii) are directly related and achieved through capacity design whose rationale is discussed in the next section. Further discussion on capacity design considerations are provided in Sect. 2.2.4.

1.2.6 The Rationale of Capacity Design Requirements

The concept of "capacity design" involves a set of rules, requirements, and verification checks defining a hierarchical designation of the types of failure modes and their location within the lateral load-resisting structural system to maximize seismic input energy dissipation through ductile behaviour (allowance of structural damage without global collapse). In general, this is achieved by judicial assignment of strength at structural members against different failure modes to control the number and sequence of plastic hinges (ductile local failures) and to minimize the probability of occurrence of brittle failures.

The rationale of capacity design can be readily visualized by means of a plain, statically determinate chain structure comprising links of different strength shown in Fig. 1.12. The strength capacity of this chain (i.e., the peak static external force that the chain can resist) is equal to the strength F of its weakest link. If this specific link is brittle, the chain fails in a brittle manner, that is, suddenly, without exhibiting any significant inelastic deformation first. However, if the weakest chain link is ductile, then the chain yields under an externally applied force F prior to breaking, exhibiting (large) plastic deformation. In the case of seismic/cyclic dynamic applied loads, such a failure entails (large) dissipation of seismic/kinetic energy.

It is noted that the non-yielding links of the chain may not be brittle: they can be ductile, similar to the weakest link, though they will remain elastic due to their higher strength and, thus, will not dissipate additional energy, under an externally applied force F (corresponding to the "design seismic action"). This is because, in the case of the above considered statically determinate chain, a single local failure results in global failure no matter what the nature of the local failure is. However, actual r/c buildings are complex structural systems with large degrees of redundancy (static indeterminacy) comprising a large number of inter-connected structural members ("links"). Therefore, a significant number of local (ideally ductile) failures is required for the development of a plastic mechanism. In this regard, in the practical case of actual r/c structures, the aims of capacity design are

- to avoid brittle modes of failure; and
- of all the possible plastic mechanisms that can be potentially developed in a given structure, to achieve the formation of the one which maximizes the dissipation of the seismic input energy.

The above aims are accomplished by appropriately detailing certain predetermined "critical" zones of r/c structural members (commonly, the ends of beams and columns and the base of shear walls) to dissipate the input seismic energy via ductile behaviour. A significant amount of energy is dissipated at each of these ductile zones (plastic hinges) by means of hysteretic mechanisms until potential local "failure" takes place due to excessive plastic deformations causing severe loss of stiffness and/or strength. The remaining "non-critical" zones of structural members are designed for sufficiently high yielding strength to remain elastic under the design seismic action. As they are not meant to yield under the

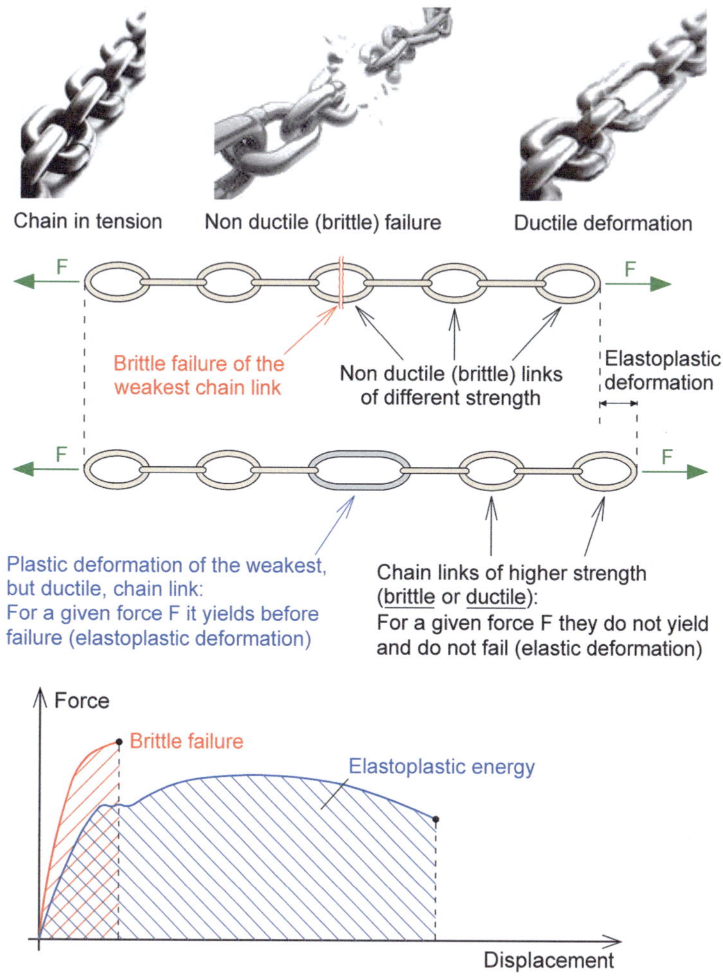

Fig. 1.12 Fundamental concept of the capacity design

design seismic action, they are typically, yet not necessarily, less ductile than the critical zones (strength "hierarchy"). Therefore, *capacity design establishes a hierarchy of zones within structural members according to their strength to "drive" yielding and plastic deformations at designated locations which:*

1. *are more capable for hysteretic energy dissipation through proper local detailing for ductile behaviour* (e.g., the ends of columns will always be less capable for ductile behaviour than the ends of beams due to the negative influence of compressive axial loads to ductility capacity- see also Sect. 2.2.3),
2. *are less important to the global structural integrity* (e.g., columns are more important structural members than beams in carrying lateral and gravitational loads. For example, a moment resisting framed r/c building will collapse if

plastic hinges form at the ends of all columns at a single storey. However, this would not be the case if plastic hinges formed at the ends of all beams at the same storey- see also Sect. 2.2.4), and
3. *are easier to inspect and repair in the aftermath of a major earthquake* (e.g., retrofitting the bottom side of beams is easier and less costly than retrofitting the top of beams).

In view of the above, the well-known capacity design rule of "strong columns-weak beams" prescribed and quantitatively verified by all contemporary seismic design codes of practice (including EC8-clause 5.2.3.3) can be readily justified and constitutes a valid example underpinning the concept of capacity design.

Further, given that, in redundant (statically indeterminate) structures, stress is redistributed within structural members whenever a new plastic hinge forms, global instability/collapse occurs after a sufficient number of plastic hinges have formed and a *plastic collapse (statically under-determined) mechanism* has been developed. In this context, capacity design ensures ultimately that, out of all the possible plastic mechanisms, the most ductile ones (i.e., the ones involving formation of the largest number of plastic hinges) develop by establishing a strength hierarchy of potential energy dissipation zones within structures according to the above three criteria. Therefore, capacity design further establishes, implicitly, a hierarchy of plastic mechanisms according to their achieved ductility, that is, their ability to dissipate the seismic input energy before collapse occurs.

In the remainder of this section, a number of important remarks are made closely related to the notion of capacity design and the underlying requirements for resisting the seismic input action by means of local and global ductile behaviour.

1.2.6.1 The Role of Plastic Hinges as the Structure's "Fuses" Against Failure

It is emphasized that, upon choosing to design a structure for a behaviour factor $q > 1$ (and especially for relatively large behaviour factors: $q >> 1$), the designated zones for energy dissipation *must* be activated (e.g., plastic hinges must form towards the development of a desirable ductile collapse mechanism) under the design seismic action following the capacity design framework. In this manner, it is ensured that the $(1-1/q)*100$ % of the input seismic energy (under the assumptions made in Sect. 1.2.4) is dissipated in a reliable fashion. Zero or partial activation/ yielding of the designated energy dissipation zones (e.g., due to accidental over-strengthening of potential plastic hinge locations during construction) indicates that the intended design for $q > 1$ was inconsistent and, thus, unsuccessful since

- the developed yielding mechanisms may be unreliable for energy dissipation,
- the inelastic deformation demands are not properly controlled and may lead to an undesirable (premature or even non-ductile) collapse mechanism, and

– the probability of global instability/collapse is not kept at sufficiently low levels for the design seismic action, let alone for the case in which the design seismic action level is exceeded (see also Sect. 1.2.6.4 below).

In this respect, plastic hinges (local mechanisms for energy dissipation through ductile behaviour entailing structural damage) of an r/c building structure exposed to the design earthquake can be viewed as the system's "fuses" which *must* "burn" (be activated). In case they do not, the probability that the whole system "burns" (i.e., the structure collapses) becomes high under the design earthquake. This is a critical issue for ensuring the life safety requirement for $q > 1$ and is similar to the way contemporary cars ensure life safety in the case of a major collision: the passengers' cabin is designed to be stiff to minimize deformations and to possess high yielding strength (strong link) compared to a designated surrounding "yielding zone" (ductile link). During severe collisions, the yielding zone is allowed to deform severely to dissipate the energy of the impact while the cabin, i.e., the critical element for life safety, remains intact.

1.2.6.2 Is Overstrength a Desirable Attribute?

It can be readily inferred from the previous discussion on the principles of capacity design that overstrength does not necessarily offer additional safety – as widely believed in the "pre-capacity-design-era" some decades ago – in case of seismic design for $q > 1$. In fact, it may actually have negative consequences. For example, placing additional longitudinal reinforcement (beyond that calculated in design) in the beams of an r/c framed building due to inadequate on-site supervision during construction may increase the strength of the beams to the point that columns at joints yield first. This cancels the intended "weak beam-strong columns" classical hierarchy of capacity design and may potentially lead to less ductile collapse mechanisms and, thus, to premature global instability.

It is thus emphasized that unevenly distributed accidental overstrength among structural members, which may occur due to factors such as poor workmanship at construction or poor quality control of material properties, *should by any means be avoided for structures designed for large q factors*. This is because it can potentially jeopardize the intended strength hierarchy of structural members and energy dissipation zones established by capacity design provisions. Specifically, unevenly distributed overstrength may render seismic design of r/c buildings:

– Inaccurate (plastic hinges may not form at the desirable ductile zones for energy dissipation),
– Inconsistent (actual seismic loads due to the design earthquake are higher than those considered in structural analysis as discussed in Sect. 1.2.4), and
– Unreliable (a plastic mechanism of reduced ductility may develop leading to premature global instability).

Therefore, strict on-site supervision is deemed essential to ensure that as-built cross-sections are consistent with capacity design considerations which, in many cases, are counter-intuitive compared to the common conception: "the stronger, the better". The latter is valid *only* for seismic design scenarios adopting $q = 1$ (linear behaviour under the seismic design loads) for which capacity design rules are not enforced.

1.2.6.3 The "Forgiving" Nature of R/C: Inherent Ductility Capacity

Arguably, ductility capacity of r/c structural members and structures plays the most crucial role in the practice of seismic design for behaviour factors q well above unity in conjunction with capacity design considerations. It renders possible the resistance of the (design) input seismic action by hysteretic energy dissipation through the development of significantly large inelastic deformations without premature local or global collapse. Associated with their inherent ductility capacity is the "forgiving" nature of r/c structures in resisting loads beyond their linear behaviour (after yielding) without collapsing, a fact that has become known to the engineering community through empirical field observations. In this context, the empirically witnessed "forgiving nature" of well-engineered r/c structures is a manifestation of their ability to redistribute high levels of stresses induced by externally applied loads whose intensity may exceed the nominal design values. This ability stems from the ductility capacity with which the design engineer has equipped structural members by means of proper local detailing rules.

Accordingly, the usefulness of making a clear distinction at design stage between *ductility demand* μ_{dem} posed by the (design) seismic input action and *ductility capacity* μ_{cap} that structures are equipped with by the design engineer, through proper local detailing of energy dissipation zones and through application of capacity design rules, is reiterated (see also Sect. 1.2.1). From the design viewpoint, ductility demand is "mapped" onto the behaviour factor $q = q_{dem}$ adopted to reduce the input seismic loads and to define the amount of energy that must be dissipated by the structure through inelastic/ductile behaviour for the design seismic action. Further, ductility capacity is "mapped" onto the highest possible behaviour factor $\max q = q_{cap}$ that could be chosen to reliably resist the design seismic action through ductile behaviour and energy dissipation. In this regard, the fact that one may choose to adopt a relatively low behaviour factor, say $q = 1.5$, to design an r/c building does not necessarily mean that the ductility capacity of this building and its ability to dissipate energy through reliable hysteretic energy dissipation mechanisms is equally low. The design engineer can still take appropriate local detailing measures to ensure ductile behaviour at "critical" zones of expected high stress demands, if so desired. In such cases, the "forgiving nature" of r/c structures (i.e., their inherent ductility capacity) can significantly contribute towards meeting the no-collapse design objective for input seismic action levels greater than the design seismic action, as is further discussed in the following section.

1.2.6.4 The Role of Ductility Capacity to Resist Seismic Loads Beyond the Design Earthquake

Although reasonably low (see Sect. 1.1.2), the probability of occurrence of an earthquake that exceeds the nominal "design earthquake" is non-negligible, as shown by recent destructive seismic events (Athens/Greece 1999, Kocaeli/Turkey, 1999, Christchurch/New Zealand 2011, L'Aquila/Italy 2012, Cephalonia/Greece, 2014). It can be argued that an r/c building structure designed for a behaviour factor $q = 1$ (i.e., linear behaviour under the design earthquake) may be capable of resisting seismic loads corresponding to *as high as an actual earthquake quite stronger than the design earthquake* without collapsing with the stipulation that some basic detailing measures for local ductility at "critical" regions of structural members are taken. This argument is based on extensive field observations and empirical evidence in the aftermath of moderate-to-major seismic events involving old under-designed (code-deficient) r/c structures. Specifically, well-engineered r/c structures, designed and constructed according to well-known "best practices", are far from being brittle. In fact, even without applying capacity design rules at a member level, *the structures possess sufficient inherent ductility capacity as a whole* (ability for hysteretic energy dissipation) which, along with the overstrength, may correspond to an "*available*" q_{cap} behaviour factor larger than 1, provided that relatively closely-spaced stirrups are placed at the ends of structural members. Although no explicit quantitative research results are provided in the international literature, the inherent ductility reserves of such designed structures may reach, in the authors' opinion, values in the range of $q_{cap} = 1.5 \div 2.0$, thus being able to avoid collapse for actual earthquakes up to $1.5 \div 2$ times stronger than the design earthquake.

On the contrary, an r/c structure designed for a large behaviour factor, say $q = 4$, will sustain significant plastic deformations under the design earthquake (e.g., up to 4 ‰ concrete compressive strain and 10 ‰ or more steel tensile strain at plastic hinges) and *it is unlikely that it will possess sufficient ductility capacity reserves to meet the ductility demand of a stronger earthquake, say* 1.5–2 times *stronger than the design one*. This is because a *capacity* behaviour factor q_{cap} equal to about $6 \div 8$ (i.e., 4×1.5 to 4×2) or more may not be readily achievable in practice.

Therefore, it can be argued that, for seismic action levels beyond the design earthquake, r/c structures designed for $q = 1$ (or 1.5, depending on the interpretation of EC8 – see end paragraph of Sect. 1.2.2 above) attain a higher probability of meeting the non-collapse requirement (provided that some basic measures for local ductility capacity are taken), compared to structures designed for large behaviour factors (e.g., $q = 4$) following the code prescribed detailing and capacity design rules for ductile behaviour. However, in the authors' opinion, it is preferable to design, apart from the selected ductility class, for low values of behaviour factor, as this provides additional strength to the structure (see also recommendations in Sects. 1.4 and 3.4), with the exception of very stiff structures. Such considerations ensure sufficient reserves of ductility capacity to resist seismic loads beyond the design earthquake within the context of the force-based design framework presented in Sect. 1.2.4. It is further emphasized that very stiff r/c building structures whose lateral load resisting system comprises a large number of strong walls follow the motion of the ground as almost non-deformable rigid bodies under earthquake excitations. In this case, ductility demand is probably lower and, thus, ductility capacity is not a major design concern. Such structures need to be designed for low behaviour factor to ensure that walls are sufficiently strong to resist the seismic loads assuming linear behaviour. In this case, most likely, the accommodation of tensile stresses at the foundation of walls becomes a critical design issue.

As a final remark, it is *recommended* that design/structural engineers bring to the attention of the owners that designing for the maximum allowed behaviour factor prescribed by current codes of practice entails the development of severe structural damage for a future "major" seismic event. Such a design achieves only "partial" protection against structural damage for the design seismic hazard and may incur considerable repair costs and downtime, while the probability for an enforced demolition in the aftermath of a seismic event exceeding the nominal design earthquake is likely. Further, they should stress that, although an "*absolute*" protection against the seismic hazard is unattainable, a "*full*" protection as defined in Sect. 1.1.3 can be achieved, if so desired, by adopting a relatively low behaviour factor value (e.g., $q < 1.5$) within the standard force-based design framework prescribed by seismic codes of practice. Reference to alternative quantitatively equivalent ways to achieve improved structural safety margins for "major" seismic events within the above framework is made in Sect. 2.3.1.3.

To conclude this section, it is emphasized that, in view of the above presented material, a thorough appreciation of the seismic design philosophy underpinning the current codes of practice is equally important to the implementation of code prescribed analysis and detailing steps in elaborating structural designs that satisfy the fundamental design objectives of Sect. 1.1.4. Developing such an appreciation allows for forming well-qualified conceptual design layouts with adequate and properly distributed stiffness, strength, and ductility properties in plan and in elevation which significantly facilitates the purpose of meeting the code specific design requirements.

1.3 The Concept of Performance-Based Seismic Design: A Recent Trend Pointing to the Future of Code Provisions

1.3.1 The Need for Performance-Based Seismic Design

In recent decades, several catastrophic earthquakes incurring significant human and economic losses (e.g., Loma Prieta/California 1989, Northridge/California 1994, Kobe/Japan 1995, Izmit/Turkey 1999, Athens/Greece 1999, L'Aquila/Italy, 2009, Fukushima/Japan 2011) have questioned the sufficiency and the reliability of the partial protection against structural damage philosophy for earthquake resistance (see Sect. 1.1.4). The latter is adopted by most current codes of practice and is implemented through the force-based design procedure discussed in Sect. 1.2.4. It involves the utilization of a considerable portion of the ductility capacity of structures to resist a nominal "design earthquake" (especially in case the largest allowable behaviour or force reduction factor is selected at design) by allowing for a certain level of inelastic deformations (structural damage) to develop which should not jeopardize the local or global structural stability. In this respect, the successful implementation of the above seismic design philosophy, especially with regards to the critical issue of collapse prevention, relies on two basic conditions:

- Sufficient site-specific seismological data exist to define the "design earthquake" in a reliable manner (within a statistical framework), and
- Sufficient quality control applies to ensure consistency and to meet best practices in all phases of the production and operation of r/c building structures, including the design process, the construction, and the maintenance.

However, certain recent seismic events have made clear that the above conditions cannot always be considered as fulfilled. Specifically,

(A) For many regions, limited or no historical evidence exists of earthquake events having occurred in the distant past which might have been of significantly higher intensity than recently recorded ones. Noticeably, several recent high intensity earthquakes took place in regions classified to be of relatively low seismic risk (e.g., Athens/Greece 1999, L'Aquila/Italy 2009, Fukushima/Japan 2011, Christchurch/New Zealand 2011).

(B) There is a lack in the required quality control of building materials used and in the on-site inspection during construction phase which ensure that cast-in-place ductile r/c structures are built-as-designed. This is especially true in several seismically prone developing countries for various historical, cultural, or even political reasons.

The above two sources of uncertainty explain, to some extent, the significant structural damage observed and human casualties incurred in recent destructive earthquakes, even in cases of structures with considerable member ductility (Fig. 1.13).

Besides the above listed points (A) and (B), there are further important practical issues which question the appropriateness of adopting a partial protection against structural damage seismic design philosophy, as detailed below:

- The high concentration of human population and activities within a relatively small number of urban centers has led to high-density of well-localized large investments in building structures which will unavoidably result in ever greater economic losses in the case of major earthquake events well beyond the design earthquake, due to expected structural damage or, even worse, to unexpected, but likely to happen, building collapses.

Fig. 1.13 Structural collapse of a parking station during the Loma Prieta earthquake (source: http://ghestalt.egloos.com/2530520)

- The rise in the World's average living standards during the last few decades and the perceived capabilities of today's level of technology have led to a steadily increasing intolerance for accepting the possibility of large economic loss due to natural disasters such as earthquakes.
- As the cost of materials decreases inversely proportional to labor costs, adopting a partial protection against structural damage seismic design approach may not be as cost-saving compared to a full protection against structural damage as in past decades. Further, the availability of novel building materials (e.g., high-strength r/c) and advancements in conceptual and architectural design provide more options for aesthetically pleasing structures designed to remain linear under the "design earthquake".
- Last but not least, it is reiterated that the State and/or pertinent regulatory agencies set, via seismic codes of practice, the *minimum allowable limits* of structural safety against the seismic hazard, leaving the choice for a higher safety level open to the structure's owner. Further, it is natural to expect that communities and non-expert individuals assume that contemporary code-compliant "earthquake resistant" structures are "earthquake-proof" and should suffer zero damage during earthquakes corresponding to the nominal "design earthquake". Therefore, why should the design structural engineer take the responsibility of applying the minimum accepted level of seismic safety to a structure by adopting the code-specified (maximum) value of behaviour factor? The choice of the achieved safety level of a structure, entailing certain cost and "risk" considerations, should normally be with the "client", that is, with the owner.

During the past two decades, the above issues and concerns have led to the development of a new earthquake-resistant structural design approach, which leaves space for more options in defining the behaviour or the *performance* of a structure against different levels of seismic excitation (Fajfar and Krawinkler 2004). Termed *performance-based seismic design* (PBSD), this emerging approach for structural design *allows for owners of structures or competent authorities to select as a design objective for a given structure a set of different structural performance levels dependent on the level of seismic hazard and on the "importance" of the structure*. PBSD further *provides tools to design engineers to ensure that these performance levels are met for each considered level of seismic input action*. PBSD has already been adopted by most contemporary codes of practice regulating the seismic assessment and upgrading of *existing* code-deficient structures but is not yet widely considered sufficiently mature for the practical seismic design of *new* ordinary structures.

The following section presents briefly the underlying philosophy of PBSD, while the subsequent section discusses the most recent guidelines adopting the PBSD approach for new structures which are expected to influence the next generation of codes of practice.

1.3.2 Early Guidelines for Performance-Based Seismic Design and Their Relation to the Traditional Design Philosophy

Performance-based seismic design (PBSD) comprises a set of organized principles, rules, methods, and criteria (qualitative and quantitative) aimed at designing structures with a specified seismic behaviour (performance) for one or more level(s) of seismic input action. The need to develop such a design approach was triggered by discussions and concerns raised in the US in the aftermath of certain destructive earthquakes that took place in the late 1980s to mid-1990s that caused unexpectedly high economic losses to major metropolitan areas (e.g., Loma Prieta/California 1989, Northridge/California 1994, Kobe/Japan 1995). It was then realized that, even though the conventional partial protection seismic design philosophy (see Sect. 1.1.4) as implemented in codes of practice (see Sect. 1.2) offers reasonable life safety against major earthquakes, the cost of damage repairs, downtime, and relocation of business and commercial activities in densely populated urban regions of developed countries is unacceptably high. To this end, a demand for a flexible design philosophy for earthquake resistance to achieve specific *structural performance* for several different levels of seismic hazard emerged. Under the supervision of the Federal Emergency Management Agency (FEMA) in the US, the above ideas were put on paper in the Vision 2000 document back in 1995 (SEAOC 1995) and have been evolving ever since in the form of guidelines and commentaries for practicing engineers e.g., ATC-40, FEMA-273 (FEMA 1997), SAC/FEMA-350 (SAC/FEMA 2000), ASCE-31 (ASCE 2002), and ASCE-41 (ASCE 2007). These series of guidelines focus on the seismic assessment of existing (older) code-deficient structures expected to exhibit severe inelastic behaviour under major earthquakes in order to rationalize rehabilitation (retrofit and/or upgrade) decisions. Nevertheless, the philosophy of PBSD is also pertinent to the design of new structures, as is further explained below.

In the context of PBSD, a series of discrete levels of (potentially desirable by owners) seismic performance of structures is defined qualitatively. These performance levels correspond to different limit states of (tolerable) structural damage described in detail. As an example, Table 1.4 lists the performance levels for the seismic assessment and rehabilitation of existing code-deficient structures defined in the FEMA-356 (FEMA 2000) pre-standard and in the Greek National Annex of EC8-Part 3 (CEN 2004c; Hellenic Organization for Standardization 2009) which will be put to force along with the already in force, Greek Code for Seismic Interventions C.S.I. (Earthquake Planning and Protection Organization (EPPO) 2013).

For visualization purposes, the seismic performance levels of Table 1.4 are pictorially represented in Fig. 1.14, for a typical r/c building with brittle (brick) infill walls on a typical base shear vs. top storey displacement graph (see also Fig. 1.5).

Further, PBSD considers appropriate structural analysis methods and prescribes qualitative verification checks and criteria to ensure that the desirable (agreed) performance level(s) are met for specific *seismic hazard levels*. The latter are quantitatively determined in a statistical/probabilistic sense for each geographic region. For example, suppose there are two site-specific seismic hazard levels of concern for the *design of new buildings*: level 1 with 10 % probability of being exceeded in 50 years corresponding to a seismic event with mean return period of about 474 years, and level 2 with 2 % probability of being exceeded in 50 years corresponding to a seismic event with mean return period of about 2500 years (see Sect. 2.3.1.1). The PBSD design approach allows for a matrix of *performance objectives* to be formed, as shown in Fig. 1.15.

Noticeably, the dual performance objective k + p essentially coincides with the "traditional" *basic design objectives* for new structures incorporated in the current codes of practice following the partial protection against structural damage delineated in Sect. 1.1.4. Specifically, the following structural performance levels can be readily mapped onto the current design objectives of code-compliant r/c buildings (see also Fig. 1.16):

(a) Negligible or light damage to non-structural members for occasional earthquakes of low to medium intensity (>Level 1)- *Immediate Occupancy performance level*,
(b) Substantial but likely to be repairable damage for the (rare, strong) "design earthquake" of high intensity (Level 1)- *Life Safety performance level*,
(c) Severe damage driving structures close to collapse for (very rare, very strong) seismic events beyond the "design earthquake" (Level 2).

Nevertheless, PBSD also allows for the explicit consideration of enhanced (more stringent) performance objectives, such as j + o or i + n, to be met either by

Table 1.4 Commonly considered seismic performance levels by codes and guidelines for existing buildings

FEMA356	EC8- part 3 (National Greek Annex/C.S.I.)
Operational (OP)	–
Very light (practically zero) damage	
Immediate Occupancy (IO)	Immediate use after the earthquake
Light damage: Practically linear structural behaviour; no residual drift; original strength and stiffness is retained.	
Life Safety (LS)	Life Safety
Substantial damage to structural members: Building may be beyond economical repair; some permanent drift; some residual lateral strength and stiffness at all storeys is retained.	
Collapse Prevention (CP)	Collapse Prevention
Extensive severe damage: Building is near collapse; large permanent drifts; vertical members can bear gravitational loads.	

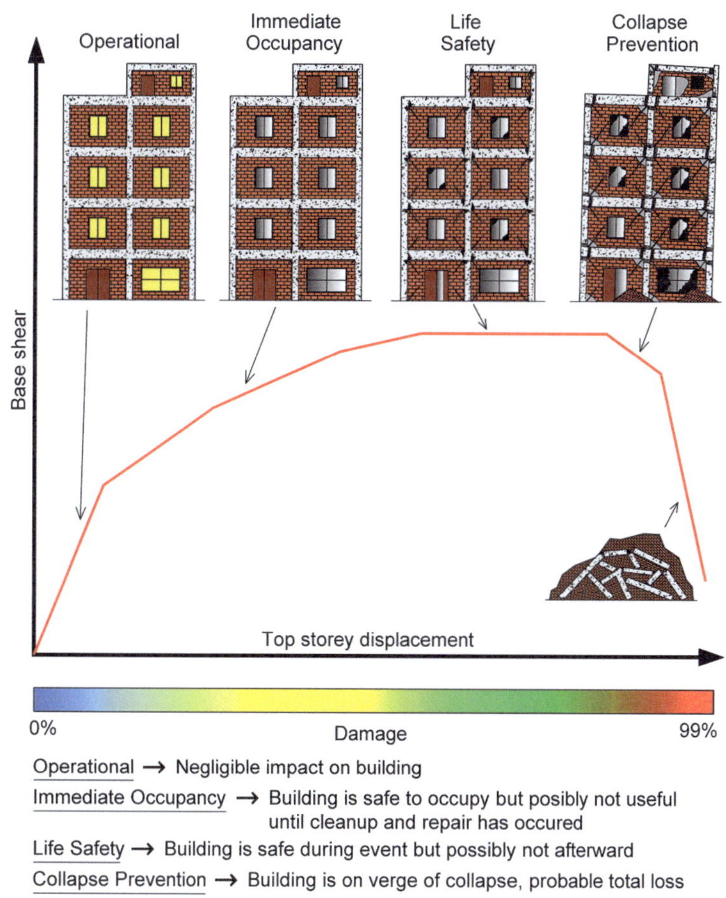

Fig. 1.14 Qualitative definition of seismic performance levels

Seismic Hazard Level			Structural Performance Level			
	Probability to be exceeded in 50 years	Mean return period	**Operational** Zero or negligible damage	**Immediate Occupancy** Damage only to non-structural members	**Life safety** Substantial, yet repairable damage to structural members	**Colapse prevention** Extensive severe damage
			→→ Gradually increased ductility demand →→			
Level 1	10%	474	i	j	k	(l)
Level 2	2%	2475	(m)	n	o	p
			→→ Gradually reduced performance →→			

Enhanced objectives: j+o or i+n to meet special building requirements (e.g., hospitals) or additional requirements of the building's owner.

Basic "traditional" objective k+p to ensure:
(a) negligible damage under minor earthquakes
(b) repairable damage under a level 1 earthquake
(c) collapse prevention under a level 2 earthquake

Fig. 1.15 Example of a performance objectives matrix for new buildings with two seismic hazard levels following the PBSD framework

important structures (e.g., schools, hospitals, large occupancy public buildings etc.), or by ordinary structures whose owners desire a higher level of safety against seismic hazard. As a final note on the example matrix of Fig. 1.15, the (m) objective is considered unattainable, while the (l) objective is unacceptable.

It is further emphasized that, as in the case of structural performance levels, the PBSD approach can accommodate an arbitrarily large number of seismic hazard levels. Consequently, an arbitrarily large number of performance objectives to be satisfied by new structures via explicit analysis and verification checks can be considered within a PBSD framework. In this regard, it is instructive to consider the performance objectives matrix shown in Fig. 1.17, set forth in the SEAOC document (SEAOC 1999) pre-standard seismic design recommendations for new structures. This matrix incorporates four seismic hazard levels and considers four performance levels which practically coincide with those identified in Fig. 1.14, though the terminology used is slightly different.

The following observations can be made with regards to the "color-mapping" used in SEAOC (SEAOC 1999):

- "Unacceptable" objectives are marked in red.
- The basic design objective (i.e., minimum for ordinary structures) requires four performance targets to be satisfied marked in green. As in the case of the performance matrix of Fig. 1.15, these "green" performance targets follow closely the conventional design objectives adopted by current codes of practice with the stipulation that a single nominal "design earthquake level" with 10 % probability of being exceeded in 50 years is assumed. Therefore, *it can be argued that they reflect the "partial" protection against structural damage* philosophy for earthquake resistance, (see also Sect. 1.1.4 and Table 1.1). *It can be further claimed that code-compliant r/c structures designed for the*

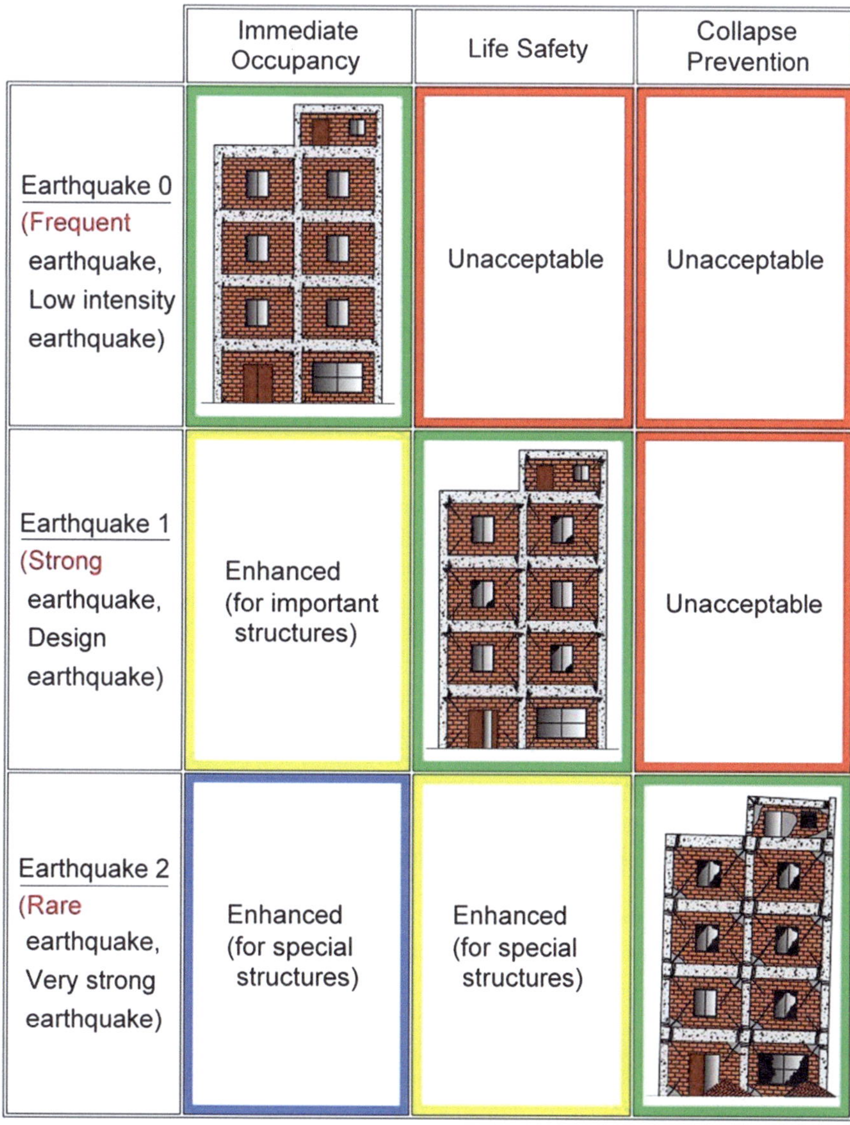

Fig. 1.16 "Traditional" seismic safety design objectives

maximum allowed behaviour factor ($q = maxq_{allow}$) *prescribed by current codes of practice (such as EC8) would, by and large, satisfy this set of basic performance targets.*

– The three performance targets marked in yellow represent we accept the change but please change "that" to "than" and can be viewed as the minimum allowed for "important" structures (see 4. in subsection 1.1.4). It can be argued that these "yellow" performance objectives correspond to roughly a "full" level of protection against structural damage assuming a nominal "design earthquake level" with 10 % probability of being exceeded for which a practically linear structural response ("Operational" performance level) is required. *Furthermore, it can be claimed based on heuristic arguments that code-compliant r/c buildings designed for a relatively low behaviour factor (e.g., $q \leq 1.50$–1.75) would satisfy this set of enhanced performance objectives provided that the usual levels of overstrength are ensured* (see also Sect. 1.2.3).

- The dual performance objectives marked in blue may be taken as the minimum ones for "special" structures whose collapse and/or downtime would affect large geographical areas and/or a considerable fraction of the total population of a country (see also Sect. 1.1.3.1).
- Marked in black is the practically unattainable objective of an "absolute" protection against seismic hazard, as described in Sect. 1.1.2. That is, no damage for (almost) any seismic action.

The above observations point to the well-recognized fact (Fardis 2009) that the traditional design objectives set by current codes of practice can be readily mapped onto appropriately defined performance objective matrices obtained via the PBSD philosophy. Moreover, they suggest, as exemplified above, that the PBSD approach offers a framework in which the choice of a value of the behaviour factor (or force reduction factor) can be rationalized and loosely mapped to expected structural performance levels. However, given the uncertainties associated with the analytical and experimental studies, it is still founded on rather qualitative criteria and engineering judgment. This is because the common force-based linear response spectrum methods of analysis and related verification checks (see Sects. 1.2.4 and 1.2.5) do not involve explicit assessment/verification of structural inelastic behaviour. Such behaviour is expected for seismic hazard levels corresponding to the "rare" earthquake of Fig. 1.17 or above for code compliant r/c buildings designed for $q = q_{allow}$, and for seismic hazard level corresponding to the "very rare" earthquake of Fig. 1.17 or above for code compliant r/c buildings designed for values of q approximating their overstrength.

The application of a full-fledged PBSD for new structures considering two, three, or more performance objectives to be simultaneously met and involving explicit verification checks and assessment of the actual structural behaviour attained in terms of (inelastic) demands for different seismic hazard levels goes beyond the current codes of practice. Therefore, this topic is not treated in this book. It is important, however, to acknowledge that, in the case of seismic assessment of existing structures, current codes of practice closely follow a performance-based approach as described in the Greek Code for Seismic Interventions (Earthquake Planning and Protection Organization (EPPO) 2013). Further, significant on-going endeavors of the earthquake engineering communities on both sides of the Atlantic to bring the PBSD approach for new structures closer to the everyday seismic design practice are being undertaken. In the past few years, two documents adopting a PBSD approach for both new and existing (code-deficient) structures

Seismic Hazard Level			Structural Performance Level			
Qualitative description	**Probability to be exceeded**	**Mean return period**	**Fully Operational**	**Operational**	**Life Safe**	**Near Collapse**
Frequent	50% in 30 years	43 Years	Basic Objective	Unacceptable	Unacceptable	Unacceptable
Occasional	50% in 50 years	72 Years	Essential Hazardous Objective	Basic Objective	Unacceptable	Unacceptable
Rare	10% in 50 years	475 Years	Safety Critical Objective	Essential Hazardous Objective	Basic Objective	Unacceptable
Very Rare	10% in 100 years	975 Years	Not Feasible	Safety Critical Objective	Essential Hazardous Objective	Basic Objective

Fig. 1.17 Performance objectives matrix for new buildings (SEAOC 1999)

became available, namely the American ATC-58 (ATC 2009) and the European Model Code (fib [fédération international du béton] 2012) following the preceding ASCE41-06 (ASCE 2007). It is expected that these documents will considerably influence the next generation of seismic codes of practice world-wide and, therefore, they are briefly discussed in the following sub-section.

1.3.3 Recent Guidelines on Performance-Based Seismic Design for New Structures (MC2010 and ATC-58)

Finalized in 2012, the so-called "Model Code 2010", produced by *fib* (Federation Internationale du Beton) (fib 2012), is the most recent document in Europe to provide recommendations on the seismic design for *both new and existing* r/c structures. It is meant to serve as a basis for the development of future codes of practice, in a similar manner as the *fib* "Model Code 1990" served as the basis for the current European code for r/c structures, namely Eurocode 2 (EC2) (CEN 2004b), which, however, does not include seismic design considerations covered within EC8. The 'model' for EC8 was CEB Model Code (CEB (Comité Euro-international du Béton) 1985). MC2010 adopts a full-fledged PBSD approach (Fardis 2013). In setting the design objectives, four structural performance levels (limit states) are considered which practically coincide with those of FEMA356 (Table 1.4). Further, four seismic hazard levels are identified which are very similar to those adopted by SEAOC 1999 (Fig. 1.17). Thus, a four-by-four performance objective matrix can be formed and, following the standard PBSD philosophy, the owner of the structure may set the performance targets to be met by the structure as a function of the level of seismic hazard and the nature or "importance" of the structure. Similar to the matrix of Fig. 1.17, the design objective for ordinary structures should observe the life safety performance for the "rare" earthquake and immediate occupancy for the "occasional" earthquake as minimum requirements. MC2010 adopts a displacement-based seismic design approach and favours dynamic response history analysis as the recommended ("reference") type of structural analysis for estimating (inelastic) deformation demands for the different hazard levels considered. Verification checks based on displacement demands and capacities of structural members are undertaken and the, possibly non-linear, behaviour of structures is assessed and quantitatively verified. Furthermore, MC2010 introduces the aspect of "time" in assessing structural performance taking into account the inevitable decay/deterioration of structures with time and aiming to provide guidance on a full life-cycle assessment of structures (Walraven and Bigaj 2011; Walraven 2013).

In 2012, the FEMA P-58 series of documents and related software tools became available in the public domain (ATC 2012). FEMA P-58 is the product of a 10 year effort undertaken by the American ATC (Applied Technology Institute) to provide guidelines on the practical implementation of the probabilistic performance-based earthquake engineering (PBEE) framework developed within the Pacific Earthquake Engineering Research Center (Moehle and Deierlein 2004). Similar to MC2010, FEMA P-58 uses a PBSD approach to seamlessly address both the design of new building structures and the assessment of existing ones, though it is not restricted to r/c buildings alone. However, the scope of FEMA P-58 goes far beyond the current structural design codes for earthquake resistance (even the so-called "first generation" performance-based codes reviewed in Sect. 1.3.2) since the "seismic structural performance/behaviour" is expressed in terms of potential losses and consequences, such as repair/replacement cost, downtime, and casualties, instead of the usual structural response terms, such as deformations and stress resultants. By adopting such a non-engineering vocabulary to define structural performance, FEMA P-58 aims to facilitate decision making by the intended stakeholders (e.g., structure owners, authorities, etc.) on the desired level of protection against seismic hazard. To accomplish this aim, standard structural analysis results are first coupled with "fragility functions" (damage analysis step) which represent the probability that a certain level of physical damage in a structural member is exhibited given specific values of structural response quantities (e.g., rotations at the end of r/c beams). Next, a loss analysis step is undertaken to estimate, statistically, the performance of a structure (interpreted as an integrated system of structural members/components) in terms of a "decision variable" (e.g., repair cost, fatalities) given the expected level of damage at each structural member/component obtained from the damage analysis step.

The practical implementation of the adopted probabilistic PBSD methodology of FEMA P-58 relies heavily on the use of a particular software tool along with an expandable database, both freely available on-line through the ATC website, which incorporates statistical data required for the damage and loss analyses steps. Interestingly, the methodology allows for seismic performance assessment of new code-compliant (r/c) buildings for a given level of seismic action represented by means of any code-specified elastic response spectrum (i.e., green curve in Fig. 1.10). This type of assessment is termed "intensity-based" and is, arguably, the most closely related to the current codes of practice.

The above brief qualitative description of MC2010 and FEMA P-58 suggests that next generation seismic codes of practice will allow for enhanced flexibility in setting case-dependent requirements and design objectives dependent on the building importance, occupancy, available resources, and level of seismic hazard. Besides technical details, an important practical aspect to be introduced is that *the choice of the desired seismic performance will be made by the owners/end-users of the structure or other stakeholders and not by the design engineer alone, taking into account life-cycle performance issues and considering consequences (in terms of replacement cost, down time, etc.) in the case in which a "rare" earthquake scenario occurs*. In support of that, analysis methods and verification checks will ensure transparency and will take advantage of the knowledge accumulated over the past few decades by researchers and field observations to assess that the intended requirements are met in an explicit manner. The latter involves treating, at least in terms of analysis methods, both newly designed buildings and existing (code-deficient) structures in the same manner. Further, the time factor will be taken into account aiming not only to assess new designs at the time they are constructed, but also to predict their future performance accounting for deterioration and addressing sustainability and life-cycle performance issues.

The final section of this chapter makes certain recommendations on the interpretation and use of current conventional codes of practice relying on the traditional force-base prescriptive methodology for the design of new structures to achieve enhanced seismic performance (i.e., level of protection against seismic hazard), beyond that minimally prescribed.

1.4 On the Selection of a Desired Performance Level in Code-Compliant Seismic Design of New R/C Buildings

As detailed in Sect. 1.2, current seismic codes of practice for ordinary r/c building structures adopt a force-based design approach based on linear types of analysis using reduced values (50 % according to EC8) for all structural elements' flexural and shear stiffnesses and considering reduced design seismic loads by a force reduction factor or behaviour factor q. The bulk of the prescribed verification checks focus on a single design objective (life safety/low performance) for a particular level of a "design seismic action" (typically having a probability of 10 % of being exceeded in 50 years and an average return period of about 475 years), while the actual structural performance (damage level) under the design earthquake is not explicitly assessed.

The *minimum* possible value of the design seismic force reduction factor is 1 (minq = 1). For q = 1, the structure is designed to behave linearly under the total (unreduced) design seismic load, that is, it (theoretically) suffers no structural damage for the design earthquake (operational/high performance). Consequently, the inherent ductility capacity and overstrength remain (theoretically) unutilized under the design earthquake and constitute safety reserves to resist higher-than-the-design levels of seismic action without collapsing.

However, as already mentioned, codes of practice define only a *maximum* allowed value for the design seismic force reduction (behaviour) factor q_{allow} depending on the (intended/targeted) ductility capacity ($\max q = \mu_{cap}$). For $q = q_{allow}$, the maximum possible (allowed) utilization of the ductility capacity takes place to resist the design seismic action by exhibiting severe inelastic behaviour without collapsing (life safety/low performance). In this regard, it is a common practice for design engineers to choose $q = q_{allow}$, without having the (written) consent of the owner of the structure and without communicating what this choice entails in case a future "design earthquake" occurs during the lifetime of the structure.

This practice is not in line with the well-established performance-based seismic design (PBSD) approach, briefly reviewed in Sect. 1.3, in which more objectives involving a set of pre-determined performance levels for different levels of seismic input action are prescribed. More importantly, in PBSD, the choice of design objectives is made by the owner in consultation with the design engineer. Still, the assumption that the design objectives of current codes of practice can be mapped onto standard performance objectives within a PBSD approach, as detailed in Sect. 1.3.2, renders possible the consideration of performance-based practices within a force-based design approach. Thus, some of the limitations of the current code-compliant seismic design practices may be circumvented. This can be better understood by emphasizing that, although current codes of practice require structural analysis to be performed explicitly for only one level of a "design" seismic action, *it is up to the design engineer in agreement with the owner to select a behaviour factor q smaller than the maximum allowed q_{allow} which corresponds to the basic objective (low performance) or to a partial level of protection against structural damage*. Therefore, in response to the very reasons that dictated the need for a PBSD philosophy in the first place, listed in Sect. 1.3.1, *it is herein recommended that design engineers opt for "full" protection against structural damage for the nominal design earthquake, unless the owner of the structure consents to adopt "partial" protection.*

In practical terms, "full" level seismic protection for ordinary r/c buildings is achieved by adopting a relatively low behaviour factor q close to the roughly estimated overstrength (e.g., $q \sim 1.5$). The lateral load resisting system of code-compliant r/c buildings designed for such values of q would practically respond linearly (only very light local damage might occur) for (future) earthquake events with 10 % probability of being exceeded in 50 years. In this regard, a need for (limited) repairs to only non-structural components may be required after such an earthquake event (immediate occupancy performance level). As a rule of thumb, according to an extensive parametric research study carried out within the framework of the Greek Seismic Code EAK2000 (Earthquake Planning and Protection Organization (EPPO) 2000) for ordinary R/C buildings of up to eight storeys (Avramidis and Anastassiadis 2002), the herein recommended enhanced level of structural safety against seismic design action can usually be accomplished by including strong concrete walls whose total length along each principal direction of the structure should be about twice the required length corresponding to a design for $q = 3.5$. Appropriate conceptual design considerations (see Sect. 2.1) need to be followed in choosing the in-plan location of these walls along with close collaboration with the architectural design team to ensure limited influence on the aesthetics and functionality of buildings (Avramidis et al. 2000, Avramidis and Anastassiadis 2002).

It is quite interesting to note that, as shown in the aforementioned research study, additional construction cost of designing different types of 4-storey to 8-storey r/c buildings for a low behaviour factor (e.g., $q = 1.5$ – Full protection against structural damage for the nominal design earthquake), as compared to the maximum allowed behaviour factor (e.g., $q = 3.5$ – Partial protection), ranges between 3 and 10 % of the total cost of the structure. For a fixed overstrength factor, this additional cost depends significantly on the site-specific level of seismic hazard (see Sect. 2.3.1.1). The additional cost will be relatively low for low seismic hazard zones. This is due to the fact that the detailing of a large number of r/c structural members is normally dominated by the minimum reinforcement requirements for structures located in low seismicity areas. Therefore, a uniform increase of seismic design loads would not impart a proportional increase to the dimensions and reinforcement of many structural members and, consequently, to the total cost of such structures. Nevertheless, for slender r/c buildings with a relatively high total height over plan dimensions ratio, as well as for buildings in high seismicity zones, significantly large footings for certain r/c (wall) elements may be required in order to achieve full protection against structural damage for the nominal design earthquake, to accommodate increased demands for compressive stresses to the supporting ground, and to control overturning due to tensile stresses at the foundation level. In case the cost of such footings become overly high and/or the bearing capacity of the supporting ground is poor, or, more generally, in case the additional construction cost to achieve a full level of protection against seismic hazard is not affordable, it is still recommended to design for as low a behaviour factor as practically possible. For instance, an "almost full" level of seismic protection against the seismic hazard can be achieved by choosing values of the behaviour factor within the range of $1.75 < q < 2.5$. Designing for such behaviour factors would still lead to significantly enhanced seismic structural performance for the nominal design earthquake compared to that achieved with the commonly prescribed maximum allowed behaviour factors (i.e., $q > 3.0$ for r/c structures).

Perhaps the most important consequence of designing for low behaviour factors to achieve high levels of structural seismic performance is that the design seismic loads are primarily resisted by means of strength (strength dominated design) without the need to utilize considerable fractions of the inherent ductility capacity of r/c structures. For example, a structure designed for $q = 1.5$ has approximately more than double the lateral strength of a structure designed for $q = 3.5$, since $3.5/1.5 \approx 2.3$. Therefore, significant reserves of ductility capacity remain to offer enhanced structural safety in cases of increased seismic demands or of reduced structural capacity due to various "accidental" and unforeseen effects not explicitly considered in codes of practice, such as:

- "Very rare" seismic events exceeding the nominal "design seismic action" level (see also the discussion in sub-section 1.2.6.4),
- Local site amplifications of the earthquake induced ground motion beyond that expected as captured by the design response spectrum due to poor soil classification assumptions (see also Sect. 2.3.1.2),
- Lateral load resisting systems of reduced capacity compared to the "as designed" structure (in terms of stiffness, strength, or ductility) due to poor workmanship of cast-in-place r/c structures and/or poor quality control and inspection during construction,
- Slab-to-slab or slab-to-column pounding of buildings with adjacent structures during major seismic events.

Further important advantages of designing for full protection against structural damage as opposed to partial protection within the common code-prescribed force-based design framework (described in Sect. 1.2.4) are:

- Standard response spectrum based linear types of structural analysis and the underlying finite element models used (see Sects. 2.4 and 2.3.2, respectively) become more reliable in predicting the actual (extreme) structural behaviour under the design seismic action.
- The construction of (strong) r/c walls, resisting the greater part of the lateral inertial seismic loads in designing for low values of the behaviour factor, is easier to achieve in practice and less prone to errors due to poor workmanship/inspection compared to the construction of (ductile) moment resisting frames.
- R/c walls are more likely to be able to resist gravitational loads after being severely damaged (near collapse stage) by extreme intensity earthquake events compared to moment resisting frames (Fintel 1991, 1995).
- The demand by contemporary societies and local communities for reduced structural damage, repair cost, and downtime in the aftermath of a seismic event corresponding to the "design earthquake" is satisfied, while the risk for human casualties becomes practically negligible.

As a final note, it is emphasized that, even if adopting a low behaviour factor q (e.g., $q = 1.5$) would normally relax the need to consider code-prescribed capacity design rules, it is still highly recommended that all the required capacity design verification checks and detailing rules are taken into account to achieve the intended level of ductility corresponding to the ductility class of choice (see also Sect. 3.1.3). Clearly, this discrete performance level result in an additional, yet quite limited and in no case prohibitive, construction cost. This recommendation is rather pertinent for moment resisting frame structural systems.

The following chapters provide further comments concerning the selection of a desired level of seismic performance or protection against structural damage by means of adopting appropriate values for the behaviour or force reduction factor q. Specifically, Sect. 3.1.4 provides a flowchart for determining the maximum allowed behaviour factor ($maxq_{allow}$) according to EC8. Further, Sects. 2.3.1 and 2.4.1 discuss the influence of the behaviour factor value on choices made at the preliminary design stage and on the structural analysis method to be adopted for design, respectively.

References

ASCE (2002) ASCE/SEI standard 31–03, seismic evaluation of existing buildings. American Society of Civil Engineers, Reston

ASCE (2007) ASCE/SEI standard 41–06, seismic rehabilitation of existing buildings. American Society of Civil Engineers, Reston

ASCE (2010) ASCE/SEI standard 7–10, minimum design loads for buildings and other structures. American Society of Civil Engineers, Reston

ATC (2009) Guidelines for seismic performance assessment of buildings, ATC-58. American Technology Council, Redwood City

ATC (2012) Next-generation methodology for seismic performance assessment of buildings, prepared by the Applied Technology Council for the Federal Emergency Management Agency, Report No. FEMA P-58, Washington, DC

Avramidis IE, Anastassiadis K (2002) Development of benchmark numerical examples for supporting of Greek Seismic Code EAK 2000 guidelines and testing relevant software packages. Proposal of a new regulatory framework for antiseismic protection including an improvement on second-order effects. Athens, Greece

Avramidis IE, Anastassiadis K, Morfidis K (2000) Full and partial antiseismic protection of buildings – proposal for a new design concept. In: Proceedings of "G. Penelis Intern. Symposium on concrete and masonry structures", Thessaloniki, Greece, 13–14 October, Ed. Ziti

Bisch P, Carvalho E, Degee H, Fajfar P, Fardis M, Franchin P, Kreslin M, Pecker A, Pinto P, Plumier A, Somja H, Tsionis G (2012) Eurocode 8: seismic design of buildings worked examples, JRC technical report. Publications Office of the European Union/Joint Research Centre, Luxembourg

Booth ED, Lubkowski ZA (2012) Creating a vision for the future of Eurocode 8. 15th world conference on earthquake engineering, Lisbon, Portugal

CEB (Comité Euro-international du Béton) (1985) Model code for seismic design of concrete structures, CEB Bull. d' Inf., 165, Paris

CEN (2004a) European Standard EN 1998-1. Eurocode 8: design of structures for earthquake resistance, Part 1: general rules, seismic actions and rules for buildings, Committee for Standardization. European Committee for Standardization, Brussels, Belgium

CEN (2004b) European Standard EN 1992-1-1. Eurocode 2: design of concrete structures, Part 1-1: general rules and rules for buildings, Committee for Standardization. Brussels, Belgium

CEN (2004c) European Standard EN 1998-3. Eurocode 8: design of structures for earthquake resistance – Part 3: assessment and retrofitting of buildings, Committee for Standardization, Brussels, Belgium

Chopra AK (2007) Dynamics of structures—theory and applications to earthquake engineering, 3rd edn. Pearson, Prentice-Hall, Upper Saddle River

De Luca F, Vamvatsikos D, Iervolino I (2013) Near-optimal piecewise linear fits of static pushover capacity curves for equivalent SDOF analysis. Earthq Eng Struct Dyn 42:523–543. doi:10.1002/eqe

Earthquake Planning and Protection Organization (EPPO) (2000) Greek Seismic Code EAK2000 (amended in 2003). Athens, Greece (in Greek)

Earthquake Planning and Protection Organization (EPPO) (2013) Greek Code for Seismic Interventions (KAN.ΕΠΕ.). Athens, Greece

Elghazouli A (2009) Seismic design of buildings to Eurocode 8. Spon Press, New York
Elnashai AS, Di Sarno L (2008) Fundamentals of earthquake engineering. Wiley, Chichester
Fajfar P, Krawinkler H (eds) (2004) Performance-based seismic design: concepts and implementation, PEER report 2004/05. Pacific Earthquake Engineering Research Center, Berkeley
Fardis MN (2009) Seismic design, assessment and retrofit of concrete buildings, based on Eurocode 8. Springer, Dordrecht
Fardis MN (2013) Performance- and displacement-based seismic design and assessment of concrete structures in the model code 2010. Struct Concr 14:215–229. doi:10.1002/suco.201300001. Submitted
Fardis MN, Carvalho E, Elnashai AS et al (2005) Designer's guide to EN 1998-1 and EN 1998-5. Eurocode 8: design of structures for earthquake resistance. General rules, seismic actions, design rules for buildings, foundations and retaining structures. Thomas Telford Ltd, London
Fardis MN, Carvalho E, Fajfar P, Pecker A (2014) Seismic design of concrete buildings to Eurocode 8. CRC Press, New York. Weiler, WA
FEMA (1997) NEHRP guidelines for the seismic rehabilitation of buildings, FEMA 273. Federal Emergency Management Agency, Washington DC
FEMA (2000) Prestandard and commentary for the seismic rehabilitation of buildings, FEMA-356. Federal Emergency Management Agency, Washington DC
fib (2013) Model code for concrete structures 2010, Federation Internationale du Beton, Ernst & Sohn, Berlin
Fintel M (1991) Shear walls – an answer for seismic resistance? Concr Int 13:48–53
Fintel M (1995) Performance of buildings with shear walls in earthquakes of the last thirty years. PCI J 40:62–80
Hellenic Organization for Standarization (2009) Standard 1498-3. Greek National Annex to Eurocode 8 – Part 3: assessment and retrofitting of existing structures (in Greek)
Lindeburg MR, McMullin KM (2014) Seismic design of building structures: a professional's introduction to earthquake forces and design details. Professional Publications, Inc, Belmont, CA
McGuire RK (1995) Probabilistic seismic hazard analysis and design earthquakes: closing the loop. Bull Seismol Soc Am 85:1275–1284
Meskouris K, Hinzen K, Butenweg C, Mistler M (2011) Bauwerke und Erdbeben. Vieweg +Teubner Verlag, Wiesbaden
Midorikawa M, Okawa I, Iiba M, Teshigawara M (2003) Performance-based seismic design code for buildings in Japan. Earthq Eng Eng Seismol 4:15–25
Miranda E (2000) Inelastic displacement ratios for structures on firm sites. J Struct Eng 126:1150–1159
Moehle J, Deierlein GG (2004) A framework methodology for performance-based earthquake engineering. Int Work Performance-Based Eng Bled, Slov
Penelis GG, Kappos AJ (1997) Earthquake-resistant concrete structures. E & FN Spon, London
Penelis GG, Penelis GG (2014) Concrete buildings in seismic regions. CRC Press/Taylor & Francis Group, London
Ruiz-García J, Miranda E (2006) Inelastic displacement ratios for evaluation of structures built on soft soil sites. Earthq Eng Struct Dyn 35:679–694. doi:10.1002/eqe.552
SAC/FEMA (2000) Recommended seismic design criteria for new steel moment-frame buildings, Report No. FEMA-350, SAC Joint Venture. Federal Emergency Management Agency, Washington, DC
SEAOC (1967) Recommended lateral force requirements and commentary. Structural Engineers Association of California, Sacramento
SEAOC (1995) Vision 2000: performance-based seismic engineering of buildings. Structural Engineers Association of California, Sacramento
SEAOC (1999) Seismic design manual. Structural Engineers Association of California, Sacramento

Veletsos AS, Newmark NM (1964) Response spectra of single degree-of- freedom elastic and inelastic systems. In: Design procedures for shock isolation systems of underground protective structures, report RTD TDR-63-3096, vol III, Albuquerque, New Mexico
Villaverde R (2009) Fundamental concepts of earthquake engineering. CRC Press, New York. Weiler, WA
Walraven JC (2013) Fib model code for concrete structures 2010: mastering challenges and encountering new ones. Struct Concr 14:3–9
Walraven JC, Bigaj AJ (2011) The 2010 fib model code for concrete structures: a new approach to structural engineering. Struct Concr 12:139–147

Chapter 2
Design of R/C Buildings to EC8-1: A Critical Overview

Abstract This chapter provides practical recommendations for the preliminary seismic design and the finite element modeling of reinforced concrete (r/c) building structures assumed to behave linearly. It also discusses and provides commentary on structural seismic analysis methods adopted by Eurocode 8 (EC8). Specifically, the main principles of conceptual design for achieving well-qualified lateral load-resisting structural systems for earthquake resistance are briefly reviewed. Further, capacity design rules and local detailing practices for enhanced ductility capacity in r/c buildings are presented. Different types of structural analysis methods commonly employed in code-compliant seismic design of structures are outlined and focus is given to the EC8-prescribed equivalent linear analysis methods for forced-based seismic design, namely, the lateral force method and the modal response spectrum method. In this context, the EC8-compatible seismic design loading combinations and the EC8 design spectrum for elastic analysis are also presented. Moreover, the most commonly used finite element modeling practices for linear analysis of r/c multi-storey buildings are detailed, including the modeling of floor slabs, frames, planar walls, cores, and footings resting on compliant soil. Finally, brief comments are included on the proper use and quality verification of commercial seismic design software using benchmark structural analysis and design example problems.

Keywords Conceptual seismic design • Capacity design • Ductile detailing • EC8 response spectrum • Design spectrum • Loading combinations • Finite element modeling • Lateral force method • Modal response spectrum method • Static inelastic pushover method • Overstrength distribution • Benchmark problems

The seismic design process of a typical building structure comprises three phases, as delineated in Fig. 2.1.

In phase A, a load-resisting structural system is defined by considering certain conceptual design principles for earthquake resistance based on the given architectural plans. Typically, this phase involves selecting the type of lateral load-resisting system (e.g., moment-resisting frame system, wall system, dual system, etc.) and finalizing its configuration. This is achieved by first considering several different feasible layouts which take into account potential architectural and structural

I. Avramidis et al., *Eurocode-Compliant Seismic Analysis and Design of R/C Buildings*, Geotechnical, Geological and Earthquake Engineering 38,
DOI 10.1007/978-3-319-25270-4_2

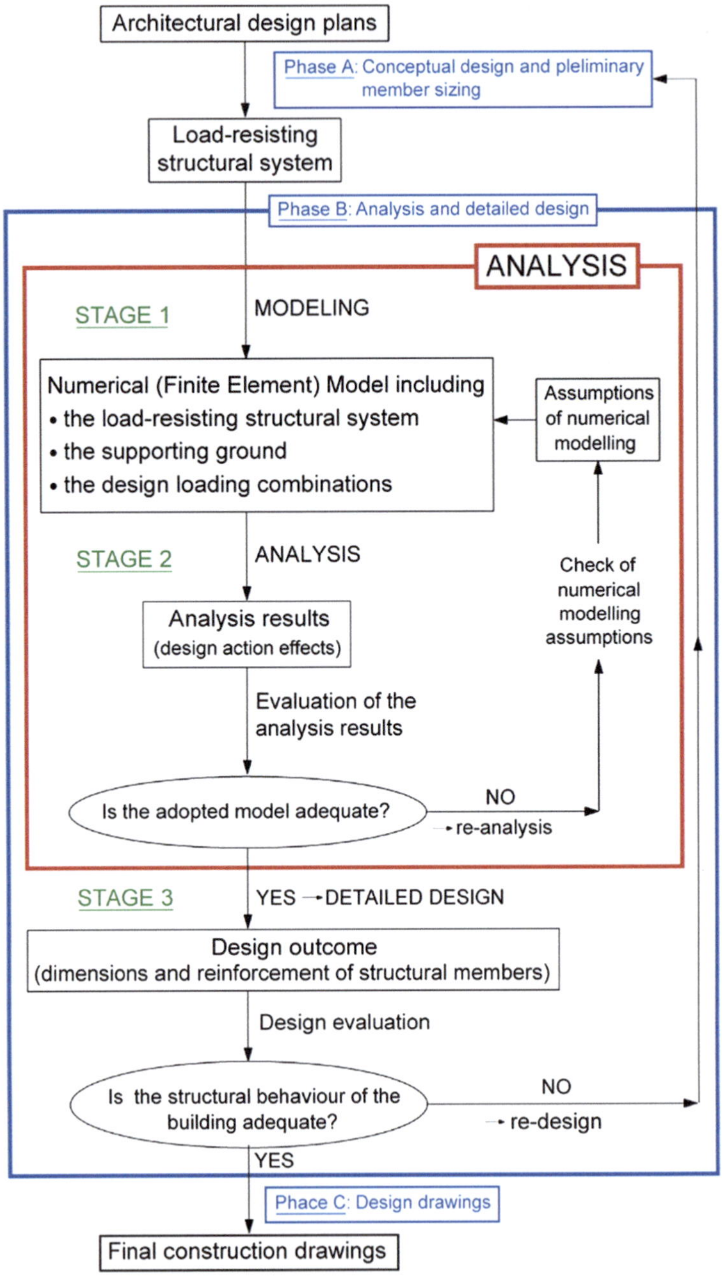

Fig. 2.1 Schematic sequence of phases and stages of the structural design process

constraints, building regulations, construction management and cost-effectiveness issues, as well as various other case-specific provisions. Next, the design structural engineer chooses a small set out of these feasible layouts for further investigation by relying heavily on his/her accumulated experience, expert judgement, and personal design preferences. The chosen layouts are examined to sufficient detail to make a quantitative comparison possible, and to finalize the configuration of the lateral load-resisting system to be considered in the second phase (phase B) of the seismic design process. To this aim, phase A involves undertaking only some preliminary (approximate) analysis steps to determine the initial sizes of r/c structural members.

In phase B, a finite element model (also called a mathematical or computational or structural analysis model; see elsewhere (Mac Leod 1995)) of the load-resisting system adopted from phase A is first developed (Stage 1: Modeling in Fig. 2.1). This is accomplished by relying on certain modeling assumptions and simplifications which are based on the in-depth knowledge of the analysis methods to be used. This model includes the building foundation system and superstructure and should take into consideration the compliance of the supporting ground, if deemed necessary. Next, the finite element model is used to calculate the "effects" (i.e., internal stress resultants/forces and deformations of structural members) of the design "actions" (i.e., design loading combinations including the seismic design loads) prescribed by the relevant design code regulations (Stage 2: Analysis in Fig. 2.1). At the end of this second stage, certain verification checks against (primarily) deformation-based criteria are made to ensure that the adopted dimensions of structural members are adequate. If these criteria are met, structural members are designed in detail (Stage 3: Detailed design in Fig. 2.1) to finalize their dimensions and the required reinforcement using the results (calculated action effects) of the analysis stage. The detailed design stage involves several verification checks to ensure that adequate levels of strength and ductility are achieved by considering appropriate longitudinal and transverse reinforcement at critical (energy dissipation) zones of structural members. Meeting these verification checks may require modifications of the adopted dimensions in a number of r/c structural members and, therefore, further re-analysis and re-design steps may be necessary to iteratively *optimize* the design of the load resisting structural system.

Finally, phase C involves the preparation of all necessary construction drawings and design plans incorporating the required reinforcement details and structural member dimensions for the practical implementation building design.

From the above brief overview of the seismic design process of a typical r/c building, it is seen that there are, at least, four stages involving critical choices and decisions to be made by the design engineer based on his/her knowledge and experience rather than on "black-box" types of calculation automated in commercial structural analysis and design software. These stages are listed below starting from those requiring more input on behalf of the designer in terms of experience and expert judgment:

Table 2.1 Mapping of seismic design process stages onto required knowledge on behalf of the design engineer and pertinent clauses of EC8-part 1

Phases and stages of the seismic design process		Required engineering knowledge	Main relevant EC8-part 1 chapters and clauses for r/c buildings
Architectural plans			
Phase A	Conceptual design of the load resisting system and preliminary member sizing	Appreciation of the seismic design "philosophy" underpinning current codes of practice	Chapter 2 Performance requirements and compliance criteria
		Selection of the desirable structural performance level	§4.2.1 Basic principles of conceptual design
			§5.2 Design concepts
Modeling of:			
	the load resisting structural system and its foundation	Knowledge and understanding of the finite element method using equivalent frame models as well as 2-D finite elements	§4.2.3 Criteria for structural regularity
	the supporting ground		§4.3.1 Modeling
			§4.3.6 Additional measures for masonry infill walls
			§4.3.1(9)P
	the design loading combinations	Access to EC1 clauses and understanding of the response spectrum concept and its use in seismic design	Chapter 3 Ground conditions and seismic action
			§4.2.4 Load combination coefficients for variable actions
			§4.3.2 Accidental torsional effects
Phase B	Structural analysis and deformation-based verification checks	Knowledge of (static and dynamic) structural analysis methods involving finite element models	§4.3.3 Methods of analysis
			§4.3.4 Displacement calculation
			Verification checks:
			§4.4.2.2(2): $\theta \leq 0.1$
			§4.4.3.2(a): $d_r v \leq 0.005\ h$
	Final detailing of structural members and verification checks	Access to EC2-part 1 clauses and knowledge of design and detailing of r/c structures	§5.4 Design for Ductility Class Medium (DCM) Buildings
			§5.5 Design for Ductility Class High (DCH) Buildings

(continued)

Table 2.1 (continued)

Phases and stages of the seismic design process		Required engineering knowledge	Main relevant EC8-part 1 chapters and clauses for r/c buildings
			§5.6 Anchorage and splices
			§5.8 Concrete foundations
			§5.9 Local effects due to masonry or concrete infill walls
			§5.10 Provision for concrete diaphragms
			§5.11 Pre-cast concrete structures
Phase C	Final design and implementation drawings	Computer-aided design software (CAD)	

- conceptual design of the lateral load-resisting system,
- development of the numerical (finite element) model,
- critical appraisal and verification of analysis results, and
- detailed design of structural members.

Further, Table 2.1 maps the required knowledge that the design engineer should possess onto the main stages of the design process as discussed above and presented in Fig. 2.1. The most relevant chapters and clauses of EC8 part-1 to be consulted in each of the identified design stages are also included in Table 2.1.

The remainder of this chapter is organized into five sections following the design process steps outlined in Fig. 2.1. In Sect. 2.1, the main principles of conceptual design for achieving well-qualified lateral load-resisting structural systems for earthquake resistance are briefly presented. Section 2.2 discusses certain (capacity) design rules and local detailing practices for enhanced ductility capacity which facilitate the preliminary sizing of structural members for ductile r/c buildings. Further, Sect. 2.3 provides details on developing appropriate (finite element) structural models and defining the EC8-compatible seismic design loading combinations to be used in the analysis stage of the design process. Next, Sect. 2.4 focuses on the different types of structural analysis methods commonly employed in the code-compliant seismic design of structures. Finally, Sect. 2.5 includes a brief discussion on the quality verification and proper use of commercial structural analysis and design software which is an essential tool for elaborating reliable designs of structures for earthquake resistance.

2.1 Conceptual Design Principles for Earthquake-Resistant Buildings

2.1.1 Desirable Attributes of the Lateral Load-Resisting Structural System and Fundamental Rules

In the recent past, the structural layout of ordinary buildings had to be kept relatively simple and straightforward from a structural analysis viewpoint to ensure that the analysis and detailing steps undertaken by structural engineers were accomplished in reasonable time in the absence of high computational power. During the past two decades, the advent of powerful low-cost computers and dependable commercial finite element-based analysis and design software have built confidence among practicing structural engineers that "almost any structural layout" can be readily and swiftly designed. In this regard, reduced time and effort is spent in the first phase of the design process of common buildings (i.e., the conceptual design of the load resisting structural system), since the second phase (i.e., modeling, analysis, and detailing) is seen as a mere "computer data input" problem. Thus, structural engineers may sometimes find themselves obliged to design structures within tight timescales based on hastily conceived structural layouts of questionable rationale and on their corresponding mathematical (computational) models using three-dimensional linear finite element structural analysis software. This approach is erroneous, especially when it comes to structures subjected to seismic excitations. The reason is that *the deficiencies of an inadequate lateral load-resisting structural system inherently vulnerable to seismic input action cannot be ameliorated or rectified at any later phase in the design and/or construction process (i.e., not even by the most consistent and detailed structural analysis and detailing steps)*. This is true irrespective of the adopted type of lateral load-resisting structural system of the structural analysis method of choice. In fact, a deficient structural layout adopted during the conceptual design stage can hardly ever be brought to the same level of seismic performance with a structural layout satisfying certain qualitative criteria and rules in line with the seismic design philosophy adopted by the current codes of practice.

In this respect, it can be readily recognized that the conceptual design stage is the one least dominated by the use of automated software. It relies heavily on the experience, expertise, and subjective preference of the design engineer to accommodate the case-dependent architectural requirements. Certain research efforts for the development of automated computational tools (relying on principles from the field of Artificial Intelligence) to support and assist design engineers in composing alternative structural layouts have been made in the past few decades (see, e.g. (Avramidis et al. 1995; Berrais 2005)). However, such "knowledge-based expert systems" have not yet reached a satisfactory level of maturity and are not considered to be capable of offering enhanced solutions beyond the average level of creativity of structural design experts. Despite being subjective in many respects, there exist qualitative rules and criteria for facilitating the conceptual design phase.

Most of these conceptual design rules have already been listed in Sect. 1.2.5 under the first of the three essential classes of requirements for ductile structural behaviour under design seismic action, namely the maximization of dispersion of the seismic input energy within the lateral load-resisting structural system. Further, some of these rules are also related to the requirements for prevention of (premature) global structural instability/collapse and for maximization of the dissipation of the seismic input energy via hysteretic behaviour. However, the last two requirements are more closely related to capacity design rules and structural member detailing for ductility which are discussed in some detail in the next section.

Focusing on the specifics of EC8, the following list of desirable attributes that structural layouts should observe to expedite the code-compliant seismic design process is included in clause §4.2.1 of EC8- Basic principles of conceptual design for earthquake resistant building structures:

- Structural simplicity,
- Uniformity, symmetry (regularity), and redundancy,
- Bi-directional resistance and stiffness,
- Torsional resistance and stiffness,
- Diaphragmatic behaviour at the storey level,
- Foundation capable of transmitting the superstructure forces to the ground.

Certain brief clarification notes highlighting the meaning and importance of the above qualitative conceptual design rules follow.

Structural simplicity

A simple load resisting structural system in plan and elevation ensures that unambiguously identifiable, continuous, and relatively short stress/load paths exist through which all external loads applied statically (gravitational loads) and dynamically (lateral inertial seismic loads) are transmitted from building superstructure to its foundation and the supporting ground. Complex or indirect load paths (e.g., due to columns supported by beams) may result in undue local stress and strain concentrations and, thus, to increased local strength and ductility demands. In the inelastic range of structural behaviour (which is expected to be severe for large values of the behaviour factor q), such local ductility demands may not be adequately captured by code-prescribed linear types of analysis methods. Consequently, the code-compliant seismic design process becomes inherently less reliable in accounting for and properly verifying the expected local ductility demands for structural layouts of increased complexity. Therefore, safeguarding simplicity and clarity of the lateral load resisting system at the conceptual design stage is essential for reducing the inherent uncertainties associated with the analysis, detailing, and construction of earthquake resistant buildings complying with the intended code-specific requirements.

Uniformity, symmetry and redundancy

It is well established through field observations, large scale experimental results, and computational/analytical research work that building structures with even (*uniform*) and *symmetric* distribution of inertial (mass), stiffness, and strength properties in plan and elevation generally exhibit favourable dynamic/vibration response to severe strong ground motions compared to *irregular* structures with non-uniform distribution of one or more of the above properties. Further, in uniform and symmetric building structures, undue local concentrations of deformation/ ductility and stress demands in a small number of structural elements are prevented. Specifically, "short column" formation is avoided by ensuring even stiffness distribution in plan and elevation, floor/slab rotations about the vertical (gravitational) axis are limited since the center of gravity lies close to the horizontal shear resistance center in plan (see also Fig. 2.2), and relative (differential) lateral and vertical displacements among structural members are minimized. Note that mass distribution is mainly related to the global geometrical shape of the building in plan and elevation, (lateral) stiffness distribution depends on the location and size of vertical structural members (columns, walls, and cores) in plan, while strength distribution is mostly associated with the longitudinal steel reinforcement ratios in structural members. Thus, in-plan symmetry does not necessarily imply in-plan uniformity (regularity), which is primarily related to the "compactness" of the building footprint (in-plan envelop). For example, H-shaped and cross-shaped plans are symmetric, but they are not uniform since they may have large in-plan recesses or elongated wings, respectively.

Redundancy allows for the development of alternative load paths upon plastic hinge formations or other local modes of failure at structural members. Therefore, redundancy is necessary to ensure redistribution of stresses which reduces the adverse effects of local (unanticipated) failures at structural members and, thus, the inherent uncertainty of the achieved seismic design. Further, redundancy increases the overall exhibited global "overstrength" of the lateral load-resisting structural system and its global ductility capacity.

Bi-directional resistance and stiffness

The horizontal design seismic action consists of two independent and simultaneously applied orthogonal components of the same order of magnitude. Therefore, the vertical structural members should be ideally arranged along two orthogonal axes ("principal axes"). Further, the overall level of lateral "resistance" of the structure against the seismic action in terms of stiffness, strength, and ductility should be similar along both principal axes.

Torsional resistance and stiffness

The torsional (rotational about the gravitational axis) seismic excitation component is typically negligible. However, building structures subjected to horizontal translational seismic excitations exhibit both translational and torsional displacements. This is due to "structural" and/or "accidental" in-plan eccentricities, that is, distances between the center of gravity and the horizontal shear resistance center at each storey level due to lack of perfect in-plan symmetry and/or non-uniform mass distribution of live gravitational loads. Therefore, the lateral load-resisting structural system should possess adequate torsional stiffness and strength. This is

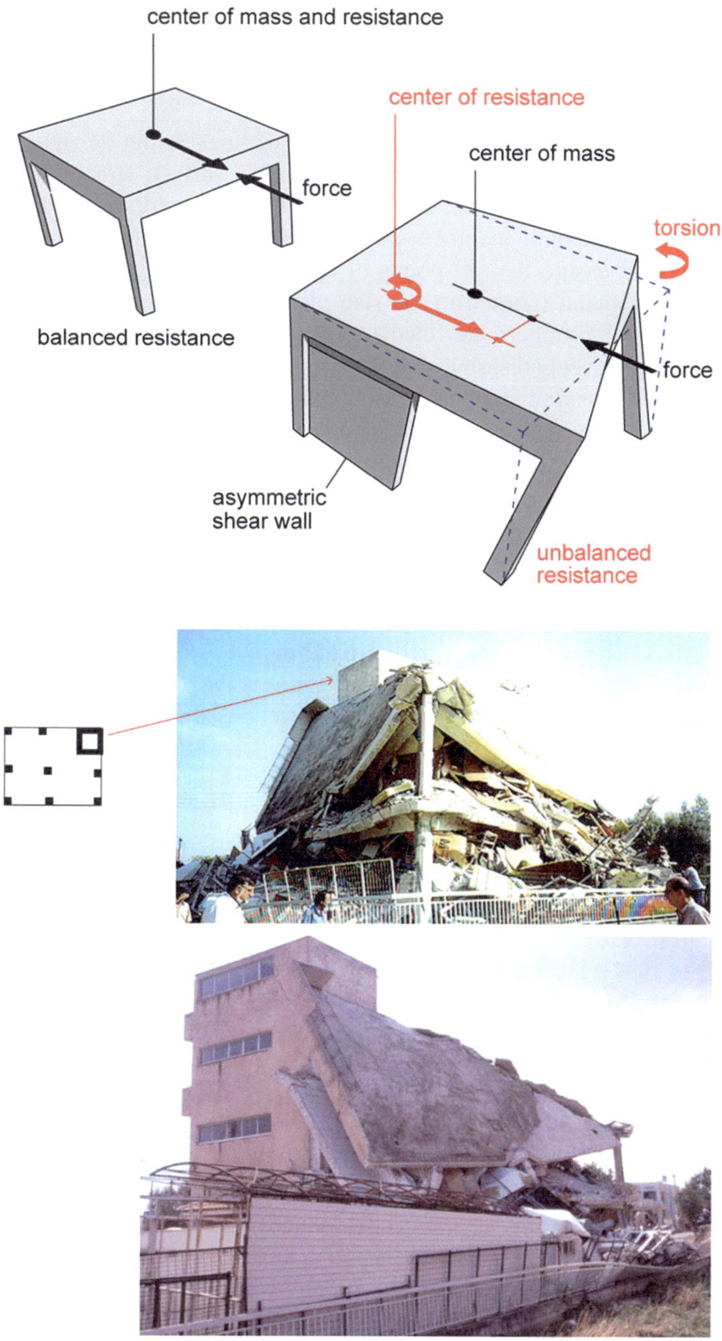

Fig. 2.2 Strongly asymmetric stiffness distribution in plan (Arnold 2006)

practically satisfied by ensuring that adequately stiff and strong vertical structural members are aligned (symmetrically) on or close to the perimeter of buildings.

Diaphragmatic behaviour at storey level

Concrete slabs at each storey level of r/c buildings acting as rigid in-plane diaphragms contribute significantly to a favourable seismic structural response behaviour. This is because they minimize horizontal relative (differential) displacements between structural members of the lateral load-resisting system at each storey level and ensure that all points of each storey undergo a single rotation about its gravitational (normal to the slab plane) axis. Further, they ensure that vertical structural members are "tied together" and that the horizontal seismic inertial forces are evenly distributed at these members according to their individual lateral stiffness. This rigid-disk like "diaphragmatic" behaviour at each storey level of buildings is achieved when slabs are compact, adequately stiff in their plane, and have relatively small in-plan aspect ratios and few/small floor openings.

Adequate foundation

A stiff and strong foundation tying the base of all vertical structural members of the superstructure together well in a grid-like layout is essential for a favourable structural response to earthquake excitations. This is because it minimizes the adverse effects of spatially incoherent ground motion, while preventing differential/relative settlements and horizontal translations at the foundation level. Further, it minimizes potential relative translations and rotations about the horizontal axes at the base of vertical structural members. Lastly, an adequately stiff and strong foundation evenly distributes the lateral seismic forces (in the form of base shears) concentrated primarily in the stiffer vertical structural members to the supporting grounds For the same reasons, the consideration of basements with perimetric r/c walls during the conceptual design stage is also recommended.

A more detailed list of practical guidelines and rules for facilitating the conceptual design for the earthquake resistance phase are provided in Table 2.2 (Penelis and Kappos 1997). Though not compulsory for code-compliant seismic design, they ensure the effectiveness and reliability of the current earthquake-resistant design philosophy adopting a partial protection against damage for design seismic action. That is, they allow for adopting relatively high values of behaviour factor q in conjunction with (equivalent) linear structural analysis methods, as detailed in Sect. 1.2.4. In general, they contribute to a favourable structural behaviour of the lateral load-resisting system for the case of severe earthquake shaking under which structures will exhibit strong inelastic behaviour, ensuring that inelastic response will only take place in the superstructure where damage can be visually detected and repaired. It is further emphasized that the above rules should be adopted even when the structure is expected to exhibit insignificant inelastic behaviour under the design seismic action (i.e., case of adopting relatively small values of behaviour factor q, e.g., $q \leq 1.75$ – high level of seismic performance for the design earthquake), since they ensure favourable static/dynamic structural behaviour for (almost) linear elastic structures as well.

Table 2.2 Summary of the main conceptual design rules and principles for earthquake resistant building layouts

Building footprint	Close to unity aspect ratio of outer building dimensions
Layout of load resisting system	Simple, clear, and highly redundant to ensure straightforward, continuous stress/load paths, and redistribution of stresses upon plastic hinge formation
Geometry in plan	Compact plan configuration to ensure relatively uniform mass distribution; approximate symmetry with respect to two orthogonal ("principal") axes; consideration of expansion joints to "isolate" unavoidable elongated wings and/or severe in-plan set-backs
Floor slabs	Avoidance of large floor openings and multi-level slabs in a single storey to ensure rigid in-plane diaphragmatic behaviour of floor slabs; avoidance of "beamless" flooring systems (slabs supported directly by columns)
Geometry in elevation	Compact building envelop in elevation without significant and abrupt set-backs to ensure uniform or smoothly decreasing mass distribution along the building height; avoidance of adversely large mass concentration at the top storeys of buildings
Lateral stiffness and strength distribution in plan	Symmetric in-plan configuration of vertical structural members (and especially of concrete walls and cores) with respect to two orthogonal ("principal") axes to ensure uniform in-plan lateral stiffness and strength distribution; arrangement of adequately stiff elements on the building perimeter and of walls along both principal axes; consideration of expansion joints to avoid substantially asymmetric in-plan layouts; avoidance of short-length beams and of beams not directly supported by columns
Lateral stiffness and strength distribution in elevation	Smooth distribution of lateral stiffness and strength along the height of buildings to ensure that no "soft" storeys (of significantly reduced lateral stiffness) and/or "weak" storeys (of significantly reduced lateral strength) exist; all vertical structural members, especially walls and cores, continue from the foundation to the top of the building without interruption avoidance of large openings in concrete walls and cores; avoidance of short columns
Foundations	Very stiff and strong foundation system tying together all elements at a single level to avoid differential displacements during seismic ground motions; use of strong tie-beams to connect isolated footings (pads) in a grillage; use of strong concrete walls to connect multi-level foundations or foundations on significantly different soil conditions; use perimetric r/c walls to ensure rigid box-like behaviour of basements; consideration of special foundations (e.g., micro-piles) to support walls on soft soils
(Masonry) infill walls	Symmetric configuration in plan and elevation; continuous arrangement in elevation with minimum offsets between storeys; continuous along the full height of each storey in order to avoid short column formation

(continued)

Table 2.2 (continued)

Building footprint	Close to unity aspect ratio of outer building dimensions
Expansion joints (seismic separation gaps)	Adequate clearance between adjacent buildings in urban environments to avoid collisions during strong ground shaking (seismic pounding) especially for corner or last-in-the-row buildings and for adjacent buildings of significantly different total height and/or different storey levels; consideration of measures to mitigate the effects of seismic pounding (e.g., use of soft material to fill in insufficient separation gaps)
Ductility	Appropriate detailing of designated energy dissipation zones of structural members for ductile behaviour; avoidance of premature/brittle local failures and instability; avoidance of forming "soft storey" collapse mechanisms; application of the "strong columns/weak beams" capacity design rule

2.1.2 *Frequently Observed Deficiencies in Structural Layouts*

Structural layouts that do not possess one or more of the attributes discussed in the previous sub-section due to poor conceptual design or unavoidable architecturally-driven constraints may possess a reduced (global) ductility capacity. Arguably, most partial or global building collapses observed in severe historical seismic events are due to adverse effects caused by not complying with one or more of the fundamental conceptual seismic design principles. In this respect, it is instructive to highlight the potential adverse effects of adopting structural layouts that do not follow or significantly deviate from the desirable conceptual design principles summarized in Table 2.2. To this aim, Table 2.3 lists the most commonly encountered deficiencies of structural layouts from the seismic design perspective and the effects that these may have during severe earthquake shaking.

Special attention should be paid to the cases of buildings with:

- excessive asymmetry in the in-plan distribution of lateral stiffness which poses extreme ductility demands on vertical structural members along the "soft" sides of the building due to torsional displacement (Fig. 2.2),
- short columns which attract significant shear stresses and, thus, under severe ground shaking, may be driven to brittle modes of failure (Fig. 2.3),
- a "soft" ground floor ("pilotis", as it is commonly referred to, mainly in Mediterranean countries) which generates considerably high and localised ductility demands that commonly lead to premature "storey" types of global plastic mechanisms of reduced global ductility (Figs. 2.4, and 2.13).

The above are considered to be the most common deficiencies observed in practice that result in significant and typically difficult to repair damages or even in partial or total building collapse depending on the level of the induced seismic action.

Table 2.3 Frequently observed deficiencies in structural layouts and potential adverse effects due to poor conceptual seismic design

Deficiencies of structural layouts	Potential adverse effects
Over-complicated load resisting systems in plan and/or elevation	Lack of clarity, simplicity, and/or continuity in the stress/load paths
Non-compact geometry in plan (elongated wings, large openings, inlets, and setbacks)	Reduced in-plane rigidity of slabs (lack of diaphragmatic action), considerable mass eccentricities
Significant inlets and/or setbacks in elevation	Unfavourable influence of higher modes of vibration due to significant deviation from a uniform mass distribution along the building height
Unduly asymmetric positioning of r/c walls and cores in plan (non-uniform stiffness distribution in plan)	Unfavourable concentration of ductility demands to a small number of structural elements due to large torsional displacement of slabs (Fig. 2.2)
Strong beams and weak columns in framed systems (lack of capacity design provision or implementation)	Plastic hinge formation at the columns leading to "column" types of collapse mechanisms of reduced global ductility (Fig. 2.14, Fig. 2.15)
Short column formation not taken into account at design stage (e.g., due to openings in or interruptions of infill walls)	Non-ductile column behaviour and/or local brittle column collapse due to large shear stress demands (Fig. 2.3)
Short beams connecting strong r/c walls without provisions for special shear/transverse reinforcement	Non-ductile beam behaviour and/or local brittle collapse due to excessive shear stress demands
Discontinuity of strong r/c walls and/or cores in elevation	Unfavourable influence of higher modes of vibration due to significant deviation from a uniform stiffness distribution along the building height
Lack of infill walls at the ground floor ("pilotis", "soft" storey) of predominantly frame load resisting systems	Undue concentration of ductility demands at the ground storey leading to a premature "floor" type of plastic mechanism (Fig. 2.4, Fig. 2.13)
R/c walls supported by columns at the ground storey (e.g., "pilotis") or at intermediate storeys	Undue concentration of ductility demands at a single "soft" and/or torsionally flexible storey leading to a premature "storey" type of plastic mechanism
Abrupt variations of lateral strength in elevation ("weak" storey)	Undue concentration of local failures at "weak" floors leading to "premature "floor" types of collapse mechanism (see also Fig. 2.13)
Slabs on columns (beamless flat-slab structural systems)	Reduced moment resisting capacity to lateral seismic loads and brittle local punch-through failures at the column-slab connections
Secondary supports of columns on beams	Undue local concentration of ductility and strength demands to the supporting beams
Slenderness of structural members	Buckling of structural members under flexure with time-varying axial loads
Multi-level foundations without strong coupling	Differential settlements and lateral displacements
Insufficient clearance between adjacent buildings in densely-built urban environments	Undue local strength demands at locations of collision/pounding during the asynchronous earthquake induced vibration of structures

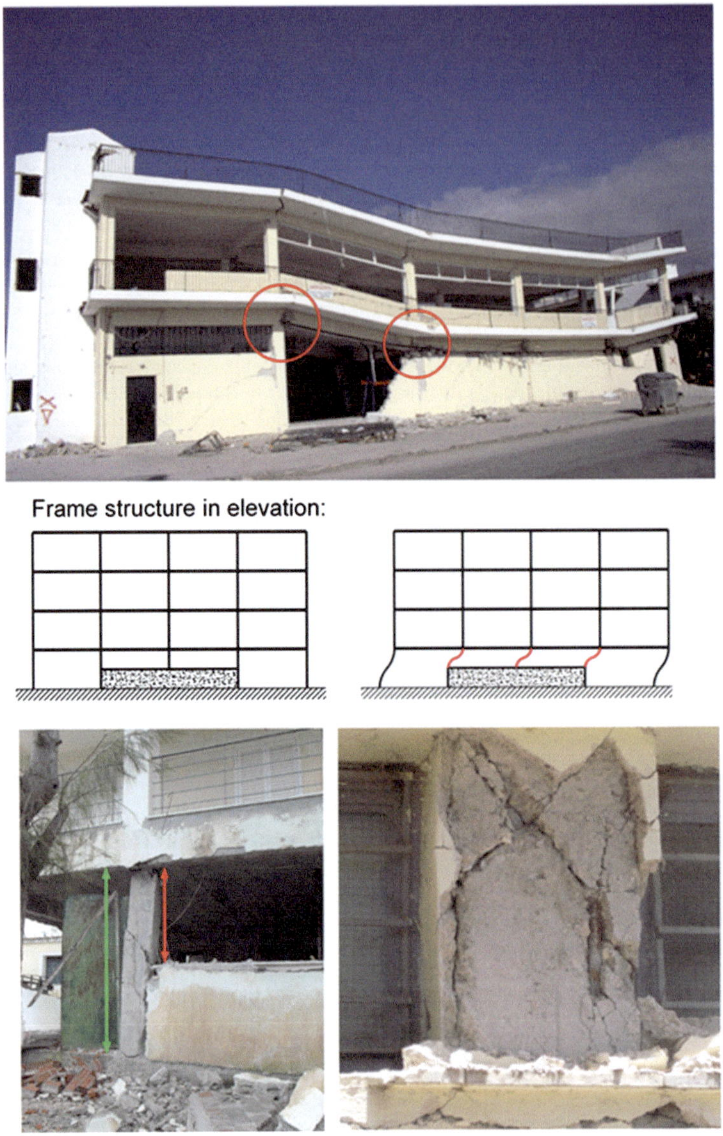

Fig. 2.3 Short column due to ex post masonry infill walls (*left*) and to initial architectural facade requirement (*right*)

Therefore, they should be avoided at first instance during the initial stages of conceptual design. A further discussion on the observed damages in the aftermath of major seismic events due to poor conceptual building design falls beyond the scope of this book. The reader is referred elsewhere for additional information (Arnold 2001, 2006; Bisch et al. 2012; Elghazouli 2009; Fardis et al. 2014; Lindeburg and McMullin 2014; Penelis and Kappos 1997; Villaverde 2009).

Fig. 2.4 Soft storeys (pilotis): r/c walls supported by columns and moment frames with no infills in the ground floor

2.2 Ductile Behavior Considerations and Preliminary Sizing of R/C Structural Members

2.2.1 The Fundamental Question at the Onset of Seismic Design: What Portion of the Ductility Capacity Should Be "Utilized"?

Arguably, the most critical decision that a design engineer needs to take in consultation with the building owner in the initial stages of seismic design, as

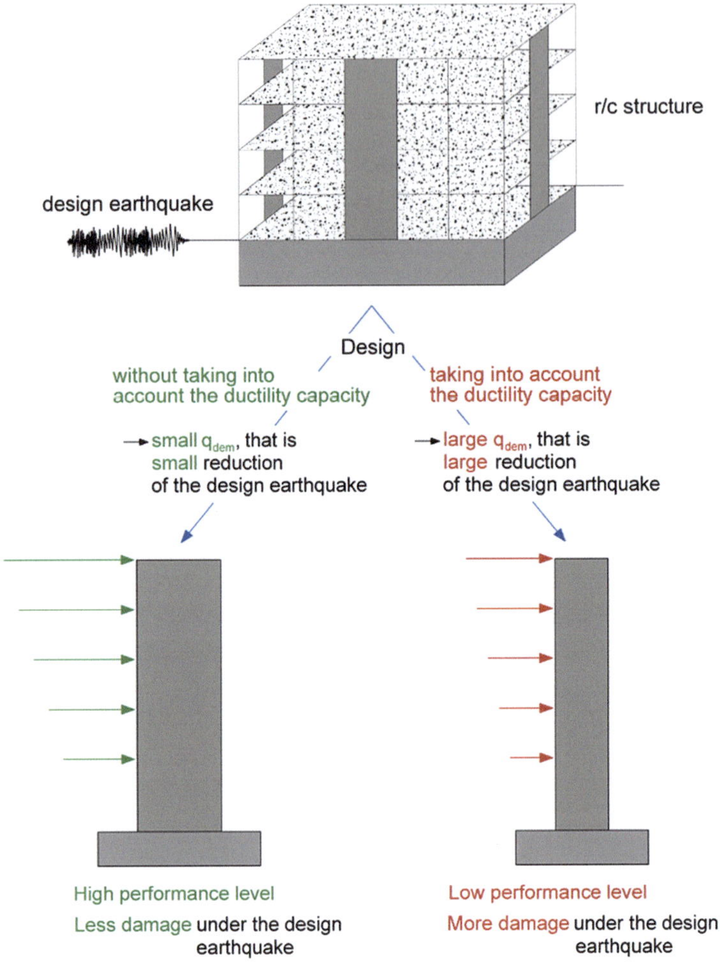

Fig. 2.5 Seismic design with and without utilization of the available ductility capacity

discussed in Sect. 1.4, concerns the desirable level of seismic protection against the nominally defined "design earthquake" (or level of seismic performance). In fact, this decision may significantly affect conceptual design considerations and preliminary member sizing. Practically speaking, the following question arises within the current framework of code-compliant seismic design (Fig. 2.5): *Will it be allowed for the utilization a significant portion of the ductility capacity of the structure to resist the nominal "design earthquake action"?*

– *A negative answer to the above question entails* that the structure will be designed for relatively small values of the force reduction factor or behaviour factor q. Therefore, design seismic loads will be relatively high and would necessitate sufficiently large sized r/c structural members (mainly r/c walls) to accommodate the required longitudinal reinforcement ratios. Overall, the structure will be relatively stiff and will resist design seismic loads mainly through its strength capacity, suffering light damages, if any. In this regard, there are two practical implications in the design process.

- Firstly, taking special measures for ductile behaviour is not mandatory, though it is reminded that a certain level of ductility capacity, inherent to all r/c structures properly designed for earthquake resistance, will be maintained. Further, an inherent level of overstrength will also be exhibited. Nevertheless, it falls to the decision of the owner, and is recommended by the authors (see also Sect. 1.4), to take certain additional measures in order to bring ductility capacity to a level above the minimum required. This ductility capacity will not be "activated" by future seismic events corresponding to the nominal design earthquake. It will be reserved to resist potential future earthquakes posing higher-than-the-design-earthquake demands.
  - Secondly, the need for sufficiently large-sized structural members (mainly r/c walls) should be taken into account in the preliminary (empirical/approximate) dimensioning and sizing step when adopting small values of the behaviour factor (high performance structure). In principle, the initially chosen dimensions (especially those of the vertical structural members: walls, cores, and columns) should be sufficient to contain the expected amounts of longitudinal reinforcement required. In this manner, the need for potential changes in structural member sizes after the analysis and verification checks is minimized.

– *An affirmative answer to the above question entails* that the structure will be designed for a large force reduction factor or behaviour factor q. Consequently, design seismic loads will be significantly lower compared to the previous case and so, potentially, will the sizes of the r/c structural members and their required longitudinal reinforcement. Overall, the structure will be more flexible, will have reduced strength against lateral loads and will be designed to suffer local damages (plastic hinges) under the design earthquake. In this case,
  - *It is mandatory that special measures for local and global ductile behaviour are taken* to ensure that the structure attains sufficient levels of ductility capacity corresponding, at a minimum, to the adopted behaviour factor value as elaborated in Sects. 1.2.2 and 1.2.3.
  - Given the significant reduction to the design seismic loads through division by the behaviour factor, equivalently reduced sizes for structural members should be assumed during the preliminary (empirical/approximate) dimensioning step.

In view of the above, it is deemed essential to re-iterate that:

– the fact that r/c structures possess an inherent level of ductility capacity (which may be readily enhanced by taking additional measures, as detailed later in this section) does not necessarily imply that this ductility capacity *should always* be utilized to resist the design earthquake through a reduction of the design seismic loads, and that

– seismic design of r/c structures for small values of q does not necessarily imply that they are non-ductile, i.e., brittle.

In the remainder of this section, certain important considerations and practical detailing rules in achieving earthquake resistant r/c building designs with ductile behaviour are presented in continuation of the general discussion on the concept of ductility included in Sect. 1.2. These rules and considerations ensure that a sufficiently high level of ductility capacity is achieved to justify the adoption of relatively large values of the behaviour factor (force reduction factor) q for r/c building structures, if so desired.

2.2.2 Local and Global Ductility Capacity

Ductility is a key-concept in the response of yielding structures subject to earthquake ground motions. As discussed in Sect. 1.1.5, ductility is qualitatively defined as the ability of a cross-section, a structural member, or a structure as a whole to exhibit significant inelastic deformations under cyclic/seismic external loads without losing large parts of its original stiffness and strength after each loading cycle. In this respect, apart from the important distinction between *ductility capacity* and *ductility demand* (see Sect. 1.2.1), it is also pertinent to distinguish between *local* ductility (capacity or demand) related to a cross-section or a critical energy dissipation zone within a structural member, and the *global* ductility (capacity or demand) related to the whole building structure or one of the substructures comprising the lateral load resisting structural system.

In this regard, it is reminded that ductility capacity is quantitatively defined in terms of displacement u (displacement ductility capacity μ_u) as the ratio of the peak attainable displacement value u_{tot} beyond which it is assumed that the structure collapses over the yielding displacement u_y signifying the onset of inelastic behaviour (Fig. 2.6). That is,

$$\mu_u = u_{tot}/u_y. \tag{2.1}$$

The above definition applies for any type of displacement/deformation, including the ductility capacity in terms of rotations θ (rotation ductility capacity)

$$\mu_\theta = \theta_{tot}/\theta_y, \tag{2.2}$$

and the ductility capacity in terms of curvature κ (curvature ductility capacity):

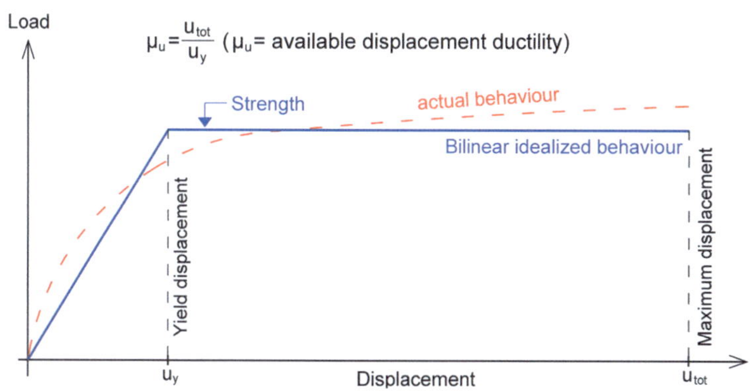

Fig. 2.6 Definition of ductility (in terms of displacements)

$$\mu_\kappa = \kappa_{tot}/\kappa_y. \tag{2.3}$$

In Fig. 2.7, a single r/c cantilevered wall laterally loaded by a point load acting at its tip is considered to clarify the difference between local (rotation) ductility capacity and global (tip-displacement) ductility capacity.

In particular, under the assumptions that strength is constant along the height of the wall (constant longitudinal reinforcement) and that no premature shear type of failure occurs, yielding initiates (plastic hinge forms) at the base of the wall where the moment diagram is maximized. Further, given that plastic deformations concentrate at the locations of yielding initiation (plastic hinges), plastic rotations θ (or curvature κ) within a certain distance L_{pl} from the base of the wall (plastic hinge length) increase significantly faster compared to those observed beyond L_{pl} (Fig. 2.7). Therefore, to resist collapse through ductile behaviour, the *local* rotation ductility capacity μ_θ (or curvature ductility capacity μ_κ) within the plastic zone length L_{pl} must be considerably higher than μ_θ (or μ_κ) beyond L_{pl}. Moreover, the local ductility capacity μ_θ (or μ_κ) within the "critical" L_{pl} length is considerably higher then the *global* tip-displacement ductility capacity μ_u. In general, the following relationship holds (Fardis 2009; Penelis and Kappos 1997; Penelis and Penelis 2014)

$$\mu_\theta = \theta_{tot}/\theta_y \gg \mu_u = u_{tot}/u_y. \tag{2.4}$$

Similarly, it can be readily understood that, in the case of laterally loaded pure moment resisting frame structures of constant strength along their height, the flexural deformation (curvature) of the beams at lower stories is more prominent than at the higher stories (Fig. 2.8). Consequently, due to increased (inelastic) deformation demands, the beam members at the lower floors yield for a significantly lower base shear (sum of external lateral forces) than the one that drives the top storey inelastic displacement to its maximum attainable value u_{tot}. In other words, plastic hinges at beam members of the first few floors form well before the

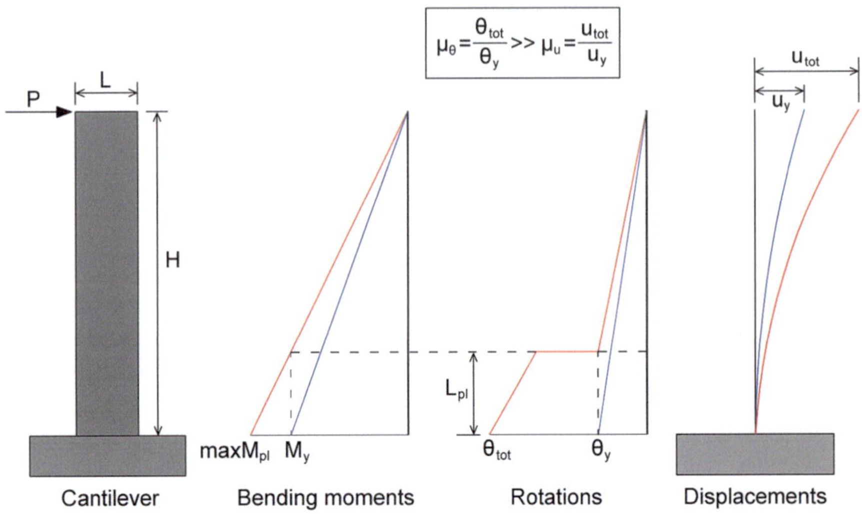

Fig. 2.7 Relation between local ductility μ_θ and global ductility μ_u for a single r/c wall

global displacement ductility capacity is reached under a gradually monotonically increasing base shear. Therefore, to achieve an overall ductile behaviour for frame building structures, the *local* curvature ductility capacity μ_κ of the beams at lower floors must be significantly higher than the *global* top-storey displacement ductility capacity μ_u. That is,

$$\mu_\kappa = \kappa_{tot}/\kappa_y \gg \mu_u = u_{tot}/u_y. \tag{2.5}$$

The following section summarizes *local detailing measures* typically taken in r/c buildings to ensure increased *local* ductility capacity at critical zones of structural members, while Sect. 2.2.4 discusses the issue of achieving high values of *global* ductility capacity of structures by means of capacity design considerations.

2.2.3 Factors Influencing the Local Ductility Capacity of R/C Structural Members

The local ductility capacity at "critical" energy dissipative zones of r/c structural members depends on the properties of the concrete and the reinforcing steel bars, as well as on the reinforcement detailing. Specifically, local ductility capacity increases by

- using concrete of higher compressive strength,
- using reinforcing steel of lower tensile strength for the longitudinal steel bars,
- using reinforcing steel of higher ductility and tensile post-yield stiffening,

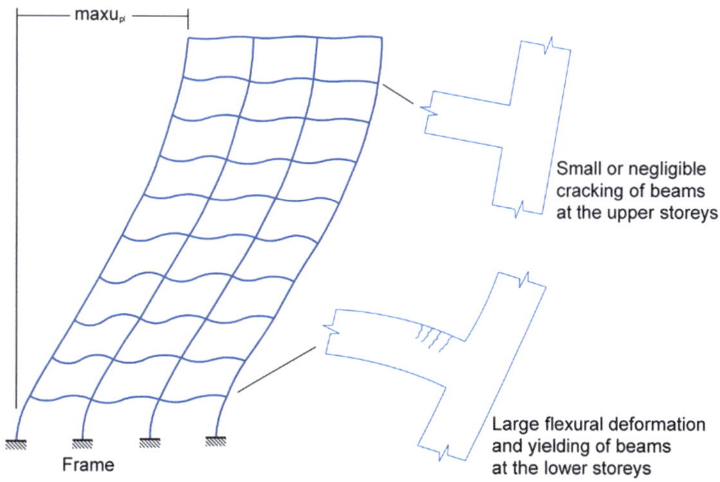

Fig. 2.8 Relation between local μ_κ ductility and global ductility μ_u for a multistorey r/c moment frame

- reducing the ratio of longitudinal reinforcement under tension,
- increasing the ratio of longitudinal reinforcement under compression,
- increasing the level/effectiveness of concrete core confinement (e.g., by using denser transverse reinforcement),
- increasing the achieved concrete-reinforcement bond,
- decreasing the axial load ratio,
- decreasing the level of sustained shear stresses.

Note that confinement of the "concrete core" (i.e., the part of the concrete inside the "cage" formed by the longitudinal and the transverse reinforcement in typical r/c structural members) to prevent its outward dilation is the most common detailing measure for ensuring local ductile behaviour of r/c structural members. It is accomplished by means of dense hoops or ties placed further to the transverse reinforcement normally required to accommodate shearing stresses (see (Penelis and Kappos 1997) for a detailed discussion on this topic). In this respect, considerable local ductility capacity in r/c columns is achieved by adhering to the following three qualitative detailing rules (see Appendix C for the complete EC2/EC8-compliant detailing requirements of r/c structural members).

1. The spacing between hoops or ties along the length of columns should be kept sufficiently small such that early buckling of longitudinal reinforcing bars is prevented (Fig. 2.9), while a desirable level of concrete core confinement is achieved.
2. The stirrups and hoops/ties used must link all the longitudinal reinforcing bars. Further, they must end in "closed hooks" (forming an angle of about 45° with the main stirrup/hoop pattern, as shown in Fig. 2.10) to prevent from opening-up

Fig. 2.9 Longitudinal steel bar buckling and concrete cover spalling due to inadequate sparse spacing of stirrups

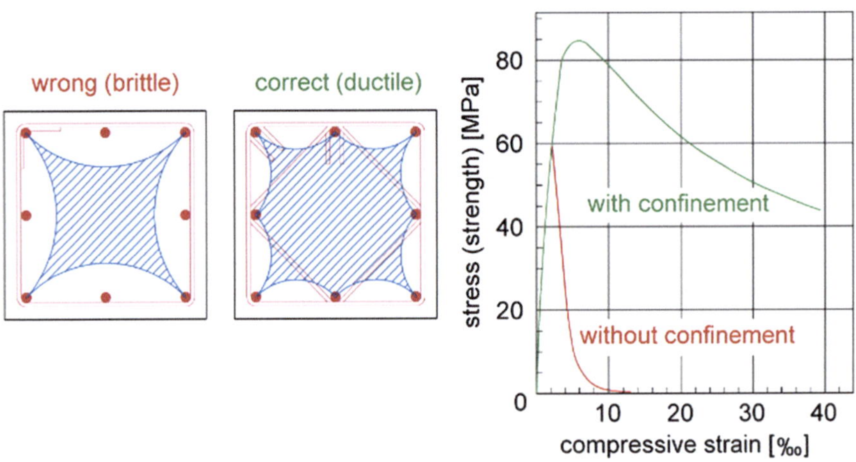

Fig. 2.10 Lateral ties arrangements in columns (hatched area: confined concrete core; white area: unconfined concrete susceptible to spalling)

under tensile forces and, thus, to prevent premature spalling of the concrete cover and buckling of the longitudinal reinforcement, as shown in Fig. 2.11. Further, closed hooks, along with a proper hoop pattern linking and tying all longitudinal reinforcing bars together, increase significantly both the strength of r/c and, more importantly, its ductility (Fig. 2.10).

Fig. 2.11 Examples of premature column failures due to opening-up of inadequately end-detailed hooks

3. The spacing between the longitudinal reinforcing bars linked by stirrups and hoops should be sufficiently small (e.g., less than 20 cm) such that the assumed confined cross-sectional area of concrete (hatched area in Fig. 2.10) is maximized. This is because the assumed confined concrete core area is defined by parabolic "arcs" of confining stresses between consecutive bars, as shown in Fig. 2.10, beyond which it is taken that concrete spalls in a similar manner as the concrete cover lying outside the stirrups (Mander et al. 1988; Sheikh and Uzumeri 1982). Clearly, smaller spacing of longitudinal bars results in parabolic confinement arcs of smaller length and, thus, in a larger area of concrete whose spalling is prevented. To this end, the use of a larger number of closely-spaced longitudinal bars of small diameter should be preferred over the use of fewer large diameter bars in practical detailing of r/c structural members for ductile behaviour (Penelis and Kappos 1997).

It is further noted that similar detailing rules apply (see Appendix C) for the critical zones of structural members where plastic hinges are anticipated to form (i.e., at the ends of beams and at the base of walls) following capacity design considerations elaborated in the following section.

As a final note, it is reminded that the above influencing factors and detailing rules for ductile behaviour presuppose that local brittle types of failure due to shearing stresses such as those shown in Fig. 2.12 do not develop and, therefore, an overall sufficient level of strength is maintained by the structure during inelastic deformations. Therefore, over-designing for flexure (e.g., by considering more longitudinal reinforcement from that required/calculated) should be avoided or should be taken into account when calculating the transverse/shear reinforcement as it increases not only the peak moment potentially developed/demanded by a future earthquake but also the demanded peak shearing force.

Fig. 2.12 Brittle (shear) failure due to inadequate lateral reinforcement

2.2.4 Capacity Design Rules for Ductile Global Collapse Mechanisms

In case a large value of behaviour/force reduction factor q is adopted in design, several structural members *must* yield and deform far into the inelastic range under the design seismic action for the structure to successfully withstand the input seismic forces, as discussed in Sect. 1.2.6. In this regard, stringent detailing rules along the lines delineated in the previous section (see also Appendix C) for local ductility capacity at "critical zones" of structural members must be observed. Furthermore, in this case, additional capacity design rules and considerations are put in place to achieve sufficient global ductility capacity. Specifically (see also Sect. 1.2.6),

– regarding individual structural members of the lateral load resisting system, energy dissipation zones are pre-specified and driven to local ductile flexural modes of failure ("plastic hinges") by application of a judicial strength hierarchy,

- regarding the lateral load resisting structural system as a whole, a certain sequence of plastic hinges is pre-specified to maximize the dissipation of the input seismic/kinetic energy via hysteretic behaviour prior to the development of "desirable" plastic mechanisms.

The aforementioned energy dissipation zones are sized and detailed appropriately to exhibit ductile behaviour. That is, to undergo large plastic deformations under the design seismic action without losing a large part of their moment bearing capacity. Cross sections of structural members outside the "critical" yielding zones, and especially cross sections neighboring these critical zones, are strengthened to ensure that they behave elastically upon plastic hinge formation.

Considering the development of plastic mechanisms, mechanisms that demand a relatively small amount of seismic energy to be dissipated in order to develop must be avoided. These are the mechanisms that require only a few plastic hinges to form. The "desirable" collapse mechanisms are those that maximize the required seismic energy dissipation in order to develop. Typically, these mechanisms require a maximum total number of potential plastic hinges to form before the structure collapses.

2.2.4.1 Plastic Mechanisms for Frame Lateral-Load Resisting Systems

In the case of pure moment resisting frame structural systems, two "extreme" examples of plastic (collapse) mechanisms are depicted in Fig. 2.13. The "desirable" plastic mechanism commonly referred to as the "beam-sway mechanism" is the one targeted via capacity design rules and requirements in code-compliant seismic design. At the other end rests the "storey-sway mechanism" due to a soft and/or weak storey which seismic codes of practice aim to avoid by relying on both capacity design and conceptual design rules (see Sect. 2.1). A third type of mechanism is shown in Fig. 2.15 called a "column-sway mechanism", which is also, theoretically, achievable but should be avoided for reasons discussed below.

Beam-sway mechanism

The desirable beam-sway plastic mechanism develops upon plastic hinge formation at the ends of all beams and at the base of the columns of the ground storey. As shown in Fig. 2.13, the required rotation θ_1 at each one of the several plastic hinges of the beam-sway mechanism is much smaller than the required rotation θ_2 at the few plastic hinges of the storey-sway mechanism for the same top-storey peak displacement u_{tot}. Clearly, local ductility demands of the beam-sway mechanism are significantly smaller. Furthermore, it is easier to accommodate ductility demands of a beam-sway mechanism, since beams of typical building structures carry negligible axial force compared to columns due to the diaphragmatic action of floors. The low axial load level positively influences the local ductility capacity of beams compared to that achieved by columns.

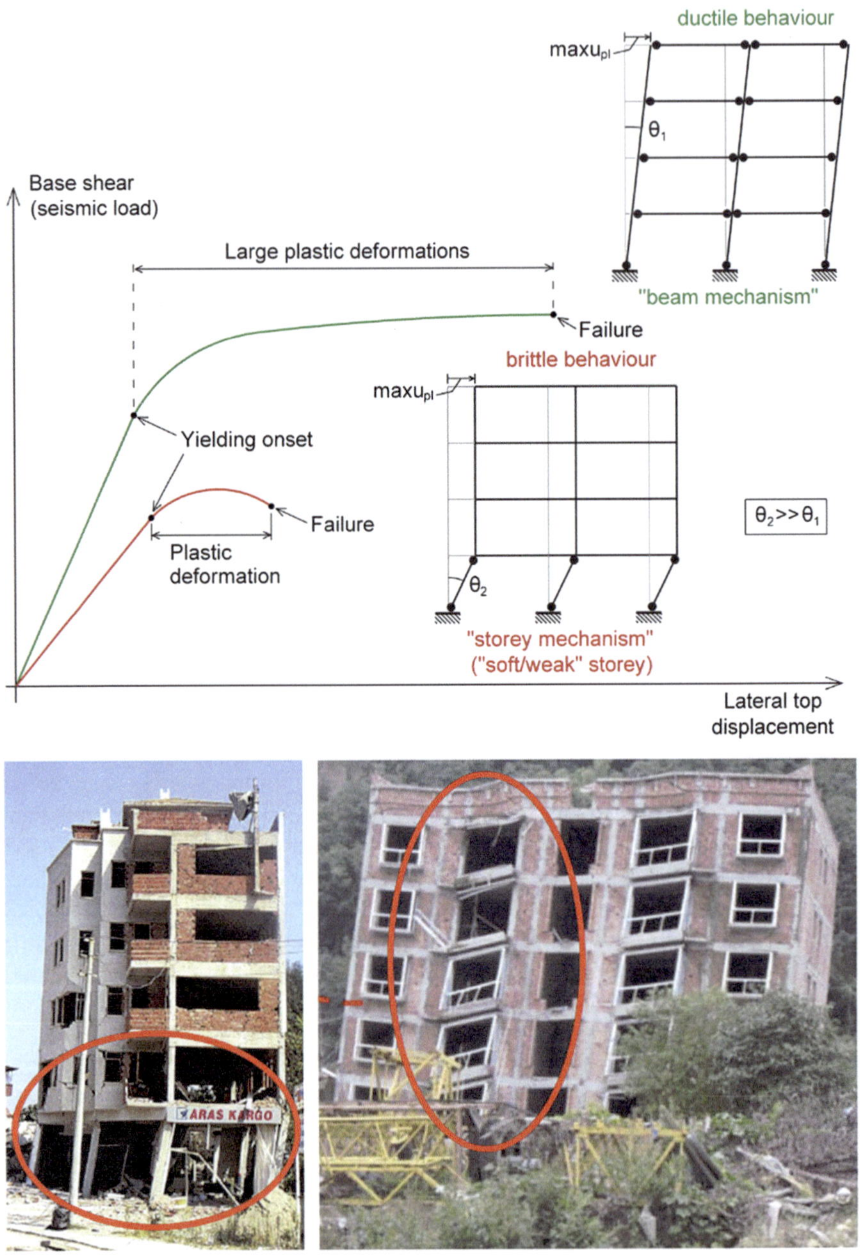

Fig. 2.13 Moment frames: unfavourable "storey mechanism" to be avoided (*left*) and favourable "beam mechanism" (*right*)

Fig. 2.14 Moment frames: consequences of non-compliance with the "strong columns – weak beams" rule

Nevertheless, it is pointed out that, in a beam-sway mechanism, the base of the columns at the ground floor will eventually yield due to unavoidable high values of locally developed moments. Therefore, it is recommended to increase the flexural strength of the columns at the ground floor (beyond the strength required to accommodate calculated moments from the structural analysis step) to "delay" the formation of plastic hinges. Ideally, plastic hinges at the base of columns should form last, upon yielding of all the beams.

Storey-sway mechanism

The "storey-sway mechanism" is avoided by application of the well-established capacity design rule of "weak beams-strong columns" (Fig. 2.14) which needs to be verified/checked quantitatively (e.g., § 4.4.3.2 of EC8). In particular, at every joint, the column longitudinal reinforcement ratios should be computed such that the sum of the flexural strength capacity (peak bending moments calculated based on the longitudinal reinforcement) of columns is higher than the flexural strength capacity of beams accounting for the potential overstrength factors.

Furthermore, a second capacity design rule applies to eliminate the possibility that a premature shear (brittle) type of local failure occurs before plastic hinges form. According to the latter rule, the transverse shear reinforcement in the "critical" zones of beams and columns is calculated based on the so-called "capacity design shear forces" (e.g., § 5.4.2 of EC8). These shear forces are computed by

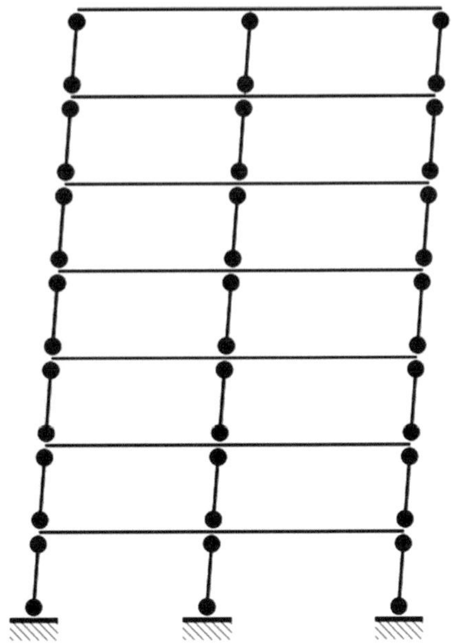

Fig. 2.15 Moment frames: unfavourable "column-sway mechanism" to be avoided

assuming that the ends of all beams and columns converging at any particular joint have yielded accounting for the potential overstrength factors. Notably, the capacity design shear forces are commonly considerably higher than the shear forces derived from the structural analysis step. As a final note, it is emphasized that the development of a storey-sway mechanism must be avoided not only for the ground storey as shown in Fig. 2.13 for the sake of exemplification, but also for each and every storey of the building.

Column-sway mechanism

The "column-sway mechanism" involves plastic hinge formation at the ends of columns at all stories (Fig. 2.15). Ensuring the reliable development of such a mechanism is very challenging at design, if not unfeasible. This is because the (time-varying during an actual earthquake) axial load carried by each column and, consequently, the flexural strength of each column changes significantly at each storey. In practice, the column-sway mechanism will most probably degenerate into a "storey-sway mechanism" at the weakest storey. Further, designing for a column-sway mechanism is not practical, since repairing plastic hinges at columns is considered to be harder and more expensive than repairing plastic hinges at beams. For these reasons, capacity design to achieve column-sway mechanism should be avoided.

Avoiding local brittle failure at beams-column joints

Further to proper detailing of the ends of beams and columns converging to joints, local and global ductile behaviour of moment resisting r/c frames involves ensuring that the joints (joint panel zones) are designed such that they do not fail

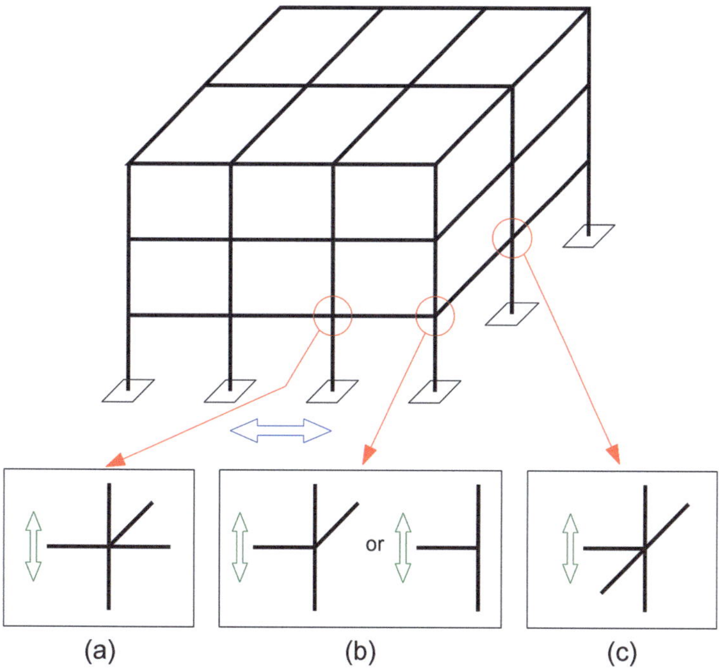

Fig. 2.16 Beam-column joint types at the perimeter of a spatial frame: (**a**) interior joint of exterior frame, (**b**) exterior (corner) joint with no or one lateral beam, (**c**) exterior joint of internal frame with two-sided lateral beams

prematurely. To this end, it is noted that the seismic behaviour of beams-columns joints depends on several factors, including their geometry and position in spatial (three-dimensional) r/c frames (see also Penelis and Kappos 1997). Special care needs to be taken for the design of the joints lying on the perimeter of r/c frames, as shown in Fig. 2.16. Specifically, exterior (corner) joints having a single or no out-of-plane converging beam (Fig. 2.16b) are particularly vulnerable to shear and, therefore, brittle failure due to crushing along the joint diagonal. Still, joints with beams converging from both out-of-plane sides such as those shown in Fig. 2.16a, c enjoy significantly increased shear capacity.

In this regard, it can be deduced that capacity design is an effectively complex procedure. Further, given that current seismic codes of practice do not distinguish among different types of joints, it can be argued that the relatively simplistic code-compliant capacity design requirements may lead to significant uncertainty in design and to potentially unpredictable structural behaviour under extreme seismic loads. In fact, recent comprehensive experimental and theoretical research work, such as (Park and Mosalam 2009) and references therein, demonstrated that current design procedures may result in severe damage to the joint regions despite the use of the "weak beam – strong column" design philosophy. The latter may be jeopardized during severe earthquakes by the premature shear failure of the joint region itself. In this respect, under certain conditions, the joints themselves might become the "weak link", even in EC2- and EC8-compliant r/c structures (Tsonos 2007). To address this issue, several procedures have been proposed in the literature, as reviewed in (Park and Mosalam 2009) and (Penelis and Penelis 2014), to ensure that the initial formation of plastic hinges as well as the subsequent extensive damage occurs at the ends of the converging beam members, while columns and joints remain intact. It is envisioned that such procedures will be incorporated into future versions of design codes to achieve improved capacity design implementation.

2.2.4.2 Collapse Mechanisms for Dual Lateral-Load Resisting Systems

From a structural design viewpoint, pure frame lateral load resisting systems are not recommended in high seismicity areas for more than three- or four-storey r/c buildings. This is because rigid-jointed (moment resisting) frames are relatively flexible and exhibit increased ductility/deformation demands under severe seismic excitation, rendering them sensitive to second-order effects. Consequently, they require careful local detailing during construction to achieve sufficient levels of ductility capacity at critical zones which may not always be readily achievable in practice. In this regard, it is usually preferable to choose lateral load-resisting systems of increased stiffness and, thus, of reduced overall ductility/deformation demands, by incorporation of r/c walls (Fintel 1991, 1995). As a rule of thumb, it is generally easier in practice to construct a wall with sufficient flexural and shear strength, rather than a ductile frame system.

A "ductile" behaviour of laterally loaded slender r/c walls (with sufficiently large height over width ratio such that flexural modes of failure prevail) is considered to be achieved when a single energy dissipation zone ("plastic hinge") forms at their base (Fig. 2.17-right panel). This is where the bending moment diagram is

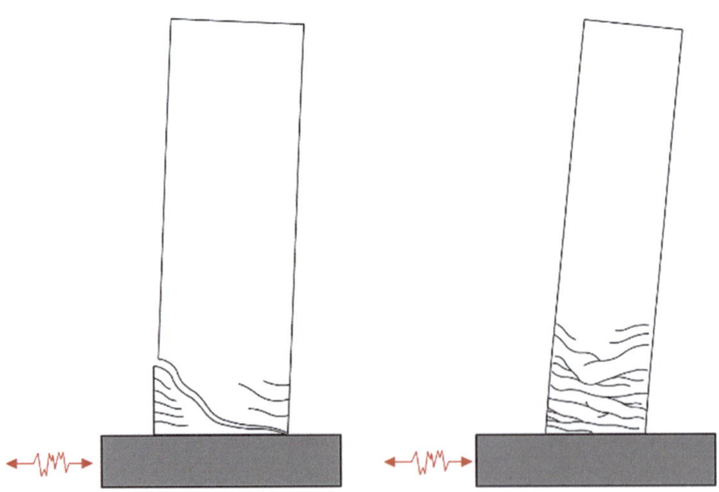

Fig. 2.17 Shear wall: brittle shear failure to be avoided (*left*) and favourable formation of plastic hinge (zone) at the base (*right*)

Fig. 2.18 Brittle shear failure of r/c walls at the ground floor of buildings to be avoided

maximized in typical slender walls. Moreover, walls should remain elastic along the rest of their height.

This desired behaviour is assured by observing the following two capacity design rules for ductile r/c walls (e.g., § 5.4.2.4 of EC8).

- Shear and flexural types of failure at storey levels above the ground storey must be avoided. This is achieved by designing/detailing the cross-sections of walls above the ground storey for higher shear and flexural strength capacity than those required from the structural analysis step.
- Shear type of failure at the base of the wall must be avoided (Fig. 2.18). This is achieved by placing sufficient transverse/shear reinforcement to sustain the levels of shearing forces developed upon the base of the wall that has yielded (capacity shear).

Figure 2.19 shows are the two possible plastic collapse mechanisms for typical combined (dual) frame-wall lateral load resisting structural systems. Notably, both require that the base of the wall yields, as well as the ends of the beams connected to the wall, where bending moment diagram is maximized due to lateral loads applied at each storey level. However, the leftmost mechanism involving the formation of plastic hinges at the ends of columns instead of beams (rightmost mechanism) is less preferable. This is because the seismic energy dissipation capacity of plastic hinges at columns is reduced due to the sustained axial force.

2.2.4.3 A Reminder of the "Limits" of Capacity Design

Conforming to all the aforementioned capacity design requirements and rules, such as the "strong columns – weak beams" rule, is perhaps the most efficient way (and certainly the code-prescribed way) to achieve the targeted "no-collapse" requirement for the design seismic action for "low seismic performance" code-compliant

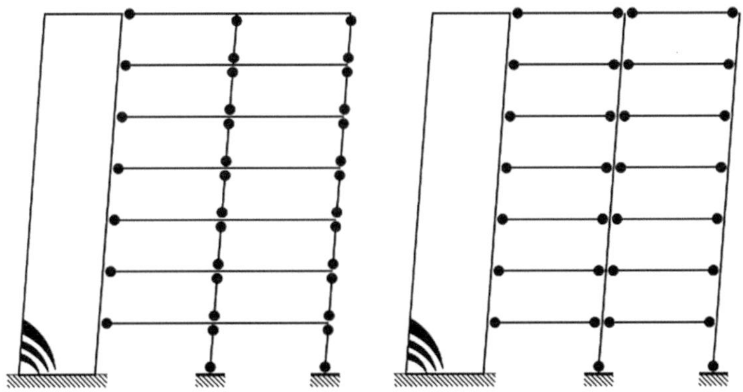

Fig. 2.19 Dual system (r/c wall – moment frame structure): unfavourable (however: acceptable for long span beams) and preferred plastic mechanisms (*left* and *right*, respectively)

building structures. This is because capacity design rules allow for utilizing the ductility capacity of the structure to resist the design seismic loads in a controlled manner, avoiding local and global instability/collapse. However, it is emphasized that designing to meet the "no-collapse" requirement via ductile behaviour (e.g., by developing plastic hinges following the beam-sway mechanism pattern) involves repairing a large number of (beam) structural members where plastic hinges form after a severe seismic event. The repair cost of these damaged structural members required to bring the structure to the same overall level of seismic performance it observed before the seismic event can be quite significant. In fact, this cost may be important even in the aftermath of more frequent (less intense) earthquakes than the "design earthquake" during which some of the designated energy dissipation zones may still yield (be damaged) to a certain extent. These issues should be accounted for (within a probabilistic context) by the owner of the structure in deciding on the targeted level of seismic performance (i.e., the adopted value of the behaviour factor q) at the onset of the design process.

The above discussion points to the question raised at the beginning of the current section and, in this regard, it closes the decision-making loop on what should be a desirable level of seismic protection against the design earthquake. The answer may not be straightforward within the current (traditional) seismic design approach followed by codes of practice (see also Sects. 1.3 and 1.4). However, it does affect conceptual design and preliminary sizing of structural members and, therefore, it was deemed essential to be included here in order to inform phase A of the seismic design process (see Fig. 2.1 and Table 2.1). The remaining sections of this chapter follow phase B of the seismic design process.

2.3 Structural and Loading Modeling for Seismic Design of R/C Buildings Using Linear Analysis Methods

Upon completion of the conceptual design phase during which the preliminary sizing of structural members takes place, the seismic design process proceeds with the analysis phase. The first stage of the analysis phase involves the development of the mathematical or computational model of the building structure to be used in the next, structural analysis stage. Routine "modeling" of ordinary building structures for seismic design includes:

- defining the design seismic loading combination comprising the design seismic action and the permanent, plus a fraction of variable gravitational loads; and
- developing a numerical finite element (FE) model which can adequately serve the purpose of determining the seismic effects (deformations and stress resultants) for the detailing of the load resisting system of the building structure.

Section 2.3.1 discusses several important issues arising in defining the EC8-compliant design seismic action. Section 2.3.2 presents the general structural modeling requirements prescribed by EC8. Lastly, Sect. 2.3.3 provides guidance and practical recommendations for the development of adequate FE models to be adopted in the context of routine seismic design of multistorey r/c buildings according to EC8, using linear methods of analysis.

2.3.1 *EC8-Compliant Loading Modeling for Seismic Design*

In the context of EC8-compliant force-based seismic design, equivalent linear types of analysis are routinely employed in which the design seismic action is defined by means of an inelastic pseudo-acceleration spectrum (see also Sect. 1.2.4). This spectrum is termed the "design spectrum for elastic analysis" in clause §3.2.2.5 of EC8, hereafter the *design spectrum*. It attains reduced ordinates compared to the elastic pseudo-acceleration response spectrum (termed the "*horizontal elastic spectrum*" in clause §3.2.2.2 of EC8) depending on the assumed behaviour (or force reduction) factor q. The lateral seismic design base shear is proportional to the design spectrum ordinates and to the inertial/mass properties of the structure. Therefore, lateral seismic loads are determined by means of the design spectrum and the nominal gravitational loading combination (gravitational permanent plus variable actions according to Eurocode 0 (CEN 2002) from which the inertial properties of the structure can be derived and accounted for in the analysis stage. The thus defined seismic loads are further combined with the nominal gravitational loading combination acting simultaneously with the lateral seismic loads. In the case of buildings with structural members susceptible to undue local vertical vibrations (e.g., horizontal cantilevered structural members or beam supporting columns; see §4.3.3.5.2 of EC8), the vertical component of the ground motion needs to be accounted for as well, yielding additional seismic loads in the vertical (gravitational) direction (§3.2.2.3 of EC8). Finally, the EC8-prescribed loading modeling for seismic design is complemented by certain additional considerations, such as the site seismic hazard from which the "design seismic action" is specified and the assumed directions along which the seismic action is applied.

In general, EC8, complemented by the National Annexes and Eurocode 0 (CEN 2002), describes most aspects of the seismic loading modeling with sufficient clarity for practitioners to follow. Therefore, the following paragraphs of this section provide only brief comments on certain issues related to (i) the definition of the design seismic action in terms of peak ground acceleration within a probabilistic context, (ii) the EC8 design spectrum, (iii) the relation between the design peak ground acceleration and the behaviour factor q, and (iv) the EC8 prescribed inertial structural properties for seismic design and the seismic loading combination. Other aspects of loading modeling (e.g., direction of seismic action, in plan points of action for the lateral seismic loads in the lateral force method, etc.) are discussed in subsequent sections and chapters focusing on the EC8 prescribed linear analysis methods and their practical implementation (Kappos 2002).

2.3.1.1 Reference Seismic Action α_{gR}, Design Seismic Action α_g and, Importance Factor γ_I

According to clause §2.1(1)P of EC8, the *"reference"* level of the seismic intensity for the *no-collapse requirement* is defined in terms of the *peak (horizontal) ground acceleration value* α_{gR} recorded on rock and having a *return period* T_R of 475 years (denoted by T_{NCR}), or a probability P_R of 10 % (denoted by P_{NCR}) to be exceeded within a time period t_e ("exposure time") equal to 50 years (see also Sect. 1.1.4). Further, for the *damage limitation requirement*, a reduced seismic action is taken, represented by a peak ground acceleration value with a return period T_{DLR} equal to 95 years, or a probability P_{DLR} of 10 % to be exceeded in 10 years.

The above correspondence between the return period T_R and the probability P_R relies on modeling the occurrence of an earthquake within a time interval as a discrete random variable following the Poisson distribution. In particular, the Poisson model involves the following three assumptions:

- the number of earthquake events in one time interval is independent of the number of earthquake events in any other past or future time interval;
- the probability of an earthquake event in a short time interval is proportional to the duration of this time interval; and
- the probability of observing more than one earthquake event during a short time interval is negligible.

Under the above assumptions, it can be shown that the probability P_R that a certain value of ground acceleration α_g will be exceeded within a given "exposure time" t_e in years (note that EC8 uses the symbol T_L to denote exposure) is expressed by

$$P_R = 1 - \exp(-P_1 t_e), \tag{2.6}$$

in which P_1 is the *probability that α_g is exceeded in one year* (*annual probability of exceedance*). Further, the return period of the considered α_g is $T_R = 1/P_1$. Therefore, by solving Eq. (2.6) for the return period, it is possible to obtain the following relationship (clause §2.1(1)P of EC8)

$$P_1 = \frac{-\ln(1 - P_R)}{t_e} \quad \text{or} \quad T_R = \frac{-t_e}{\ln(1 - P_R)}. \tag{2.7}$$

By substituting $t_e = 50$ years and $P_R = P_{NCR} = 10\ \%$ in the last equation, one obtains a return period of $T_R = T_{NCR} \approx 475$ years (no-collapse requirement of EC8). The value of the peak ground acceleration on rock ground conditions corresponding to the T_{NCR} return period (reference peak ground acceleration α_{gR} of EC8) is site specific and can be obtained from National (or regional) seismic hazard maps (clause §3.2.1 of EC8). Such maps are developed by relying on probabilistic seismic hazard analysis (McGuire 1995). For illustration purposes, the seismic hazard maps included in the National Annexes to EC8 of Greece and of The Netherlands are given in Fig. 2.19, characterized by significantly different levels of seismicity. Note that the reported peak ground acceleration values in these hazard maps correspond to the reference return period $T_{NCR} = 475$ years of EC8 for rock ground conditions (ground type A as of EC8), and, thus, they can be directly used in conjunction with EC8 code. A more detailed Seismic Hazard Map of Europe has also been produced in the framework of the European Project SHARE, though still with no direct reference to the corresponding National Annexes (Giardini et al. 2013).

However, seismic hazard maps may be developed for peak ground acceleration corresponding to any return period T_R. Therefore, caution needs to be exercised by practitioners to ensure that an EC8-compatible reference seismic action is adopted in undertaking EC8-compliant seismic design, especially in regions outside the European Continent (Fig. 2.20).

The EC8 *design seismic action*, α_g, is defined as the product of the reference seismic action α_{gR} times the *importance factor* γ_I. That is,

$$a_g = \gamma_I a_{gR}. \tag{2.8}$$

Practically, the value of the importance factor γ_I is associated with the intended use and occupancy of building structures and their potential consequences for collapse due to severe ground shaking (see also Sect. 1.1.4). Based on such criteria, buildings are classified into four classes given in Table 4.3 of EC8 (§4.2.5). *Conventional ordinary occupancy r/c buildings fall into importance class II* for which $\gamma_I = 1$ *by definition* (clause §4.2.5(5)P of EC8). Therefore, *for "ordinary buildings", the design seismic action equals the reference seismic action having 10 % probability of being exceeded in 50 years or a return period of 475 years.* Lower or higher than $\gamma_I = 1$ values are used for less critical or more critical

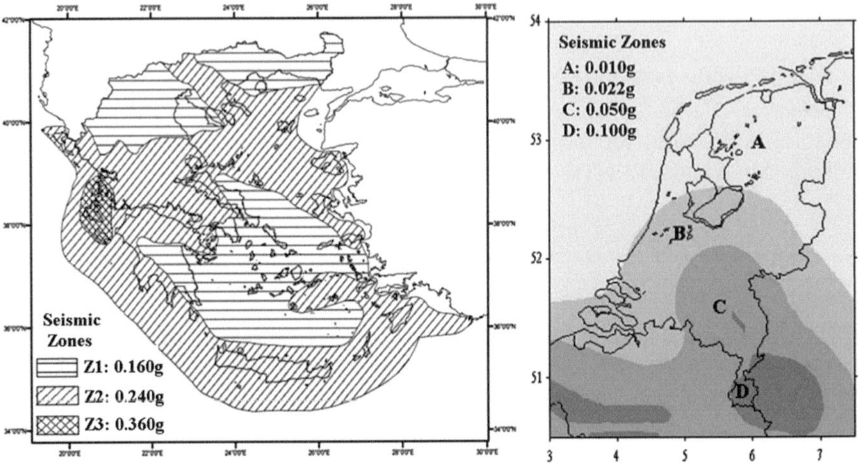

Fig. 2.20 Seismic hazard maps included in the National Annexes to EC8 of Greece (*left*) and of The Netherlands (*right*)

structures, respectively. As a final remark, it is noted that, from a theoretical viewpoint, a change to the value of the importance factor is equivalent to changing the return period of the considered design seismic action α_g compared to the reference seismic action α_{gR} (clause §2.1(4) of EC8).

2.3.1.2 The Design Spectrum for Elastic Analysis

The *horizontal elastic response spectrum* S_e is defined analytically in clause §3.2.2.2(1)P of EC8 as a function of the natural structural period T. Its shape is characterized by four distinct period dependent branches demarcated by the corner periods T_B, T_C, and T_D as follows (Fig. 2.21a and Fig. 3.1 of EC8):

- Short natural period branch ($T \leq T_B$) corresponding to very stiff structural systems where spectral ordinates increase with increasing T;
- Medium natural period branch ($T_B \leq T \leq T_C$) of constant spectral ordinates ("plateau");
- Long natural period branch ($T_C \leq T \leq T_D$) of decreasing spectral ordinates with increasing T; and
- Very long natural period branch ($T \geq T_D$) corresponding to very flexible structural systems of constant peak relative displacement ordinates $S_e \cdot (2\pi/T)^2$.

Apart from the design peak ground acceleration α_g, the amplitude of the EC8 elastic response spectrum depends on the soil amplification factor, S, and on the linear (viscous) damping dependent factor, η. This is indicated in the peak value (plateau) of the elastic response spectra normalized by the design peak ground acceleration α_g included in Fig. 2.21. *For 5 % ratio of critical viscous damping commonly*

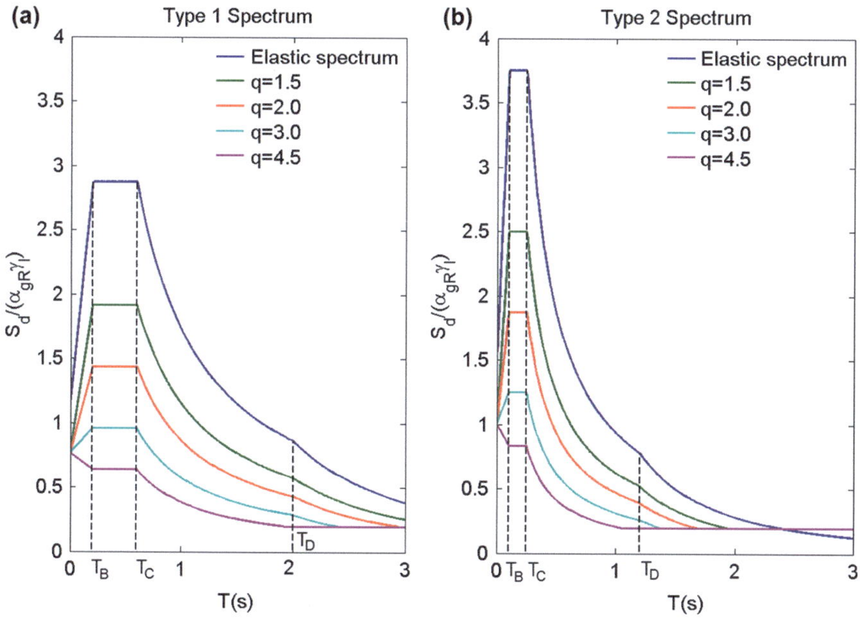

Fig. 2.21 Normalized horizontal EC8 elastic spectrum recommended for ground type "C" (Type 1 (a) and Type 2(b)) and corresponding design spectra for various values of the behavior factor q

assumed for r/c structures and adopted hereafter, the damping factor takes on the "reference" value $\eta = 1$ (clause §3.2.2.2(3) and (4) of EC8).

The values of the soil amplification factor S and the corner periods T_B, T_C, and T_D are site-specific and depend on the properties of the local supporting ground and seismological considerations. Specifically, EC8 classifies the supporting ground into five different "basic" *ground types* ("A" to "E") plus two "special" ones (Table 3.1 in clause §3.1.2 of EC8) based on certain quantitative local soil related criteria. Ground type "A" corresponds to rock and is the "reference" ground type for which $S = 1$.

Further, two different *spectrum types* (type 1 and 2) are defined accounting for the surface-wave magnitude M_s of the earthquake that contributes the most to the site seismic hazard within a probabilistic seismic hazard analysis (clause §3.2.2.2 (2)P of EC8). Type 2 spectrum corresponds to $M_s < 5.5$. *Recommended* values for all five basic ground types and for types 1 and 2 spectra are provided in Tables 3.2 and 3.3 in clause §3.2.2.2 of EC8. In Fig. 2.21a, b, the type 1 and 2 elastic response spectra normalized to the design peak ground acceleration for ground type "C" are included, respectively. It is seen that type 2 spectra representing typical intra-plate seismo-tectonic environments observe higher spectral ordinates within a narrower band of natural periods shifted towards shorter periods compared to type 1 spectra. The latter spectra correspond to typical interplate seismic events dominating seismic structural design in most regions of South Eastern European countries (e.g., Italy, Greece, Turkey) and many other high seismicity areas worldwide.

The *EC8 design spectrum for elastic analysis* S_d is analytically expressed as (§3.2.2.5 of EC8)

$$S_d(T) = \begin{cases} \gamma_I \alpha_{gR} S\left[\frac{2}{3} + \frac{T}{T_B}\left(\frac{2.5}{q} - \frac{2}{3}\right)\right], & 0 \le T \le T_B \\ \gamma_I \alpha_{gR} S\frac{2.5}{q}, \quad T_B \le T \le T_C & \\ \gamma_I \alpha_{gR} S\frac{2.5}{q}\left(\frac{T_C}{T}\right) \ge \beta\gamma_I \alpha_{gR}, & T_C \le T \le T_D \\ \gamma_I \alpha_{gR} S\frac{2.5}{q}\left(\frac{T_C T_D}{T^2}\right) \ge \beta\gamma_I \alpha_{gR}, & T_D \le T \end{cases}, \tag{2.9}$$

where:

- $\gamma_I \alpha_{gR}$ ($=\alpha_g$) is the peak (horizontal) ground acceleration corresponding to the design seismic action (for importance factor $\gamma_I = 1$, the value of α_g has 10 % probability of being exceeded in 50 years, as discussed in Sect. 2.3.1.1);
- S is the soil amplification factor (reference value S = 1 for rock ground "A");
- T_B and T_C are the corner periods defining the second constant acceleration branch ("plateau") of the design spectrum;
- T_D is the corner period signifying the beginning of the constant peak relative displacement ($S_d(2\pi/T)^2$) response range;
- β denotes a lower bound for the design spectrum normalized by α_g, having a recommended value of 0.20 and
- q is the behaviour (or force reduction) factor (see Sect. 1.2.2).

The EC8 design spectrum S_d in Eq. (2.9) is a piecewise continuous function of T obtained by dividing the three branches of the elastic spectrum S_e for $T \ge T_B$ by the behaviour factor q and by reducing the theoretical "zero period" spectral ordinate $S_e(T=0)$ by 2/3. Additionally, a minimum bound is applied to the two right-most branches along the natural period axis.

Therefore, by setting q = 1 and by replacing the 2/3 ratio of the first branch in Eq. (2.9) by 1, one retrieves the EC8 elastic response spectrum for 5 % damping. This is further illustrated in Fig. 2.21 where four EC8 design spectra obtained by the type 1 and type 2 elastic response spectra for ground "C" are plotted for four values of the behaviour factor q. Note that these design spectra are "pinned" at a $S_d(T=0)/\alpha_g = 2/3 \cdot S_e(T=0)/\alpha_g$ value which remains the same for any behaviour factor. In fact, for large behaviour factors, the first branch of the design spectrum may have a negative slope (decreasing spectral ordinates with increasing natural period) which is never the case for the elastic response spectrum. It is further noted that a lower bound of $\beta = 0.2$ applies for very long natural periods.

The significance of the behaviour factor q in code-compliant seismic design has been highlighted in Sect. 1.4 and further comments and recommendations are included in the next chapter on the practical implementation of EC8 code. The following sub-section clarifies the notional and practical difference between increasing the design seismic action and reducing the behaviour factor in EC8-compliant seismic design.

2.3.1.3 Modification of the Design Seismic Action vis-a-vis the Behaviour Factor

By examining the analytical expression of the EC8 design spectrum in Eq. (2.9), it is seen that, for a given soil amplification factor S, the amplitude of the design seismic loads (which are proportional to the design spectrum ordinates) increases

- (a) either by increasing the importance factor γ_I and, therefore, the design seismic action $a_g = \gamma_I\, a_{gR}$,
- (b) or by reducing the behaviour factor q.

Interestingly, the above two operations may have the same quantitative (numerical) impact on the stress resultants computed from a linear analysis for which the structure needs to be designed (i.e., for which structural members are detailed). *However, operations (a) and (b) bear a completely different qualitative meaning*. In case (a), the design seismic loads increase by increasing the linear seismic demand (design seismic action) either by increasing the exposure time of the structure to the seismic hazard t_e, or, equivalently, by increasing the considered return period T_R (see also discussion in Sect. 2.3.1.1). Nevertheless, the level of the expected inelastic deformations that the structure will undergo under the increased design seismic action level remains the same. In case (b), the structure undergoes smaller inelastic deformations and, therefore, achieves a higher seismic performance under the same design seismic action level.

The above qualitative difference can perhaps be better understood in the context of performance-based seismic design (Sect. 1.3). Focusing on Fig. 1.17, case (a) corresponds to a downwards column-wise change to the structural performance: the structure performs the same but for a more intense (less frequent) earthquake event. Case (b) corresponds to a row-wise change towards higher seismic performance for the same seismic intensity level. Interestingly, in both cases, enhanced seismic structural behaviour is achieved within the performance based seismic design framework.

2.3.1.4 Inertial Properties for Seismic Design and Seismic Loading Combination

Lateral seismic forces imposed on structures due to strong ground motion are mass proportional. Therefore, apart from the intensity of the ground motion (expressed in terms of the design spectrum), nominal (design) mass/inertial structural properties need to be specified as well to determine seismic effects. According to EC8 (clause

§3.2.4), the mass/inertial properties for seismic design are obtained by considering the characteristic value G_k of all j permanent gravity loads (e.g., self-weight of structural and permanent non-structural elements, fixed equipment, finishing, etc.), but only a fraction of the characteristic value Q_k of all i variable gravity loads (e.g., loads due to occupancy, movable equipment, etc.) in the following combination of actions:

$$\sum_j (G_{k,j})\text{“}+\text{”}\sum_i (\psi_{E,i} Q_{k,i}), \tag{2.10}$$

where the symbol $\sum_k$ implies “the combined effect of all k actions” and the symbol “+” implies “to be combined with” following the standard notation adopted within the Eurocode series. In the above expression, $\psi_{E,i}$ is a combination coefficient for variable action i given as (clause §4.2.4 of EC8)

$$\psi_{E,i} = \varphi \psi_{2i}, \tag{2.11}$$

where, ψ_{2i} is the combination coefficient of the *quasi-permanent value* of the i variable action, $\psi_{2i} Q_{k,i}$, and $\varphi \leq 1$ is a coefficient that may further reduce the quasi-permanent value of variable action depending on the type of variable action and the storey occupancy in a building structure.

The combination coefficients ψ_{2i} are given in Eurocode 0 (CEN 2002) Annex A1, and may be as low as 0.3 for ordinary occupancy residential and office buildings recognizing that, during an earthquake, a relatively small fraction of the characteristic value of the variable actions will be acting combined with the action of the full permanent loads. In fact, the quasi-permanent value of a variable action $\psi_2 Q_k$ is considered to be “almost always” exceeded during the life-time of a structure within a probabilistic/statistical context. Moreover, *recommended* values for the φ coefficient are included in Table 4.2 of EC8 (§4.2.4) which can be as low as 0.5 allowing for up to 50 % reduction of the considered mass of the building contributed by the quasi-permanent gravity variable loads.

As a closure to this section, it is deemed essential to note that the seismic action is classified as an “accidental” action to Eurocode 0 (CEN 2002) and that the “total” design action combination, which includes permanent and variable actions together with the seismic action, is given by (clause §6.4.3.4 of EN1990:2002)

$$\sum_j (G_{k,j})\text{“}+\text{”}P\text{“}+\text{”}A_{ED}\text{“}+\text{”}\sum_i (\psi_{2,i} Q_{k,i}), \tag{2.12}$$

in which A_{ED} is the design seismic (accidental) action and P denotes the prestressing action, if it exists. In view of Eqs. (2.10), (2.11), and (2.12), it is important to note that the φ enters only in the definition of the gravity loads used to obtain the inertia property of the building structure and therefore influences the value of A_{ED}. However, it is not used in combining the quasi-permanent variable actions in the total seismic loading combination of Eq. (2.12). In other words, if $\varphi \neq 1$ is adopted, the mass of the building used to define the design seismic action A_{ED} will not be consistent with the gravity loads applied to the structure in combination with the design seismic action in the analysis stage.

2.3.2 EC8-Compliant Modeling of Superstructure, Foundation, and Supporting Ground

The gravity and lateral load-resisting (or load-bearing) structural system of r/c building structures comprises the superstructure and the foundation. Its mission is to safely transfer externally applied loads (e.g., gravity loads primarily applied to floor slabs, lateral loads due to wind pressure, lateral inertial loads due to horizontal seismic excitations, etc.) to the supporting ground. Apart from *floor slabs*, the superstructure of a typical contemporary r/c building designed for earthquake resistance can include *beams*, *columns*, *walls*, and *cores*, as schematically shown in Fig. 2.22. Further, the foundation system may include *simple pad footings*, *strip footings*, *deep beam grillages*, or even *monolithic mat-slab (raft) foundations* to support the vertical structural members (columns, walls, and cores).

One-dimensional finite elements (e.g., two-node Euler-Bernoulli beam/column element) and sometimes *two-dimensional finite elements* (e.g., four-node rectangular shell elements for plate bending or plane stress) together with appropriate *finite element meshing* schemes are routinely incorporated to represent the material, structural, and inertial properties of the above structural members and to define their *topology and connectivity* in typical *numerical/computational finite element (FE) models*. Furthermore, for the case of ordinary r/c building structures subject to seismic excitation, soil compliance can be accounted for, if deemed necessary, by FE models by introducing elastic support conditions at the foundation level. More sophisticated numerical/modeling techniques for capturing explicitly the dynamic soil-foundation-structure interaction phenomenon (e.g., use of two-dimensional/three dimensional finite elements or boundary elements to model the supporting ground) are rarely considered in the seismic design of structures concerned in this book and, thus, are left out of the ensuing discussions.

Focusing on the superstructure FE modeling of a typical r/c building, a better insight into the involved considerations and requirements can be gained by examining structural members and structural sub-systems on an individual basis. In a nutshell, beams are horizontally oriented and exhibit primarily uni-directional flexural behaviour along the vertical (gravitational) plane due to vertical and lateral loads. Due to their orientation and monolithic connection with the floor slabs, beams bear negligible loads along their longitudinal direction (axial loads). Further, they are taken to have flanged cross-sections, either T-shaped or L-shaped.

On the antipode, columns bear significant axial loads, since they are vertically oriented, and undergo significant bi-directional flexure under seismic excitation.

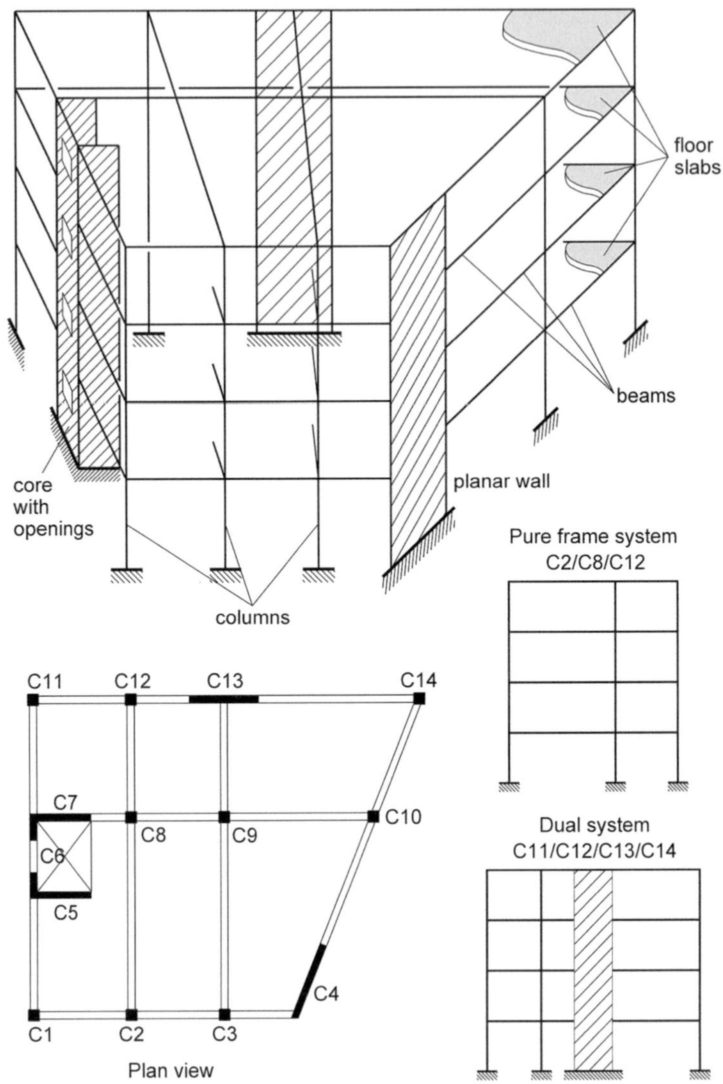

Fig. 2.22 Typical structural components of multistory r/c buildings

The cross-sectional dimensions of beams and columns are an order of magnitude smaller than their longitudinal dimension (length) and, therefore, are viewed as "one-dimensional" members in space.

Reinforced concrete walls are vertically oriented planar ("two-dimensional") structural members with elongated cross-sectional dimensions (aspect ratios >1/4) such that their width (longest cross-sectional dimension) is of the same order of magnitude with the storey height. They exhibit primarily either flexural behaviour (*flexural-type/slender walls*) or shear behaviour (*shear-type/squat walls*).

Cores are assemblies of two or more planar walls having either open-loop or closed-loop (tube-like) cross-sections. Vertically oriented lintel braces may be included along one open side of cores with open-loop cross-sections. Cores can be viewed as spatial ("three-dimensional") structural members in space, since their "envelop" cross-sectional dimensions are usually of the same order of magnitude with the storey height.

It is noted that planar r/c walls and r/c cores are commonly used in areas of high seismicity to increase the lateral stiffness of building structures and, when existent, are important members of the seismic load-resisting system of a building structure designed to exhibit ductile behaviour. Therefore, they should be distinguished from *infill* or architectural walls which exhibit non-ductile material behaviour and are normally considered as "non-structural" elements in code-compliant seismic design of structures. Notably, the contribution of infill walls in resisting seismic forces is neglected in the majority of seismic codes.

Moment resisting frame (MRF, *pure* frame or simply *frame*) structural sub-systems form due to the monolithic (though not necessarily perfectly rigid) connection of beams and columns achieved in cast-in-place and in properly engineered pre-cast r/c buildings (see Figs. 2.22 and 2.33).

Further, *dual* structural sub-systems form by coupling walls and/or cores together with columns and/or frames via beam members at each floor level (see Figs. 2.22 and 2.41).

Moreover, planar *coupled wall* (or simply *wall*) structural sub-systems form by coupling together two or more planar walls via strong coupling beams, not necessarily at (only) the floor levels.

Several such sub-systems of the same or different types linked together in space via beams monolithically embedded within floor slabs (at least in cast-in-place r/c structures) form the superstructure of the (gravitational *and* lateral) load resisting system in r/c buildings.

In light of the above, it is seen that the development of a typical computational (FE) model to efficiently represent/capture the properties of the combined superstructure-foundation-soil system involves several simplification steps and assumptions with regards to

- the material behaviour of concrete and steel (e.g., linear-elastic or elasto-plastic stress-strain relationships, etc.);
- the structural behavior of r/c structural members (e.g., axially inextensible members, perfectly rigid in-plane slabs or diaphragms, etc.)
- the connectivity of r/c structural members (e.g., perfectly/semi- rigid frame connections, level of in-plane stiffness of floor slabs, etc.);
- the cross-sectional properties of r/c structural members (e.g., second moment of area for flanged beams/planar walls/cores, torsional stiffness of cores, etc.);
- the distribution of inertia/mass properties in the structure (e.g., mass concentration at the center of gravity of floors, distribution across many nodes, etc.); and
- the soil properties (e.g., spring constant values of elastic foundations, inclusion of rotational springs, etc.).

However, codes of practice provide only limited guidance for practitioners with regards to the development of proper FE models for seismic analysis, as this is considered to be a subjective issue of personal preference and cumulative experience, while it is closely related to the capabilities of available commercial software. EC8 is no exception. It only provides brief, primarily qualitative, comments and requirements addressing the above modeling issues for building structures (clause §4.3.1 of EC8). These general EC8 requirements are listed in the following paragraphs and complemented by some additional comments. Further, a summary of recommended assumptions and modeling techniques made to address the most common modeling requirements for EC8 compatible *linear analysis* is given in Table 2.4.

Table 2.4 Common requirements and assumptions in structural (elastic and inertial) modeling of r/c building structures

Common modeling aspects and requirements	Common modeling assumptions made
Full or partial diaphragmatic action of r/c floor slabs	Perfectly rigid, within their plane (either horizontal or inclined), slabs for full diaphragmatic action
	Deformable, within their plane, slabs for partial diaphragmatic action (see Sect. 2.3.3.1)
R/c moment resisting frames	Frames with semi-rigid joints
	Perfectly rigid-jointed frames (see Sect. 2.3.3.2)
Planar r/c walls and r/c cores above the ground level (in the superstructure) and below the ground level (basement)	Use of equivalent frame models comprising one-dimensional linear finite elements
	Use of alternative equivalent linear models
	Use of linear two dimensional finite elements (see Sects. 2.3.3.3 and 2.3.3.4)
Concrete cracking effects in r/c members in equivalent linear analyses	Modification of structural members properties (stiffness reduction of beams, columns, walls, and cores) using empirical reduction factors-§4.3.1(7) of EC8 (see Sect. 2.3.2.1)
Infill wall contribution to lateral load resistance	Completely ignore contribution
	Modeling by means of equivalent linear bracing bars (see Sect. 3.3)
Supporting ground compliance	Foundation beams and slabs resting on
	linear elastic springs (Winkler springs)
	a continuous elastic medium
	two or three-dimensional finite elements (see Sect. 2.3.3.5)
Inertial/mass properties concentrated at the level (height) of floor slabs	Floor masses lumped at a single node (center of mass) of each floor with two horizontal translational plus one rotational about the vertical axis dynamic degrees of freedom
	Floor masses lumped at many nodes on each floor with two horizontal translational dynamic degrees of freedom

Adequate modeling of stiffness and inertial property distribution (§4.3.1(1)P of EC8)

The need to ensure that the adopted structural building model "adequately" represents the distribution of its elastic properties (stiffness) and its inertial properties (mass) such that all significant deformation shapes and inertia forces are properly accounted for under the seismic action is highlighted. This is an important consideration, especially in the case of adopting the modal response spectrum method of analysis which takes into account the higher than fundamental modes of vibration (see Sects. 2.4.2 and 2.4.3). The additional requirement for adequate strength distribution representation if a non-linear analysis method is adopted for design purposes is also posed. A brief discussion on material non-linear FE modeling options is included in Sect. 2.4.1.2 where inelastic methods of analysis are reviewed. In the remainder of this section, the assumption of linear material behaviour is made in developing FE models for EC8-compliant linear types of static or dynamics analysis (see also Sect. 2.4.2).

Joint rigid offsets (§4.3.1(2) of EC8)

It is a *requirement* that the structural model accounts for the contribution of "joint" regions (where two or more structural members are connected) to the deformability of the building. Thus, the end zones of beams and columns in frame structural systems must be explicitly modeled as rigid depending on the geometry of the joint. A detailed discussion on this issue is provided in Sect. 2.3.3.2.

Diaphragmatic action of floor slabs and mass modeling/discretization (§4.3.1 (3) and (4) of EC8)

"In general the structure may be considered to consist of a number of vertical and lateral load resisting systems, connected by horizontal diaphragms." (§4.3.1(3)). This consideration is in perfect alignment with previous discussions in view of Fig. 2.22. Floor slabs can be modeled as horizontal diaphragms "binding together" vertical structural members (e.g., uncoupled walls and cores) and sub-systems (e.g., frames and dual systems) at the level of each floor with the aid of beams.

"When the floor diaphragms of the building may be taken as being rigid in their planes, the masses and the moments of inertia of each floor may be lumped at the center of gravity." (§4.3.1(4)). Apart from being quite favourable in terms of seismic structural response of buildings (see Sect. 2.1), the in-plane stiffness of floor diaphragms significantly facilitates modeling and analysis, as it allows for lumping all inertial properties of the building at the center of gravity of floors. Therefore, only three dynamic degrees of freedom per floor need to be considered: two translational along two orthogonal horizontal axes and one torsional about the gravity axis (see Sect. 2.3.3.1). This consideration expedites the modal analysis step and the interpretation of mode shapes. Nevertheless, a rigorous verification of diaphragm rigidity according to EC8 requires FE modeling using two-dimensional finite elements, since "The diaphragm is taken as being rigid, if, when it is modeled with its actual in-plane flexibility, its horizontal displacements nowhere exceed those resulting from the rigid diaphragm assumption by more than 10 % of the corresponding absolute horizontal displacements in the seismic design situation." (§4.3.1(4)). Modeling details and practical recommendations on this issue are included in Sect. 2.3.3.1.

Use of planar structural models (§4.3.1(5) of EC8)

The use of (two) planar structural models along (two) "principal" directions of buildings instead of spatial (three-dimensional) models is allowed for the seismic design of regular in plan buildings or of buildings conforming to the, largely qualitative, conditions of clause §4.3.3.1(8) of EC8. A detailed discussion and practical recommendations on this issue are provided in Sect. 2.3.2.2.

Accounting for the effect of concrete cracking (§4.3.1(6) of EC8)

The adopted values of the stiffness properties of r/c structural members should correspond to the initiation of yielding of the reinforcement. Therefore, appropriately reduced stiffness values compared to those corresponding to uncracked structural members should be adopted in the analysis stage. A detailed discussion and practical recommendations on this issue are provided in Sect. 2.3.2.1.

Assumption of reduced stiffness properties for r/c structural members (§4.3.1 (6) and §4.3.1(7) of EC8)

In clause §4.3.1(6) of EC8, it is stated that the adopted values of the stiffness properties of r/c structural members should account for the effect of concrete cracking and should correspond to the initiation of yielding of the reinforcement. "Unless a more accurate analysis of the cracked elements is performed, the elastic flexural and shear stiffness properties of concrete and masonry elements may be taken to be equal to one-half of the corresponding stiffness of the uncracked elements." (§4.3.1(7)). The use of "effective" stiffness properties corresponding to cracked r/c members (at the onset of reinforcement yielding) is interweaved with the use of "equivalent" linear analysis for seismic design (see also Sect. 1.2.4). However, it is noted that EC8 does not make any particular reference to the need for reducing the axial and torsional stiffness of structural members, while it suggests the same level of reduction of flexural and shear stiffness for all different types of r/c members. Both these assumptions need careful consideration and are discussed in some detail in Sect. 2.3.2.1.

Accounting for infill walls effect in structural models (§4.3.1(8) of EC8)

In clause §4.3.1(2) of EC8, the need to account for non-structural elements influencing the response of the load-resisting structural system of r/c buildings is emphasized. Masonry infill walls, though significantly stiff within their plane, are not normally taken to *contribute* to load-resistance under the design seismic action, as they are brittle and usually prone to non-ductile failure at even lower than the nominal design earthquake levels of seismic action (at least for an adopted q factor equal or close to the maximum allowable value). However, the adverse effects of non-uniform distribution of infill walls in plan and elevation need to be taken into account in certain cases, as specified in clauses §4.3.1(8) and §4.3.6 of EC8. Further details on this matter are provided in Sect. 3.3.

Foundation and supporting ground compliance (§4.3.1(9) of EC8)

"The deformability of the foundation shall be taken into account in the model, whenever it may have an adverse overall influence on the structural response."; "Foundation deformability (including the soil-structure interaction) may always be taken into account, including the cases in which it has beneficial effects." (§4.3.1 (9)). The issue of whether and in which cases the effects of soil compliance and, even more, the soil-structure interaction phenomenon have positive effects on the seismic response of structures is open to research and is certainly not readily predictable in advance. Therefore, it is generally recommended to include the foundation and soil compliance in the overall structural model with due consideration of the distinct features of static and dynamic soil-structure interaction. More details on practical ways to accomplish this are provided in Sect. 2.3.3.5.

Inertial/Mass properties for seismic design (§4.3.1(10) of EC8)

"The masses shall be calculated from the gravity loads appearing in the combination of actions indicated in 3.2.4. The combination coefficients ψ_{Ei} are given in 4.2.4(2)." (§4.3.1(10)). This issue has already been discussed in Sect. 2.3.1.4 and is only included here for the sake of completeness of the EC8 modelling requirements.

Guidance and recommendation on typical discretization FE schemes and modeling techniques for ordinary r/c buildings in accordance with the general EC8 modeling requirements are provided in Sect. 2.3.3. In the remainder of this section, special attention is focused on two practical issues arising in the modeling stage of the seismic design, as these can have a significant impact on the (linear) analysis results and, therefore, on the final design. The first issue relates to the stiffness reduction in r/c structural members due to concrete cracking. The second issue relates to whether two-dimensional (planar) FE models can be adopted instead of three-dimensional (spatial) models in the context of equivalent linear analysis. In the following two paragraphs, certain comments and practical recommendations are given in relation to the above two issues.

2.3.2.1 Stiffness Reduction of R/C Members for Linear Analysis (§4.3.1 (6) and (7) of EC8)

As already discussed in Sects. 1.1.5.4 and 1.2.4 and noted in clause §4.3.1(6) of EC8, it is reasonable to assume that r/c structural members are cracked in performing "equivalent" linear types of analysis using an inelastic spectrum to define the seismic input action. Therefore, "effective" or "secant" stiffness values should be considered which are typically found (e.g., by lab testing) to be significantly lower than the theoretical stiffness properties calculated from the uncracked gross section properties of r/c members. That is,

- the flexural rigidity (EI),
- the shear rigidity (GA_s),
- the axial rigidity (EA), and
- the torsional rigidity (GJ),

where

E is the modulus of elasticity,
G is the shear modulus,
A is the member cross-sectional area,
A_s is the member shear area,
I is the moment of inertia (second moment of area), and
J (or I_T) is the torsional moment of inertia (polar moment of area).

It is further noted that, in the context of (linear) seismic analysis and design, the overall stiffness of the r/c load-resisting structural system (and of its adopted computational model) significantly influences its dynamical properties and ultimately the design seismic effects. In particular, structural stiffness has a profound effect on the value of the fundamental and the higher natural periods of the structure which, in turn, influence the seismic loads obtained from the design spectrum of Eq. (2.9), for which the structure needs to be designed. Further, the stiffness of the load resisting structural system depends heavily on the stiffness properties of its individual constituent r/c structural members. Therefore, care should be exercised in the adopted values of effective/reduced r/c member stiffness properties accounting for concrete cracking. These values should not be unrealistically low, since this will significantly increase the natural period of the structure and, thus, may underestimate the design seismic loads (see Fig. 2.21), leading to non-conservative designs. Conversely, if relatively large values for secant stiffness properties are adopted, excessively high design seismic loads may be reached, leading to cost-ineffective designs.

According to EC8 (clause §4.3.1(7), the flexural rigidity (EI) and the shear rigidity (GA_s) should be reduced by 50 % in cracked members compared to the uncracked values (unless a rigorous analysis is undertaken), assuming that such reduced stiffnesses correspond to the initiation of yielding of the reinforcement (clause §4.3.1(6)). The above default reduction to stiffness properties is considered to be relatively small compared to what is observed in relevant experimental tests (Fardis 2009). As such, it yields conservative (safe-sided) designs in the context of force-based seismic design, though it may underestimate deformations. The latter issue is not considered to be important, as EC8-prescribed deformation-based verification checks (see Sect. 3.2) are usually not critical for the majority of ordinary r/c building structures.

Furthermore, the assumption that the same level of flexural and shear stiffness reduction, i.e., 50 %, applies to all different types of r/c members, combined with the additional assumption that axial and torsional member deformations have negligible influence on bending and shear stiffness, significantly expedites the design process from a practical viewpoint, as the same structural (FE) model used for "equivalent" linear seismic analysis (assuming cracked r/c members) may also be used for (linear) analysis under gravity loads (assuming uncracked r/c members), i.e., for the ultimate limit state basic design combination for permanent and variable actions (no accidental/seismic). This is because stress resultants obtained from linear analysis for static externally applied loads (i.e., support settlements or temperature effects are considered) depend only on the *relative* stiffness contribution of each structural member (i.e., relative stiffness values assigned to each finite element in the model), and not on the absolute stiffness values.

Nevertheless, despite being convenient for design purposes, it should be noted that considering a uniform flexural and shear stiffness reduction for all types of structural members is not realistic, since the extent of concrete cracking during cyclic inelastic deformation is smaller for members carrying large axial forces (e.g., columns) compared to members under (almost pure) flexure (e.g., beams). In this regard, EC8 (clause §4.3.1(7)) does allow for the adoption of more elaborate stiffness properties if a more accurate analysis of the cracked elements is performed without providing any further suggestions with regards to the nature of such an analysis. To this end, it would be rational to relate the level of stiffness reduction to the targeted seismic performance level, as this is expressed via the behaviour (force reduction) factor q within the EC8-prescribed force based seismic design. In particular:

- if a high seismic performance level is targeted (i.e., a relatively low value of the behaviour factor q is adopted in design), the structure undergoes insignificant, if any, inelastic deformation under the design seismic action and, thus, relatively small stiffness reduction/degradation at structural members can be assumed due to limited extent of concrete cracking; while
- if a low seismic performance level is targeted (i.e., a relatively large value of the behaviour factor q close to the maximum allowable is adopted in design), the structure undergoes significant inelastic deformation under the design seismic action and, thus, significantly higher, compared to the previous case, stiffness reduction/degradation at structural members needs to be assumed.

The above reasoning suggests that the adopted values for effective stiffness properties corresponding to cracked members (i.e., flexural, shear, axial, and torsional rigidity) should depend on the value of the behaviour factor q. However, at present, no research work along these lines is found in the open technical literature. In this respect, the cracked member (secant) stiffness at yield for several types of r/c members reported in (Priestley et al. 2007) can be adopted as being a function of axial loading.

As a final remark, it is noted that EC8 does not make any recommendation with regards to effective/secant values for the axial and torsional rigidities of structural members. It is *recommended* that the axial rigidity of cracked members remains the same as that for the uncracked members. It is further emphasized that the torsional stiffness of structural members has a non-negligible influence on seismic effects, assuming that spatial (three-dimensional) FE models are used in the analysis, as recommended in the next section. For example, the assumed value of torsional stiffness in the modeling of r/c cores may influence design seismic effects in all structural members of an r/c building (see Sect. 2.3.3.4). Given that torsional stiffness is significantly reduced due to cracking, a reduced value in the uncracked torsional stiffness (i.e., to the member torsional moment of inertia, I_T) of the order of 90 % is herein *recommended* (Fardis 2009).

2.3.2.2 On the Use of Planar Structural FE Models for Linear Analysis (§4.3.1 of EC8)

Due to the practically unavoidable asymmetries in stiffness and inertial property distribution within building structures, a certain level of "coupling" between translational and rotational response will always occur. The practical implications of this coupling may be better understood by considering a three-dimensional linear FE model of a regular building structure with vertical structural members and sub-systems aligned parallel along two orthogonal horizontal axes X and Y. Assuming that floor slabs behave as rigid diaphragms, each floor is assigned two horizontal translational dynamic degrees of freedom along X and Y and one rotational (torsional) degree of freedom about a vertical (gravitational) axis passing through the floor center of gravity (§4.3.1(4) of EC8). An asymmetric stiffness distribution and/or mass distribution in-plan would cause translational-rotational mode coupling (see also Appendix B). Specifically, floor diaphragms would rotate about a vertical axis under horizontal seismic action along the X or Y direction. Consequently, these floor rotations will result in translations of structural members along the Y or X directions, respectively. Clearly, such mode coupling effects (or "torsional effects"), that is, horizontal translations along the perpendicular direction to the direction of the seismic action, can only be accounted for in an explicit manner by considering a spatial (three-dimensional) FE model. The severity of the torsional effects depends heavily on the level of observed in-plan asymmetry (or eccentricity, as defined in Appendix B).

Still, in clause §4.3.1(5) of EC8, the consideration of two planar models (one along each "principal direction" X and Y) is allowed to be used for "equivalent" linear types of analysis for

- (a) regular in-plan buildings satisfying the conditions of §4.2.3.2; *and for*
- (b) buildings of up to 10 m high having uniformly distributed infill walls and rigid in plane floor slabs (clause §4.3.3.1(8) and (9) of EC8).

In case (b), the analysis results (seismic effects) are multiplied by a "corrective" factor equal to 1.25. Further, in both cases, the torsional effects are implicitly taken into account by multiplying the seismic design effects (stress resultants) in each structural member determined by the planar models by a factor dependent on the location of structural members in-plan and on the type of linear analysis method used.

It should be recognized that allowing for the use of planar models together with the above semi-empirical modifications aims to facilitate practical EC8 compliant seismic design. However, the adoption of two planar models vis-à-vis a single spatial model may underestimate significantly the seismic effects obtained by means of linear dynamic analyses (Anastassiadis et al. 2003; Athanatopoulou and Avramidis 2008). Furthermore, nowadays, the majority of contemporary structural analysis software packages can readily perform static linear and/or modal response spectrum analysis for linear three-dimensional models. In fact, the case of planar (two-dimensional) models is commonly treated as a "special case" for which additional restraints along the transverse direction are imposed or, equivalently, elimination of dynamic degrees of freedom along one horizontal direction needs to be defined. Moreover, a proper verification check for torsional effects involves, either way, the consideration of linear static analysis in spatial three-dimensional structural models, since locations of accidental eccentricities need to be included in the model even for building structures with two orthogonal horizontal axes of symmetry (see also Sect. 3.1.1.1). Even more, the very development of planar models along two "principal" directions may be quite challenging, or even non-feasible, in the case of vertical structural members or structural sub-systems not aligned along two orthogonal horizontal axes. For all the above reasons, *it is recommended that the use of planar (two-dimensional) FE models is avoided for the seismic design of r/c buildings. Instead, spatial (three-dimensional) models are adopted which can explicitly account for torsional effects and coupling of torsional-translational response.*

2.3.3 Common Structural FE Modeling Practices of Multistorey R/C Buildings for Linear Methods of Analysis

In the context of force-based seismic design using "equivalent" linear analysis methods, structural (FE) modeling does not necessarily aim for a "realistic" representation of the actual behaviour of the structural system subject to the nominal design seismic action (*after all, under such a level of excitation, unless a small behaviour factor is adopted, the actual structure is expected to yield and to respond inelastically, while the considered analysis and FE models assume linear-elastic behaviour*). Rather, FE models aim to achieve dependable (and conservative to a certain extent) analysis results (seismic effects) such that EC8-compliant detailing of structural members can take place to achieve the prescribed design objectives. Still, the FE models used in the analysis need to comply with the EC8 modeling requirements presented in the previous Sect. 2.3.2.

In this context, a better appreciation of the actual requirements for an adequate structural FE model for equivalent linear analysis can be gained by considering an overview of the common structural analysis steps taken for routine seismic design of ordinary r/c buildings. To this aim, consider a typical multistorey r/c building comprising floor slabs and various load-bearing structural sub-systems, as shown in

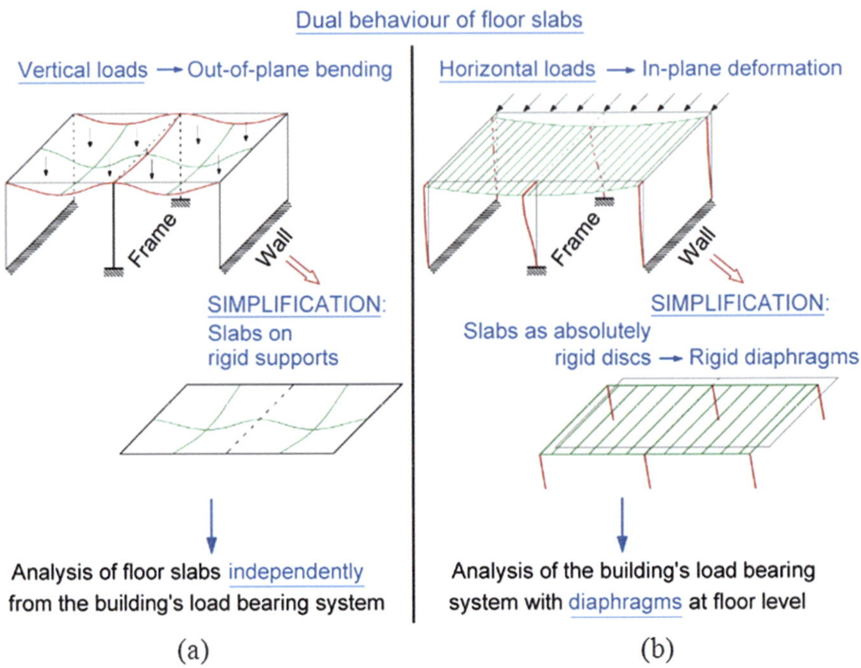

Fig. 2.23 Bending (**a**) and diaphragmatic (**b**) behavior/function of floor slabs in r/c buildings

Fig. 2.22. Most gravity loads (permanent and variable as defined in Sect. 2.3.1.4) and all seismic inertial loads (assuming mass properties are lumped in one or more locations on each slab) are acting upon the floor slabs. Therefore, floor slabs in r/c building structures perform a dual function (see also Fig. 2.23):

– they transfer via bending (flexural) behaviour externally applied gravity loads to the supporting beams, columns, walls, and cores, being monolithically connected to them; and
– they transfer via diaphragmatic (membrane) behaviour in-plane applied loads, including the lateral seismic inertial loads to the load-resisting sub-systems (e.g., frames, walls, etc.).

Based on the above two independent functions of floor slabs (i.e., bending/ flexural and diaphragmatic/membrane structural behaviour), FE modeling and analysis of conventional r/c buildings is commonly undertaken in two independent steps:

– (1) At first, slabs are analyzed only for gravity loads independently from other structural members, assuming they are supported by non-deformable (rigid) members (see Fig. 2.23a). For simple slab geometries, simple methods of linear structural analysis are considered, often involving the use of tables given in standard structural engineering handbooks. For more complicated slab geometries involving in-plan setbacks and openings, linear static FE analysis is

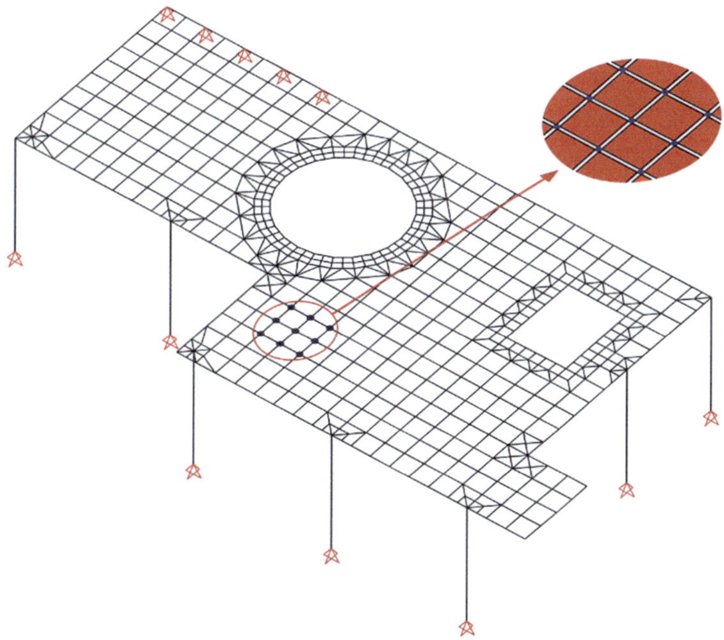

Fig. 2.24 Floor slab of complex geometry discretized using plate/shell finite elements

required using two-dimensional plate-bending or shell finite elements (see, e.g., Fig. 2.24). The obtained slab reactions from the above analysis are applied in the next step as vertical loads acting on the supporting structural members and sub-systems.

- (2) In the second step, the remaining load-resisting structural system is loaded by the slab reactions plus all other actions of the considered design load combination, such as: gravity loads from infill walls acting directly onto beams, lateral seismic loads distributed among structural sub-systems based on their relative lateral stiffness (assuming perfectly rigid diaphragmatic slab action), temperature effects, etc. A typical load-resisting system comprises various vertical members (i.e., columns, walls, and cores) coupled together by beams. The latter are assumed to have flanged T-shaped or L-shaped cross-sections, being monolithically connected to slabs. The above load-resisting system is "solved" under the considered loads to determine deformations and stress resultants in all structural members using any of the two EC8-prescribed linear analysis methods (see Sect. 2.4.2).

Note that in the previous step (2), floor slabs are assumed to act as rigid diaphragms within their plane. The case of modeling in-plane flexible slabs is briefly discussed in Sect. 2.3.3.1 below. Regarding the modeling of flanged beams and columns, linear one-dimensional Euler-Bernoulli beam finite elements in three-dimensional space are used. Certain details on beam-column connectivity

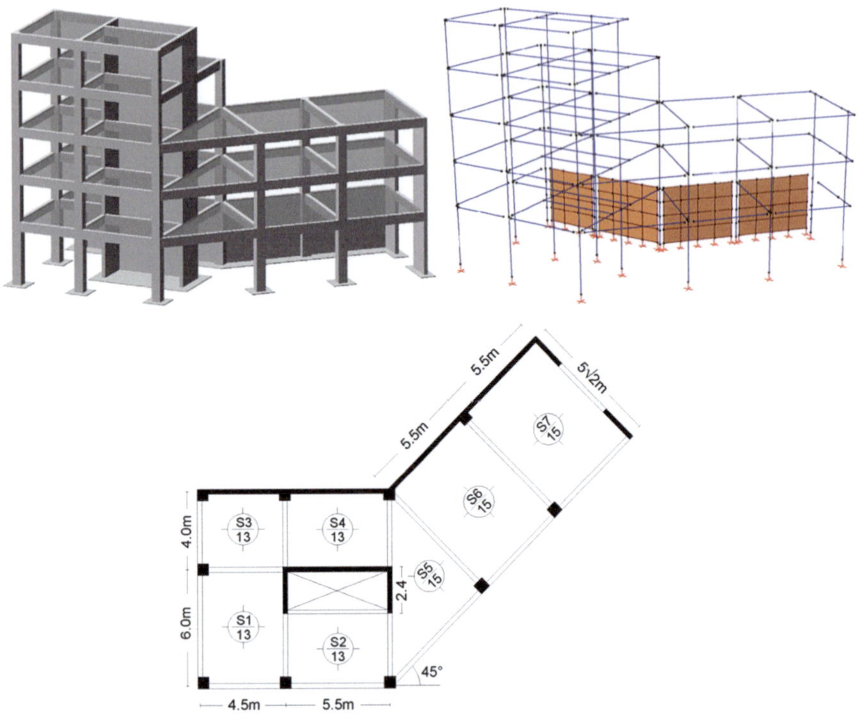

Fig. 2.25 Modeling of strong shear-type r/c walls using 2D finite shell elements and of a U-shape r/c core using 1D beam/column finite elements according to the equivalent frame modeling technique

issues in frame systems are included in Sect. 2.3.3.2. However, walls and cores may only be adequately modeled via one-dimensional beam elements under certain conditions. In case these conditions are valid, walls and cores are modeled via equivalent frame models relying on the wide-column analogy and, thus, the resulting three-dimensional FE model of the load-resisting system of the building superstructure comprises only one-dimensional beam elements. In routine seismic design practice, the use of equivalent frame models to represent walls and cores is acceptable in terms of accuracy for structural members with primarily flexural behaviour (i.e., height to width ratio greater than 2) and without having significant geometrical setbacks and openings or exhibiting significant torsional deformations. In case of walls and cores for which one or more of the above empirical conditions are violated, two-dimensional (planar) finite elements are employed in the modeling. In such cases, the resulting three-dimensional FE model of the load-resisting system of the building superstructure comprises both one-dimensional beam elements and two-dimensional shell or plane stress elements (see, e.g., Fig. 2.25). Refined conditions on the range of applicability of equivalent frame models to represent walls and cores are discussed in Sects. 2.3.3.3 and 2.3.3.4.

In the following paragraphs of this section, guidance and remarks are provided on common modeling practices followed and practical issues arising in the development of linear finite element (FE) models to be used in the seismic design of multistorey r/c buildings. When possible, the recommended modeling approaches are illustrated by means of self-explanatory figures for the proper modeling of

- floor slabs acting as in-plane rigid or flexible diaphragms;
- r/c rigid or partially rigid jointed frames comprising beams and columns;
- r/c planar walls and planar dual systems (coupled walls with frames);
- r/c cores; and
- r/c footings and foundation beams resting on compliant soil.

2.3.3.1 Modeling of Floor Slabs

Floor slabs in cast-in-place r/c buildings are monolithically connected with the supporting beams, columns, walls, and cores. They distribute and transfer horizontal seismic (inertial) loads to the lateral load-resisting system of the building. Furthermore, they contribute in maintaining the floor geometry in-plan and ensure that sufficient horizontal in-plane stiffness exists such that beams are stressed under bending within only one plane (the vertical but not the horizontal).

Floor slabs are commonly assumed to act as rigid diaphragms in their plane. However, in reality, floor slabs have neither infinite in-plane stiffness nor strength, so care must be exercised when using the rigid diaphragm assumption. Therefore, the level of in-plane flexibility of floor slabs needs to be verified and, if deemed appropriate, to be appropriately accounted for.

In practice, the rigid diaphragm assumption is heuristically considered to be valid for floors of "compact" geometry/shape in-plan (e.g., Fig. 2.26). For

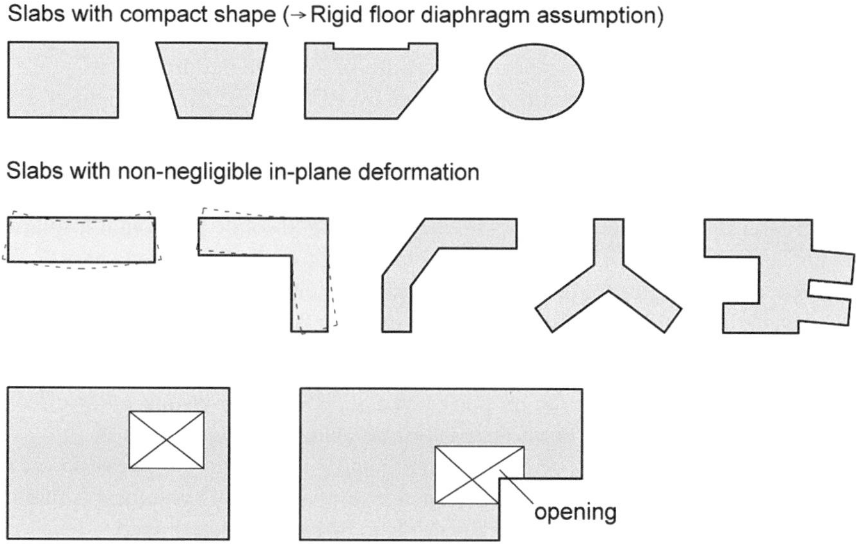

Fig. 2.26 Compact and non-compact plan configurations of floor slabs

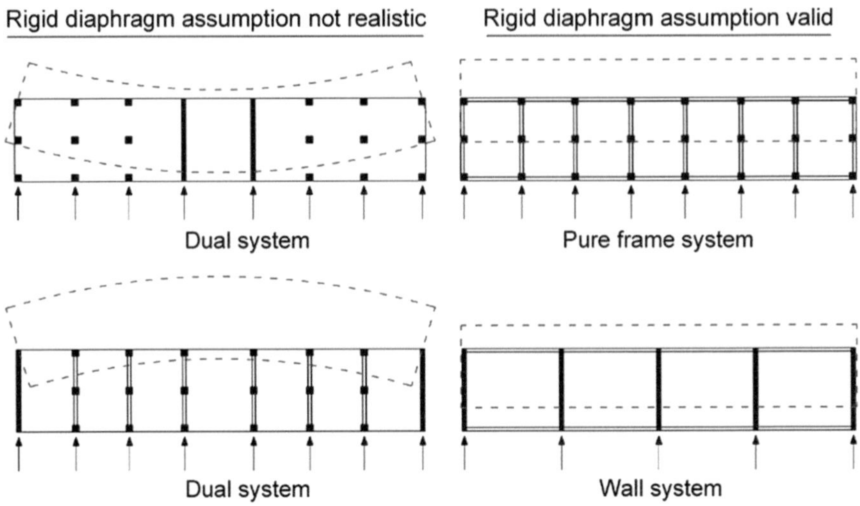

Fig. 2.27 In-plane deformability (*left*) and diaphragmatic behaviour (*right*) in elongated plan configurations

significantly elongated, "winged", or "flanged" floor plans or for slabs with large openings, the rigid diaphragm assumption may not be realistic and may lead to inaccurate or even non-conservative values of seismic effects in certain structural members.

However, such conditions based on purely geometric considerations need to be complemented and informed by structural criteria. In this regard, it is noted that the rigid diaphragm assumption may be accurate even for buildings with elongated floor plans as long as all the vertical structural members attain the same lateral stiffness as shown in Fig. 2.27. It is also worth noting that floor slab in-plane deformability primarily influences the response of low-rise buildings.

The only verification condition specified by EC8 on the floor slab rigid diaphragm assumption is included in clause §4.3.1(3), stating, "The diaphragm is taken as being rigid, if, when it is modeled with its actual in-plane flexibility, its horizontal displacements nowhere exceed those resulting from the rigid diaphragm assumption by more than 10 % of the corresponding absolute horizontal displacements in the seismic design direction.", which is of limited practical use. In practice, design engineers decide on the basis of their personal expertise whether the rigid diaphragm assumption holds or the in-plane flexibility of floor slabs should be explicitly accounted for. The latter should always be the case if

- floor slabs are *required* to be proportioned and detailed for seismic action effects and, therefore, in-plane stress distribution needs to be computed, or if
- part of the slabs are supported by monolithically connected pre-stressed beams and, therefore, the stress field developed in the vicinity of these beams due to significant axial compressive forces needs to be reliably determined.

Modeling of floor slabs as rigid diaphragms

In modern commercial FE structural analysis software, the rigid diaphragm action of floor slabs is modeled by considering appropriate kinematic constraints applied to nodes belonging to a particular diaphragm (slab). These kinematic constraints are expressed by means of a set of equations "coupling" nodal displacements, or equivalently, nodal *static* degrees of freedom (DOFs), to ensure that all nodes translate within the diaphragm plane and rotate about an axis normal to this plane such that the distances among them remain the same. In this manner, the diaphragm moves as a rigid body in its plane without any in-plane deformations (strains) developing. However, out-of-plane diaphragm deformations related to transverse flexure/bending of the slab are free to develop under transverse (e.g., gravity) loads.

The above rigid diaphragm constraint is commonly implemented to any single slab in a FE model of a building structure by introducing an "auxiliary" (virtual) node at the center of gravity of the slab known as the "master" node. This node is assigned only three independent DOFs (out of the possible six nodal DOFs in space) with respect to the local x,y,z orthogonal coordinate system of the slab, with z axis being normal to the slab plane, as shown in Fig. 2.28.

Specifically, three DOFs, two translational along axes x and y and one rotational about the z axis, are assigned to the master node. All other ("slave") nodes of the FE mesh belonging to the slab are assigned all six independent DOFs. Let U_{xm}, U_{ym}, and R_{zm} be the nodal displacements of the master node under some external loading that corresponds to the assigned DOFs, as indicated in Fig. 2.28. The corresponding (in-plane) nodal displacements due to the considered external load of all slave

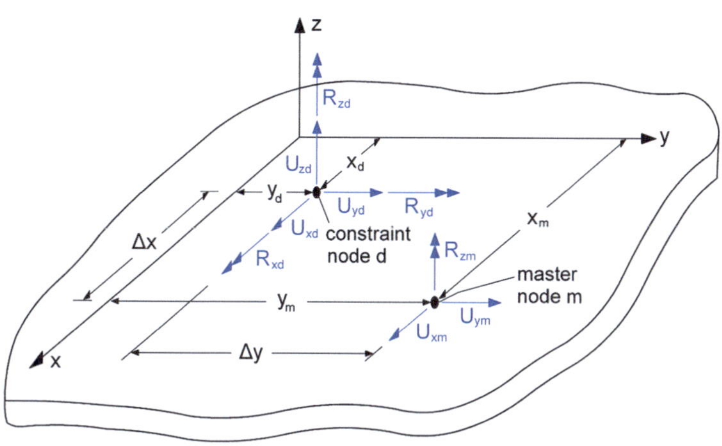

$$U_{xd} = U_{xm} + \Delta y \, R_{zm}$$
$$U_{yd} = U_{ym} - \Delta x \, R_{zm} \quad \text{with:} \quad \Delta x = x_m - x_d, \quad \Delta y = y_m - y_d$$
$$R_{zd} = R_{zm}$$

Fig. 2.28 The rigid diaphragm kinematic constraint concept

nodes belonging to the slab are related to those of the master node through the following set of equations:

$$U_{xd} = U_{xm} + \Delta y R_{zm} \quad , \quad U_{yd} = U_{ym} - \Delta x R_{zm} \quad , \quad R_{zd} = R_{zm}, \quad (2.13)$$

where $\Delta x = x_m - x_d$ and $\Delta y = y_m - y_d$, and (x_m, y_m), (x_d, y_d) are the coordinates of the master node and the slave node, respectively. *The above equations implement the rigid diaphragm kinematic constraint* and need to be satisfied simultaneously with the nodal equilibrium equations within the stiffness method formulation of matrix structural analysis. In typical r/c building structures, floor slabs are horizontally aligned (the local z axis coincides with the gravitational axis), and thus all DOFs refer to the global orthogonal coordinate system X, Y, Z where Z lies along the gravitational axis. Each floor slab is assigned a master node and the dependent DOFs (the ones coupled by the set of rigid diaphragm equations at each floor) can be eliminated from the system of nodal equilibrium equations expressed in terms of the global coordinate system using standard *static condensation* techniques (Chopra 2007). Eventually, only the DOFs of the master nodes remain and, therefore, the rigid diaphragm constraint concept leads to a substantial reduction in the total DOFs or, equivalently, to the number of independent equilibrium equations that governs the deflection of the building for static lateral loads or the motion of the building for horizontal strong ground motion. Hence, it significantly expedites dynamic analysis, as the size of the required eigenvalue problem to be solved is only 3 N, where N is the number of storeys.

In case of *inclined slabs* in which the local z axis does not coincide with the global gravitational axis Z (see, e.g., Fig. 2.29a), the diaphragm constraint equations can still be written in the local coordinate system as above, but then an appropriately defined rotation matrix needs to be applied to transform the master nodal displacements in space from the local to the global coordinate system. As a final remark, it is noted that slabs may not always span the full plan of a building and, therefore, "partial floor diaphragms" need to be considered (see, e.g., Fig. 2.29b).

Fig. 2.29 Inclined diaphragm (*left*), partial diaphragm (*right*)

Accounting for in-plane flexibility of floor slabs

In case the in-plane flexibility of a floor slab needs to be accounted for, two-dimensional finite elements along with an appropriate FE mesh needs to be considered. Typically, simple plane stress finite elements would suffice to capture the membrane (diaphragmatic) behaviour of slabs. However, shell finite elements have to be used in case the transverse flexural (plate bending) behaviour due to loads acting normal on the slab plane needs to be accounted for as well.

In flexible diaphragm modeling, the coarseness of the FE mesh to be adopted is case-specific and depends heavily on the intended scope of the undertaken analysis and the results/outcomes sought. A fine discretization of each particular slab is required in order to capture in detail the in-plane stress distribution developed within the slab under seismic (horizontal) excitation (see, e.g., Figs. 2.30 and 2.31b). This requirement arises in the case in which the identification and quantification of local potentially undue in-plane stress concentrations are sought at critical regions of slabs, such as close to slab openings (e.g., Fig. 2.31b), near the inner edge of in-plan setbacks (e.g., Fig. 2.30), along the common edge of two building "wings", or in the vicinity of supporting pre-stressed concrete beams. It is noted that such fine FE meshing necessitates the consideration of compatible (fine) FE discretization of the supporting beams, as depicted in Fig. 2.30. By noting that beams are modeled by one-dimensional beam elements (see Sect. 2.3.3.2), special care is needed to ensure the proper "coupling" (connection) of the beam elements with the two-dimensional plane stress elements at common nodes of the FE grid (Fig. 2.30).

It is emphasized that seeking to determine in-plane stress concentrations in floor slabs with high accuracy via significantly fine FE meshing, unless sufficiently justified, is generally unnecessary in routine code-compliant seismic design using linear analysis methods. In fact, it may hamper the seismic design process, as it increases the computational demands: each additional node in the FE mesh introduces 6 DOFs. Furthermore, due to the unavoidable fine discretization of the slab

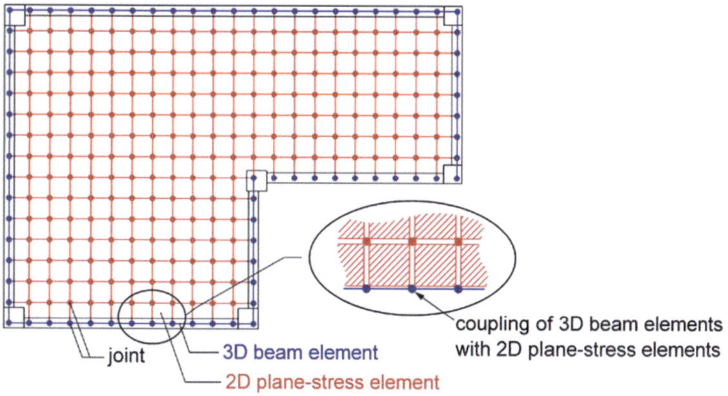

Fig. 2.30 Plane stress finite element discretization of in-plane flexible floor slab

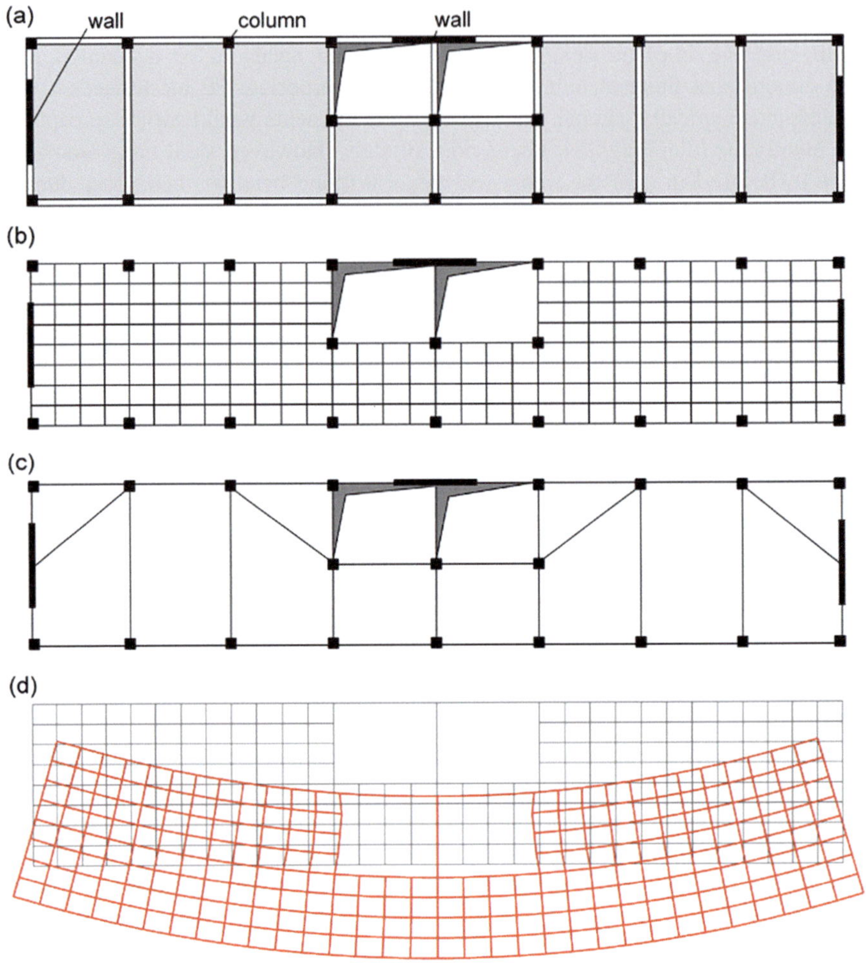

Fig. 2.31 Detailed and simplified finite element models for in-plane flexible diaphragm modeling. (**a**) Sectional plan of structural system. (**b**) Discretisation of the diaphragm with a fine mesh of finite elements. (**c**) Discretisation of the diaphragm according to a proposed simplified model. (**d**) Deformed shape of the floor slab

supporting beams, the subsequent proportioning/detailing procedure of each beam becomes involved and requires special post-processing subroutines. In this respect, there is scope in adopting less refined modeling approaches which can capture the influence of the in-plane slab flexibility at a minimum increase of computational cost and post-processing effort. One such approach is to discretize the slab using a very coarse mesh of two dimensional plane stress finite elements considering only the nodes of the vertical elements at the slab level, as shown in Fig. 2.31c. It has been demonstrated that such coarse meshing yields acceptable results in the context of seismic analysis (Doudoumis and Athanatopoulou 2001).

Further, it is important to note that, in flexible diaphragm modeling, the mass at each floor level needs to be distributed to match the actual distribution of mass over the plan area. This is in contrast to rigid diaphragm modeling, in which lumped mass properties at the center of rigidity of each level can be used.

As a final note, it is worth mentioning that the output from FE models are element stresses, and forces and displacements at the element nodes. However, not all FE analysis programs allow for designing of slabs on the basis of stress fields and only a few of them allow the user to define sections across a diaphragm so that resultant design forces are calculated over a series of nodes.

2.3.3.2 Modeling of Beams, Columns and Frames

Beams and columns are modeled using classical one-dimensional two-node finite elements with six degrees of freedom per node characterized by the cross-sectional area A, the moments of inertia I_{22} and I_{33} with respect to the local axes 2 and 3 of the cross-section, respectively, the effective shear areas A_{S3} and A_{S2} along the direction of the local axes 3 and 2, respectively, and the torsional moment of inertia I_{11} with respect to the centroidal axis 1 (Fig. 2.32). Care is needed to account for the orientation of beams and columns in space by appropriately defining their local 1, 2, 3 axes which do not usually coincide with the global X-Y-Z reference coordinate system.

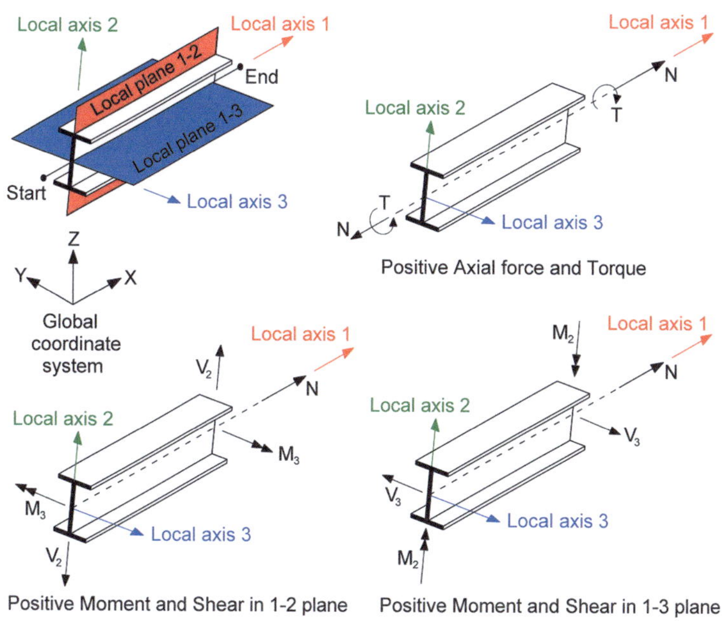

Fig. 2.32 Classical one-dimensional two-node finite beam/column element (Positive sign convention shown is the one adopted in numerical example problems of Chap. 4)

The cross-sectional geometry and mechanical properties of columns are straightforward to determine. However, the beams of cast-in-place r/c buildings are monolithically connected to floor slabs and, therefore, they are taken to have a flanged cross-section, either T-shaped (if the slab extends to both sides of the beam) or L-shaped (if the beam lies at the perimeter of the floor slab). The "effective" flange width of r/c beams is specified in detail in EC2 and is usually assumed to be constant throughout the length of the beam. Therefore, mechanical properties, constant along the length of beams, can be readily determined for flanged T- or L-shaped cross-sections. It is worth noting that the flexural rigidity EI_{22}, the shear rigidity GA_{S3}, and the axial rigidity EA are theoretically "infinite" for beams supporting perfectly rigid slabs. This condition is achieved in a numerically stable manner by enforcing the rigid diaphragm constraint equations to the beam nodes, that is, treating the nodes of beams belonging to a certain rigid diaphragm as "slave" or "constraint" nodes, as shown in Fig. 2.28.

The beam-column connections (joints) of cast-in-place r/c structures are characterized by a high level of rigidity which depends on the shape and dimensions of the joint and, in turn, on the shape and dimensions of the cross-sections of the converging members. *Appropriately representing the rigidity of joints* in rigid-jointed frames (or moment resisting frames) is an important modeling issue for the purpose of seismic design and is *required to be accounted for* (clause §4.3.1 (2) of EC8). This is because joint rigidity significantly influences the overall lateral stiffness of frame lateral load-resisting structural systems and, therefore, the natural periods and seismic design forces of buildings.

To this end, a reasonable modeling approach is to assume the whole joint as being rigid. This implies that the part of a one-dimensional beam finite element used to model a beam lying within the physical region of a joint with a column, i.e., the part from the centerline of the column to the outer face of the column, is considered rigid ("*wide-column*" model). Similarly, the part of a one-dimensional beam finite element used to model a column lying within the physical region of a joint with a beam, i.e., the part from the centerline of the column to the outer face of the column, is considered rigid ("*deep-beam*" model). Modern structural analysis software offers the option of one-dimensional beam element with rigid ends of arbitrary length ("rigid offsets") which significantly facilitate the implementation of wide-column and deep-beam models. However, considering rigid offsets for all converging structural members assuming that the whole joint is rigid (see, e.g., Fig. 2.33) may lead to an overestimation of the actual lateral stiffness of a frame subject to the design earthquake, because it ignores the significant non-linear deformations and stiffness degradation of the joint region that are expected to occur under extreme shaking.

In this respect, a modeling option that is often deemed preferable is *to assign rigid end offsets only to the weakest* (most flexible) *structural member converging to each joint*. Capacity design considerations suggest that the weakest members converging to each joint would normally be the beams as shown in Fig. 2.34a, and thus the "wide-column" model. For the more unusual case of frame systems with

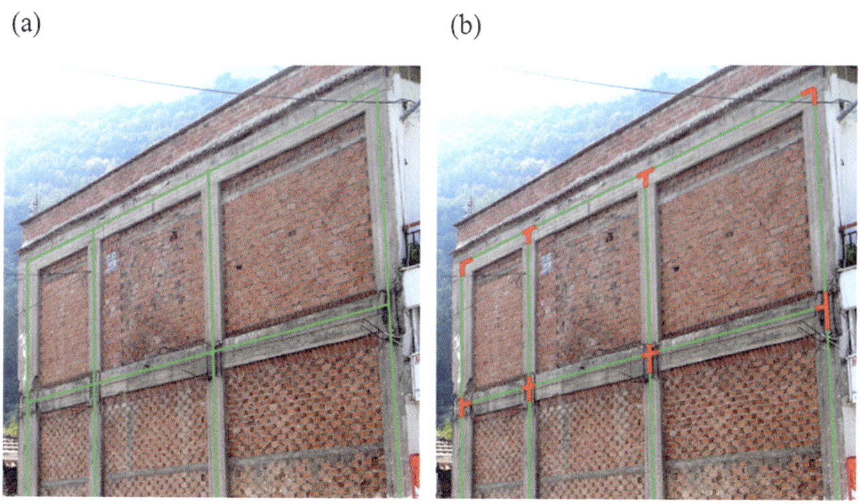

Fig. 2.33 Rigid-joint concept for r/c frames

deep beams, rigid offsets can be assigned to the end of columns, as illustrated in Fig. 2.34b ("deep-beam" model).

An alternative modeling option accounting for frame joint rigidity is to assign rigid offsets to all converging structural members to a joint, but of reduced ("effective") length compared to the actual joint geometry by the same factor, say ¼, as shown in Fig. 2.34c ("wide-column, wide-beam" model). Considering case-specific parametric analysis to quantify the sensitivity of the overall stiffness of the lateral load-resisting system of a building (and, thus, the code-prescribed design seismic loads) is recommended to ensure a reasonable value of the aforementioned reduction factor (see also Sect. 2.4.4.2). Notably, "effective" rigid end offsets are incorporated into ASCE/SEI 41-06 Supplement No.1 based on the proposed ratio of column to beam moment strength, $\Sigma M_c/\Sigma M_b$, (Elwood et al. 2007).

In this context, useful insights on the influence of the length of rigid offsets to the overall stiffness of frames can be gained by considering the flexural stiffness of a simple beam element under reverse bending as a function of the rigid offset length b/2 in both ends (Fig. 2.35a). The relationship between end moment M and end rotation θ is $M = 6EI\theta/[L(1 - b/L)^3]$, where L is the distance between the center-lines of the two supporting columns and b is the column width, i.e., twice the rigid offset length. For $b = 0$, the well-known moment-rotation relationship $M_{b=0} = 6EI\theta/L$ is retrieved for a beam without any rigid offset. Therefore, by plotting the ratio $\gamma = M/M_{b=0} = [1/(1 - b/L)^3] \times 100\ \%$ versus (Fig. 2.35b, the significant influence of the length of rigid offsets to the flexural stiffness of the beam can be readily quantified. Similar case-specific plots can be readily devised by practitioners to build confidence in fine-tuning rigid offset lengths of beams and columns (and other) important modeling parameters.

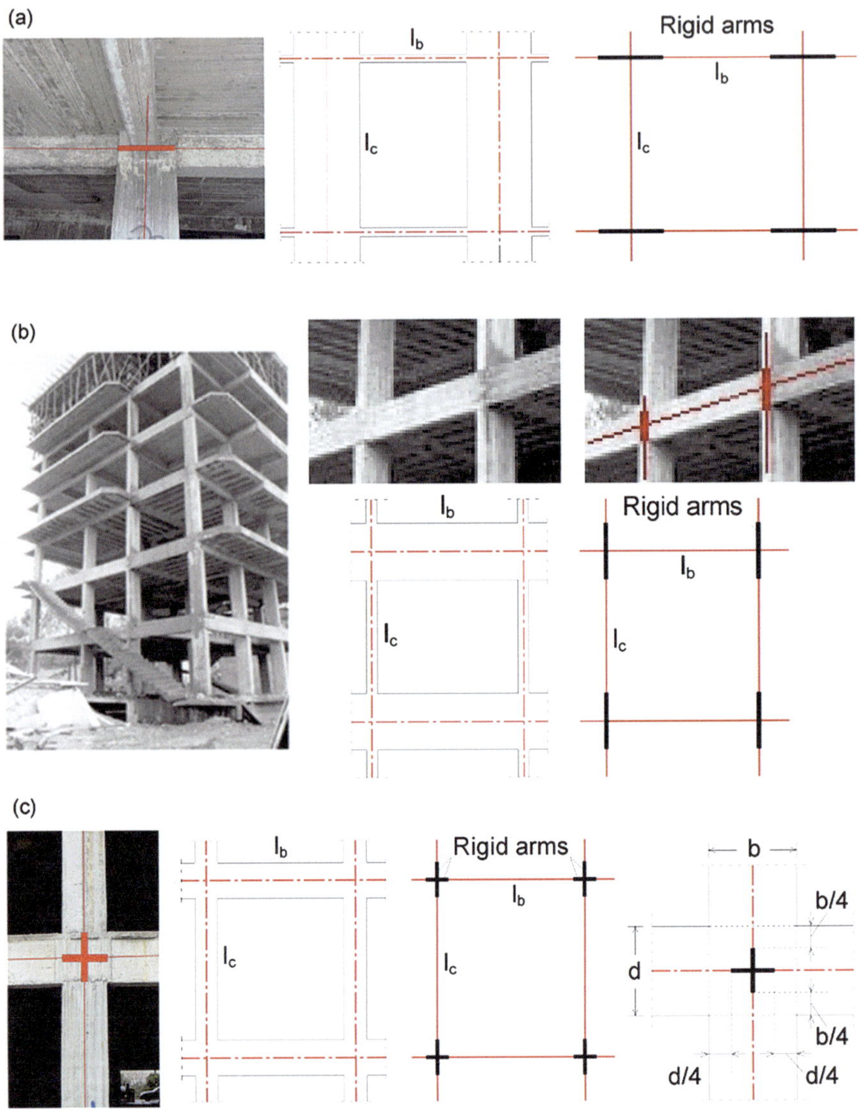

Fig. 2.34 Modeling of rigid joints in frame structures (to simplify the sketch of the floor, slabs are omitted). (**a**) Beam joined by bulky columns: "wide-column model" (preferable seismic design). (**b**) Columns joined by deep beams: "Deep-beam model" (unfavourable connection of perimeter beams to the weak axes of the columns). (**c**) Columns and beams of similar stiffness: "Wide-column, deep-beam model"

Special attention in frame joint modeling is also needed in the case in which the centerlines of two beams or/and two columns converging to a joint do not intersect. In this case, the common node is placed on the centerline of one of the intersecting

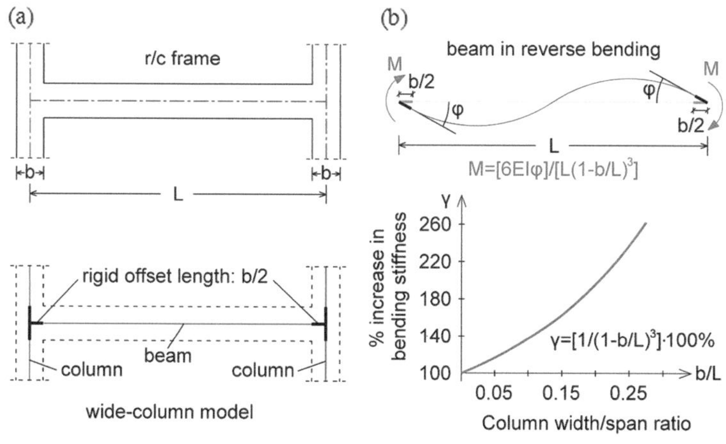

Fig. 2.35 Effect of rigid arm length on bending moment

members and the rigid ends of the elements are connected to this joint with an eccentricity (see Fig. 2.36). To this end, note that most modern analysis programs offer rigid-end offset beam elements and, therefore, allow beams and columns to be modeled by a single finite element shown in Fig. 2.36b. In the case of eccentric connections in space, rigid offsets should be allowed to be non-collinear with the axis of the beam element. A possible finite element to address this problem is schematically shown in Fig. 2.36c.

Further, Fig. 2.37 illustrates the way such an element is incorporated into joint modeling. Note that, as distributed loads on a beam (e.g., the reactions of supported slabs) may act all the way from its left to its right node, rigid offsets non-collinear with the element axis should be capable of being assigned loads (Fig. 2.36 c).

From the structural analysis viewpoint, it is worth noting that the computer implementation of rigid offsets is achieved either by using beam element stiffness matrices which already incorporate the rigid-end offsets within their element stiffness matrix, or by imposing appropriate sets of constraint equations (coupling the otherwise independent DOFs) analogous to the one considered in Eq. (2.13) for the rigid diaphragm implementation. Modeling of the rigid end offsets manually using virtual short beam elements of very large ("infinite") values for the stiffness properties should be avoided, as it may cause conditioning problems and numerical errors in the solution of the underlying system of nodal equilibrium equations (i.e., inversion of the global stiffness matrix) for nodal displacements.

As a final remark, it is noted that, when end rigid offsets are present in a member, design effects (cross-sectional internal forces and moments) are determined at the end-sections of the elastic length of the member, i.e., at the faces of the left and right supports for beams, and at the base and top of the clear height of columns.

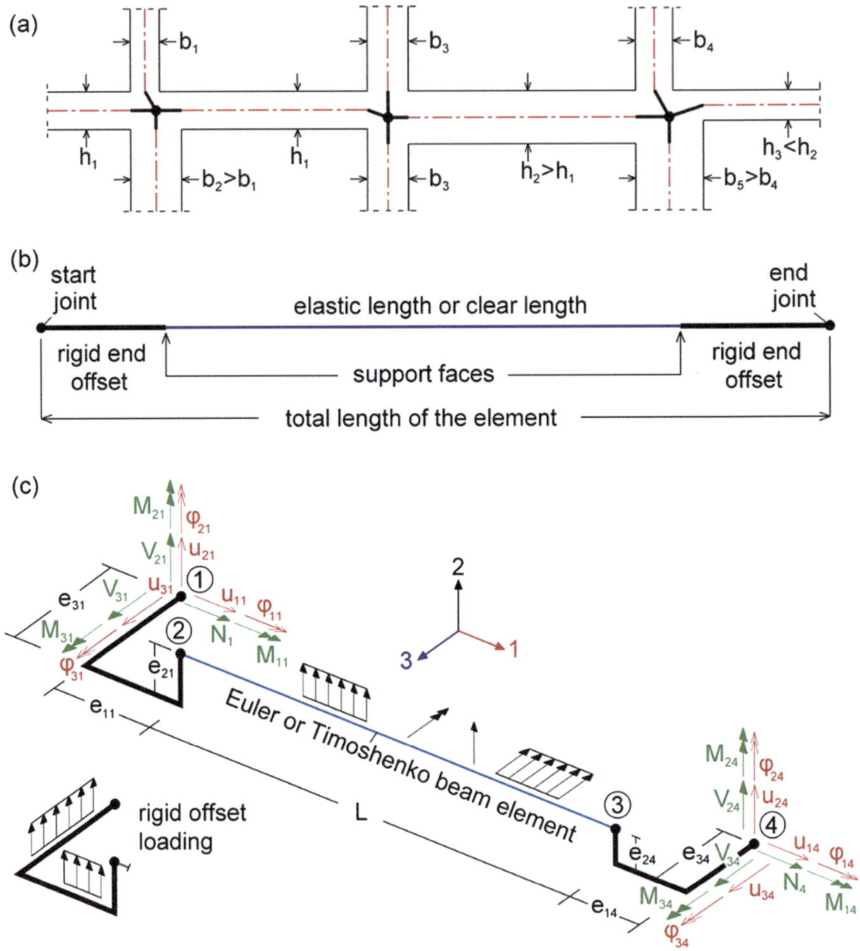

Fig. 2.36 Modeling connection eccentricities in frame structures (to simplify the sketch of the floor, slabs are omitted). (**a**) Connection eccentricities in plane frame. (**b**) Simple beam/column element with collinear rigid offsets. (**c**) Generalized beam/column element with 3D rigid offsets

2.3.3.3 Modeling of Planar Walls

Arguably, the most accurate finite element (FE) modeling practice to capture the response of structural walls subject to static and/or dynamic loads involves the use of two-dimensional plane stress or even shell elements arranged in a properly defined FE mesh (Fig. 2.38b). In this manner, all important degrees of freedom at connecting nodes are explicitly accounted for. The required mesh density is case-dependent and is typically chosen by means of a "convergence" analysis to ensure that the desired level of accuracy is achieved. However, in routine seismic design

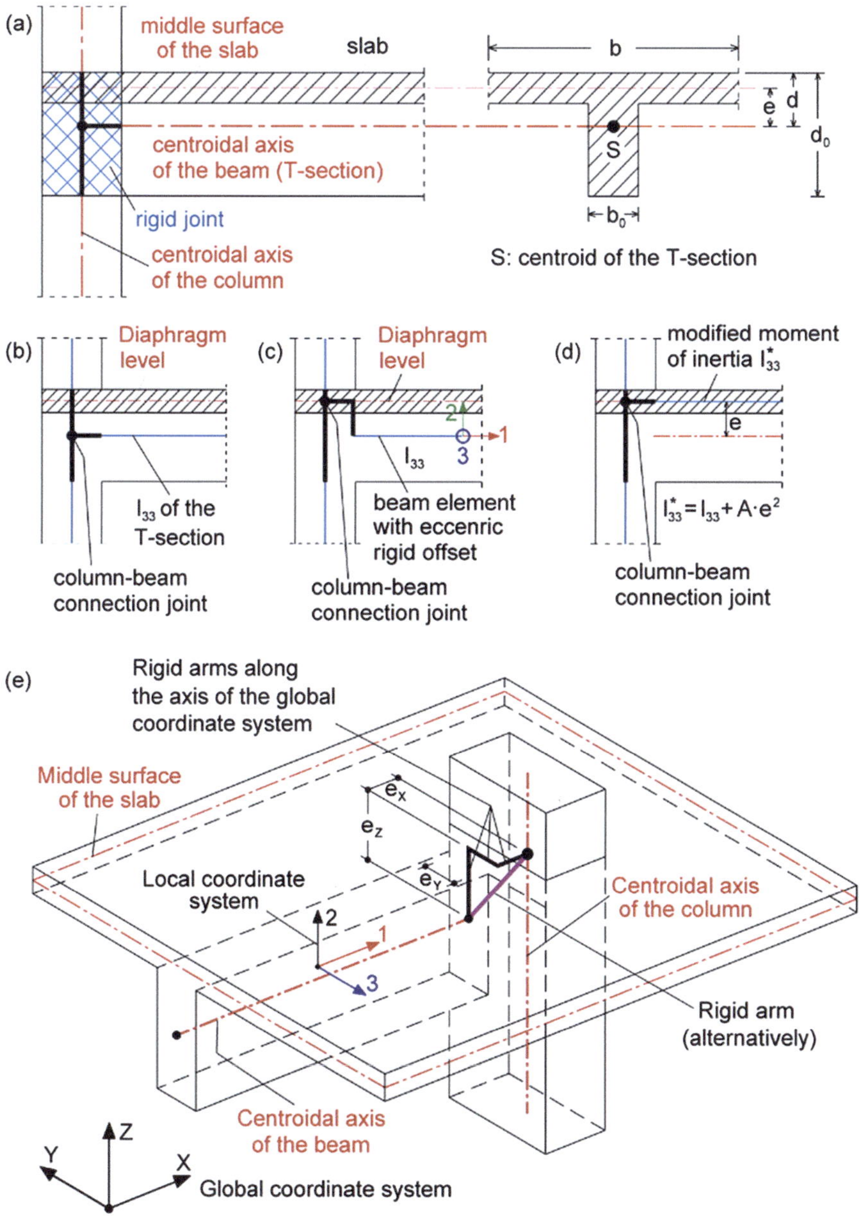

Fig. 2.37 Three-dimensional model of column-beam connection in r/c buildings with floor slabs. (**a**) Geometry of column-beam connection. (**b**) Incorrect model. (**c**) Beam with I_{33} of T-section. (**d**) Beam with $I^*_{33} = I_{33} +$ Steiner. (**e**) 3D model of the column-beam connection

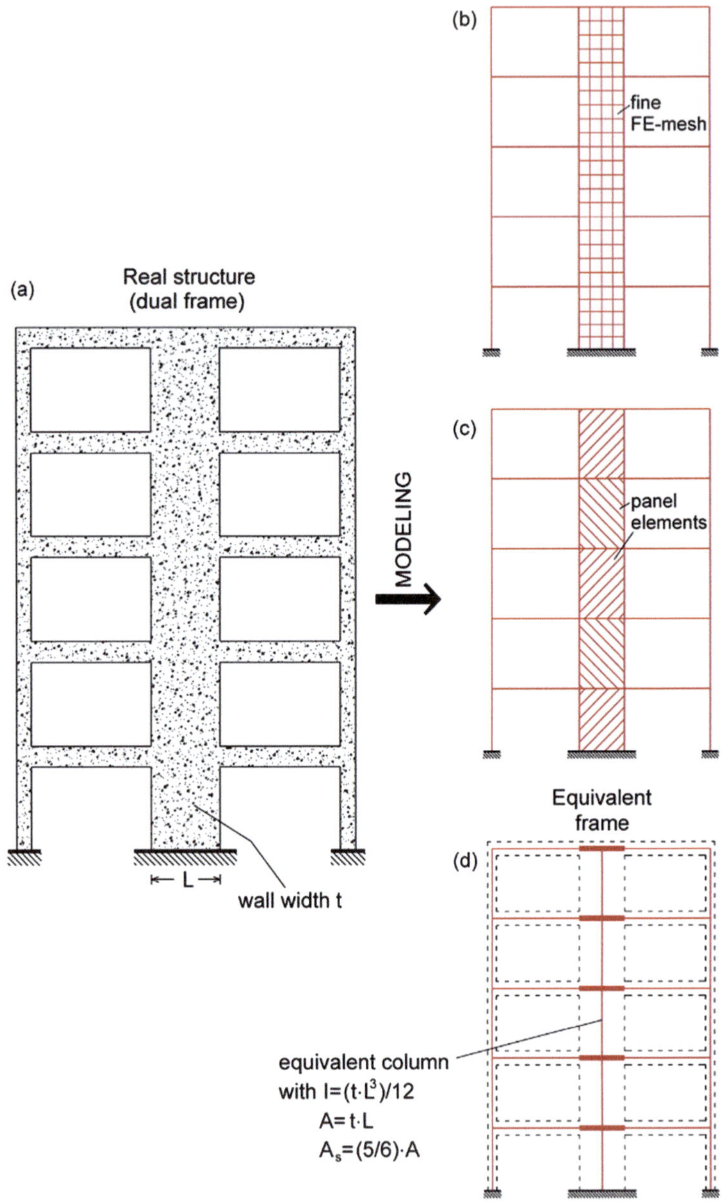

Fig. 2.38 Finite element, panel, and equivalent frame models for a planar wall

practice of ordinary/conventional building structures, there is scope in avoiding using such refined FE models since these models typically involve:

– Excessively tedious and time-consuming input data preparation process compared to the alternative option discussed below,
– Post-processing difficulties concerning wall proportioning and detailing on the basis of stress fields rather than on stress resultants, especially in the case of multi-modal response spectrum analysis,
– Inconsistent (and unjustifiably high) level of accuracy given the large number of over-simplifying/gross assumptions commonly made in code-prescribed seismic analysis and modeling of ordinary r/c buildings, and
– Challenges arising in connecting the one-dimensional elements modeling the beams at each floor level to the two-dimensional shell elements which may require special local meshing to avoid the development of spurious local stresses.

At a preliminary (conceptual) design stage, some of the above issues may be bypassed using simple "panel element" models involving coarse FE mesh with only one finite element per storey to model structural walls (Fig. 2.38c). However, such models are too rough to be used for the main structural analysis stage. To this end, simplified FE meshes comprising only one-dimensional (frame/beam) elements, often called "equivalent frame models" (Fig. 2.38d), have been introduced by (MacLeod 1977) to model structural walls as an alternative to more refined FE models incorporating two-dimensional shell elements (MacLeod 1990; Stafford-Smith and Girgis 1984). These simplified models are widely used, as they strike an acceptable balance between accuracy and efficiency in capturing the behaviour of slender walls exhibiting predominantly flexural/bending deformation (Fig. 2.39). Conveniently, the majority of structural walls in earthquake-resistant buildings are designed and expected to exhibit such behaviour under seismic loads.

The basic concept underlying the equivalent frame model consists in the replacement of planar walls by one-dimensional frame/beam elements, often called "equivalent columns", positioned at the centerline (i.e. vertical centroidal axis) of the actual wall and assigned section properties corresponding to the true geometry of the wall (Fig. 2.38d). These equivalent columns are connected to the beams at the floor levels by rigid arms (virtual perfectly rigid in flexure and in tension/compression one-dimensional beam elements) whose length reflects the actual width of the wall. Similar modeling techniques as described in the previous section can be applied to define the rigid arms, that is, by using either beam elements incorporating end rigid offsets to their element stiffness matrix or by applying nodal constraints.

Clearly, the fundamental underlying principle of the above modeling technique for walls is the Bernoulli assumption that plane sections remain plane and normal to the deformed neutral axis. This is indicated for a simple coupled wall system in Fig. 2.39 in which separate equivalent columns are used to model each wall individually. Evidently, this assumption tends to become less realistic as the height to width ratio h/b of the wall decreases. Therefore, as a general rule of thumb, the equivalent frame model may be applied with acceptable accuracy for high-rise, slender, bending/flexural type of walls with a ratio h/b of at least 4 (i.e. $h/b > 4$). It

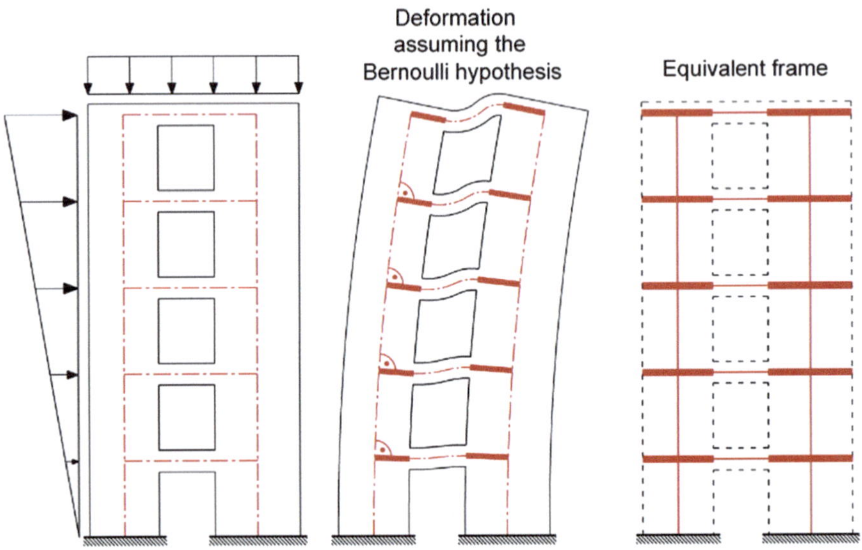

Fig. 2.39 Equivalent frame model for two walls coupled by lintel beams

should not be used to model low-rise, squat, shear types of walls with h/b $\leq$ 2. For intermediate geometries, a local parametric sensitivity analysis is recommended to be undertaken to gauge the accuracy of equivalent frame models (see also Sect. 2.4.4.2).

In Fig. 2.40, an equivalent frame model example is shown for a typical "stepped wall", while in Fig. 2.41, the application of the equivalent frame modeling technique to various real-life structural systems is exemplified.

2.3.3.4 Modeling of Cores

Cores in r/c buildings are spatial substructures composed of two or more (typically three or four) planar shear walls commonly positioned in-plan around staircases and/or elevators. Cores are, in general, desirable in earthquake resistant design of buildings providing significant lateral and torsional stiffness to the overall load resisting structural system. However, r/c cores exhibit a considerably different mechanical behaviour compared to planar walls, let alone beams and columns and, therefore, use of one-dimensional FE beam elements to achieve reasonable models for routine seismic design of structures is a challenging task. In fact, modeling of r/c cores may involve significant assumptions and, thus, may become an important source of uncertainty in the analysis of multistorey buildings.

Accurate computational modeling for cores involves discretization using two-dimensional shell finite elements, including all six degrees of freedom at each node (Fig. 2.42). The density of the finite element mesh depends on the accuracy desired in each particular case and is typically determined by means of standard converging ("sensitivity") FE analyses

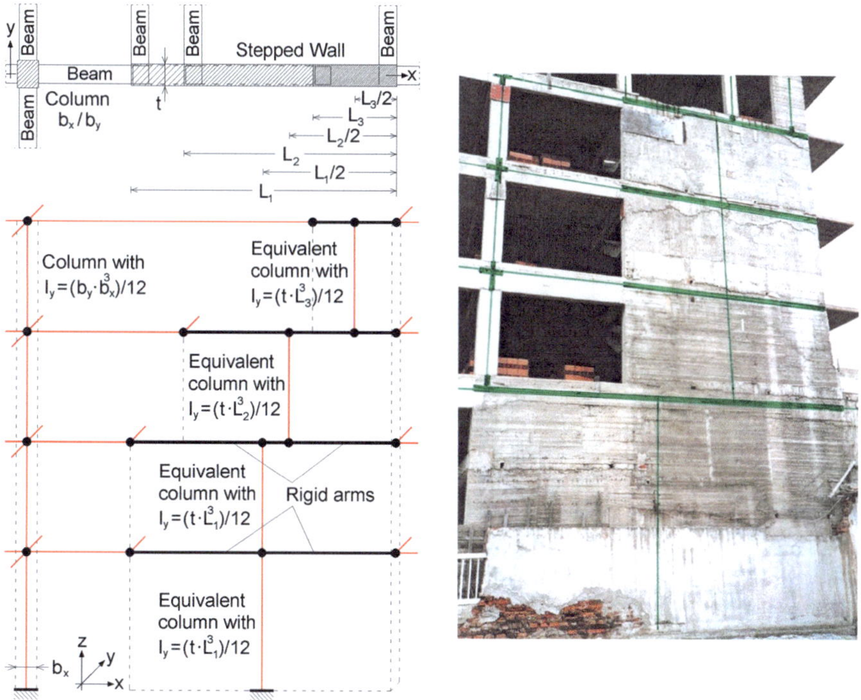

Fig. 2.40 Equivalent frame model for stepped walls

However, for the same reasons listed in the previous Sect. (2.3.3.3), the use of detailed finite element models for the seismic analysis of conventional buildings is not preferred in the every-day practice of seismic design. Therefore, as for planar walls, simplified "equivalent frame models" comprising one-dimensional beam elements are frequently used in the context of code-prescribed linear analysis methods (see Sect. 2.4.2). In fact, different equivalent frame models for r/c cores have been proposed in the literature and integrated into various structural analysis software used extensively by practitioners worldwide (Mac Leod and Hosny 1977; Xenidis et al. 1993). Arguably, the most widely accepted equivalent frame model for U-shaped cores is shown in Fig. 2.43. It treats the three "wings" of the core as individual planar frames and relies on the Bernoulli hypothesis by considering the core to be a cantilever thin-walled beam with a U-shaped cross-section. It is based on the following practical rules:

– each individual wing ("flanges" and "web") of the core is replaced by an equivalent column positioned at the centerline of the wing and assigned properties, as shown in Fig. 2.43; and
– the equivalent columns are linked together at the floor slab levels by "virtual" rigid arms, i.e., beam elements perfectly rigid in flexure *but not in torsion*.

It is emphasized that the above model becomes less accurate as the ratio of the core outer dimensions in-plan over its total height increases (Avramidis 1991).

Fig. 2.41 Equivalent frame models for structural walls in r/c building (Rigid joints in column-beam connections are omitted)

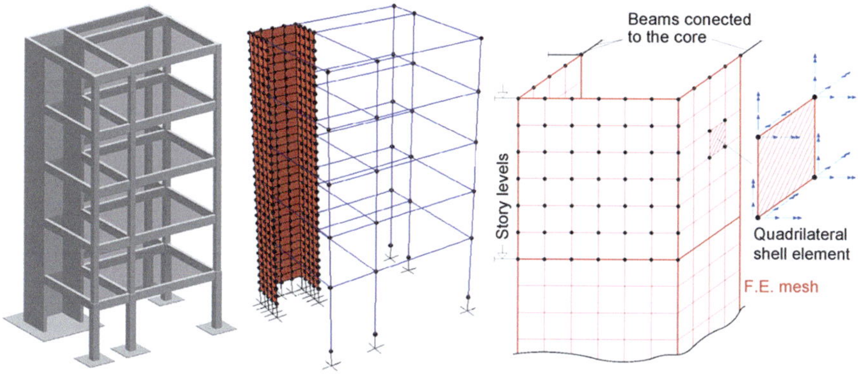

Fig. 2.42 Shell finite element model of a U-shaped r/c building core

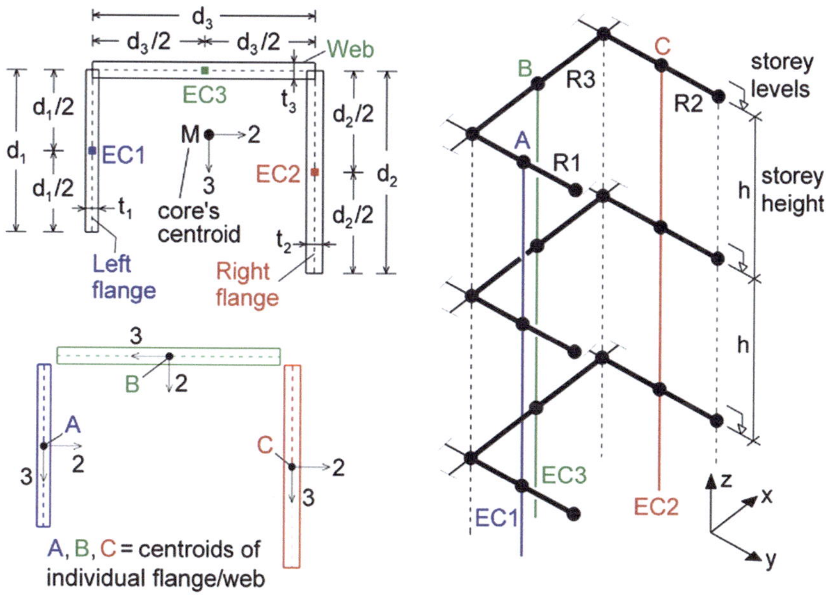

Properties of equivalent columns

	Left Flange EC1	Right Flange EC2	Web EC3
A	$d_1 \cdot t_1$	$d_2 \cdot t_2$	$d_3 \cdot t_3$
I_2	$(d_1^3 \cdot t_1)/12$	$(d_2^3 \cdot t_2)/12$	$(d_3^3 \cdot t_3)/12$
I_3	$(d_1 \cdot t_1^3)/12$	$(d_2 \cdot t_2^3)/12$	$(d_3 \cdot t_3^3)/12$
A_{s2}	$5/6 \cdot (d_1 \cdot t_1)$	$5/6 \cdot (d_2 \cdot t_2)$	$5/6 \cdot (d_3 \cdot t_3)$
A_{s3}	$5/6 \cdot (d_1 \cdot t_1)$	$5/6 \cdot (d_2 \cdot t_2)$	$5/6 \cdot (d_3 \cdot t_3)$
J	$\alpha \cdot (d_1 \cdot t_1^3)$	$\alpha \cdot (d_2 \cdot t_2^3)$	$\alpha \cdot (d_3 \cdot t_3^3)$

where $\alpha = \frac{1}{3} - 0.21 \cdot \frac{t_i}{d_i} \cdot \left[1 - \frac{1}{12} \cdot \left(\frac{t_i}{d_i}\right)^4\right]$, i=1,2,3

Properties of rigid arms

	R1	R2	R3
A	∞	∞	∞
I_2	∞	∞	∞
I_3	∞	∞	∞
A_{s2}	∞	∞	∞
A_{s3}	∞	∞	∞
J	J_{R1}	J_{R2}	J_{R3}

$J_{Ri} = \beta \cdot (h \cdot t_i^3)$

where $\beta = \frac{1}{3} - 0.21 \cdot \frac{t_i}{h} \cdot \left[1 - \frac{1}{12} \cdot \left(\frac{t_i}{h}\right)^4\right]$

Fig. 2.43 Equivalent frame model of a U-shaped r/c building core

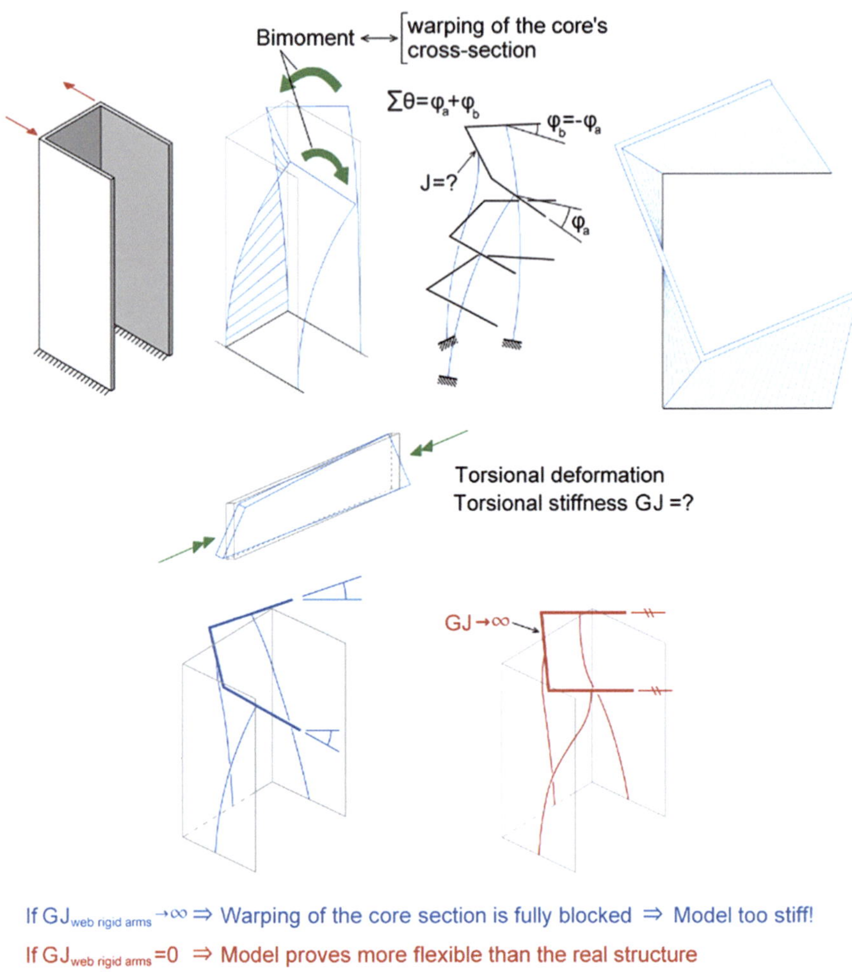

Fig. 2.44 Torsional rigidity GJ of the rigid arms of typical equivalent frame model for a U-shaped r/c building core

Furthermore, particular attention must be focused on assigning the torsional stiffness property GJ of rigid arms such that the overall torsional behaviour of the core acting as an open-loop thin-walled cantilever beam is properly captured. For example, in order to adequately model the torsional stiffness of the U-shaped core of Fig. 2.42, *the torsional stiffness of the web rigid arms needs to be assigned a finite value that corresponds to the actual geometry of the core*, as illustrated in Fig. 2.43 (as opposed to all other stiffness properties of the rigid arms, which should be assigned very large artificial values) such that the independent deformation of the rigid arms corresponding to the two "flanges" of the core is only partially restricted, as shown schematically in Fig. 2.44. In this manner, the actual out-of-plane torsional warping of the core can be reasonably captured. If, by mistake, the

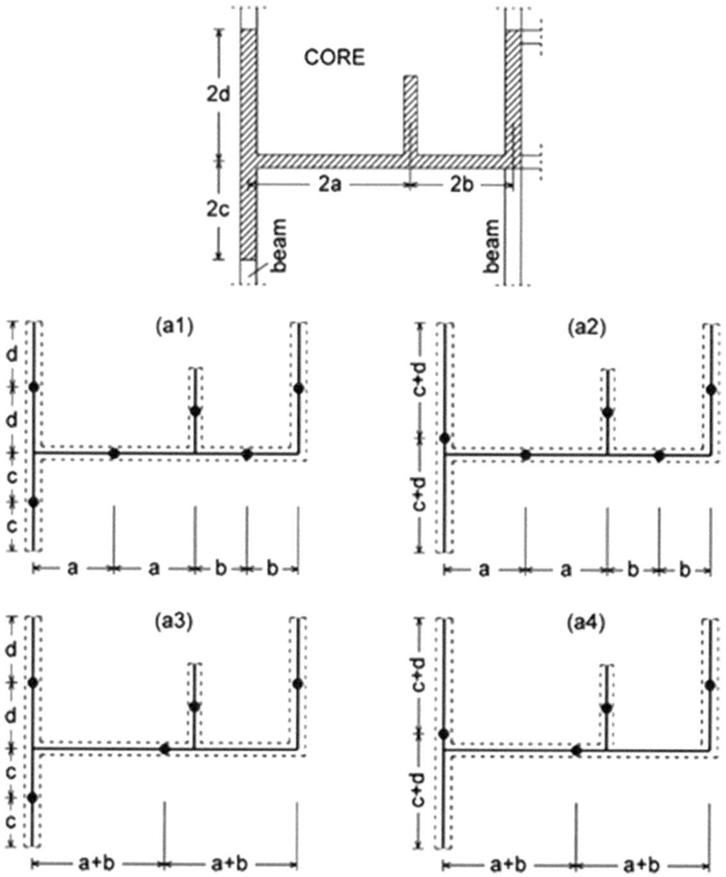

Fig. 2.45 Possible alternatives for the equivalent frame modeling of a complex building core

rigid arms of the web are modeled as perfectly rigid in torsion, all three rigid arms at each storey level will remain in the same plane, thus making the frame model unrealistically stiff against torsion. This would further cause significant errors to the estimated values of bending moments developing at the beams linked to the core flanges at each floor level.

It is worth noting that topology issues with regards to the potential positioning of the columns in equivalent frame models for multi-cell cores are often encountered in practice. In particular, for a given multi-cell core, more than one alternative configuration of an equivalent frame model may be defined with regards to the location and number of column elements used and, consequently, the mechanical properties of each column. For example, in Fig. 2.45, four possible choices for the number and locations of the equivalent columns for an open two-cell core are shown.

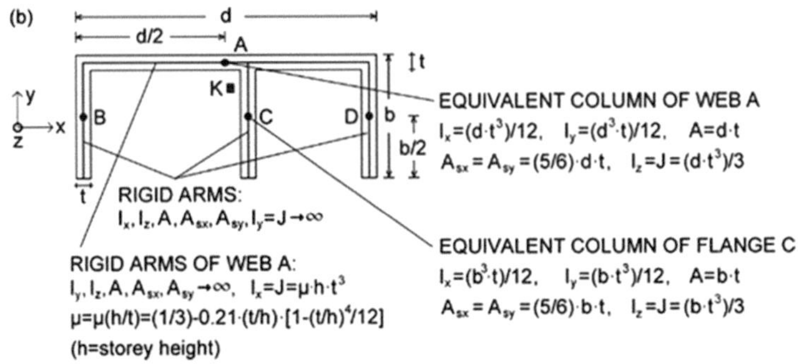

Fig. 2.46 Suggested beam element properties for a 4-column equivalent frame model of a typical open two-cell core

Further, possible reasonable member properties for a particular four-column model representing a simple open two-cell core are given in Fig. 2.46 (Xenidis et al. 2000). In this simplified model, a finite value is assigned to the torsional stiffness of the rigid arm representing web A to account for the warping of the core section in an approximate manner. In such cases, it is recommended that sensitivity analysis is undertaken with regards to the topology and the properties of the elements of equivalent frame models for r/c core as further discussed in Sect. 2.4.4.2.

2.3.3.5 Modeling of Footings and Foundation Beams on Flexible Ground

As previously discussed, it is recommended that foundation and the influence of the supporting ground are included in the numerical (FE) model of the superstructure used to determine the gravitational and seismic effects. In routine earthquake resistant design of conventional structures, soil compliance is typically modeled by introducing elastic support conditions at the relevant support nodes. This consideration relies on the so-called *Winkler model* in which the supporting ground is represented by means of point springs for simple pad footings (e.g., Figs. 2.47 and 2.48) or by means of continuously distributed springs underneath strip footings, horizontal foundation beams and mat-slabs (e.g., Figs. 2.49 and 2.50).

The linking of the superstructure finite elements to the support nodes relies on the rigid-joint-idealization approach already discussed in detail for the case of beam-column connection modeling (see Sect. 2.3.3.2). Typical examples of connectivity in FE meshes of vertical structural elements to foundation nodes are given in Figs. 2.47, 2.48, 2.49, and 2.50 without additional comments.

Further, Fig. 2.51 shows a generalized finite element for modeling beams resting on flexible soil allowing for connecting to vertical structural elements with arbitrary eccentricity (Morfidis and Avramidis 2002). As a final remark, it is noted that, in case the effects of fully coupled dynamic soil-structure interaction need to be

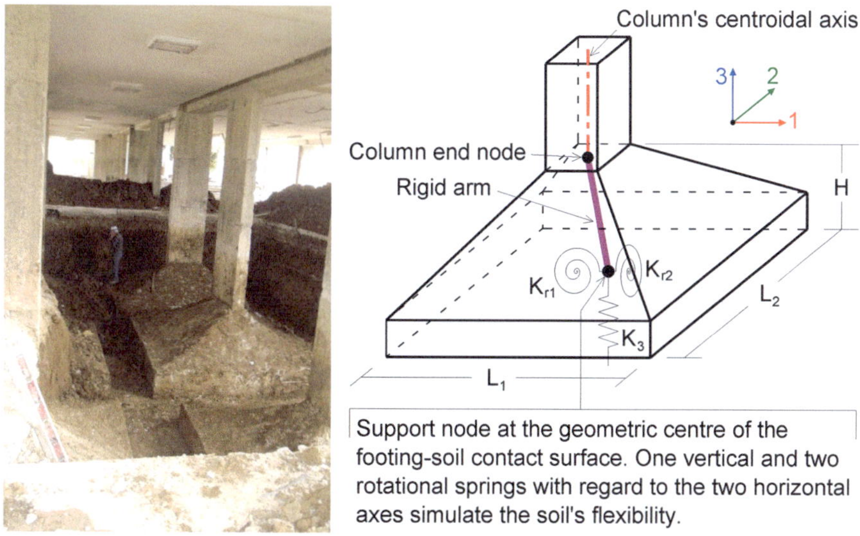

Fig. 2.47 Modeling of column-footing connection on flexible soil

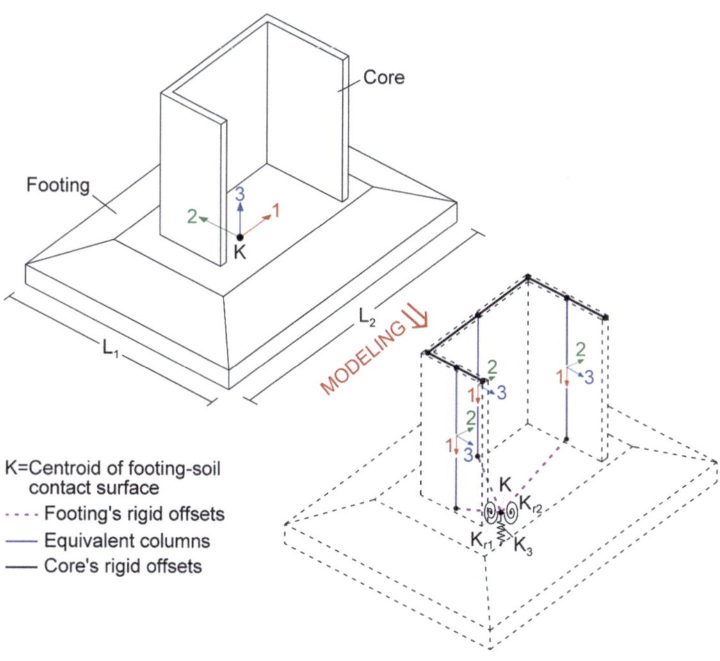

Fig. 2.48 Modeling of core-footing connection on flexible soil

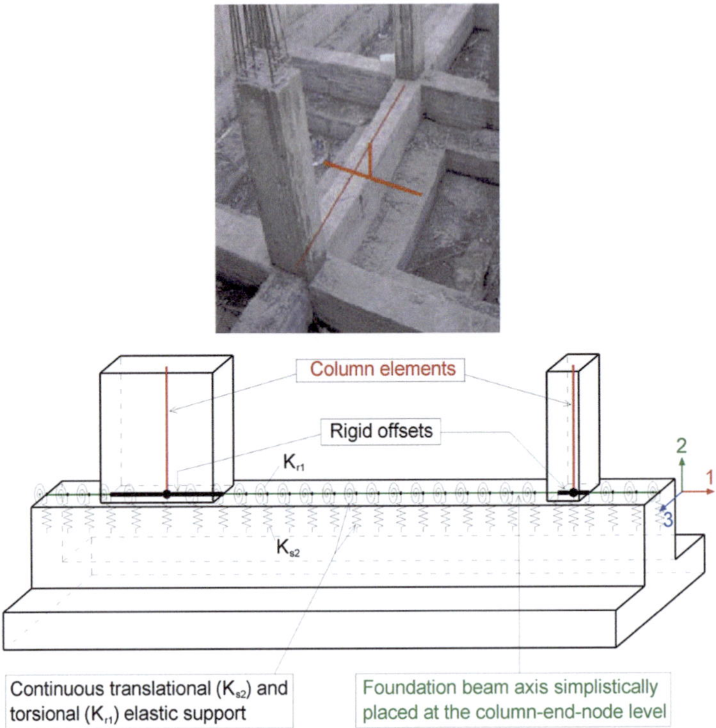

Fig. 2.49 Modeling of column/wall-foundation beam connection on flexible soil using beam elements on continuous elastic support

considered, the soil strata below and around the foundation is modeled explicitly using the finite element or other numerical techniques used in geotechnical engineering to represent explicitly the soil properties. Such considerations fall beyond the scope of this text and are not treated.

2.4 Structural Analysis Methods for Seismic Design of R/C Building Structures

Stress and deformation analysis involving numerical (finite element) structural models is accomplished using well-established computational methods. For static (time-invariant) external loads, the standard direct stiffness method of structural analysis is commonly used. For dynamic/seismic (time-varying) external loads, analysis methods of structural dynamics such as the response spectrum-based mode superposition method and the various numerical schemes for direct integration of the dynamic equations of motion are employed (Chopra 2007). Further, depending

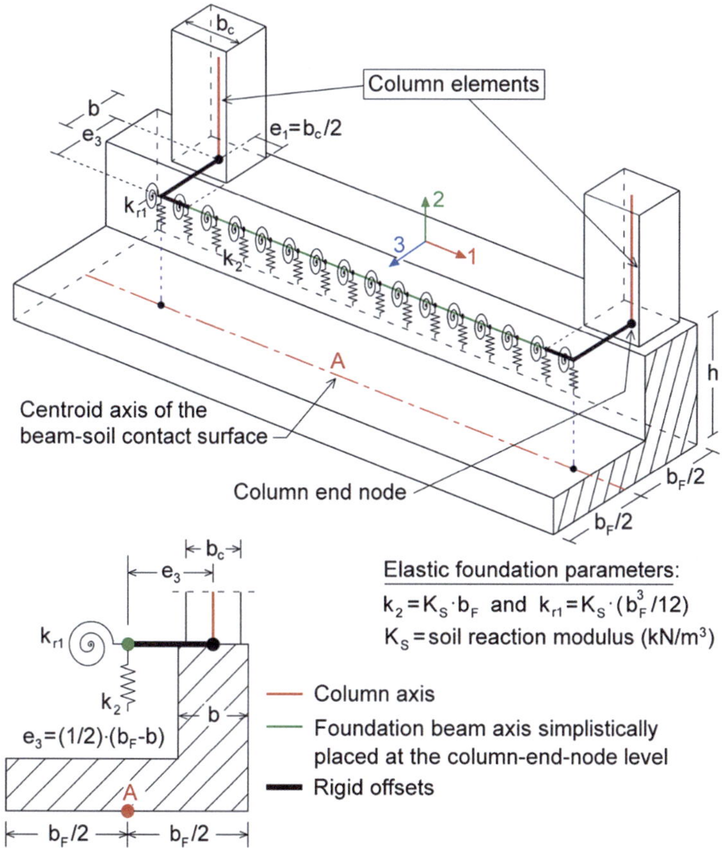

Fig. 2.50 Simplified modeling of column-foundation beam eccentric connection

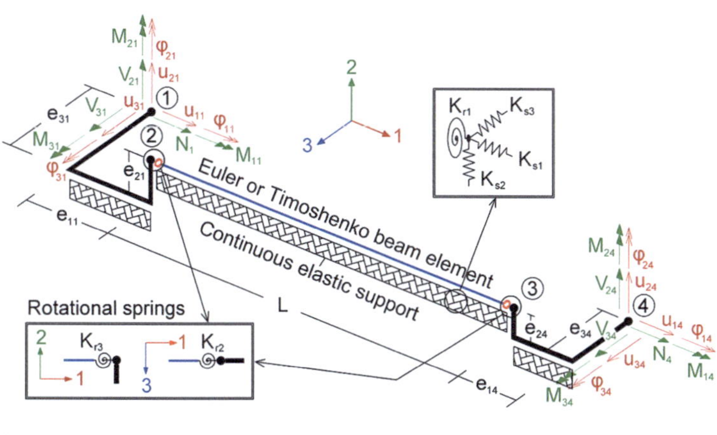

Fig. 2.51 Generalized beam finite element on continuous elastic (Winkler) foundation with 3D rigid offsets for modelling eccentric column-foundation beam connections

on the intensity of the considered external loads, a given structure may deform either elastically or elasto-plastically. In the latter case, the structure exhibits non-linear inelastic (hysteretic) behaviour due to material non-linearity. Therefore, the following four different types of analysis are readily identified depending on the nature of the external loads (static or dynamic) and the material behaviour (elastic or inelastic):

- Elastic static analysis (e.g., lateral force method)
- Elastic dynamic analysis (e.g., modal response spectrum method)
- Inelastic static analysis (e.g., non-linear pushover method)
- Inelastic dynamic analysis (e.g., non-linear response history method)

Moreover, the severity of the axial loads carried by structural members and the level of structural deformations may require the consideration of second-order theory (or at least inclusion of the P-Δ effects) to approximately account for the potential geometrically non-linear behaviour.

For each of the above listed types of analysis, which may or may not account for geometric non-linear phenomena, one or more analysis methods may be applicable. In this regard, the question of "*what is the most appropriate analysis method for r/c buildings exposed to seismic loads?*" becomes pertinent. Section 2.4.1 addresses this question by taking into account the expected structural seismic performance (i.e., severity of inelastic behaviour under a specific level of the seismic action) beyond code-compliant seismic design. Section 2.4.2 lists the types of analysis allowed by the EC8 for the design of new ordinary (conventional) structures. Next, Sect. 2.4.3 includes remarks and recommendations for the practical implementation of the most common analysis methods of EC8, including the inelastic static analysis method (detailed in Appendix A). Finally, Sect. 2.4.4 discusses the need to verify the distribution of overstrength across a structure in support of capacity design approaches and to conduct parametric sensitivity analyses to quantify potentially important sources of uncertainty in finite element modeling.

2.4.1 Selection of Structural Analysis Methods for Seismic Design

It can be argued that, since code-compliant r/c building structures are usually designed to yield under a design earthquake, an inelastic dynamic analysis method such as the *non-linear response history analysis* (NRHA) should always be used for seismic design. However, this type of analysis poses a number of challenges to the everyday practice of seismic design which may not be easily addressed by practitioners. Specifically, NRHA involves direct integration of the equations of motion in the time domain for a given input seismic ground motion record in terms of acceleration (accelerogram). These equations of motion are derived upon appropriate modeling of the non-linear hysteretic (inelastic) material behaviour under cyclic loading (see further Sect. 2.4.1.2) and of any potential geometric non-linearity due to gaps, discontinuities, friction, and second-order phenomena. In this regard, the following main practical difficulties arise in applying NRHA for the seismic design of new ordinary structures:

- Given the significant variability of the peak response of yielding structures subject to recorded seismic accelerograms, there may be a lack of a sufficient number of site-representative seismic records for design purposes. Note that this issue can be addressed by using various record selection algorithms, e.g., (Katsanos and Sextos 2013) and/or by employing scaling/modification schemes applied to recorded or artificial accelerograms, e.g., (Giaralis and Spanos 2009). However, such considerations extend beyond the usual capacity of structural design engineers as they require access to sufficiently large databases of recorded accelerograms and to specialized software, along with the expertise and experience to use such software properly.
- There exists a limited number of dependable hysteretic laws capable of adequately capturing the inelastic behaviour of r/c structural members and of the supporting soil in spatial (three-dimensional) finite element models available in commercial software. Further, such hysteretic relationships involve a plethora of parameters for each class of structural members (and soil types) which may not be readily known/available to the design engineer (see also Sect. 2.4.1.2)
- There is limited knowledge and guidance for practitioners with regards to allowable seismic demand limits for r/c structural members under different dynamic load combinations. In more general terms, the verification, interpretation, and utilization of NRHA numerical results (i.e., time traces of structural response quantities such as inelastic rotations at plastic hinges of critical zones within various structural members) for design purposes requires considerable experience and specialized expertise. The cost of such specialized consultancy services in the routine seismic design of ordinary structures is not practically justified.
- Dependable commercial software which may undertake NRHA within reasonable time and monetary constraints applicable for the seismic design of ordinary r/c building structures is scarce.

Furthermore, as has been discussed in Sects. 1.3 and 1.4, a building owner may decide, in consultation with the design engineer, to adopt a higher seismic performance level for the nominal design seismic action than the minimum life safety (LS) performance level commonly targeted by the current (traditional) codes of practice. This is achieved by ensuring different levels of stiffness, strength, and ductility and, therefore, a given structure may be allowed to exhibit lower levels of inelastic behaviour for the design seismic action up to "almost" elastic or even purely elastic behaviour. Clearly, in the last two cases, undertaking NRHA is unecessary, as a linear elastic dynamic analysis would be sufficient.

In this respect, for new structures, the selection of the most appropriate analysis method, and, consequently, of a suitable computational (finite element) structural model, depends primarily on the targeted/desired level of seismic performance to

Table 2.5 Range of applicability of structural analysis methods for seismic design and assessment of ordinary structures

	High performance existing structures (OP or IO) or code-compliant structures	Low performance existing structures (LS or CP)
Linear dynamic analysis methods (e.g., response spectrum method)	Of general use (the "reference" method of EC8-part 1)	Can be used, but inelastic methods are more suitable
Linear static analysis methods (e.g., lateral force method)	Can be used for low-rise regular buildings with a dominant translational fundamental mode shape	Can be used, but inelastic methods are more suitable
Inelastic dynamic analysis methods (e.g., non-linear response history)	Mostly used for important or low performance structures, but:	
	involves limitations/uncertainties in modeling	
	is computationally demanding	
	requires special expertise, experience, and software	
Inelastic static analysis methods (e.g., inelastic pushover analysis)	Recommended to be used for low performance structures	
	It is less involved/demanding than inelastic dynamic analyses, but:	
	is reliable only for planar regular structures of low to medium height	

be achieved through design. Accordingly, the selection of an appropriate analysis method for the seismic assessment of existing structures depends on the level of seismic performance expected to be attained by the structure. Therefore, the selection of a suitable structural analysis method for seismic design (and assessment) is closely related to the performance-based seismic design philosophy and considerations discussed in Sect. 1.3. In this respect, certain comments and practical recommendations follow on the selection of analysis methods accounting for the desired (for new structures) or expected (for existing structures) level of seismic performance. Further, Table 2.5 summarizes certain key points discussed and justified in the remainder of this section in a matrix form.

2.4.1.1 Linear Methods for Seismic Analysis

Typical linear *multi-mode (dynamic) analysis methods* used in the practice of seismic design of structures include the classical (modal) response spectrum based analysis and the linear response history analysis. The *response spectrum based* analysis involves modal (eigenvalue) analysis of the adopted elastic finite element (FE) structural model to derive natural frequencies and modes of vibration. Next, peak structural response quantities of interest (e.g., building floor displacements, stress resultants at critical cross-sections of structural members, etc.) are computed separately for each identified mode of vibration by means of seismic response spectra used to represent the input seismic action. These "modal" peak response quantities are then "combined" using appropriate modal combination rules to obtain the total values of peak response used for design purposes (Chopra

2007). The *linear response history analysis* considers seismic accelerograms to represent the input seismic action in the same manner as the non-linear (inelastic) response history analysis (NRHA) previously discussed. However, non-linear phenomena are not accounted for in the development of the FE models and, therefore, the numerical integration of the underlying equations of motion may involve modal analysis based techniques.

Such linear dynamic methods of analysis should normally be used for structures exhibiting relatively high seismic performance levels (e.g., "operational" or "immediate occupancy" in Fig. 1.15) or structures whose structural members exhibit a low demand capacity ratio (DCR). The latter is defined as the ratio of seismic demand over the capacity of a particular (critical) cross-section. These structures are expected to suffer very little, if any, structural damage under the design seismic action and, thus, they respond in an "almost linear" fashion.

Further, linear *single-mode (fundamental-mode) static analysis methods* can be used in the special case of high performance low-rise regular building structures. These structures have a dominant translational fundamental mode of vibration and, therefore, the influence of higher modes of vibration on their overall dynamic response is insignificant. An example of a typical single-mode linear elastic static analysis method is the "lateral force method" prescribed in the EC8 (§ 4.3.2.2 of EC8) which is further discussed in Sects. 2.4.3 and 3.1.5.2 (see also flowchart 3.8). The seismic action is represented by means of a seismic response spectrum which is used to define the seismic design base shear force (see, e.g., Fig. 1.5) corresponding to the first (fundamental) natural period of the structure. This base shear force is distributed along the height of a building following an inverted triangular pattern in the form of lateral static loads applied to floor masses lumped at pre-defined locations on each storey level.

At this point, it is important to note that EC8 treats "equivalent" linear response spectrum based dynamic or static (for regular buildings) analysis methods as the "preferred" ones for the design of new structures, even though they may exhibit a relatively low, "life safety", seismic performance level (i.e., in case of adopting the maximum, or close to the maximum, allowed value of the behaviour factor q). As already discussed in Sect. 1.2.4, this is accomplished by adopting an "inelastic" design (response) spectrum of reduced spectral ordinates compared to the elastic response spectrum in defining the input seismic action. In this manner, the range of applicability of linear elastic analysis methods is broadened to include the case of moderate-to-low seismic performance levels in undertaking code-compliant seismic design of ordinary building structures.

It is further noted that the use of linear methods of analysis for the case of building structures exhibiting locally strong inelastic behaviour is also allowed in the pertinent American pre-standards within the performance-based seismic design framework [e.g., ASCE 41-06] (ASCE 2007). However, in these documents the unreduced elastic response spectrum is considered in the determination of the seismic input loads and, instead of a universal reduction behaviour factor q applied to the loads, a local reduction factor m is applied to the stress resultants at the critical cross-sections. The latter factor accounts for the local ductility capacity of cross-sections and its value is given in tabular form for each type of structural member and assumed level of seismic performance. For example, for r/c columns $m = 1.25 \sim 2.00$ for immediate occupancy performance level, $m = 1.75 \sim 3.00$ for life safety performance level, and $m = 1.75 \sim 4.00$ for collapse prevention performance level. Further, for the case of cross-sections of reduced ductility capacity bearing significant compressive or shearing stresses, $m = 1$. Interestingly, the m factor is applied to stress resultants due to both the seismic and the gravitational load combination.

The above discussion highlights the rationale of using response spectrum based (dynamic or static) linear methods of analysis within a performance-based seismic design and assessment philosophy for yielding structures. Nevertheless, it is emphasized that the use of such methods outside the prescriptive seismic codes of practice and/or the relevant pre-standards is not recommended for the case of structures exhibiting strongly inelastic behaviour. Instead, inelastic methods of analysis should be preferred which account for the non-linear inelastic behaviour of structures in an explicit manner.

2.4.1.2 Non-Linear Methods for Seismic Analysis

The dynamic *non-linear response history analysis* (NRHA) method is the most refined and detailed type of inelastic analysis and can be theoretically applied to any type of structure whether it exhibits inelastic behaviour or not under a given seismic ground motion excitation. Focusing on r/c building structures, NRHA accounts for the seismic/cyclic response of r/c structural members explicitly by considering either distributed (fiber) or lumped (discrete) plasticity non-linear FE structural models.

The use of lumped plasticity models is a long established approach involving elastic one-dimensional "beam/column elements" with end member point plastic hinges. These plastic hinges are represented by inelastic rotational springs connected in series to the elastic beam/column elements. In this regard, a set of assumptions needs to be made regarding the length of the plastic hinge and the adoption of an appropriate hysteresis law in the form of moment-curvature or moment-rotation relationships for the rotational springs of each r/c structural member. These assumptions are often associated with non-negligible uncertainty. Specifically, the length of the assumed plastic hinges is one controversial issue; typically, this is to some extent compensated by either using semi-empirical expressions found in the literature (Penelis and Kappos 1997) or by modeling the critical zones for energy dissipation of structural members with multiple small-length elements having adequately closely spaced rotational springs (Kappos and Sextos 2001). Further, numerous hysteretic material laws are available for capturing the post-yielding behaviour of r/c members under cyclic loading, including stiffness degradation, strength deterioration, and the pinching effect (Katsanos et al. 2014). However, such advanced material models are hardly ever used in the design of ordinary structures, as they require a large number of input parameters which usually are not readily available. More importantly, the current state-of-the-art suggests that lumped plasticity models can be used to perform reliable and numerically stable NRHA only in planar (two-dimensional) structural models. Lumped plasticity hysteretic laws for spatial (three-dimensional) FE models that are sufficiently simple, general, and numerically robust for practical design purposes are not available. Thus, in the case most needed in design (e.g., for asymmetric in-plan building structures with significant torsional response exposed to two horizontal components of the seismic ground motion), lumped plasticity models do not offer a reliable analysis option.

An alternative to the lumped plasticity approach is the so-called distributed (fiber) plasticity models which lend themselves to more reliable NRHA for spatial FE models. In distributed plasticity models, the cross-section of each structural member, including the reinforcing rebar, is discretized into a set of longitudinal fibers. Each fiber is assigned an appropriate non-linear uniaxial stress-strain relationship. Next, integration operations lead to the desired stress-strain relationship of the entire section (Fardis 2009). An advantage of this approach is that the hysteretic behaviour of cross-sections is explicitly derived based on the inherent material constitutive relationships, while the time-varying axial load in structural members is also accounted for during run time. However, the required computational time for NRHA using distributed plasticity spatial FE models for a sufficiently large number of representative strong ground motions is practically prohibitive for the design of ordinary r/c structures.

Overall, in view of the above brief presentation on available options for non-linear material modeling, it can be concluded that the required modeling effort, computational demand, and experience in defining the non-linear properties of the structural FE models, in selecting and appropriately scaling the input ground motions, and in the proper interpretation and validation of numerical results, renders the use of NRHA for the seismic design of new structures impractical. To this end, a plethora of *inelastic static analysis methods* have been proposed to circumvent some of the above challenges and difficulties associated with the application of NRHA, while aiming to capture/predict more realistically the response of structures designed for relatively low seismic performance levels (e.g., "life safety") than the code-prescribed "equivalent" linear elastic methods discussed in the previous section. It is noted that these methods are not based on any rigorous theoretical background and rely on assumptions and methodologies that, in many cases, contradict basic principles of structural mechanics (e.g., invoking the principle of superposition for non-linear structures). Still, they are being extensively used in practice, especially for evaluating the seismic vulnerability of existing (code-deficient) structures.

Of these inelastic static analysis methods, the least involved, namely the *non-adaptive single-mode inelastic pushover analysis* or simply the *standard pushover analysis* (SPA), has been adopted for the design of new structures and the assessment of existing structures by current codes of practice, including the EC8 (see also Sects. 2.4.2 and 2.4.3.4). In brief, SPA is a modern variation of the classical plastic 'collapse' analysis adequately adapted to seismically excited structures aiming to predict the hierarchy of (localized) structural damages up to the onset of collapse (Krawinkler and Seneviratna 1998). It considers monotonically increasing lateral loads applied statically to the structure following a particular distribution along the height of the building resembling the distribution of the fundamental modal forces that the structure would be subjected to under horizontal dynamic support excitation. As these loads increase, a series of plastic hinges develop gradually at critical sections of the structure, leading to force redistribution and eventually to a failure mechanism (see also Fig. 2.13). Therefore, through this analysis, it is possible to obtain the non-linear relationship (pushover curve) between the sum of the lateral applied force (base shear) and the deformation of the structure monitored at a specific location. Based on this curve, various approaches exist for the definition of a surrogate ("equivalent") single-degree-of-freedom (SDOF) oscillator which may be non-linear/inelastic, e.g., EC8-1 Annex B (CEN 2004), FEMA356 (FEMA 2000), or linear heavily damped, e.g. ATC-40 (Applied Technology Council 1996). This SDOF-oscillator is used in conjunction with seismic response spectra for seismic performance assessment (and/or design) of the considered structure, see also FEMA440 (FEMA 2005).

Several variants of the SPA exist which account for the influence of the computed "equivalent" SDOF oscillator properties (Manoukas et al. 2011) for the influence of higher modes of vibration, such as the *modal pushover analysis* (Chopra and Goel 2002; Goel and Chopra 2005; Manoukas et al. 2012; Paraskeva et al. 2006; Paraskeva and Kappos 2010), and/or for the fact that seismic load distributions change as the structure yields, such as the *adaptive pushover analysis* (Antoniou and Pinho 2002; Elnashai 2001). Clearly, these and other existing variants of the SPA come at an increase in computational cost and modeling effort, while their availability in commercial structural analysis software is currently quite limited. In terms of material non-linearity modeling, both lumped and distributed plasticity models can be used in conjunction with inelastic static analyses. Compared to the case of NRHA, the required (monotonic rather than cyclic) moment-curvature relationships for the definition of plastic hinge properties at critical cross-sections of r/c buildings can be derived in a much easier fashion. This is achieved by means of readily available software performing conventional fiber model analysis given a reasonable estimate of the member axial load. Alternatively, tables of plastic hinge properties aiming to facilitate the practical implementation of inelastic static analyses are available in the public domain, e.g., ASCE-41 (ASCE 2007), and can be adopted for undertaking SPA or its variants. Still, the application of inelastic static (pushover) analyses to spatial FE models with gradually increasing lateral loads in two orthogonal directions is computationally sensitive, and it is not recommended for practical seismic design and assessment purposes. Thus, static inelastic analysis methods share the same practical limitation with the NRHA in terms of applicability: they can be reliably applied only to planar FE models.

Collectively, compared to the NRHA, the inelastic static analysis methods

- are computationally significantly less demanding,
- require much less experience in modeling the material non-linear behaviour for which sufficient guidance for design engineers is provided by relevant pre-standard documents while dependable commercial software also exists,
- do not involve representing the seismic input action by means of properly selected and scaled recorded (or artificial) accelerograms, and
- yield numerical results (pushover curves) which can be readily interpreted from structural engineers and used in conjunction with prescriptive semi-empirical methodologies found in seismic codes of practice and in relevant guidelines for practitioners for the seismic assessment of new and existing structures.

However, being "static" in nature, these analysis methods cannot account for

- the part of the input seismic energy dissipated during the dynamic response of structures through mechanisms modeled via viscous damping (typically 5 % of critical viscous damping is used in r/c structures), and
- dynamic effects influencing the seismic energy dissipation demands in structures such as the duration of the ground motion which is well related to the total number of loading-unloading-reloading response cycles.

Compared to the equivalent linear response spectrum based analysis methods discussed in the previous section, inelastic static analysis methods provide a more realistic prediction of the actual (non-linear) seismic performance of new code-compliant structures targeting a "life safety" performance under the design seismic action. This is because they can, approximately, take into account the redistribution of stresses among structural members due to plastic hinges (post-yield behaviour) expected to form at critical cross-sections under the nominal design earthquake. Therefore, inelastic static analysis methods can provide useful estimations and insights on

1. the level of stress demands in structural members which may potentially exhibit non-ductile behaviour (e.g., deep beams due to high shearing force demands),
2. the severity of the local inelastic deformation demands at designated (and detailed for ductile behaviour) hysteretic energy dissipation zones within structural members,
3. the influence of the gradual local stiffness and strength reduction at individual structural members to the global behaviour/performance of the structure,
4. the potential formation of undesirable collapse mechanisms (e.g., due to "weak" or "soft" floors in building structures, as discussed in Sect. 2.2.4), and
5. the severity of inter-storey drifts and floor rotations in building structures.

In this regard, such methods are useful in the identification of local critical cross-sections in r/c building structures where undue concentrations of inelastic ductility demands occur, necessitating member re-sizing or the taking of special detailing measures (see also Sects. 2.2.2 and 2.2.3). Further, they are also useful in informing whether accounting for the influence of second-order "P-Δ" phenomena (i.e., geometric non-linearity) should be an issue of concern at design (see also Sect. 2.4.3.3). In case of severe influence of geometric non-linearity (i.e., large inelastic storey drifts and/or inter-storey rotations), increasing the size of vertical r/c structural members, especially of r/c walls, is commonly considered.

2.4.2 Overview of EC8 Structural Analysis Methods

EC8 specifies four different analysis methods that can be used in the seismic design of new structures (§ 4.3.3.1). These are:

1. the "*modal response spectrum analysis*" method, taken as the "reference" method (§ 4.3.3.1 (2)P) applicable to all types of (conventional) structures;
2. the "*lateral force method of analysis*", applicable only to structures characterized by a single dominant translational mode of vibration along each "principal direction" (§ 4.3.3.2.1 (1)P);
3. the "*non-linear static (pushover) analysis*"; and
4. the "*non-linear time history (dynamic) analysis*", hereafter response history analysis.

The first two analysis methods are linear elastic and are by far the most commonly used for the seismic design of new r/c building structures. The seismic action is represented via the EC8 (inelastic) design spectrum of reduced spectral ordinates by the behaviour factor q compared to the EC8 elastic spectrum (see Sect. 1.2.4). In this regard, the non-linear structural behaviour assumed in design is indirectly accounted for through the adoption of a behaviour factor $q > 1$ in the (linear) analysis step and by performing deformation checks against structural displacement results multiplied by q. The modal response spectrum method is of general use and accounts for a sufficiently large number of vibration modes such that at least 90 % of the total building mass is activated along the (principal) direction of the seismic action (§ 4.3.3.3.1(3)). The lateral force method of analysis is the least demanding in computational power and complexity, as it accounts for only the fundamental mode of vibration. In principle, it involves a series of four separate static analyses with horizontal lateral seismic loads triangularly distributed along the height of buildings acting at the level of each floor along two orthogonal principal directions. Accordingly, the lateral force method is allowed to be applied to buildings whose response is not significantly affected by contributions from modes of vibration higher than the fundamental mode in each principal direction (§ 4.3.3.2.1(1)P). Details on the implementation of these two methods are included in Sect. 3.1.5 and described in Flowcharts 3.7 and 3.8.

EC8 allows for the use of a specific form of "pushover" analysis as an alternative analysis method to the two above (equivalent) linear elastic methods (§ 4.3.3.4.2.1 (1)). It is noted, however, that in various National Annexes of the EC8, such as in the one applicable to Greece (highest seismicity country within the EU), the non-linear static (pushover) analysis of EC8 is only allowed to be used *in conjunction with* the modal response spectrum analysis and not as an exclusive stand-alone method for the case of designing new structures. A discussion on the limitations of this method is provided in Sect. 2.4.3.4 below, while a description of its practical implementation steps are given in Appendix A. Finally, the non-linear time history analysis (i.e., non-linear response history analysis-NRHA) is generally allowed, but its use is restricted by the issues raised in the previous section. In the case of practical designing of ordinary new structures, it is rarely used in practice. However, it is more commonly used in the special case of structures whose non-linear response must be explicitly accounted for, such as for base-isolated buildings covered in chapter 10 of EC8. The consideration of the seismic design of such structures falls outside the scope of this book and, thus, the NRHA will not be discussed in what follows. As a final remark, it is noted that EC8 does not consider at all the case of linear response history analysis, which, however, can prove useful in the case of "irregular" buildings designed for stringent seismic performance levels (i.e., IO).

2.4.3 Discussion and Recommendations on EC8 Analysis Methods

This section provides critical discussion on a number of practical issues arising in the application of the three most commonly used analysis methods prescribed by the EC8 for the design of new buildings. Listed in order of increasing complexity, these methods are: the lateral force method, the modal response spectrum method and the non-linear static method. In the interest of guiding the reader through this section, the main conclusions and recommendations derived from the ensuing detailed discussion are summarized:

1. The use of the lateral force method of analysis, as prescribed in EC8, is not recommended, except for in the ideal case of up to medium height buildings whose lateral load-resisting structural systems are symmetric with regard to two orthogonal horizontal directions (Sect. 2.4.3.1). Instead, the modal response spectrum analysis method should be preferred even in the case of "regular" buildings, as defined in clause §4.3.2 of EC8.
2. The use of the percentage combination rules of clause §4.3.3.5.1(3) of EC8 to estimate the most unfavourable effects due to the simultaneous action of two or three components of the seismic action should be avoided (see Sect. 2.4.3.2). Instead, the "square root of the sum of the squared values" (SRSS) or other more accurate combination rules should be considered.
3. Caution should be exercised in the calculation of the interstorey drift sensitivity coefficient defined in clause §4.4.2.2(2) of EC8. When possible, the use of commercial computer software capable of performing reliable second-order static analyses is recommended to determine the above coefficient (Sect. 2.4.3.3).
4. The version of the non-linear static analysis prescribed in EC8 should be used with caution in the case of designing new asymmetric in-plan buildings. For such

buildings, it is recommended that pushover analysis is not used as a stand-alone analysis method but, rather, in conjunction with the modal response spectrum analysis method (Sect. 2.4.3.4).

2.4.3.1 The Range of Applicability of the "Lateral Force Method of Analysis"

In clause §4.3.3.2.1(1)P of EC8, it is stated that the lateral force method of analysis "…may be applied to buildings whose response is not significantly affected by contributions from modes of vibration higher than the fundamental mode in each principal direction." This prescription raises the following two practical issues:

- "Principal directions" can be unambiguously defined only for buildings whose lateral load-resisting structural systems (i.e., moment resisting frames and shear walls in the case of r/c buildings) are aligned along two orthogonal horizontal axes in-plan (see clause § 4.3.3.1(11)P of EC8, where, however, the term "relevant directions" is used). There is no provision in EC8 on how the principal directions should be defined in the case of buildings whose vertical structural members do not follow the above ideal layout. Such cases may arise in practice due to space restrictions within an urban environment and/or architectural requirements. It is noted that, in these cases, a viable definition of "principal axes" may be accomplished by considering the theory of "virtual elastic axes for multi-storey buildings" discussed in the technical literature (Athanatopoulou and Doudoumis 2008; Makarios and Anastassiadis 1998). However, commonly used commercial software for code-compliant structural design does not usually offer capabilities of defining the virtual elastic axes and their associated principal directions and, thus, their definition requires additional computational effort and technical substantiation on behalf of practicing engineers.
- Further, suppose that, under some reasonable engineering assumptions, the "principal horizontal directions X and Y" of a building can be identified. In the general case where the lateral load-resisting structural elements are not aligned along two orthogonal horizontal axes, the determination of, say, the fundamental mode along the X direction requires undertaking modal analysis considering a three-dimensional finite element model of the structure with all degrees of freedom along the Y direction restraint. Similarly, in determining the fundamental mode along the Y direction, modal analysis needs to be performed for a different model of the structure where all degrees of freedom along the X direction are restraint. Thus, even though the lateral force method is supposed to offer a computationally less demanding analysis option compared to the modal response spectrum method, it requires, in general, two additional modal analyses to be performed for two different appropriately modified (restrained) computer models to check whether its use is allowable according to EC8.

More importantly, clause §4.3.3.2.1(2) of EC8 states that higher modes of vibration do not significantly contribute to the seismic response of building structures and, thus, the lateral force method can be applied if the following two conditions are met:

1. The fundamental periods of vibration in the two principal directions of a building are lower than 2 s or $4T_c$, where T_c is the corner period between the constant spectral pseudo-acceleration and the constant spectral pseudo-velocity segments of the EC8 elastic spectrum (see §3.2.2.2 of EC8 and Sect. 2.3.1.2), and
2. The building is regular in elevation, as defined in clause §4.2.3.3 of EC8.

However, it is noted that there does not appear to exist any rigorous research study in the technical literature suggesting the dominance of the fundamental natural mode of vibration for buildings satisfying these two conditions. In this respect, the use of the lateral force method may lead, in some cases, to non-conservative designs compared with the use of the more general and modal response spectrum analysis, which is based on a sound theoretical background. Examples of non-conservative designs obtained by application of the lateral force method for the case of buildings that meet the above two conditions but are not regular in-plan are reported elsewhere (Paglietti et al. 2011).

In conclusion, the use of the lateral force method of analysis for design purposes should be avoided and the modal response spectrum analysis method should be preferred except for cases of up to medium height r/c buildings with symmetric layouts where moment resisting frames and/or shear walls are aligned along two orthogonal directions in plan.

2.4.3.2 Spatial Combination of Peak Response Quantities from Individual Components of the Seismic Action

There exist two distinct cases where the issue of accounting for the effects of the two or three independent components of the seismic action acting simultaneously arises in performing any of the two methods of linear elastic analysis considered in EC8. These are:

- The case of determining the expected peak value of a single response quantity (e.g., the peak moment at a critical cross-section of a beam element), and
- The case of estimating the most probable value of a response quantity that occurs concurrently with the expected peak value of each one of the two or more internal stress resultants acting simultaneously at a particular cross-section of a structural element (e.g., two bending moments about two orthogonal horizontal axes and an axial force acting concurrently at the base of a column).

The first case is addressed in clause §4.3.3.5.1(2)b of EC8, which puts forth the use of the square root of the sum of squared values (SRSS combination rule) of the considered response quantity evaluated separately for each component of the seismic action (see also §4.3.3.3.2 of EC8). However, it is noted that, in the context of the modal response spectrum analysis method, the application of the SRSS rule is valid only when the same value of behaviour factor q is adopted along the considered directions of the seismic action components. If different values of q are adopted in defining the design spectrum along the considered directions, a consideration allowed by EC8 (§ 3.2.2.5(3)P), then a more general rule for spatial combination should be applied (Anastassiadis et al. 2002), as discussed in more detail in Sect. 3.1.4.4.

In addressing the second case, EC8 allows for the application of "percentage combination rules" (§4.3.3.5.1(3)), as an alternative to the SRSS rule. However, it is well-established in the literature that percentage combination rules are not founded on any rigorous theoretical basis and may lead to erroneous results (Menun and Der Kiureghian 1998; Wilson et al. 1995) Thus, it is recommended that the SRSS rule is utilized in all cases, unless the use of more accurate models is deemed essential (§4.3.3.5.1(2)c), e.g. (Gupta and Singh 1977).

2.4.3.3 Geometric Non-Linearity: Second-Order Theory and "P-Δ" Effects

For relatively flexible structures, second-order effects (geometric non-linearity caused by axial loads acting on the deformed configuration of the structure) may need to be considered in undertaking structural analyses for design purposes. The necessity for this consideration is governed by the intensity of the axial forces carried by structural members and the severity of deflections observed by structural members. In the case of ordinary r/c buildings adequately designed to resist earthquake induced lateral loads, the influence of second-order effects is not usually a critical design factor. However, EC8 prescribes a specific verification check to assess the potential influence of second-order effects for all structures. To this end, it is deemed essential to discuss certain assumptions made in EC8 related to second-order effects and to provide some practical recommendations.

In EC8, second-order effects refer only to the additional destabilizing moments induced to vertical structural members (columns and shear walls in r/c buildings) due to a relative translation of their ends by Δ (see Fig. 2.52). Under the common assumption of rigid floor diaphragms in r/c multi storey buildings, Δ becomes equal to the interstorey drift at each storey and, thus, the additional destabilizing moment acting on a vertical structural member carrying an axial load P becomes P·Δ ("P-Δ" or "structure P-Δ" or "P-Big Delta" effect). The additional destabilizing moments due to the flexure/curvature of structural members between their nodes, commonly referred to in the literature as "P-δ" (or "member P-δ" or "P-little delta") effects, are assumed to be negligible in EC8. This assumption is reasonable for ordinary multi-storey buildings, especially in the case of r/c buildings. In Fig. 2.52a, a simple portal frame is considered to illustrate the negative influence of P-Δ effects on the lateral load resistance capacity of structures. Specifically, P-Δ effects result in:

1. A reduction in the lateral stiffness K_E of the structure, assuming linear elastic behaviour, by the so-called geometric stiffness K_G. In the case of the portal

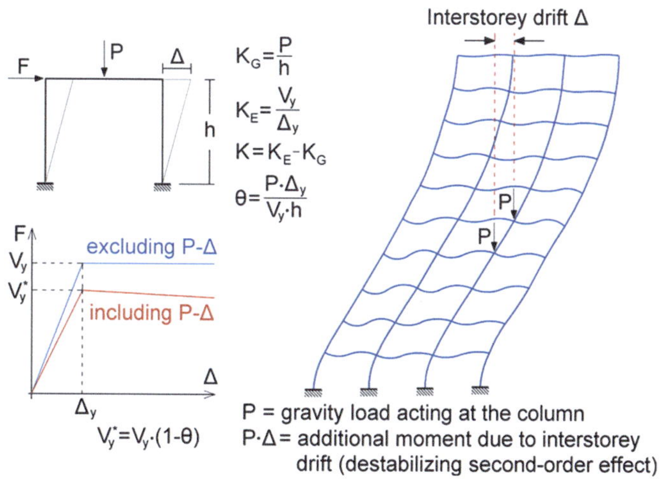

Fig. 2.52 Influence of P-Δ effects on the lateral stiffness K and the lateral resistance capacity V_y of a single storey frame (*left*); destabilizing P-Δ effects in a multi storey frame (*right*)

frame considered in Fig. 2.52a, K_G is equal to P/h (vertical load over the height of the frame) and, thus, the reduction in stiffness is proportional to the level of the gravity load, i.e., total stiffness $K = K_E\text{-}K_G = K_E\text{-}P/h$.

2. A reduction in the seismic base shear V_y (strength) at the yielding displacement Δ_y by a factor of 1-θ as shown in Fig. 2.52b, where θ is defined in Fig. 2.52a. Clearly, this reduction in strength is proportional to both the level of the applied vertical load and to the yielding displacement.
3. An increase in the negative slope of the post-yield segment in the base shear-floor deformation diagram (V-Δ diagram), as shown in Fig. 2.52b. The latter eventually leads to structures of reduced ductility.

Along these lines, EC8 (§ 4.4.2.2(2)) defines the "interstorey drift sensitivity coefficient" θ_i at a storey i by the expression

$$\theta_i = \frac{P_{tot,i}\Delta_i}{V_{tot,i}h_i} \tag{2.14}$$

where:

$P_{tot.i}$ is the sum of the axial forces carried by all the vertical structural members of the storey i due to the gravity loads of the seismic design combination of actions;
$V_{tot.i}$ is the total seismic storey shear at the storey i;
$\Delta_i = q\Delta_{el,i}$ is the average interstorey drift of the storey i;
$\Delta_{el,i}$ is the *average* interstorey drift of the storey i determined by a response spectrum based linear elastic analysis;
q is the behaviour factor used in the analysis undertaken to compute Δ_{el};
h is the height of storey i.

- If $\theta \leq 0.10$, the P-Δ effects are ignored.
- If $0.10 < \theta \leq 0.20$, the P-Δ effects are accounted for in an approximate manner by multiplying the response quantities (stress resultants and displacements) by a factor of $1/(1\text{-}\theta)$.
- If $0.20 < \theta \leq 0.30$, an exact second-order analysis must be undertaken.
- If $\theta > 0.30$, a re-design of the structure is required to increase its lateral stiffness such that θ is reduced to below 0.30.

Note that, in determining the interstorey drift sensitivity coefficient θ using Eq. (2.14), the values of $P_{tot.i}$, $V_{tot.i}$, and Δ_i should occur *simultaneously* and correspond to the same load combination. Thus, in the context of modal response spectrum analysis, θ should be determined by application of Eq. (2.14) for each considered direction of seismic action and each vibration mode followed by application of a modal and spatial combination rule (see also Sect. 3.2.1). Clearly, this is not a trivial task if automated structural design software is not available.

Further, it is noted that the ratio θ as determined by Eq. (2.14) is accurate only for planar frames with a predominantly shear-type of deflected shape in elevation under static loads (McGregor and Hage 1977). The application of the above formula for coupled frame-shear wall structural systems may yield large non-conservative deviations from the true value of θ, especially when the product λH is less than 6, where H is the total height of the building, $\lambda^2 = GA_s/EI$, and GA_s, EI are the total shear and flexural stiffness, respectively, of the considered coupled system treated as a continuous structure comprising a purely shear sub-system and a purely flexural sub-system. For example, if $\lambda H = 1$, the value of θ computed from Eq. (2.14) will deviate by 50 % from the true value θ. In general, Eq. (2.14) is not valid for three-dimensional asymmetric in plan building structures.

The aforementioned problems of accuracy related to the determination of the interstorey drift sensitivity coefficient θ defined in Eq. (2.14) can be readily circumvented by making use of finite element structural analysis software capable of performing reliably static second-order analysis. In case such software is available, it is herein proposed to use an elastic stability index expressed by

$$\theta_e = \frac{P}{P_{cr,el}} \tag{2.15}$$

as a basis of an alternative verification check against second-order effects. In the above equation, P is the total gravity load of the seismic design combination of actions and $P_{cr,el}$ is the critical buckling load of the structure following the well-known theory of stability of structures under static loads.

The expression in Eq. (2.15) applies universally to all types of lateral load resisting structural systems. Further, the index θ_e is a property uniquely defined for each structure since it depends only on the stiffness of the structure and on the distribution of the gravitational loads, while it is independent of the assumed horizontal seismic loads. Finally, the influence of the inelastic behaviour (material non-linearity) regulated by the behaviour factor q in the context of the EC8 can be taken into account by considering the product θ_e·q as is prescribed for all cases of displacement response quantities (§ 4.3.4(1)P of EC8). Hence, it is herein proposed to consider the following criterion for checking against second-order phenomena instead of that of Eq. (2.14)

$$\theta = q\theta_e \leq 0.10 \tag{2.16}$$

where θ_e is defined in Eq. (2.14).

It is recognised that the above criterion relies on a simple static second-order analysis which, however, can be considered sufficiently accurate for ordinary r/c buildings. Undertaking more sophisticated types of second-order analyses (i.e., accounting for material non-linearity and/or the dynamic nature of the input seismic action) would not be reasonable within the context of an "equivalent linear" type of analysis adopted by codes of practice like EC8, where material non-linearity is crudely taken into account through a single scalar (i.e., the behaviour factor q) and a simple flat reduction of all structural elements' stiffness properties.

2.4.3.4 Material Non-Linearity: The Inelastic Static "Pushover" Analysis

For all practical reasons discussed in Sect. 2.4.1, the use of the inelastic response history analysis is rarely considered in "routine" seismic design or assessment of ordinary r/c building structures. However, inelastic static analysis methods are extensively used in practice and, in fact, they are considered the preferable methods for the seismic assessment of existing (code-deficient) structures. Notably, EC8 prescribes the use of a particular inelastic static analysis method not only for seismic assessment of existing structures, but also for determination of the overstrength ratio involved in the calculation of the q factor value for design using equivalent linear analysis methods (see also Sect. 3.1.5). Furthermore, it is the first seismic code of practice worldwide to allow for the use of such an analysis method as an alternative to the standard linear elastic methods of analysis (clause §4.3.3.4.2.1(d) of EC8) to account for material post-yield behaviour (see also Sect. 2.4.1.2). To this end, it is herein deemed necessary to discuss in some detail the EC8 prescribed static inelastic method, although it is not utilized in the ensuing chapters of this book.

The so-called "non-linear static (pushover) analysis" is described in clause §4.3.3.4.2 and in the Informative Annex B of EC8. It is an inelastic static analysis method developed by Fajfar (Fajfar 2000). The detailed steps of the method are presented in Appendix B of this book. In accounting for material non-linearity, a bilinear force-deformation (or moment-rotation/curvature) relationship with no strengthening in the post-yield range (§4.3.3.4.1(2) and (3)), that is, an elastic-perfectly plastic model, is allowed to be used (at minimum) at the structural member level (see, e.g., Fig. 1.9). Conveniently, this model requires knowledge of only the pre-yield stiffness (cracked section is assumed as per Sect. 2.3.2.1), the yield strength, and the ultimate displacement (collapse) capacity of critical cross-sections. The latter two can be found either from relevant tables provided by pre-standards and guidelines to practitioners, e.g., FEMA356 (FEMA 2000), ASCE-41 (ASCE 2007), or from standard fibre section analysis, as discussed in Sect. 2.4.1.2.

At the core of the method lies the application of single-mode non-adaptive pushover analysis, that is, the standard pushover analysis (SPA) discussed in Sect. 2.4.1.2. Therefore, the inelastic static method of EC8 comes with all the drawbacks related to the SPA. Specifically, it does not account for the contribution of higher-than-the-fundamental modes of vibration and it does not consider the change of the load pattern along the height of a building (e.g., according to the fundamental natural period) as the building undergoes gradual plastic deformations and, hence, a change in its structural (stiffness) and modal properties. In this respect, the non-linear static (pushover) analysis of EC8 yields dependable results for planar structural systems with compact envelop in elevation and high uniformity in mass distribution which are expected to exhibit a relatively uniform distribution of inelastic demands along their height. Further, the above method is also reliable for spatial structural systems with insignificant torsional response under horizontal ground excitation (e.g., having an, almost, double-symmetric in-plan configuration). The application of the method to spatial FE structural models of arbitrary geometry may be problematic. Moreover, it cannot take into account the influence of the vertical component of the ground motion.

Apart from the above limitations, common to all single-mode, non-adaptive inelastic static analysis methods, there are certain additional issues of concern with regards to the practical applicability of the EC8 prescribed inelastic static analysis. These are listed and discussed to some extent below.

Distribution of Horizontal Loads along the Structure's Height

In clause §4.3.3.2.1(1) of EC8, it is stated that the SPA should be applied by considering at least two different lateral load distributions along the height of any given building. These comprise a "uniform" floor mass proportional pattern and a "modal" pattern proportional to the lateral seismic loads found by application of the standard linear elastic analysis methods along the horizontal (principal) direction under consideration. EC8 implies that either the lateral force method (see flowchart 3.8 in Sect. 3.1) or the modal response spectrum analysis (see flowchart 3.7 in Sect. 3.1) can be used to define the "modal" pattern, of horizontal loads for the pushover analysis. However, in the case of the "modal" distribution pattern it is not clear (i) whether the horizontal loads refer only to forces along one or two principal directions or include moments about the vertical axis, and (ii) which mode of vibration should be taken in case the multi-modal response spectrum analysis is considered to define the modal distribution pattern. Conveniently, taking for granted that the non-linear inelastic static analysis of EC8 is reasonably applicable only to symmetric structures and planar structural systems, it can be argued that solely horizontal forces and solely the first fundamental mode shape needs to be considered in defining the "modal" load distribution pattern. In every case, design engineers should be well aware of the fact that pushover analysis results (i.e., the shape of the pushover curve, the target displacement, and the sequence of plastic hinge development) are strongly dependent on the adopted distribution of the applied lateral loads, e.g. (Krawinkler and Seneviratna 1998).

Influence of Torsional Effects

Increased seismic deformation demands are expected for vertical structural members lying close to the perimeter of buildings with in-plan asymmetry. This is because of the coupling (combined) effect of translational with torsional components of vibration modes. In conducting non-linear static (pushover) analysis, EC8 mandates that the lateral seismic loads are applied at 4 specific locations about the center of gravity of each floor to account for accidental eccentricity (§4.3.3.4.2.2(2) P), as shown in Fig. 3.4. However, in clause §4.3.3.4.2.7 of EC8, it is stated that this consideration may underestimate the deformation demands on structural members lying in the stiff/strong side of torsionally sensitive buildings (see also Appendix B). It is therefore proposed to increase the thus determined deformation demands by applying an amplification factor "based on the results of an elastic modal analysis of the spatial structural model" (§ 4.3.3.4.2.7(2)). Although research work along these lines has been conducted (Fajfar et al. 2005; Marusic and Fajfar 2005) and some practical recommendations are outlined in research oriented texts (Fardis 2009; De Stefano and Pintucchi 2010), it is pointed out that this recommendation needs further justification and explanation within EC8.

Furthermore, in the case of adopting a spatial (three dimensional) model, the requirements for considering 4 different sets of points where lateral loads need to be applied (corresponding to 4 different mass positions to account for accidental eccentricity) times the two different lateral load distributions (a uniform and a modal) along both principal axes and both directions (positive and negative) separately leads to performing a total of 32 inelastic static pushover analyses. Evidently, the required computational and modeling effort seems to well exceed what is practically feasible for the case of routine seismic design of conventional structures.

Spatial Combination

In clause §4.3.3.5.1(6) of EC8, it is allowed to use the same spatial combination rules as in the case of linear analysis to account for the simultaneous action of two orthogonal components of the ground motion, namely the SRSS rule and the percentage combination rules (Sect. 2.4.3.2). However, it is noted that these rules have been analytically derived and numerically verified in the published scientific literature assuming linear structural behaviour. For example, the SRSS spatial combination rule can be shown to provide exact peak results by assuming that the two simultaneous horizontal orthogonal components of the seismic ground motion are statistically uncorrelated and that a linear response spectrum based analysis is used to derive peak structural responses along each direction of the seismic action (Smebby et al. 1985). Therefore, the aforementioned clause of EC8 contradicts the assumptions made in deriving the recommended spatial combination rules, and the issue of whether their adoption in conjunction with non-linear static types of analysis lead to conservative results has not yet been adequately investigated in the open literature.

Overall, the above discussion on certain challenges in applying the inelastic static pushover analysis method demonstrates that it has not yet reached an appropriate level of maturity and dependability for the everyday practice of seismic design of new structures. This is especially true for the case of spatial structures with asymmetric stiffness, strength, and mass distribution in plan and/or elevation whose higher modes of vibration have a non-negligible influence on their overall seismic response. In fact, this method has been initially proposed for planar structures under a single component of excitation with the aim of being a practically useful and less computationally demanding analysis method compared to the NRHA. As such, it is not recommended to be used as an alternative to the equivalent linear methods (modal response spectrum based and lateral force method), especially for in-plan asymmetric structures until future corrective steps and developments take place. It is noted, however, that such improvements may most likely come at an increased computational cost and level of sophistication which may contradict the purpose for which non-linear static analyses have been proposed in the first place.

Still, this method may certainly serve well to *complement* the linear methods of analysis for design purposes as it can provide a useful estimate of the performance of a newly designed structure. For example, it may be used as a verification tool for the achieved performance level within a performance-based seismic design framework for new structures. It is reminded that such verification is not provided by the force-based equivalent linear design methods dominating EC8 (among many other seismic codes of practice). In this context, the Greek National Annex to EC8 (Greece is by far the highest seismic risk country within the current EU) allows for the use of the inelastic pushover analysis method only in conjunction with the modal response spectrum analysis. The same document does not allow for reducing action effects in case the pushover analysis yields smaller demands than the modal response spectrum analysis with the exception of the overstrength a_u/a_i ratio (see Sect. 3.1.4). In any case, design engineers should develop a thorough appreciation of the inherent assumptions and limitations of the EC8 prescribed pushover analysis such that, combined with adequate experience in earthquake resistant design, they are in a position to critically appraise and use the pushover analysis results for seismic design purposes.

2.4.4 Overstrength Distribution Verification and Sensitivity Analyses

As a closure to this section on analysis methods, it is deemed important to further discuss the issue of ensuring that plastic hinges towards desirable collapse mechanisms will actually develop under the design seismic action in the case of adopting relatively large behaviour factors q (i.e., for new relatively low seismic performance structures). Furthermore, a second important issue is discussed with regards to the need for parametric sensitivity analyses using a series of FE structural models at the design stage to investigate the influence of uncertain parameters involved at the modeling stage.

2.4.4.1 Verification of Overstrength Distribution

As already pointed out in Sects. 1.2.6.1 and 2.2.4, a sufficient number of plastic hinges at designated (ductile) zones for energy dissipation must form under the design seismic action for structures designed for relatively large values of the behaviour factor q (e.g., for the maximum allowed by EC8 q_{allow}). This consideration epitomizes the capacity design approach and ensures life safety performance of code-compliant structures under the design earthquake. However, accidental overstrengthening of critical cross-sections of r/c members designed to yield first can lead to an insufficient number of plastic hinges and to failures at cross-sections with inherently low ductility capacity (Sect. 1.2.6.2). Therefore, due to undesirable/unpredicted overstrength, the very capacity design approach is jeopardized and the underpinning design objectives of the partial protection against the seismic hazard design philosophy may not be successfully met. In practice, this situation may arise in case of non-uniform or even "arbitrarily" distributed overstrength among r/c structural members.

In the context of seismic design using equivalent linear analysis methods, the distribution of overstrength within a structure can be verified by monitoring the demand capacity ratio (DCR) introduced in Sect. 2.4.1.1 (ratio of seismic demand over capacity). This can be readily accomplished by use of structural analysis/design software capable of automatically returning the DCR upon analysis and detailing of each critical cross-section. An even distribution of the DCR among all structural members should be sought. This may require additional re-design steps or local "manual" fine-tuning of the longitudinal reinforcement at certain cross-sections. Ideally, but practically hard to achieve, the value of the DCR at all critical cross-sections designed to yield (designated zones for energy dissipation) should be equal to or slightly lower than one ($DCR < \sim 1$) for the design seismic action. Furthermore, the distribution of overstrength can also be monitored at the design stage by application of the non-linear static (pushover) analysis of EC8, though its use is practically limited to planar structural models (Sect. 2.4.3.4).

2.4.4.2 The Need for Parametric Sensitivity Analyses of Structural Models

Due to various sources of uncertainty (e.g., to the mechanical properties of building materials or to the fixity conditions of structures to the supporting ground at the foundation level), finite element (FE) models developed to perform static or dynamic analysis for design purposes can only approximate the actual behaviour of r/c structures. Therefore, even if the uncertainty associated with the externally applied loads is left aside (e.g., by adopting code-compliant loading design actions which have already accounted for the uncertain nature of environmental and man-made loads), *no single structural model, no matter how "refined" it may be regarded, can fully and accurately represent the properties of a given structure in a deterministic context. Consequently, the actual response of a structure to a given set of (design) loads cannot be reliably predicted by one structural model.*

Furthermore, FE computer models are based on simplified modeling assumptions, whose appropriateness and reliability in capturing the actual behaviour of structures subject to external loads depends heavily on, and bears the signature of, the structural analyst/design engineer or even the software developer in cases of using commercial software with automated modeling capabilities (e.g., automated FE meshing). Therefore, the subjective nature of modeling assumptions contributes an additional source of uncertainty. Such modeling uncertainties due to incomplete knowledge of structural properties and assumptions made during FE model development are recognized by (seismic) codes of practice and guidelines for practitioners. These uncertainties are implicitly treated by ensuring that a sufficiently high degree of conservatism is achieved in code-compliant (seismic) design of structures. Still, there are certain circumstances where it is common (and, in fact, recommended) that design engineers consider a series of different/alternative FE models to represent the same load-resisting structural system during the analysis stage of the seismic design procedure. Such needs for parametric/sensitivity analyses to crudely quantify the aforementioned uncertainties arise in practice in the cases of

1. requiring consideration of a range of values for a certain modeling parameter or group of modeling parameters;
2. pursuing the calibration of simplified structural models achieving a small trade-off in terms of accuracy compared to more refined models.

The first case is encountered in accounting for phenomena of significant inherent uncertainty which are not specifically covered by codes of practice and cannot be readily (or inexpensively) quantified. A typical example is the consideration of the effects of the supporting ground compliance at the foundation level of building structures. This issue may become critical for relatively stiff structural members (e.g., r/c walls and cores) founded on relatively soft soil. As discussed in Sect. 2.3.3.5, supporting ground compliance is commonly modelled via the "Winkler" model involving single (e.g., Fig. 2.47) or distributed (e.g., Fig. 2.49) linear springs. Given that only a crude qualitative description of the supporting ground type is usually available in the case of ordinary structures, a relatively wide range of spring constant (stiffness) values may be adopted in the analysis. It is thus reasonable to consider performing a parametric sensitivity analysis for a sufficient number of different sets of spring constants lying within a soil-dependent range of values to trace the "envelop" (peak values) of deformations and stress resultants developing at critical cross-sections. In this manner, structural members can be detailed based on a "worst-case scenario" approach chosen out of a series of analyses with varying sets of Winkler spring constants.

The second case may be dictated either by a requirement to expedite analysis through the development of an FE model computationally efficient to analyze, or by limitations in the analysis/modeling/post-processing capacities of the commercial structural design software available. A typical example is the modeling of r/c cores using equivalent frame systems, as discussed in Sect. 2.3.3.4. Such simplified models are commonly considered in practice, as they are computationally inexpensive and easy to develop, while they can be treated by simple structural design software packages using only one dimensional beam elements. However, the topology of equivalent frame systems and the mechanical properties assigned to certain beam/frame members are not unique for any given (multi-cell) core (see, e.g., Fig. 2.45). In such cases, it is recommended to consider two or more alternative FE models to gauge the sensitivity of deformations and stress resultants at critical cross-sections against different equivalent frame topologies used to capture the behaviour of (multi-cell) r/c cores.

In view of the above discussion, it can be further argued that design engineers and structural analysts should *resist the temptation of using today's capabilities of commercial software and the availability of computational power to develop very detailed and computationally demanding linear or non-linear FE models,* as these typically require many input parameters for which realistic/representative values are usually not available. In fact, such detailed models may be erroneously regarded as "accurate", while they do not contribute to a better understanding of the underlying inherent modeling uncertainties. Instead, structural engineering practitioners should be taking advantage of the enhanced capabilities of modern structural analysis software and computers *to run thorough investigations of the structural/mechanical behaviour of a given load-resisting structural system by developing series of alternative FE models within the context of parametric/sensitivity analyses.* Such an analysis/design approach is not only useful for the identification and quantification of critical sources of uncertainty in structural FE modeling, but it also contributes to the development of the all-important skill/ability to detect errors in the structural analysis step (e.g., due to errors in input data) by inspection of numerical results (e.g., displacement and stress demand concentrations, stress/load paths, etc.). Note that the development of this ability among practitioners has been hindered in recent decades by the availability of automated "user-friendly" computer software used as a "black-box" in (seismic) structural design. Nevertheless, the ability to check empirically the correctness of structural analysis results through possession of a good sense ("instinct") of expected stress paths and local strain concentrations can significantly facilitate, if not ensure, the elaboration of well-balanced, economical, and functional structural designs from a conceptual design stage up to the local detailing stage. A further discussion on the need for verification of analysis results obtained by means of structural analysis/design software follows in the last section of this chapter.

2.5 On the Use of Commercial Software for Routine Seismic Design

In recent decades, the advent of low-cost high-performance computer hardware and automated structural analysis commercial software made it possible to undertake computationally demanding structural analyses and design/detailing procedures in a user-friendly and cost-effective manner. Contemporary seismic codes of practice, EC8 inclusive, *implicitly presume* the availability of adequate computational resources and automated software to structural design engineers. Consequently, code-compliant seismic design of r/c structures can only be practically accomplished with the aid of appropriate structural analysis and design/detailing software run on sufficiently powerful computers. Typically, such software facilitates finite element modeling, performs the required structural analysis steps, undertakes and reports on the code-prescribed verification checks, and automates design/detailing of r/c structural members to code requirements. Further, it safeguards productivity by ensuring transparent and accurate data input and by offering capabilities for visual and end-results checking and interpretation. Clearly, special care is needed to ensure that specialized commercial software for code-compliant seismic analysis and design addressing engineering consultants and practitioners (assumed to be knowledgeable structural engineers) performs as intended with an acceptable *degree of reliability*. In this respect, this section provides brief practical comments on effective ways to verify the reliability of typical professional structural analysis and design software via benchmark problems and on the attributes of quality software packages and their proper usage.

2.5.1 *Verification of Commercial Structural Analysis Software via Benchmark Structural Analysis and Design Problems*

The need and importance of verifying the reliability of automated structural analysis and design software has attracted the attention of the research and engineering community since the early 1980s (Borri and Vignoli 1984; Gifford 1987; Lurie and Wells 1988; Melosh and Utku 1988; Pixley and Ridlon 1984). In those days, personal computers became relatively affordable, offering the opportunity for computer-aided structural analysis and design. This was achieved by developing purpose-made software capitalizing on advances in computational structural mechanics (statics and dynamics), which had taken place in previous decades.

Due to the complexity of modern seismic codes of practice, a stringent verification against all possible sources of errors and "bugs" of code-compliant seismic design software is not practically feasible. Such consideration would require full-fledged testing for all different code-prescribed modeling, analysis, and design/detailing options and alternative logic paths given any type of structural load-resisting system. In this respect, commercial software developers aim to *minimize the probability of systematic errors* occurring and, therefore, to ensure that structural analysis and design software packages are *reliable to a certain acceptable level*. This is achieved by considering *well-established computer programming techniques* (e.g., structured programming, modular programming, and fault-tolerant programming), which *minimize the risk* of coding errors (Chemuturi 2010; Lyu 1996; Mills et al. 1987; Stavely 1999). Furthermore, standard quality control practices are normally applied for all pertinent stages of software development. The thus developed software packages are usually copyrighted. As such, they are made available to users as "black boxes", since gaining access to, let alone changing, the source code by a third party is prohibited. Consequently, verification and debugging of the source code of commercial structural analysis and design software is undertaken by software developers (who may not necessarily have a structural engineering background) rather than by the end-users (structural engineers).

Nevertheless, apart from verification at the source code level, considering appropriate benchmark problems, testing a sufficiently broad spectrum of software functionality is regarded as an effective way to assess any software package (Lyu 1996). In this respect, the quality and reliability of commercial software for code-compliant seismic design may also be verified by means of *benchmark analysis and design example problems*. Such benchmark problems should go beyond simple "academic" exercises commonly included in software documentation to illustrate the accuracy of certain modeling techniques and analysis algorithms. They should ideally involve complete code-compliant modeling, seismic analysis, and design/detailing of sufficiently complex structures (e.g., multistorey buildings comprising different types of load-resisting structural systems). Still, they should also be judicially defined, documented, and worked out such that:

- they are not unnecessarily complex and, therefore, they can be promptly input, run, and checked in a straightforward manner by practitioners; and that
- each individual problem of a certain set of benchmark problems tests a different but well-targeted source of potential errors/bugs and/or of software functionality accounting for the underlying code requirements.

Conveniently, appropriate sets of benchmark structural analysis and design problems possessing the above attributes allow for undertaking software quality and reliability assessment not only by software developers, but also by regulatory State agencies at a regional, national, or international level, or even by individual software users (i.e., professional structural engineers). Furthermore, these benchmark problems can further serve as:

- tutorial examples to gain familiarity with the proper usage, the functionality, the capabilities, and the limitations of any commercial structural analysis and design software; and as
- case-study examples of code-compliant structures designed/detailed to the required specifications illustrating proper seismic code implementation.

However, it should be recognized that the development of appropriate sets of benchmark structural design problems is a formidable task that requires knowledge of computational mechanics, structural analysis and dynamics, design and detailing of r/c structures for earthquake resistance, and computer programming, along with a thorough appreciation of the requirements of codes of practice and their intended practical implementation. Moreover, their solution requires the use of proper software and computer coding. In this regard, dependable benchmark problems are difficult to develop for individual structural design engineers alone This task is usually accomplished by the involvement of the research community and/or of experienced and respectable engineering consultants under the auspices of national or international regulatory agencies.

Along these lines, the three numerical examples provided in Chap. 4 of this book can be viewed as prototypes of benchmark seismic analysis and design problems of r/c building structures compliant with the EC8 and EC2. Therefore, they may be used not only as instructive case-studies to illustrate EC8-compliant seismic analysis and design but also as benchmark problems to partially check the "correctness" of commercial seismic structural analysis software and design of r/c building structures according to Eurocodes. In fact, these three example problems are based partly on a large set of purpose-developed benchmark structural analysis problems according to the late Greek Seismic Code EAK2000 (Earthquake Planning and Protection Organization (EPPO) 2000a) in the frame of a research project funded by the Hellenic Earthquake Planning and Protection Organization (Avramidis et al. 2005) and partly on a set of benchmark problems (Sextos et al. 2008) compliant with the latest Greek Reinforced Concrete Code ΕΚΟΣ2000 (Earthquake Planning and Protection Organization (EPPO) 2000b). Incidentally, the above sets of benchmark problems are free to view and download from www.ec8examples.gr (in Greek).

2.5.2 Desirable Attributes and Use of Good Quality Software

The quality of professional software for code-compliant seismic design of r/c structures can be verified and gauged against different criteria, including:

(a) The *correct* implementation (coding) of structural analysis algorithms and code-prescribed design verification checks and detailing of structural members;
(b) The *completeness* in accounting for all particular verification checks and relevant requirements prescribed by structural codes of practice;
(c) The rationale, validity, and applicability of the *default simplification assumptions* made in the modeling and analysis stages; and
(d) The *ease of use*, including the graphical interface, the visualization of input data and output results, and the accompanied documentation.

As discussed in the previous section, the "correctness" of any software package in terms of analysis results, data post-processing, implementation of code-prescribed verification checks, and detailing of r/c members according to seismic code rules and requirements can be effectively assessed by relying on standard sets of purpose-developed benchmark analysis and design problems. In cases where only a particular (partial) software verification check is deemed necessary, targeted "hand-calculations" or calculations readily automated in a spreadsheet format may suffice. Further, the level of completeness by which a particular structural design software package covers and fulfils the requirements of a given seismic code of practice can also be assessed by case-dependent benchmark design examples. However, for such an assessment, a thorough appreciation of seismic code prescriptions is required as well. To this aim, the implementation flowcharts included in Chap. 3 provide a solid basis for commercial software verification in terms of completeness and coverage of the EC8 code for the seismic design of ordinary r/c structures.

Typical commercial software comes with various default simplification assumptions pertaining to the modeling and analysis stages aiming to expedite routine code-compliant seismic design. Therefore, unlike the case of research-oriented fully customizable software, analysis results obtained by means of commercial structural design programs may be biased and dependent on the inherent assumptions made. In this regard, it is normal to expect that different commercial software packages yield different analysis results, which might impact detailing of certain critical structural members, for the same input structural system and loading (Fig. 2.53a). For "good-quality" software (in terms of correctness, completeness, and default assumptions), the observed differences in the final structural designs (e.g., required longitudinal reinforcement ratios at critical cross-sections) should be well within an acceptable engineering precision covered by the inherent conservatism of seismic codes of practice. Still, in principle, all default modeling and analysis assumptions made should be explicitly stated and described with sufficient detail in the accompanying documentation of quality commercial structural analysis/design programs, even if these assumptions are not amenable to user-defined changes. This consideration ensures transparency and safeguards software users (structural engineering practitioners) from developing erroneous presumptions on the level of accuracy achieved by certain modeling and analysis techniques which may be acceptable for routine seismic design of ordinary structures but may not be applicable to more demanding/specialized seismic design scenarios.

Lastly, the "ease of use" criterion is rather subjective and, to a great extent, a matter of personal preference. Nevertheless, the current consensus suggests that certain standard features, such as user-friendly graphical interface for data input, visualization tools to inspect output analysis and detailing results, transparency in assumptions made and steps followed, completeness of accompanying manuals and software documentation including tutorials and benchmark problems, are quite desirable and beneficial to practising engineers.

Apart from the quality of structural analysis and design software itself, it is important to note that commercial seismic design software must always be used

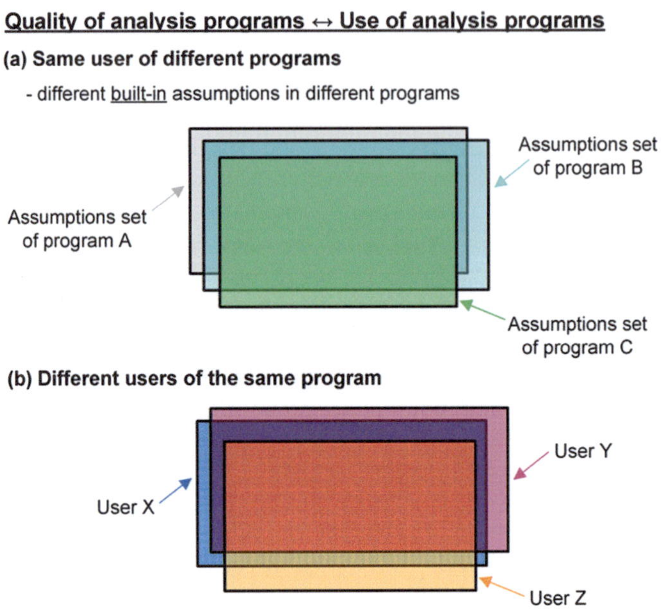

Fig. 2.53 Different default modeling and analysis assumptions in different commercial software packages, (**a**), and different analysis results and designs produced by different users of the same software package, (**b**)

appropriately. Indeed, a high-quality and well-verified seismic design software package is not sufficient to ensure the quality of the end product: erroneous analysis results and unreasonable structural design and detailing will be observed if a high-quality software is not properly used by well-qualified practising engineers with sufficient experience and knowledge of the adopted seismic code of practice. In every case, the so-called GiGo (*Garbage in – Garbage out*) "rule" applies, implying that the achieved quality and appropriateness of the output product (i.e., r/c building designed and detailed for earthquake resistance consistent with the desired level of seismic protection) cannot be higher than the quality and appropriateness of the input data considered (e.g., design loading scenarios, geometry and topology of the adopted finite element structural model, choice of the adopted analysis method, etc.) and of the related modeling assumptions made (e.g., structural and loading modeling, mass discretization, etc.). Unreasonable modeling assumptions, poor appreciation of load paths, errors/typos in inserting data input and other possible user mistakes will lead to poor quality structural designs. In fact, it is not uncommon that two different users (structural designers) of the same software may reach different structural designs given the same structural model at the conceptual design stage (Fig. 2.53b).

References

Anastassiadis K, Avramidis IE, Panetsos P (2002) Concurrent design forces in structures under three-component orthotropic seismic excitation. Earthq Spectra 18:1–17

Anastassiadis K, Avramidis IE, Athanatopoulou AM (2003) Critical comments on Eurocode 8, Draft no.5, Part I, Sections 3 and 4. fib symposium on concrete structures in seismic regions, 6–8 May, Athens/Greece

Antoniou S, Pinho R (2002) Advantages and limitations of adaptive and non adaptive force-based pushover procedures. J Earthq Eng, ATC and EPPO 8:497–522

Applied Technology Council (ATC) (1996) Seismic evaluation and retrofit of concrete buildings, ATC-40, Redwood City

Arnold C (2001) Architectural considerations. In: Naeim F (ed) The Seismic Design Handbook, 2nd edn. Kluwer Academic Services, Boston, pp 275–326

Arnold C (2006) Seismic issues in architectural design. Designing for earthquakes. A manual for architects. FEMA 454/Dec 2006, risk management series. NEHRP, Oakland, CA, pp 1–54

ASCE (2007) ASCE/SEI standard 41–06, seismic rehabilitation of existing buildings. American Society of Civil Engineers, Reston

Athanatopoulou AM, Avramidis IE (2008) Evaluation of EC8 provisions concerning the analysis of spatial building structures using two planar models. 5th Eurropean workshop on the seismic behaviour irregular complex structure, Catania, Italy, pp 285–297

Athanatopoulou AM, Doudoumis IN (2008) Principal directions under lateral loading in multistorey asymmetric buildings. Struct Des Tall Spec Build 17:773–794. doi:10.1002/tal

Avramidis IE (1991) Zur Kritik des äquivalenten Rahmenmodells für Wandscheiben und Hochhauskerne. DIE BAUTECHNIK 68:275–285

Avramidis IE, Manolis GD, Andreadakis MG (1995) PADEX – a preliminary antiseismic design expert system. In: Pahl PJ, Werner H (eds) Computing civil building engineering. Balkema, Rotterdam, pp 215–220

Avramidis IE, Anastassiadis K, Athanatopoulou A, Morfidis K (2005) Benchmark numerical examples to support correct application of the Greek anti-seismic code and facilitate control of professional software (in Greek). Aivazis Editions, Thessaloniki, Greece

Berrais A (2005) A knowledge-based expert system for earthquake resistant design of reinforced concrete buildings. Expert Syst Appl 28:519–530

Bisch P, Carvalho E, Degee H, Fajfar P, Fardis M, Franchin P, Kreslin M, Pecker A, Pinto P, Plumier A, Somja H, Tsionis G (2012) Eurocode 8: seismic design of buildings worked examples, JRC technical report. Publications Office of the European Union/Joint Research Centre, Luxembourg

Borri A, Vignoli A (1984) Micro-computer-aided design: a critical discussion of the reliability of programs of current use. In: Proceedings of the international conference on engineering software microcomputing. Pineridge Press, Swansea, pp 239–250

CEN (2002) European Standard EN1990. Eurocode 0: basis of structural design. Committee for Standarization. Brussels

CEN (2004) European Standard EN 1998-1. Eurocode 8: design of structures for earthquake resistance, Part 1: general rules, seismic actions and rules for buildings. Committee for Standarization. European Committee for Standardization, Brussels

Chemuturi M (2010) Software quality assurance: best practices, tools and techniques for software developers. J. Ross Publishing, Fort Lauderdale

Chopra AK (2007) Dynamics of structures—theory and applications to earthquake engineering, 3rd edn. Pearson, Prentice-Hall, Upper Saddle River

Chopra AK, Goel RK (2002) A modal pushover analysis procedure for estimating seismic demands for buildings. Earthq Eng Struct Dyn 31:561–582. doi:10.1002/eqe.144

De Stefano M, Pintucchi B (2010) Predicting torsion-induced lateral displacements for pushover analysis: influence of torsional system characteristics. Earthq Eng Struct Dyn 39:1369–1394

Doudoumis IN, Athanatopoulou AM (2001) Code provisions and analytical modeling for the in-plane flexibility of floor diaphragms in building structures. J Earthq Eng 5:565–594
Earthquake Planning and Protection Organization (EPPO) (2000a) Greek seismic code (EAK2000) (amended in 2003), Athens, Greece (in Greek)
Earthquake Planning and Protection Organization (EPPO) (2000b) Greek reinforced concrete code (EKOΣ2000), Athens, Greece (in Greek)
Elghazouli A (2009) Seismic design of buildings to Eurocode 8. Spon Press, New York
Elnashai AS (2001) Advanced inelastic static (pushover) analysis for earthquake applications. Struct Eng Mech 12:51–69
Elwood KJ, Matamoros AB, Wallace JW, Lehman DE, Heintz JA, Mitchell AD, Moore MA, Valley MT, Lowes NL, Comartin CD, Moehle JP (2007) Update to ASCE/SEI 41 concrete provisions. Earthq Spectra 23:493–523
Fajfar P (2000) A nonlinear analysis method for performance-based seismic design. Earthq Spectra 16:573–592
Fajfar P, Marusic D, Peru I (2005) Torsional effects in the pushover-based seismic analysis of buildings. J Earthq Eng 9:831–854
Fardis MN (2009) Seismic design, assessment and retrofit of concrete buildings, based on Eurocode 8. Springer, Dordrecht
Fardis MN, Carvalho E, Fajfar P, Pecker A (2014) Seismic design of concrete buildings to Eurocode 8. CRC Press, New York/Weiler
FEMA (2000) Prestandard and commentary for the seismic rehabilitation of buildings FEMA-356. Federal Emergency Management Agency, Washington, DC
FEMA (2005) Improvement of nonlinear static seismic analysis procedures FEMA-440. Federal Emergency Management Agency, Washington, DC
Fintel M (1991) Shear walls – an answer for seismic resistance? Concr Int 13:48–53
Fintel M (1995) Performance of buildings with shear walls in earthquakes of the last thirty years. PCI J 40:62–80
Giaralis A, Spanos PD (2009) Wavelet-based response spectrum compatible synthesis of accelerograms—Eurocode application (EC8). Soil Dyn Earthq Eng 29:219–235. doi:10.1016/j.soildyn.2007.12.002
Giardini D, Woessner J, Danciu L, Crowley H, Cotton F, Grünthal G, Pinho R, Valensise G, Consortium S (2013) SHARE European seismic hazard map for peak ground acceleration, 10% exceedance probabilities in 50 years. doi:10.2777/30345
Gifford JB (1987) Microcomputers in civil engineering – use and misuse. J Comput Civ Eng ASCE 1:61–68
Goel RK, Chopra AK (2005) Role of higher-"mode" pushover analyses in seismic analysis of buildings. Earthq Spectra 21:1027–1041. doi:10.1193/1.1851547
Gupta AK, Singh MP (1977) Design of column sections subjected to three components of earthquake. Nucl Eng Des 41:129–133
Kappos AJ (2002) Dynamic loading and design of structures. Spon Press, London
Kappos AJ, Sextos AG (2001) Effect of foundation compliance on the lateral load response of R/C bridges. J Bridg Eng 6:120–130
Katsanos EI, Sextos AG (2013) ISSARS: an integrated software environment for structure-specific earthquake ground motion selection. Adv Eng Softw 58:70–85. doi:10.1016/j.advengsoft.2013.01.003
Katsanos EI, Sextos AG, Elnashai AS (2014) Prediction of inelastic response periods of buildings based on intensity measures and analytical model parameters. Eng Struct 71:161–177. doi:10.1016/j.engstruct.2014.04.007
Krawinkler H, Seneviratna GDPK (1998) Pros and cons of a pushover analysis of seismic performance evaluation. Eng Struct 20:452–464. doi:10.1016/S0141-0296(97)00092-8
Lindeburg MR, McMullin KM (2014) Seismic design of building structures: a Professional's introduction to earthquake forces and design details. Professional Publications Inc, Belmont
Lurie PM, Wells BD (1988) Computer-assisted mistakes. J Civ Eng 58(12):78–80

Lyu MR (1996) Handbook of software reliability engineering. IEEE Computer Society Press/McGraw-Hill, New York
Mac Leod IA (1995) Modern structural analysis – modelling process and guidance. Thomas Telford Ltd, London
Mac Leod IA, Hosny HM (1977) Frame analysis of shear wall cores. J Struct Div ASCE 103:2037–2047
MacLeod (1977) Analysis of shear wall buildings by the frame method. Proc Inst Civ Eng 55:593–603
MacLeod IA (1990) Analytical modelling of structural systems. Ellis Horwood, London
Makarios TK, Anastassiadis K (1998) Real and fictitious elastic axis of multi-storey buildings: theory. Struct Des Tall Spec Build 7:33–55
Mander JBJB, Priestley MJN, Park R (1988) Theoretical stress-strain model for confined concrete. J Struct Eng ASCE 114:1804–1826
Manoukas GE, Athanatopoulou AM, Avramidis IE (2011) Static pushover analysis based on an energy-equivalent SDOF system. Earthq Spectra 27:89–105
Manoukas GE, Athanatopoulou AM, Avramidis IE (2012) Multimode pushover analysis for asymmetric buildings under biaxial seismic excitation based on a new concept of the equivalent single degree of freedom system. Soil Dyn Earthq Eng 38:88–96
Marusic D, Fajfar P (2005) On the inelastic seismic response of asymmetric buildings under biaxial excitation. Earthq Eng Struct Dyn 34:943–963
McGregor JG, Hage SE (1977) Stability analysis and design of concrete frames. J Struct Div ASCE 103:1953–1970
McGuire RK (1995) Probabilistic seismic hazard analysis and design earthquakes: closing the loop. Bull Seismol Soc Am 85:1275–1284
Melosh RJ, Utku S (1988) Verification tests for computer-aided structural analysis. microcomputers in civil engineering, vol 3, pp 289–297
Menun CA, Der Kiureghian A (1998) A replacement for the 30%, 40%, and SRSS rules for multicomponent seismic analysis. Earthq Spectra 14:153–163. doi:10.1193/1.1585993
Mills HD, Dyer M, Linger RC (1987) Cleanroom software engineering. IEEE Softw 4:19–24
Morfidis K, Avramidis IE (2002) Formulation of a generalized beam element on a two-parameter elastic foundation with semi-rigid connections and rigid offsets. Comput Struct 80:1919–1934
Paglietti A, Porcu MC, Pittaluga M (2011) A loophole in the Eurocode 8 allowing for non-conservative seismic design. Eng Struct 33:780–785
Paraskeva TS, Kappos AJ (2010) Further development of a multimodal pushover analysis procedure for seismic assessment of bridges. Earthq Eng Struct Dyn 39:211–222. doi:10.1002/eqe
Paraskeva TS, Kappos AJ, Sextos AG (2006) Extension of modal pushover analysis to seismic assessment of bridges. Earthq Eng Struct Dyn 35:1269–1293. doi:10.1002/eqe.582
Park S, Mosalam KM (2009) Shear strength models of exterior beam-column joints without transverse reinforcement. Pacific Earthquake Engineering Research Center report. University of California, Berkeley, 2009/106
Penelis GG, Kappos AJ (1997) Earthquake-resistant concrete structures. E & FN Spon, London
Penelis GG, Penelis GG (2014) Concrete buildings in seismic regions. CRC Press/Taylor & Francis Group, London
Pixley RA, Ridlon SA (1984) How to check out an engineering computer program. In: Proceedings of the 3rd conference computing in civil engineering. American Society of Civil Engineering (ASCE), San Diego, CA, pp 583–593
Priestley MJN, Calvi GM, Kowalsky MJ (2007) Displacement based seismic design. Building 23:1453–1460. doi:10.1016/S0141-0296(01)00048-7
Sextos AG, Nasiopoulos G, Katityzoglou K, Avramidis IE (2008) Benchmark numerical examples for dimensioning r/c structures according to the codes. In: Proceedings of the 3rd panhellenic conference earthquake engineering and engineering seismology, Technical Chamber of Greece & Hellenic Society for Earthquake Engineering, 5–7 November, Athens, Greece

Sheikh SA, Uzumeri SM (1982) Strength and ductility of tied concrete columns. J Struct Div ASCE 106:1079–1102
Smebby W, Der Kiureghian A, Smeby W (1985) Modal combination rules for multicomponent earthquake excitation. Earthq Eng Struct Dyn 13:1–12. doi:10.1002/eqe.4290130103
Stafford-Smith B, Girgis A (1984) Simple analogous frames for shear wall analysis. J Struct Eng ASCE 110:2655–2666
Stavely A (1999) Toward zero-defect programming. Addison-Wesley, Reading
Tsonos AG (2007) Cyclic load behaviour of reinforced concrete beam-column subassemblages of modern structures. ACI Struct J 194(4):468–478
Villaverde R (2009) Fundamental concepts of earthquake engineering. CRC Press, New York/Weiler
Wilson E, Suharwardy I, Habibulah A (1995) A clarification of the orthogonal effects in a three-dimensional seismic analysis. Earthq Spectra 11:659–666
Xenidis H, Athanatopoulou AM, Avramidis IE (1993) Modeling of shear wall cores under earthquake loading using equivalent frames, 21–23 June 1993, 01-910. EURODYN'93, 2nd European conference on structural dynamics, Trondheim, Norway, pp 901–910
Xenidis H, Morfidis K, Avramidis IE (2000) Modeling of two-cell cores for three-dimensional analysis of multi-story buildings. Struct Des Tall Build 9:343–363

Chapter 3
Practical Implementation of EC8 for Seismic Design of R/C Buildings – Flowcharts and Commentary

Abstract This chapter delineates all the required computational and logical steps involved in the analysis stage of seismic design of ordinary reinforced concrete (r/c) buildings according to Eurocode 8 (EC8) by means of detailed self-contained flowcharts and pertinent comments. Special focus is given to verification checks for structural regularity in plan and elevation, on the classification of building structures based on torsional sensitivity, on the determination of the maximum allowed behavior factor by EC8, and on the selection and implementation of the two equivalent linear methods of analysis considered by EC8, namely the lateral force method and the modal response spectrum method. Furthermore, additional flowcharts and comments are included for the implementation of EC8-prescribed post-analysis verification checks based on deformations, that is, verification check of second-order effects via the interstorey drift sensitivity coefficient and verification check for maximum interstorey drift to ensure that damage limitation requirements of EC8 are met. Practical recommendations expediting the implementation of EC8-compliant analysis of ordinary r/c buildings are provided. Finally, the required detailing and verification checks for the design of r/c structural members according to Eurocode 2 and Eurocode 8 are presented in the form of self-explanatory flowcharts, along with the special requirements for the determination of seismic design bending moment and shear force diagrams.

Keywords EC8 implementation flowcharts • Structural regularity • Torsional sensitivity • Behavior factor • Lateral force method • Modal response spectrum method • Design seismic effects • Second-order effects • Interstorey drift sensitivity coefficient • Interstorey drift • Infill walls • Seismic detailing or r/c structural members

In Chap. 2, the following three phases of the seismic design process for typical r/c building structures have been identified and discussed (Fig. 2.1 and Table 2.1)

- Phase A: Conceptual design and preliminary sizing of structural members;
- Phase B: Analysis and detailed design;
- Phase C: Design drawings.

I. Avramidis et al., *Eurocode-Compliant Seismic Analysis and Design of R/C Buildings*, Geotechnical, Geological and Earthquake Engineering 38,
DOI 10.1007/978-3-319-25270-4_3

At the core of the seismic design process lies Phase B, comprising three stages:

- Stage 1: Structural (finite element, FE) and seismic loads modeling;
- Stage 2: Structural analysis and deformation-based verification checks;
- Stage 3: Detailing of r/c members.

This chapter focuses on stages 2 and 3 of phase B, presenting all prescriptive computational steps and verification checks involved in the practical implementation of the current EC8 (CEN 2004a), along with additional explanatory remarks and comments. It is assumed that equivalent linear analysis methods are adopted for the analysis stage, as this is the current state of practice.

Prior to dwelling on the details of structural analysis and detailing according to EC8, it is deemed useful to highlight certain practical issues arising in Phase A of the seismic design of code-compliant r/c buildings. In Phase A, it is assumed that architectural plans of the building to be designed are available. Based on these drawings, the outer dimensions and geometry of the lateral load-resisting structural system in plan and elevation are determined. At this preliminary point of seismic design, the structural/design engineer needs to set a target as to the desired seismic performance level of the structure, *upon consultation with the building owner.* Detailed discussions on this matter have been included in various sections of Chap. 1. The main points are conveniently summarized in Fig. 3.1, with the aid of simplified (bilinear) force-deformation (or base shear – top storey deflection) diagrams (see also Figs. 1.5 and 1.8) corresponding to three different structures (A, B, and C) subject to four different levels of seismic action (see also Fig. 1.17). Specifically,

- for frequent earthquake events, all three structures A, B, and C behave linearly (points A_1, B_1, C_1- no structural damage occurs);
- for occasional earthquake events, the higher strength structures B and C behave linearly (points B_2 and C_2), while the lower strength structure A undergoes some plastic deformation (point A_2);
- for the rare (design) earthquake event, the higher strength structures B and C remain linear (points B_3 and C_3), while the lower strength structure A undergoes severe plastic deformation, without, however, collapsing, since it attains sufficient ductility capacity to withstand the design seismic loads (point A_3). This significant level of utilization of the ductility capacity of structure A entails extensive structural damage at critical cross-sections;
- for the very rare earthquake event, well beyond the nominal design earthquake, structure C, which attains *both* high strength and sufficient (reserved) ductility capacity, has a significantly higher survival (no-collapse) probability (point C_4) than structures A and B.

It is reminded that the decision on the desired seismic performance level significantly influences the preliminary sizing of structural members and, therefore, the initial finite element (FE) model to be developed in the subsequent Phase B, as shown in Fig. 3.2.

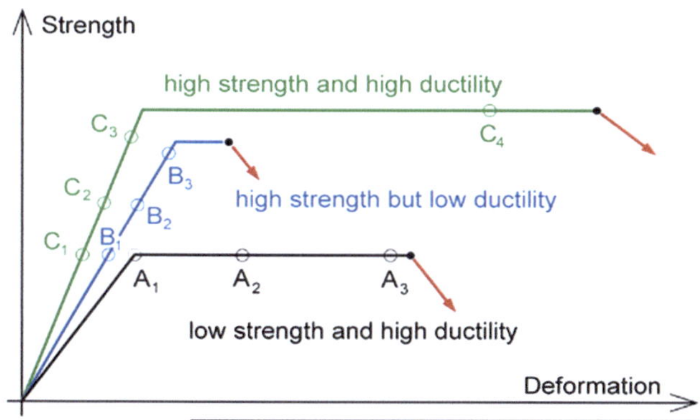

Structure	behavior factor $q=q_{dem}$ available ductility $\mu=\mu_{cap}$	Seismic response low intensity (frequent)	moderate intensity (occasional)	high intensity (rare) [design earthquake]	very high intensity (very rare) [> design earthquake]
A low strength, but ductile	q: large μ: large	**A1** no yielding (no damage)	**A2** minor yielding (minor damage)	**A3** major yielding (major damage)	--- Failure
			→Partial to full utilization of the ductility capacity to dissipate the design earthquake input energy		
B high strength, but low ductility	q: small μ: small	**B1** no yielding (no damage)	**B2** no yielding (no damage)	**B3** no yielding (no damage)	--- Failure
			→No utilization of the ductility capacity to dissipate the design earthquake input energy (The structure remains elastic)		
C high strength, and high ductility	q: small μ: large	**C1** no yielding (no damage)	**C2** no yielding (no damage)	**C3** no yielding (no damage)	**C4** minor to major damage
			→No utilization of the ductility capacity to dissipate the design earthquake input energy (The structure remains elastic)		**Partial to full** utilization of the ductility capacity

Fig. 3.1 Alternative choices of the desired seismic performance level (schematically)

In general, high seismic performance targets (e.g., corresponding roughly to a "immediate occupancy" performance under the design seismic action) require larger r/c member sizes compared to aiming at the basic design objectives (e.g., corresponding roughly to a "life safety" performance under the design seismic action). Nevertheless, it is noted that there are hardly any prescriptive procedures

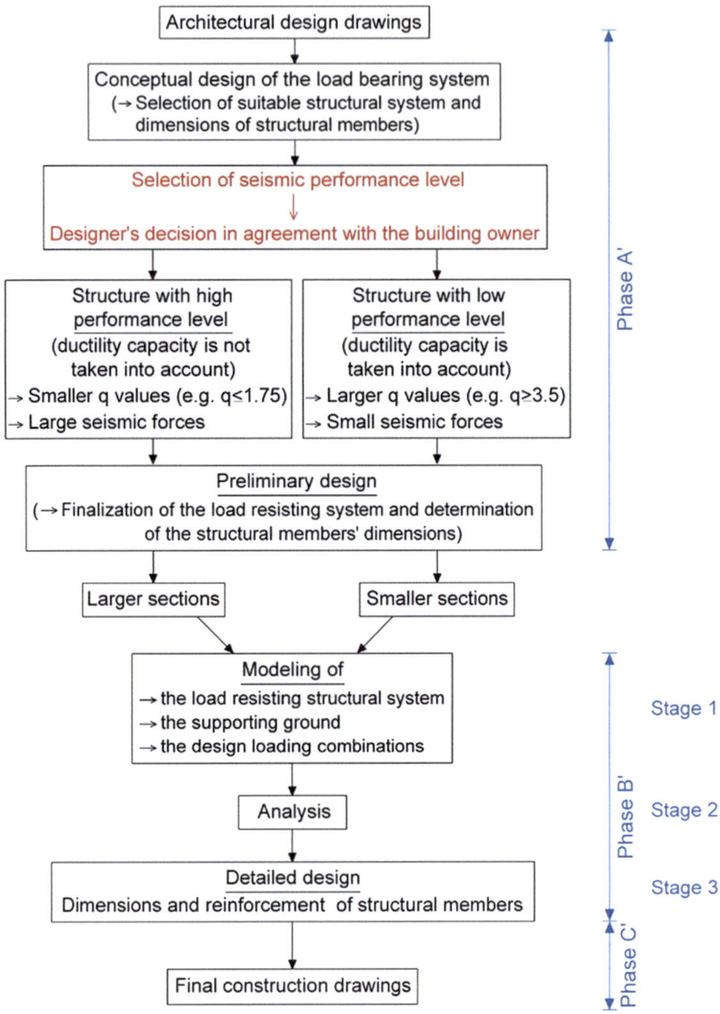

Fig. 3.2 Schematic sequence of design phases (see also Fig. 2.1) – influence of selected seismic performance level on preliminary design

and guidelines for preliminary r/c structural member sizing. In practice, this is accomplished by relying on the designer's accumulated experience with the seismic design of similar types/geometry of building structures given a specific level of seismic performance and/or by undertaking preliminary (rough and approximate) calculations (commonly of static rather than dynamic nature) using simple/crude FE models of the structure under design or of its critical substructures. Furthermore, sizing of r/c structural members is typically an iterative procedure. The adequacy of the preliminary sizes of structural members selected to carry the prescribed gravitational and seismic loads is verified via code-prescribed deformation-based and stress-based checks relying on analysis results. In case some of the selected cross-section sizes are under-designed (e.g., are not sufficient to carry loads or to fit the required reinforcing bars) or are over-designed (e.g., significantly below than the minimum required reinforcement), judicial re-sizing of members followed by analysis of the updated FE model needs to take place. In any case, the demand capacity ratio (DCR) defined in Sect. 2.4.1.1 is a valuable quantitative measure for assessing the adequacy and efficiency of the lateral load-resisting structural system to take the nominal design gravitational and seismic loads. The design engineer should aim to ensure a uniform -as far as possible- distribution of DCR values across the critical cross-sections of all structural members, which should certainly be below unity (see also Sect. 2.4.4.1).

In the context of EC8 compliant seismic design, the seismic performance level can only be indirectly set by means of a single scalar which leverages the amplitude of the design seismic forces: the behaviour factor q (see Sect. 1.4 for a detailed discussion). Specifically,

- the designer has to first adopt one of the three (i.e., Low, Medium or High) Ductility Classes (DCL, DCM, DCH) as described in Sect. 3.1.3. Depending on the structural system of the building and its regularity in plan and height, the upper ($maxq_{allow}$, see Sect. 3.1.4 below) and lower bounds ($q = 1.5$ or 1.0 depending on the interpretation of the relevant provision) of the behaviour (force reduction) factor q are defined.
- given the selected Ductility Class and the energy absorption potential provided by each Class through the corresponding design (geometric, reinforcement and detailing) rules, the designer needs to adopt a target performance level by selecting the desirable value of the behavior factor q within the aforementioned prescribed limits.
- in case a low seismic performance level is desired, i.e., a high portion of the ductility capacity is to be utilized to resist the design seismic action, a large value of the behaviour factor q is chosen *at the analysis stage* which, as already explained, cannot be larger than the maximum allowed value ($maxq_{allow}$)
- in case a high seismic performance level is desired, i.e., a small portion of the ductility capacity is to be utilized to resist the design seismic action, a lower than the $maxq_{allow}$ behaviour factor q may be adopted *at the analysis stage*, close or identical to the minimum permissible value of q (1.5 or 1.0).

It is worth noticing that, in the terminology introduced in Sect. 1.2, the q-value adopted in analysis is the "behaviour factor demand" q_{dem}. Further, as a ductility class is adopted at the beginning of the design process, a corresponding energy absorption potential (i.e., available ductility) is provided (even if $q = 1$ is used in analysis). This simply means -again in the terminology introduced in Sect. 1.2 – that a larger value of q is realized *at the design stage*, namely the "behaviour factor capacity" q_{cap} ($q_{dem} \leq q_{cap} \leq maxq_{allow}$).

Once preliminary sizes of structural members comprising the adopted (lateral) load resisting structural system are selected, the structural/design engineer proceeds with the structural (FE), the supporting ground, and the loading modeling. Although

EC8 includes only generic guidance in terms of structural and supporting ground modeling (Sect. 2.3.2), it specifies in detail the seismic design loading (Sect. 2.3.1). In any case, Sect. 2.3.3 provides certain practical recommendations for facilitating FE structural modeling for linear seismic analysis of multistorey r/c buildings. Upon development of the computational model of the structure (including structural, supporting ground, and loading modeling), the seismic design procedure proceeds with:

- the structural analysis of the computational model to determine internal stress resultants and deformations of the load resisting structural system for all EC8 loading combinations;
- the verification checks in terms of deformations and stresses; and
- the finalization of all structural member sizes and reinforcement (upon one or more sizing-modeling-analysis-verification checks-detailing iterative cycles).

In the following sections, all the EC8-prescribed computational and logical steps involved in undertaking the above three tasks are presented taking a sequential step-by-step approach. Specifically, Sect. 3.1 utilizes a series of self-contained flowcharts, supported by comments when deemed necessary, to illustrate appropriate and efficient ways to undertake EC8-compliant force-based equivalent linear types of analysis. Further flowcharts and comments related to post-analysis verification checks based on deformations and the influence of infill walls are included in Sects. 3.2 and 3.3. Section 3.4 summarizes practical recommendations for expediting the implementation of EC8-compliant analysis. Next, Sect. 3.5 delineates procedures for determining stress resultants for the seismic design of r/c walls. Finally, Sect. 3.6 walks through all required detailing and verification checks for r/c structural members in the form of self-explanatory flowcharts.

3.1 EC8-Compliant Seismic Analysis Steps and Flowcharts

In this section, all required computational procedures involved in the analysis stage of EC8-compliant seismic design of r/c buildings are presented by means of 12 self-contained flowcharts (logic diagrams). Each flowchart is assigned a unique number "x". Reference to flowcharts in the text and within flowcharts is made by the symbol "FC-x". Further, all references to particular clauses of EC8 part-1 within flowcharts are indicated by blue colored fonts. Despite being self-explanatory, certain additional remarks and comments are provided to clarify and facilitate the practical implementation of flowcharts as necessary.

The first flowchart (FC-3.1) provides an overview of all the steps involved in the seismic analysis stage (stage 2 of phase B in Figs. 2.1 and 3.1) of the FE structural model developed in the modeling stage (stage 1 of phase B in Figs. 2.1 and 3.1).

Almost every step indicated in FC-3.1 requires a series of computations and/or verification checks (either logical or numerical) to be undertaken as prescribed by the relevant EC8 clauses. These computations and verification checks are

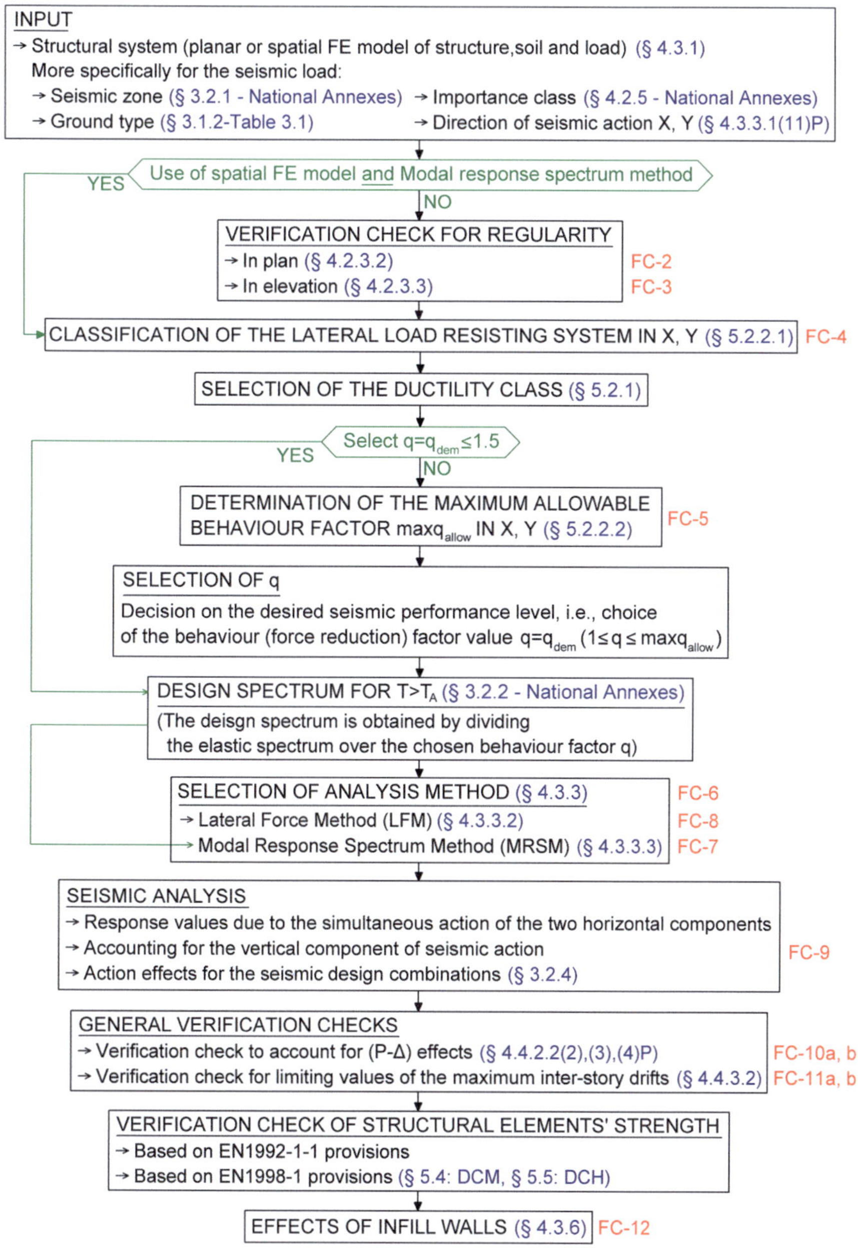

Flowchart 3.1 Global implementation flowchart

delineated in subsequent flowcharts included in this Chapter as indicated in FC-3.1. Therefore, FC-3.1 serves as a "table of contents" for all flowcharts to follow, taking a sequential step-by-step approach.

It is emphasized that it may not always be required to undertake all steps included in FC-3.1 for the seismic design of a given r/c building. Certain steps can be omitted depending on the choices made with regards to the adopted FE model properties and analysis method. Specifically, if

1. a spatial (three-dimensional) FE structural model is *a priori* adopted,
2. the modal response spectrum analysis is applied, and
3. a value ≤ 1.5 is *a priori* selected for the behavior factor $q = q_{dem}$ to be used in analysis,

then the regularity checks can be bypassed, as indicated by the alternative green colored paths shown in FC-3.1.

3.1.1 *Conditions and Verification Checks for Structural Regularity*

EC8 distinguishes between "regular" and "non-regular" building structures *separately* in plan and elevation according to certain structural regularity criteria. Regular buildings are taken to have an inherently favourable behaviour to strong ground motions due to their geometric, structural, and dynamic properties. For the same reason, their (inelastic) dynamic response to seismic excitations can be predicted with higher confidence. Consequently, favourable choices are allowed to be made in the modeling and analysis stages of regular structures with regards to (see § 4.2.3.1(4)):

- the adopted FE structural model (planar – two dimensional or spatial – three dimensional),
- the adopted equivalent linear analysis method (lateral force method or modal response spectrum method),
- the maximum allowable value of the behaviour factor.

Following Table 4.1 of EC8, Table 3.1 demonstrates the "penalization" of structural "irregularity" in plan and elevation separately in terms of the aforementioned modeling and analysis aspects.

It is noted that regularity in plan has implications only for the adopted FE structural model, while regularity in elevation influences the choice of the adopted equivalent linear analysis method and the level of the maximum allowable reduction in the design seismic loads. Nevertheless, it is emphasized that the consideration of planar FE models (typically one planar model along each principal axis) is not a *requirement* for regular in plan structures; a single spatial FE model can be adopted, leading to more reliable analysis results (see Sect. 2.3.2.2). Along similar

Table 3.1 Implications of structural regularity on modeling, analysis, and maximum allowable behaviour factor

Regularity				
in plan	in elevation	Allowed FE model	Allowed (linear) analysis method	Maximum allowed behaviour factor
Yes	Yes	Planar	LFM[a]	Reference value
Yes	No	Planar	MRSM[b]	Reduced by 20 % value
No	Yes	Spatial	LFM[a]	Reference value
No	No	Spatial	MRSM[b]	Reduced by 20 % value

[a]Lateral force method (under the additional condition in §4.3.3.2.1(2)a of EC8- see also FC-3.6)
[b]Modal response spectrum method

lines, adopting the lateral force method for regular buildings in elevation which conform to the additional condition of §4.3.3.2.1(2)a of EC8 is not a requirement. The modal response spectrum method (MRSM) can be used for the analysis of such structures, yielding more accurate results than the single mode lateral force method (LFM). Finally, irrespective of the prescribed 20 % reduction in the "basic" or reference value of the maximum allowable behaviour factor q_o imposed for non-regular in elevation buildings, it is highlighted that a reduced q_o value by 20 % or more can be adopted even for regular in elevation buildings in case a higher-than-the-basic-seismic-performance-design is sought (see also discussion in Sect. 1.4).

A series of structural regularity conditions in plan and elevation are prescribed in clauses §4.2.3.2 and §4.2.3.3 of EC8. Some of these conditions are *purely quantitative*, relying on numerical calculation and verification checks of certain geometric and dynamic properties of interest. Other conditions are *purely qualitative*, requiring the expert judgment of the designer. There are also "*hybrid conditions*" which employ certain numerically verifiable criteria to inform engineering judgment. Table 3.2 categorizes the different EC8 prescribed regularity conditions according to the above three categories. Additional comments on some of these conditions are included in the following two subsections discussing regularity in plan and elevation conditions separately.

3.1.1.1 Verification Checks for Regularity in Plan: FC-3.2

A recommended sequence of logic and computational steps to check for regularity in plan of a particular r/c building is provided in FC-3.2. Arguably, the most challenging condition to be verified in terms of required calculations relates to the issue of "*torsional sensitivity*" (*Note*: For reasons explained in detail in Appendix B, the term "torsional sensitivity" is preferred over the quite misleading term "torsional flexibility" adopted in EC8) according to clause § 4.2.3.2(6) of EC8. The latter states that, in a regular in plan building structure, the structural eccentricity e_o and the torsional radius r shall satisfy the following conditions at each floor level and along each direction of analysis (principal axes) X and Y

Table 3.2 Qualitative and quantitative regularity conditions in plan and in elevation

	Regularity in plan	Regularity in elevation
Purely quantitative regularity conditions	In-plan slenderness: $\lambda = L_{max}/L_{min} \leq 4$ [§ 4.2.3.2(5)]	Setbacks along building height criteria [§ 4.2.3.3(5)]
	Static eccentricities along two principal orthogonal directions: $e_o \leq 0.30r$ [§ 4.2.3.2(6)]	
	Torsional sensitivity: $r \geq l_s$ [§ 4.2.3.2(6)]	
Purely qualitative regularity conditions	"Almost" symmetrical plan distribution of lateral stiffness and mass [§ 4.2.3.2(2)]	–
	Adequacy of in-plan floor stiffness and diaphragmatic behaviour for "L"-,"Π"-,"H"-,"I"-, and "X"-shaped floor plans [§ 4.2.3.2(4)]	
Hybrid regularity conditions	Compactness of plan configuration [§ 4.2.3.2(3)]	"Smooth" variation of floor stiffness and mass in elevation [§ 4.2.3.3(3)]
		Uninterrupted vertical structural members of the lateral load resisting system [§ 4.2.3.3(2)]
	Infill walls plan distribution [§ 4.3.6.3.1]	"Smooth" variation of resistance along height of frame buildings [§4.2.3.3(4)]
		Infill walls distribution in elevation [§4.3.6.3.2]

$$e_{oX} \leq 0.30 r_X \quad \text{and} \quad e_{oY} \leq 0.30 r_Y \,, \tag{3.1}$$

$$r_X \geq l_s \quad \text{and} \quad r_Y \geq l_s \,. \tag{3.2}$$

where

- e_{oX} and e_{oY} are the distances between the center of stiffness (or shear center) and the center of mass, measured along the X and Y directions, respectively, normal to the direction of analysis considered;
- r_X and r_Y are the torsional radii with respect to the center of stiffness given by the square root of the ratio of the torsional stiffness to the lateral stiffness in the Y and X directions, respectively; and
- l_s is the radius of gyration of the floor mass in-plan given by the square root of the ratio of the polar moment of inertia of the floor mass in-plan with respect to the center of mass of the floor over the floor mass (see also Appendix B).

It is noted that the center of stiffness (shear center) can be uniquely defined at the floor levels of only single-storey buildings or of multistorey buildings with vertical structural members observing consistent lateral deformations along the direction of analysis. This is the case for multistorey buildings with frame or wall lateral load-resisting structural systems (defined in Sect. 3.1.2) along the principal axes in which

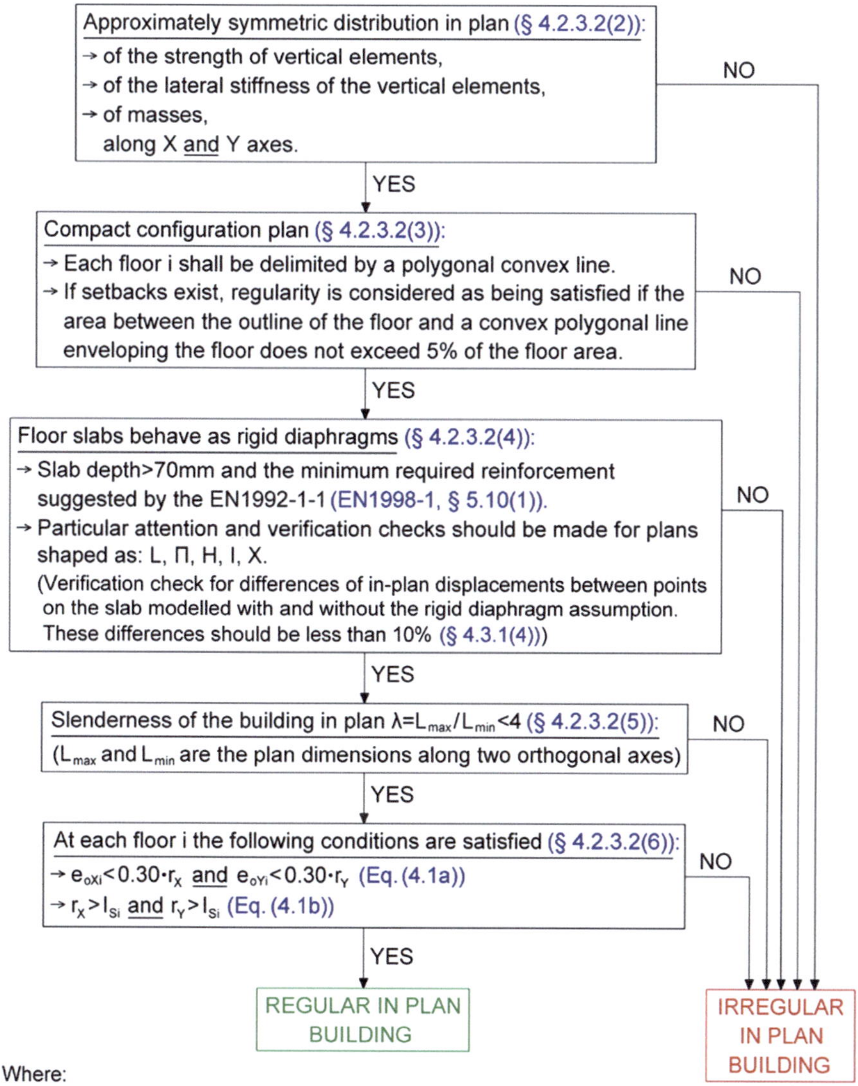

Where:

e_{oXi}, e_{oYi} : the distance between the centre of mass and the centre of stiffness at floor level i along X and Y axis respectively.

r_X, r_Y: torsional radius.

$l_{Si}=\sqrt{J_{mi}/m_i}$: radius of gyration of the floor i.

J_{mi} : mass moment of inertia of the floor i.

m_i : mass of the floor i.

Flowchart 3.2 Evaluation of regularity in plan

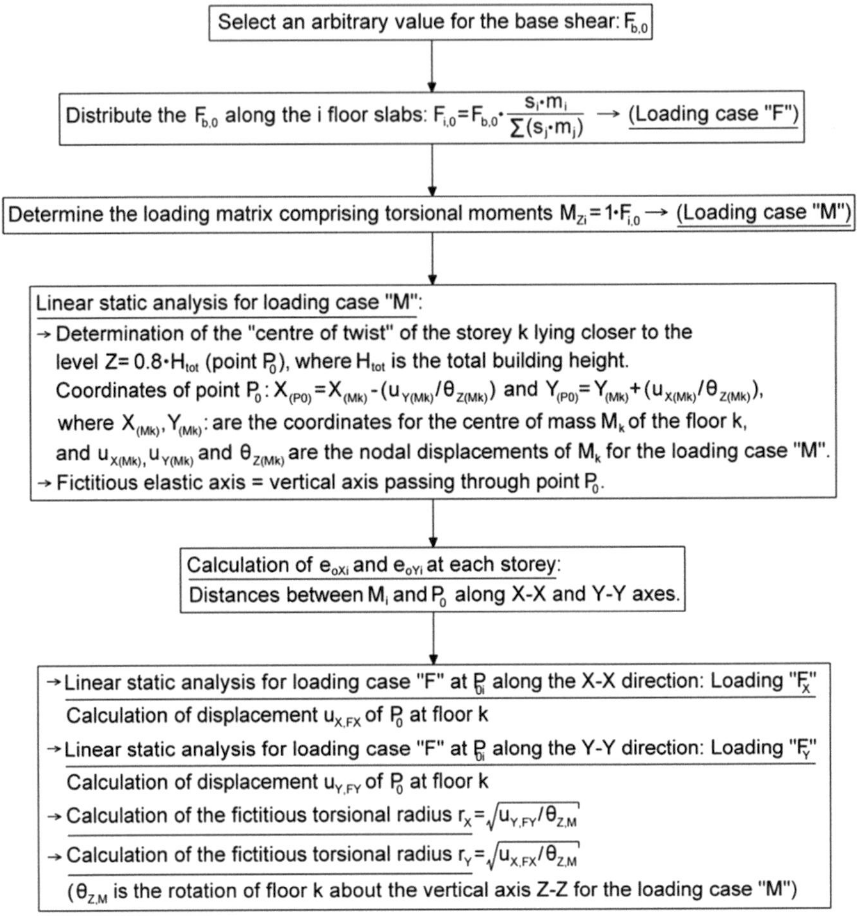

Flowchart 3.2a Determination of fictitious elastic axis and torsional radii

the vertical structural members run without interruption from the foundations to the top of the structure (see also clause §4.2.3.2(8) of EC8). Consequently, structural eccentricities (e_{oX} and e_{oY}) and torsional radii (r_X and r_Y) can be unambiguously defined only for one-storey buildings and for pure frame or wall (bending-type) multistorey buildings (Kan and Chopra 1977; Rosman 1984). In the general case of multistorey r/c buildings with lateral load-resisting structural systems comprising both walls and frames (dual systems), the above quantities can only be approximately defined, except those belonging to a special class called "isotropic buildings" (Athanatopoulou et al. 2006; Hejal and Chopra 1989).

In FC-3.2a, a rational approach is outlined for determining structural eccentricities and torsional radii for any type of multistorey building relying on the concept of the "fictitious elastic vertical axis" whose traces at each floor level observe, approximately, the property of the shear center as defined in the Greek National

Annex to EC8 (Hellenic Organization for Standarization 2009) and elsewhere (Makarios and Anastassiadis 1998). The thus defined structural eccentricities and torsional radii can be used to check for regularity in plan conditions of Eqs. (3.1) and (3.2).

As a final note, it is reminded that – as reported at the beginning of Sect. 3.1 – by adopting a spatial (three-dimensional) FE structural model and applying the modal response spectrum method of analysis, the need to consider the regularity (in plan and elevation) verification checks can be circumvented.

3.1.1.2 Verification Checks for Regularity in Elevation: FC-3.3

A recommended sequence of logic and computational steps to check for regularity in elevation of a particular r/c building is provided in FC-3.3. As shown in Table 3.1, regularity in elevation influences the method of equivalent linear analysis to be considered (LFM or MRSM) and the maximum allowable value of the behaviour factor which can be used in the analysis. In this regard, it is noted that undertaking the largely qualitative and empirically-based verification checks for regularity in elevation can be omitted by adopting the MRSM analysis in conjunction with a behaviour factor q_o reduced by at least 20 % of the maximum allowable basic value (see also Sect. 3.1.4).

3.1.2 Classification of a Lateral Load-Resisting Structural System

According to clause §5.2.2.1(1)P of EC8, r/c building structures are classified as one of the following structural system types for lateral load-resistance:

(i) Frame systems;
(ii) Ductile wall systems,

  (a) with coupled walls, or
  (b) with uncoupled walls;

(iii) Dual (coupled frame-wall) systems,

  (a) being frame-equivalent, or
  (b) being wall-equivalent;

(iv) Systems with large lightly reinforced walls;
(v) Inverted pendulum systems;
(vi) Torsionally sensitive[1] systems.

---

[1] See Note in the first paragraph of Sect. 3.1.1.1.

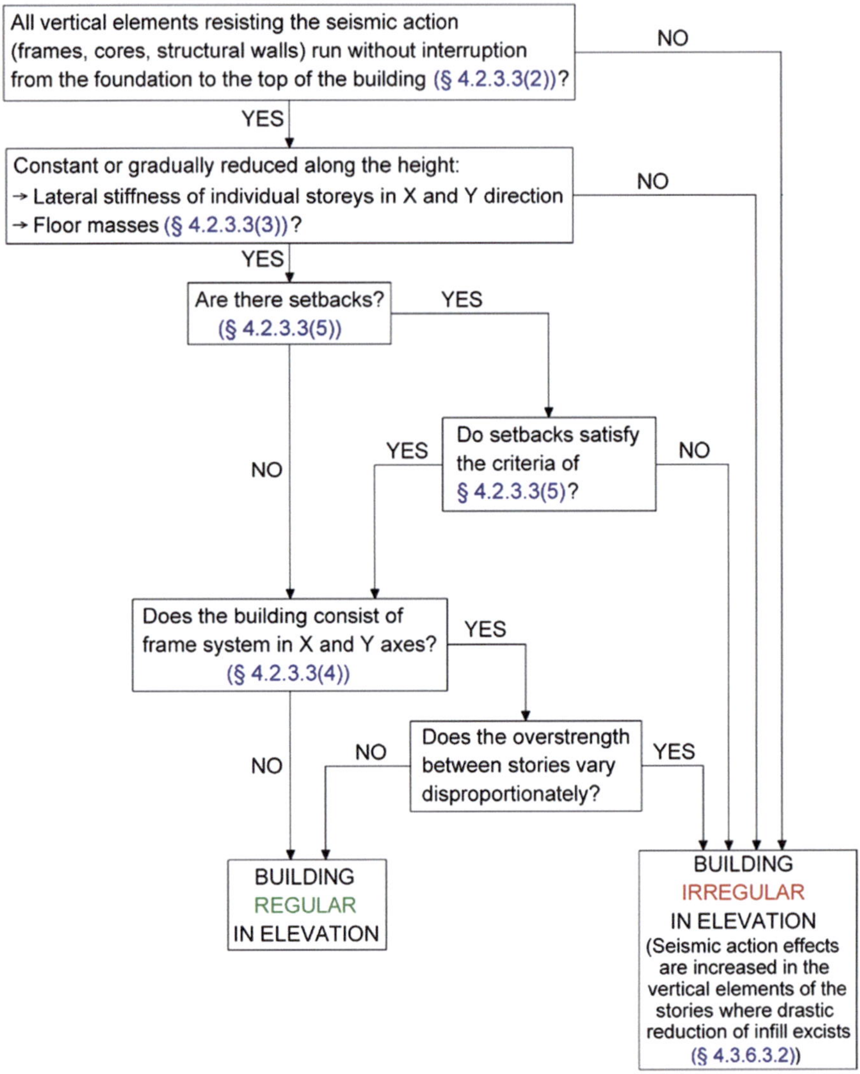

Flowchart 3.3 Evaluation of regularity in elevation

The above classification significantly influences the maximum allowable value of the behaviour factor that can be adopted within the context of an equivalent linear analysis (i.e., maximum allowed level of reduction of design seismic forces – see also Sect. 3.1.4). This implication accounts for the fact that the level of ductility capacity which can be potentially achieved by following capacity design rules and by proper local detailing of structural members varies significantly among different structural systems. Moreover, additional special provisions on verification checks and detailing apply to some of the structural systems.

Table 3.3 Definitions of structural system types according to EC8

Structural system type	Fraction of total base shear taken by spatial frames	Fraction of total base shear taken by walls	External load resistance	Torsional sensitive structures	Inverted pendulum structures
(i) Frame	>65 %		Gravitational and lateral loads taken primarily by spatial frames	All types of building structures for which $r < l_s$ along any horizontal (principal) axis.	All types of multistorey building structures with 50 % of their total mass located in the upper third of the building height.
(ii) Ductile wall		>65 %	Gravitational and lateral loads taken primarily by walls		
(iiia) Dual frame-equivalent	≤65 % and >50 %		Gravitational loads taken primarily by spatial frames		
(iiib) Dual wall-equivalent		≤65 % and >50 %			
(iv) Large lightly reinforced walls system	(a) comprises at least two walls of horizontal length $L \geq \min\{4.0 \text{ m}, 2\,h_w/3\}$, where h_w is the total wall height, which resist at least 20 % of the total gravity load of the design seismic combination				
	(b) has fundamental natural period $T_1 \leq 0.5$ s.				

The above listed types of structural systems are defined in clauses §5.1.2 and §5.2.2.1 of EC8. The key properties for each type are briefly described in the following paragraphs and are collected in Table 3.3. It is important to note that, with the exception of torsionally sensitive systems, any single r/c building may be classified differently along different horizontal directions (practically along the two principal axes), as specified in clause § 5.2.2.1(2) of EC8.

(i) **Frame system [Clause §5.1.2(1) of EC8]**

Structural system in which gravitational and horizontally applied loads are resisted primarily by spatial moment resisting frames whose shear resistance at their base should exceed 65 % of the total shear resistance at the base of the building.

(ii) **Ductile wall system [Clause §5.1.2(1) of EC8]**

Structural system in which gravitational and horizontally applied loads are resisted primarily by vertical ductile structural walls, either coupled or uncoupled, whose shear resistance at their base should exceed 65 % of the total shear resistance at the base of the building. It is noted that a structural wall is designated as *ductile* if it is rigidly fixed at its base and is designed and detailed to dissipate energy in a flexural plastic hinge zone free from openings and large perforations developed just above its base. Further, a coupled wall system comprises two or more single walls connected via adequately ductile "coupling beams" in a regular pattern such that the sum of the bending moments developing at the base of the system is at least 25 % smaller than the sum of the bending moments at the base of all walls if they were not coupled together.

(iiia) **Frame-equivalent dual system [Clause §5.1.2(1) of EC8]**

Structural system comprising spatial moment resisting frames and walls in which the gravitational loads are primarily resisted by the frames, and the shear resistance at the base of the frames should exceed 50 % of the total shear resistance at the base of the building.

(iiib) **Wall-equivalent dual system [Clause §5.1.2(1) of EC8]**

Structural system comprising spatial moment resisting frames and walls in which the gravitational loads are primarily resisted by the frames, and the shear resistance at the base of the walls should exceed 50 % of the total shear resistance at the base of the building.

(iv) **System with large lightly reinforced walls [Clause §5.2.2.1(3) of EC8]**

Wall structural system comprising at least two walls of horizontal dimension greater or equal to 4.0 m or to 2 $h_w/3$, where h_w is the total wall height, along the horizontal direction of interest, which: (a) collectively resist at least 20 % of the total gravity load of the design seismic combination, and (b) have a fundamental natural period less or equal to 0.5 s.

(v) **Inverted pendulum system [Clause §5.1.2(1) of EC8]**

Structural system in which 50 % or more of its mass is located in the upper third of the height of the structure, or in which seismic energy dissipation takes place mainly at the base of a single vertical structural member. A further special note is made to exempt from this class single storey buildings with columns carrying axial load normalized to the cross-sectional area of 30 % or less.

(vi) **Torsionally sensitive system [Clause §5.2.2.1(6) of EC8]**

Frame, wall, or dual structural system which does not satisfy the condition "torsional radius $\geq$ radius of gyration" of Eq. (3.2) [that is, Eq. (4.1b) in clause §4.2.3.2(6) of EC8] along at least one (any) horizontal direction. It is reminded that, throughout this text, the term "torsionally sensitive" is used in place of the term "torsionally flexible" adopted by EC8 for structures not satisfying the condition "torsional radius $\geq$ radius of gyration". This is because the above condition is not uniquely related to torsional flexibility (perceived as the reciprocal of torsional stiffness), as discussed in detail in Appendix B.

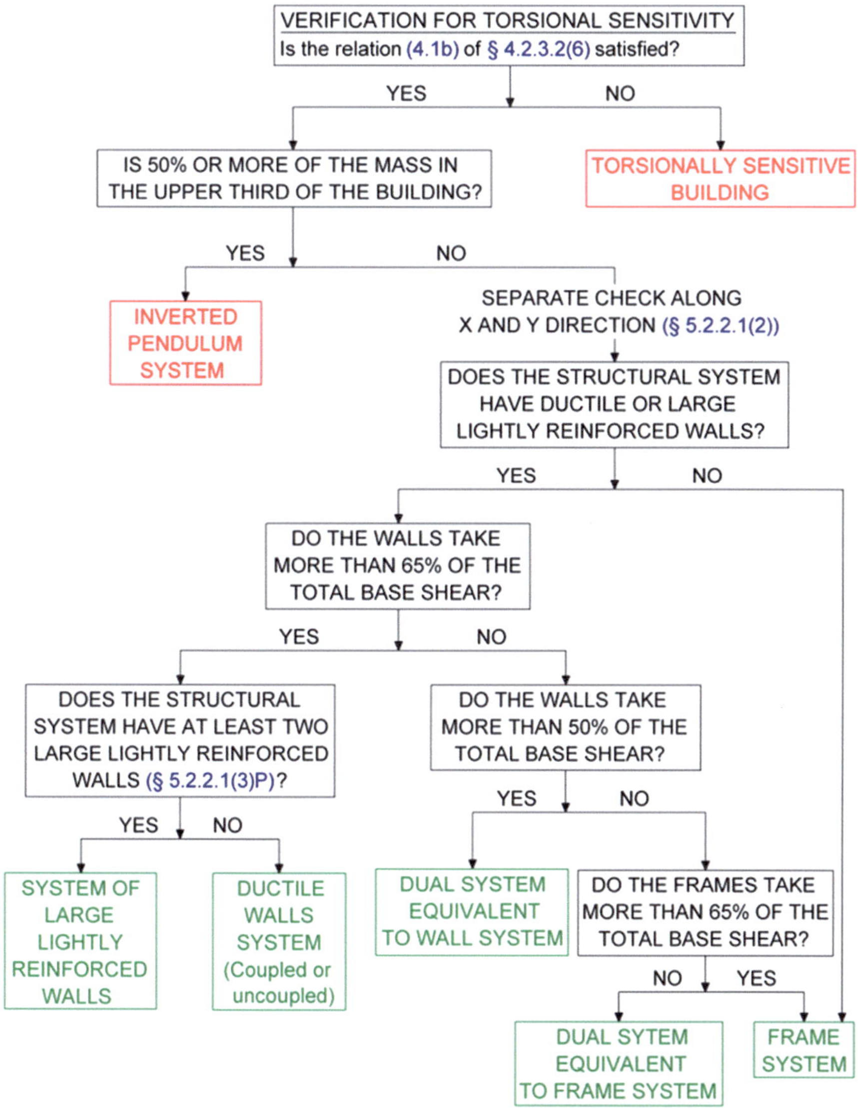

Flowchart 3.4 Classification of the structural system to a structural type

Based on the above definitions, any given r/c building structure can be classified into one (or two along the two principal axes X and Y) structural type following the sequence of logical steps and calculations shown in FC-3.4. Arguably, the most challenging steps of the above procedure are the:

– verification check of the condition "torsional radius ≥ radius of gyration" of Eq. (3.2) (§4.2.3.2(6) of EC8), and

– calculation of the fraction of the total base shear resisted by the various vertical structural elements (i.e., frames/columns and walls/cores) along the two principal axes X and Y.

The first step requires the calculation of the torsional radius and, thus, of the floor shear (resisting) center (see also Appendix B). This can be accomplished in the case of multistorey dual building structures using the procedure outlined in FC-3.2a. The second step requires linear analysis of a (preliminary) finite element model of the building structure for lateral (seismic) loads along the two principal axes X and Y. No particular reference is made within EC8 as to what type of analysis to undertake for the purpose (e.g., static analysis along the lines of the lateral force method or dynamic analysis along the lines of the modal response spectrum method), and as to whether two independent analyses need to be undertaken considering the seismic action acting along the X and Y axes separately. Moreover, in case a static analysis is used, the in-plan position of the horizontal seismic forces is not explicitly prescribed, although it significantly affects the base shear fraction resisted by frames and walls (see numerical Example A). To this end, it is recommended to undertake two independent linear static analyses along the X and Y principal axes involving lateral loads distributed along the height of the building according to the lateral force method. An arbitrary value of the total base shear can be assumed, as it is only sought to determine the distribution of the base shear among the various vertical structural elements (relative/fractional values) and not its value in absolute terms. The lateral loads may be applied at the center of mass of each floor. For each individual analysis along X and Y axes, the respective shear forces V_X and V_Y at the base of each vertical structural member are first determined and normalized by the value of the total base shear assumed in the analysis. Normalized base shears corresponding to columns are summed together and verification checks against the percentages included in FC-3.4 and in Table 3.3 are performed to classify the building into structural system types along each principal axis separately, as per EC8.

3.1.3 Selection of Ductility (Capacity) Class

According to clause §5.2.1 of EC8, r/c buildings can be designed

– either for low energy dissipation capacity: *Ductility Class Low* (DCL),
– or for adequate capacity to dissipate energy without substantial reduction in their overall ability to carry horizontal and vertical loads.

In the latter case, r/c buildings are designed and detailed to behave in a ductile manner (i.e., to be able to undergo large inelastic deformations without losing a substantial part of their initial strength and stiffness under cyclic/seismic loading). Depending on the desired level of ductility capacity, two different classes of ductile r/c structures are defined: *Ductility Class Medium* (DCM) and *Ductility Class High* (DCH).

Ductility Class Low (DCL): non-ductile r/c structures

In this case, r/c buildings are designed according to Eurocode 2 (CEN 2004b) provisions for r/c structures without any additional requirement for seismic design/ detailing except for the use of reinforcing steel of class B or C as defined in Table C.1 of EN1992-1-1:2004 (clause §5.3.2 of EC8). Design of DCL r/c buildings are recommended only in geographic regions of low seismicity (clauses §5.3.1 and §3.2.1(4) of EC8), and the maximum allowable behaviour factor q is 1.5 ($maxq_{allow} \leq 1.5$) for all structural systems.

Ductility Classes Medium and High (DCM and DCH): ductile r/c structures

All rules and concepts discussed in the first two chapters of this book for achieving ductile r/c structures apply in the case of EC8-compliant DCM and DCH buildings. Specifically, adequate (either "medium" or "high") ductility capacity, and, thus, capacity for seismic energy dissipation through inelastic behaviour, is aimed for by ensuring that local ductile failure modes (dominantly flexural) precede brittle failure modes (dominantly shear) with sufficient reliability and that ductility demands are uniformly distributed in plan and elevation across designated "critical" zones of structural members detailed for enhanced ductility capacity.

The design of DCM r/c building structures involves satisfying the requirements and provisions included in clause §5.4 of EC8, while for DCH r/c buildings, the additional (more stringent) requirements included in clause §5.5 of EC8 must be satisfied. The different levels of ductility capacity achieved by DCM and DCH structures reflect the different maximum allowable behaviour factors q prescribed for each class in clause §5.2.2.2 of EC8, as discussed in the following section.

Clearly, choosing to design between medium and high ductility classes is related to the desired level of ductility capacity that the designer seeks to achieve. Further, the higher the ductility capacity, the larger the *potential* plastic (inelastic) structural deformations which *may* be allowed and, thus, the higher the value of the maximum allowable behaviour factor $maxq_{allow}$ which *may* be adopted in seismic design. Therefore, the fact that higher $maxq_{allow}$ values are prescribed for DCH structures compared to DCM structures by EC8 is readily justifiable. However, it is reminded that whether and to what extent the available ductility (capacity) is utilized to actually resist the nominal design seismic action is a decision to be made by the designer in consultation with the building owner and is not necessarily related to the level of available ductility (see also Sect. 1.4). Rather, this decision relates to the desired seismic performance level to be targeted and depends on the selection of the behaviour factor q at the analysis step, that is, on the reduction of the seismic forces that the structure will be designed for. Design for high seismic performance buildings and, thus, for small probability of structural damage to occur under the nominal design earthquake *requires* adopting a relatively small behaviour factor (e.g., $q \leq 1.75$). This holds *irrespective* of the level of available ductility (capacity) of the structure which might allow for a higher behaviour factor to be adopted. For example, opting for a DCH structure requires that the most stringent requirements capacity design and local detailing requirements are adopted to maximize ductility capacity for a given structural system, no matter what would be the behaviour factor used in the analysis step. In case this high level of ductility capacity is not fully utilized to resist the design earthquake (by adopting a behaviour factor smaller than the $maxq_{allow}$ value), it remains as a reserve to ensure higher structural safety against collapse in case of an earthquake imposing higher demands than the design earthquake.

In general, the ductility class influences (among other important factors) the maximum allowed behaviour factor q ($maxq_{allow}$) which *may* be adopted in seismic design to reduce the design seismic loads. The lower the behaviour factor q actually adopted in analysis compared to the $maxq_{allow}$ (i.e., the less utilization of the ductility capacity to resist the design seismic action is made), the higher the seismic performance of the structure will be *for the given ductility class chosen.*

3.1.4 Determination of the Maximum Allowed Behaviour Factor

The behaviour (force reduction) factor $q \geq 1$ is the parameter by which the seismic action corresponding to linear structural behaviour is divided (and, thus, is reduced to allow for inelastic behaviour) in the context of EC8-compliant equivalent linear analyses (see also Sects. 1.2.4 and 2.3.1.2). This is achieved by dividing/reducing the ordinates of the EC8 elastic response spectrum (note, however, that the first branch of the EC8 design spectrum is not directly divided by q). In this manner, the adopted value of the behaviour factor q controls the extent of plastic deformation that a structure is allowed to exhibit under the design seismic action. Therefore, the value of $q = q_{dem}$ represents the *demand* of utilization of the ductility capacity (and overstrength) attained by the structure. Full utilization of the ductility capacity (and overstrength) corresponds to a limiting value of the behaviour factor equal to q_{cap} for which EC8 prescribes a maximum allowable value ($maxq_{allow}$) based on certain criteria. In this regard, the following relations hold (see also Sect. 1.2.2):

$$\min\ q = 1 \leq q = q_{dem} \leq \max\ q_{allow}(EC8) \leq q_{cap} = \max\ q. \tag{3.3}$$

According to clause §5.2.2.2(1)P of EC8, the upper limit or maximum allowable value $maxq_{allow}$ (for which the symbol q is used in EC8) of the behaviour factor which can be used to reduce the design seismic action along each direction of analysis (principal axes) is given as (Eq. (5.1) of EC8)

$$maxq_{allow} = q_o k_w \geq 1.5, \tag{3.4}$$

Table 3.4 Basic value of the behaviour factor q_o for buildings regular in elevation (EC8, Table 5.1)

Structural system type	Medium ductility class (DCM)	High ductility class (DCH)
Frame	$3.0\ \alpha_u/\alpha_1$	$4.5\ \alpha_u/\alpha_1$
Dual frame-equivalent		
Dual wall-equivalent		
Coupled wall		
Uncoupled wall	3.0	$4.0\ \alpha_u/\alpha_1$
Torsionally sensitive	2.0	3.0
Inverted pendulum	1.5	2.0

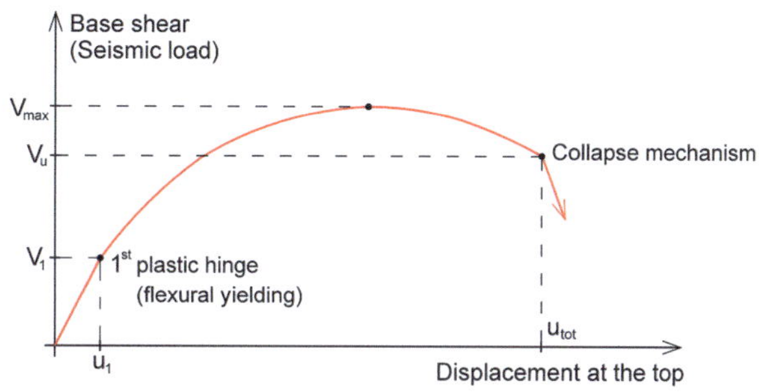

Fig. 3.3 Seismic loads corresponding to coefficients α_1 and α_u

where

- q_o is the "basic" behaviour factor value (see Table 3.4 below) which depends on
  - the adopted ductility class (see Sect. 3.1.3);
  - the lateral load-resisting structural system type (see Sect. 3.1.2);
  - the regularity in elevation (see Sect. 3.1.1.2);
  - the ratio (a_u/a_1) representing the "distance" between the first plastic hinge formation and structural collapse, as can be readily visualized in terms of a pushover curve (see Fig. 3.3).
- k_w is a reduction factor ($0.5 < k_w < 1$) reflecting the prevailing failure mode for lateral load-resisting systems which include walls (see Table 3.6 below).

A recommended sequence of the steps required to determine the maximum allowable value of the behaviour factor ($maxq_{allow}$) is provided in FC-3.5. Further notes on some of these steps follow.

Basic (reference) value q_0 of the maximum allowable behaviour factor

Included herein as Table 3.4, for convenience, is the Table 5.1 of EC8 (§5.2.2.2 (2)), which provides the basic value q_o of the maximum allowable behaviour factor for regular in elevation building structures. In the case of non-regular in elevation

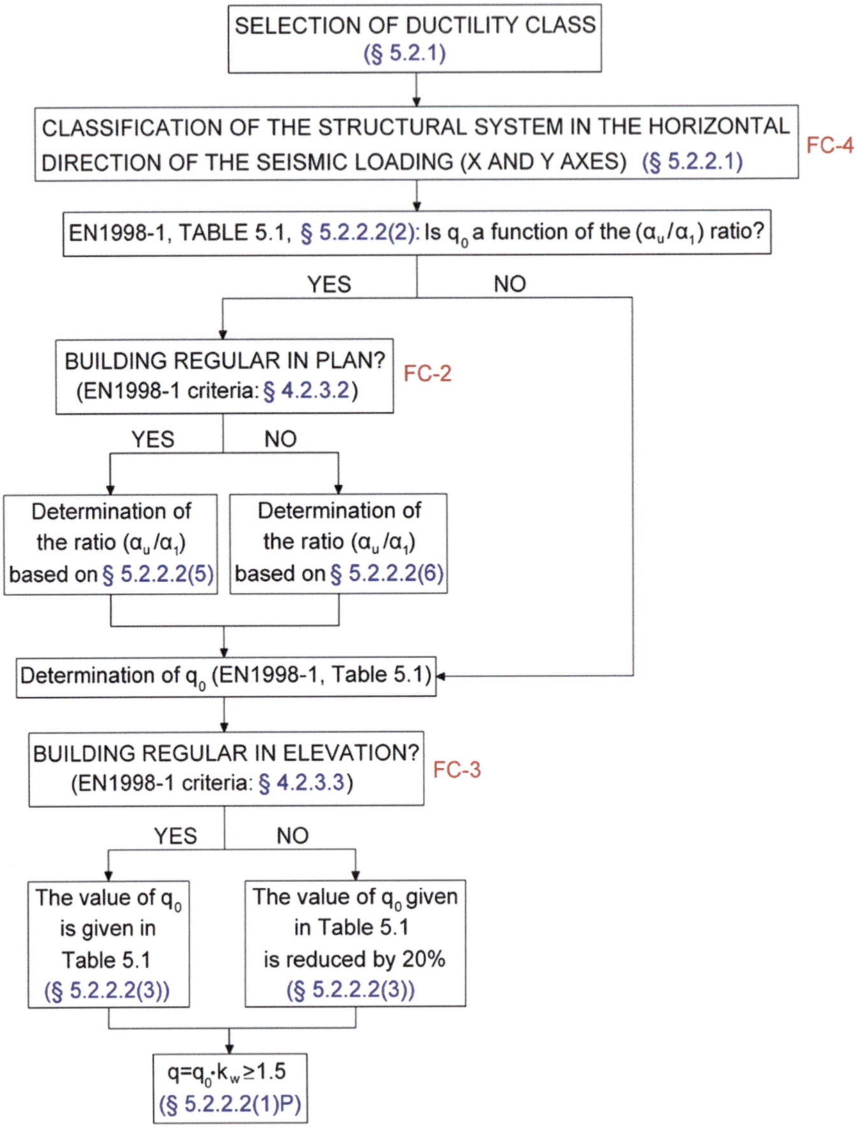

Flowchart 3.5 Determination of the maximum allowable value of the behaviour factor q

buildings, the values of q_o should be reduced by 20 % (clauses §4.2.3.1(7) and §5.2.2.2(3) of EC8).

Ratio (a_u/a_1) multiplying the basic value q_o

The quantities a_u and a_1 appearing in Table 3.4 are defined in clause §5.2.2.2 (4) of EC8 as follows:

Table 3.5 Approximate values of the (α_u/α_1) ratio for regular in plan buildings (clause §5.2.2.2 (5) of EC8)

Structural system type	α_u/α_1
Frame or dual frame-equivalent systems	
Single storey buildings	1.1
Multistorey, one-bay frames	1.2
Multistorey, multi-bay frames or dual frame-equivalent systems	1.3
Wall or dual wall-equivalent	
Wall systems with only two uncoupled walls per horizontal direction	1.0
Other uncoupled wall systems	1.1
Dual wall-equivalent, or coupled wall systems	1.2

Table 3.6 Reduction factor k_w (clause §5.2.2.11(P) of EC8)

Structural system type	k_w
Frame systems	1.0
Dual frame-equivalent systems	
Wall systems	$0.5 < (1 + \alpha_o{}^a)/3 < 1$
Dual wall-equivalent systems	
Torsionally sensitive systems	

[a] α_o is the prevailing aspect ratio (height/length) of the walls within the structural system

- α_1 is the multiplication factor applied to the horizontal seismic design action such that the first plastic hinge forms (cross-section failure under flexure) at any structural member in the structure, while all other design actions (e.g., gravitational loads) remain constant.
- α_u is the multiplication factor applied to the horizontal seismic design action such that a sufficient number of plastic hinges forms at various structural members in the structure for the development of a collapse mechanism (global structural instability), while all other design actions (e.g., gravitational loads) remain constant.

The ratio (a_u/a_1) can be determined by means of a non-linear static (pushover) analysis (pushover curve), as depicted in Fig. 3.2 (see also Sect. 2.4.3.4 and Appendix A), though its value cannot exceed 1.5 (clause §5.2.2.2(8) of EC8). Therefore, it holds that $1 \leq a_u/a_1 < 1.5$. Alternatively, the empirical values included in Table 3.5 can be used in the case of regular in plan buildings (clause §5.2.2.2 (5) of EC8), while for non-regular in plan buildings, an average (a_u/a_1) value between 1.0 and the value given in Table 3.5 can be used (clause §5.2.2.2(6) of EC8).

Reduction factor k_w for structural systems with walls

The value of the k_w factor is given in Table 3.6 (clause §5.2.2.11(P) of EC8) and accounts for the expected dominant failure mode in systems with walls.

The k_w factor recognizes that slender walls with high aspect ratios h_w/l_w, where h_w is the height of the wall and l_w is the length of the cross-section of the wall along the direction of the seismic action, are expected to fail predominantly in flexure and to have a relatively low shear stress influence. Consequently, they are expected to possess enhanced ductility capacity. If the aspect ratios of all walls of a structural system do not significantly differ, the prevailing aspect ratio α_o may be determined from the following expression (clause §5.2.2.2(12) of EC8):

$$a_0 = \sum_i h_{wi}/l_{wi}, \tag{3.5}$$

where h_{wi} and l_{wi} are the height and the cross-sectional length of wall *i* along the considered direction of the seismic action.

In the remainder of this section, additional comments and discussion on certain aspects involved in the determination of the maximum allowable behaviour factor value are provided.

3.1.4.1 Minimum Value of $\max q_{allow}$ (Maximum Allowable Behaviour Factor)

According to Eq. (5.1) of EC8 (see Eq. (3.4) above), the upper limit value of the behaviour factor q (i.e., the maximum allowable behaviour factor $\max q_{allow}$) cannot be smaller than 1.5. This lower bound limiting value of $\max q_{allow}$ obviously relates to the generally expected value of *overstrength* f that r/c structures possess (see Sects. 1.2.2 and 1.2.3). Indeed, it is reminded that the behaviour factor q (i.e., q_{cap} in the terminology introduced in Sect. 1.2) is the product $\mu_{cap} \cdot f$ of the ductility capacity μ_{cap} and the overstrength f (Eq. (1.13) in Chap. 1). Therefore, the lower bound limiting value 1.5 of $\max q_{allow}$ can be interpreted as being equal to the value of the overstrength f in the theoretical case of a perfectly brittle structure having ductility capacity $\mu_{cap} = 1$. In this theoretical case, using the value 1.5 at the analysis step to reduce the seismic loads (i.e., setting $q_{dem} = 1.5$ or $q_{dem} = q_{cap}$ in the terminology introduced in Sect. 1.2) would mean that all overstrength resources will be utilized to resist the design seismic action. Of course, the value of the "behaviour factor demand" q_{dem} (force reduction factor) is chosen equal to the value of the "behaviour factor capacity" q_{cap} only in the case of full utilization of the structure's capacity to resist the design earthquake. Otherwise, a smaller value (down to 1) may be chosen for q_{dem} to achieve higher seismic performance, if so desired.

3.1.4.2 Range of Values of the Maximum Allowable Behaviour Factor

Based on Tables 3.4, 3.5 and 3.6, the following values of the maximum allowable behaviour factor $\max q_{allow}$ can be readily computed (see Table 3 in (Fardis 2006)):

- for inverted pendulum systems: $1.5 \div 2.0$
- for torsionally sensitive systems: $1.6 \div 3.0$
- for all other systems: $2.4 \div 5.85$

where the low values correspond to ductility class medium (DCM) and the high values correspond to ductility class high (DCH).

3.1.4.3 On the Reduced Maximum Allowable Behaviour Factor for Torsionally Sensitive Structures

Building structures not satisfying the condition in Eq. (3.2) (clause § 4.2.3.2(6) of EC8) are classified as torsionally sensitive (a more accurate characterization than the term "torsionally flexible" used in EC8; see Note in the first paragraph of Sect. 3.1.1.1) and are "penalized" by imposing a significantly reduced maximum allowable behaviour factor value (see Table 3.4). This is a reasonable consideration provided that torsionally sensitive buildings will actually exhibit significant rotations about a vertical axis under horizontal ground excitations. Typically, such rotations yield non-uniformly distributed in-plan deformations to vertical structural members imposing significantly high (and possibly hard to accommodate) ductility demands on members lying at or close to the perimeter of buildings.

However, as discussed in detail in Appendix B in view of numerical examples, a torsionally sensitive structure can be, in fact, torsionally stiff and, therefore, it may not develop excessively large rotations to justify the significant penalty imposed by EC8 in terms of the maximum allowable behaviour factor. Conversely, a torsionally non-sensitive building may, in fact, be torsionally flexible and, thus, develop significant rotations under horizontal ground motion excitations. In this respect, the aforementioned penalty for torsionally sensitive structures appears to be unjustifiable, at least for the case of horizontal ground excitations. Still, it is important to note that torsionally sensitive structures are indeed prone to exhibit undue rotations about a vertical axis *for rotational ground excitations*. Nevertheless, such excitations are not treated/considered at all throughout EC8.

3.1.4.4 On the Use of Different Behaviour Factor Values Along Different Horizontal Directions of the Seismic Action (Principal Axes)

Any given r/c building needs to be classified under a single, globally applicable, ductility class (see Sect. 3.1.3). However, it is possible that the same building observes significantly different structural properties along its two principal axes, say, X and Y, such that it is classified under different structural types along X and Y. In such cases, different values for the behaviour factor, q_X and q_Y, may be adopted along the X and Y directions, respectively. Consequently, the design response spectra used to represent the seismic action along the X and Y directions observe different spectral ordinates proportional to the ratio q_X/q_Y, a consideration that EC8 does allow. Such consideration does not involve any implication in the seismic analysis step commonly undertaken separately along directions X and Y. *However, care needs to be exercised in the application of spatial (directional) combination rules to determine the extreme seismic response effects under simultaneous seismic action along directions X and Y.*

It is noted that, in the common case of considering the same design spectrum (in terms of shape and amplitude) along the X and Y directions, the extreme seismic response effect R due to simultaneous seismic action along both directions is determined via the SRSS rule as $\max R^2 = R_X{}^2 + R_Y{}^2$ (see Sect. 2.4.3.2 and clause §4.3.3.5.1(b) of EC8), where R_X and R_Y denote the maximum response of the effect R due to seismic excitation along the X and Y directions, respectively. Further, the maxR value is independent of the incident angle of the seismic action. In other words, it will be the same no matter what the incident angle that the two orthogonal horizontal components of the seismic action form with respect to the direction X. Nevertheless, if different response spectra are used to undertake (equivalent) linear analysis along any two orthogonal horizontal directions, then the extreme exR value of any response deformation or stress resultant quantity R will depend on the considered incident angle. In the special case of considering proportional design spectra (i.e., spectra with same shape but different amplitude) to represent the seismic action along two orthogonal horizontal directions, the extreme exR value (and the corresponding "critical" incident angle for which this value is attained) is given by a more involved expression than the simple SRSS (Anastassiadis et al. 2002):

$$extrR^2 = \left(1+\lambda^2\right)\left(R,_{Xa}^2 + R,_{Ya}^2\right)/2 + \left(1-\lambda^2\right)\sqrt{\left[\left(R,_{Xa}^2 - R,_{Ya}^2\right)/2\right]^2 + R_{XY,a}^2}, \quad (3.6)$$

where

$R,_{Xa}$ is the maximum response value of the effect R for seismic excitation along the X-axis represented by the spectrum S^a,

$R,_{Ya}$ is the maximum response value of the effect R for seismic excitation along the Y-axis represented by the spectrum S^a,

$R_{XY,a} = \sum_i \sum_j \varepsilon_{ij} R_{i,Xa} R_{j,Ya}$,

$S^b = \lambda\, S^a$, $0 < \lambda \leq 1$, and

ε_{ij} are the correlation coefficients (see Eq. (3.11) in Sect. 3.1.5.1).

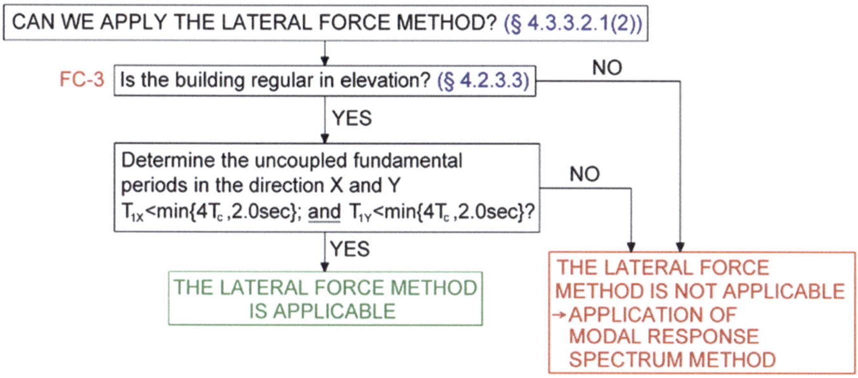

Flowchart 3.6 Choice of analysis method (MRSM or LFM)

3.1.5 Selection and Implementation of Equivalent Linear Methods for Seismic Analysis

In the analysis stage of the seismic design process of an r/c building, a qualified structural analysis method is applied to an appropriate FE model of the lateral load-resisting structural system. Specific (design) seismic effects are computed and used, next, to perform certain verification checks and to detail all r/c structural members comprising the lateral-load resisting system, with the exception of r/c floor slabs. The latter are commonly designed and detailed under flexure due to gravitational loads independently from the rest of the structure. The considered FE model is developed using standard modeling techniques discussed in Sect. 2.3.3 and should account for the diaphragmatic action of the slabs.

The most commonly used structural analysis methods for EC8-compliant seismic design of r/c buildings are the modal response spectrum method (MRSM) and the lateral force method (LFM), some aspects of which have been presented in Sects. 2.4.2 and 2.4.3. It is reminded that the MRSM is the basic (reference) method of analysis applicable to all r/c building structures, while the LFM can only be applied under certain conditions, as shown in FC-3.6.

The sequence of computational and logical steps required to apply the MRSM and the LFM are given in FC-3.7 and FC-3.8, respectively. Further, certain notes and clarification remarks on the implementation of these two methods are provided below.

3.1.5.1 Modal Response Spectrum Method (MRSM): FC-3.7

The MRSM relies on the use of standard tools and techniques of linear structural dynamics, such as the modal (eigenvalue) analysis in conjunction with the response spectrum concept, to determine the various peak modal structural response quantities and their subsequent appropriate combination of design interest (seismic

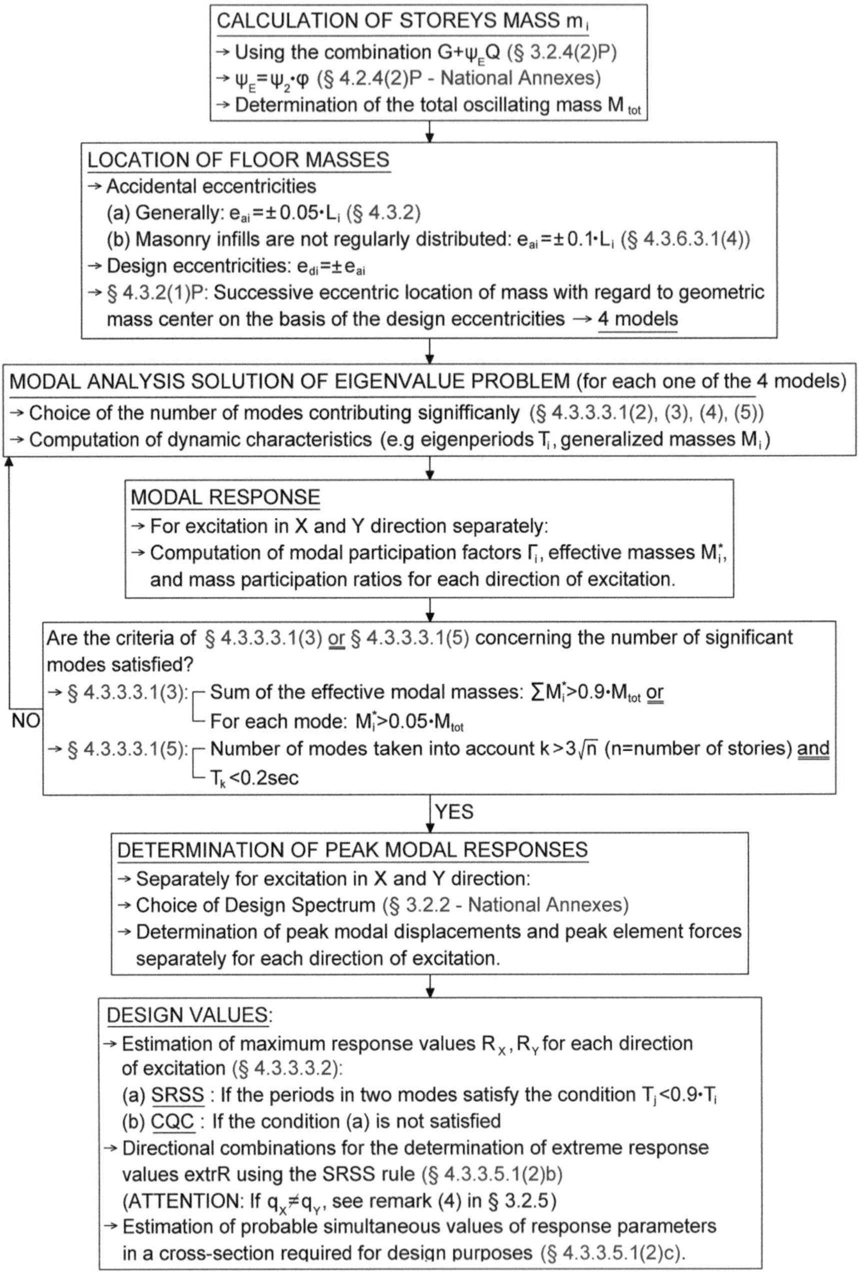

Flowchart 3.7 Modal response spectrum analysis method (MRSM)

DETERMINATION OF BASE SHEARS IN X AND Y DIRECTION (§ 4.3.3.2.2)

→ Relation (4.5): $F_{bX}=S_d(T_{1X})\cdot M_{tot}(G+\psi_E Q)\cdot\lambda_X$

$F_{bY}=S_d(T_{1Y})\cdot M_{tot}(G+\psi_E Q)\cdot\lambda_Y$

$\lambda_j=0.85$ if $T_{1j}<2T_c$ and $n>2$, (j: X,Y), (n=number of stories)

$\lambda_j=1$ in other cases

↓

VERTICAL DISTRIBUTION OF HORIZONTAL SEISMIC FORCES (§ 4.3.3.2.3)

→ Basic relation (4.10): $F_i=F_b\cdot\frac{s_i\cdot m_i}{\sum(s_j\cdot m_j)}$

→ Simplified relation (4.11): $F_i=F_b\cdot\frac{z_i\cdot m_i}{\sum(z_j\cdot m_j)}$

where: F_i is the horizontal force acting on floor level i.

F_b the seismic base shear in accordance with the relation (4.5).

s_i, s_j the displacement of masses m_i, m_j in the fundamental mode shape.

m_i, m_j storey masses.

z_i, z_j the heights of the masses above the level of application of seismic action.

↓

IN PLAN LOCATION OF SEISMIC FORCES AT FLOOR LEVELS:

→ Accidental eccentricities

(a) In general: $e_{ai}=\pm 0.05\cdot L_i$ (§ 4.3.2)

(b) Non uniform arrangements of infills in plan: $e_{ai}=\pm 0.1\cdot L_i$ (§ 4.3.6.3.1(4))

→ Desing eccentricities: $e_{di}=\pm e_{ai}$

→ Succesive application of seismic forces F_i at 4 different locations that are displaced from the center of mass on the basis of design eccentricities.

↓

ANALYSIS CASES:

(a) Case 1: Forces F_{iX} / Eccentricities $e_{d1i}=-e_{aYi}$

(b) Case 2: Forces F_{iX} / Eccentricities $e_{d2i}=+e_{aYi}$

(c) Case 3: Forces F_{iY} / Eccentricities $e_{d3i}=-e_{aXi}$

(d) Case 4: Forces F_{iY} / Eccentricities $e_{d4i}=+e_{aXi}$

↓

DESIGN VALUES:

→ Directional combination using SRSS rule (§ 4.3.3.5.1(2)b) for the determination of extreme response values produced by the following combinations:

(a) Combination 1: Case 1 "+" Case 3

(b) Combination 2: Case 1 "+" Case 4

(c) Combination 3: Case 2 "+" Case 3

(d) Combination 4: Case 2 "+" Case 4

→ Estimation of probable simultaneous values of response parameters using scientifically substantiated methods (§ 4.3.3.5.1(2)c).

Flowchart 3.8 Lateral force method of analysis

effects). The underlying theory of MRSM can be found in standard structural dynamics textbooks (Chopra 2007). In the following paragraphs, focus is given on certain practical issues arising in the implementation of the MRSM as prescribed by EC8 and listed within FC-3.7.

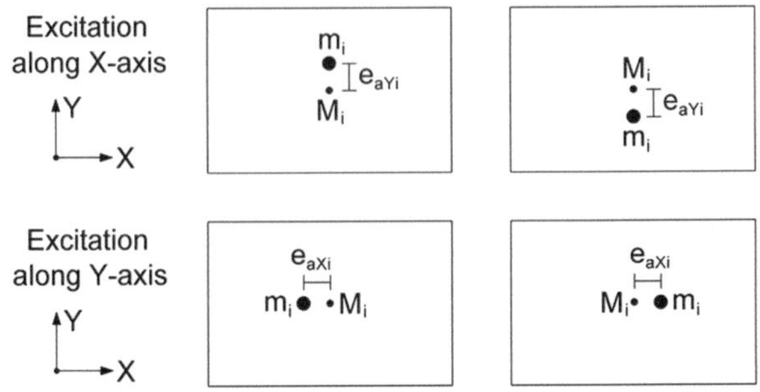

Fig. 3.4 The four analysis cases: storey masses displaced by the accidental eccentricities e_{aXi}, e_{aYi} at each storey level i with respect to the storey center of mass M_i

Inertial discretization and accidental torsional effects

In typical FE modeling/discretization practices for the seismic design of r/c buildings, masses m_i are attributed to each floor i and are distributed at the level (height) of each floor slab. These masses are determined from the gravity loads of the seismic combination ($G + \psi_E Q$) as in clause §3.2.4 of EC8 (see also Sect. 2.3.1.4). Discretization of the m_i masses at each floor and inertial modeling is commonly achieved by taking any of the following three different approaches, (A), (B), and (C), in spatial (three-dimensional) FE models. These approaches need to accommodate the additional requirements for accidental torsional effects, as in clause §3.2.4 of EC8, partly due to uncertainties in the floor mass distribution during a future seismic event.

(A) Under the assumption of perfectly rigid in-plane slab behaviour (diaphragmatic floor action), the location of the center of mass M_i at the level of each floor slab i is first identified. Next, the masses m_i and the polar moments of inertia J_{mi} with respect to M_i of each floor i are computed (see also Appendix B). Finally, an auxiliary virtual "node" at each floor level is considered located at distances equal to the accidental eccentricities, e_{aXi} and e_{aYi}, from M_i along the two principal axes X and Y of the structure, respectively, and for both (positive and negative) directions (clause §3.2.4 of EC8). This consideration is pictorially depicted in Fig. 3.4, in which the virtual node at an arbitrary floor i is noted by the heavy dot. Clearly, four different FE structural models need to be considered, each one having the virtual node displaced at a different set of locations (same "sign" and axis) along the height of the building. The virtual node at floor i belongs to the diaphragm of the floor and is assigned two horizontal masses m_{Xi} and m_{Yi} ($m_{Xi} = m_{Yi} = m_i$) along the principal (analysis) directions X and Y and a polar moment of inertia $J'_{mi} = J_{mi} + m_i\ e^2_{ai}$ about the gravitational axis.

(B) In this second approach, floor masses are not lumped at a particular point of the floor slab as in approach (A). Rather, they are distributed across n nodes of the FE model located at the level of each floor i. Specifically, two horizontal masses m_{nXi} and m_{nYi} are assigned at each node n along the principal (analysis) directions X and Y, such that $m_{nXi} = m_{nYi} = m_{ni}$ and $m_i = \sum_n m_{ni}$ where m_{ni} is the attributed mass to node n of floor i (nodal mass). The latter mass corresponds to an influence area around each node n defined by means of geometrical criteria for the gravity loads of the seismic combination ($G + \psi_E Q$). Upon definition of all nodal masses, the center of mass can be readily found. Notably, this inertial modeling approach is applicable even in the case of not perfectly rigid in-plane behaviour of slabs, while the rotational inertial properties do not need to be explicitly accounted for (i.e., it does not require assigning separate polar moments of inertia at each node).

(C) The third approach closely follows the principles of approach (B) in that floor masses $m_{nXi} = m_{nYi} = m_{ni}$ are assigned to a number of n nodes at each floor level along the principal (analysis) directions X and Y. It further takes advantage of the fact that, in a typical FE model of lateral-load resisting structural systems, slabs are not explicitly modeled using 2D finite elements (see, e.g., Fig. 2.24). Consequently, nodes at each floor are located exclusively at the nodes of the structure's equivalent frame model. Therefore, in this approach, the nodal masses m_{ni} are computed from the axial load divided by the acceleration of gravity carried at the top of each vertical structural member (columns and walls) of the i-1 floor due to the ($G + \psi_E Q$) gravity loading combination. As in the previous approach, the center of mass of each floor M_i is determined upon computing of all nodal masses. It is noted that this third approach is commonly implemented by commercial purpose-made software for seismic design of building structures as it is quite easy to code.

It is important to note that each of the above three inertial modeling approaches yield different locations for the center of mass at each floor M_i. However, these differences are usually small. Further, the development of the four different spatial FE models to account for accidental torsional effects can be readily defined using the approach (A). Nevertheless, placing the M_i at displaced locations around the nominal M_i (Fig. 3.3) in the context of the (B) and (C) approaches is a more challenging task that needs careful implementation, as additional "fictive" (auxiliary) nodes must be used and appropriately connected to the "real" nodes of the equivalent frame model. This issue can be addressed by means of the alternative approach to account for torsional effects discussed in the following paragraph.

Alternative approach to account for accidental torsional effects

In clause §4.3.3.3.3(1) of EC8, an alternative procedure to account for the accidental torsional effects is offered requiring static analyses of a single FE model for four different loading cases, as opposed to considering four different FE models required in the previously discussed approach (Fig. 3.4). In particular, this alternative approach

– considers the use of a single three-dimensional FE model in the context of the MRSM steps of FC-3.7 where the concentrated (lumped) floor masses are placed at the geometrical center of gravity of each floor without any accidental eccentricity; and
– requires additional linear static analyses to be undertaken for a set of moments applied about the gravitational axis at each floor level (or for an equivalent set of coupled horizontal forces) equal to

$$M_{ai} = \pm\ e_{ai}\ F_i, \tag{3.7}$$

where

M_{ai} is the torsional moment applied at storey i about its vertical axis;
e_{ai} is the accidental eccentricity of storey mass i as defined in clause §4.3.2 of EC8 for all considered directions of the seismic action; and
F_i is the horizontal force acting on storey i, as derived in clause §4.3.3.2.3(2)P of EC8 for all considered directions of the seismic action.

The above approach requires undertaking separate static analyses for eccentricities $\pm\ e_{aXi}$ and $\pm\ e_{aYi}$ along principal axes X and Y, respectively. Given its approximate nature (note that F_i are the horizontal loads considered in the LFM), it is reasonable to consider undertaking only one analysis for the moment $M_{ai} = \pm \max(e_{aXi}, e_{aYi})F_i$ referring to the maximum accidental eccentricity. In this case, the adoption of the percentage combination rules for estimating seismic effects due to simultaneous action of two orthogonal horizontal components of the seismic action can be considered sufficiently accurate.

Estimation of seismic effects due to the two orthogonal horizontal seismic components acting simultaneously ("spatial combination")

Maximum values of any seismic response quantity (seismic effect) $\max R_{,x}$ and $\max R_{,y}$ (denoted in EC8 as E_{EdX} and E_{EdY}) are computed independently for the two orthogonal horizontal components of the seismic action along principal axes X and Y, respectively. However, the two horizontal components of the seismic action are taken to act simultaneously (clause §4.3.3.5.1(1)P of EC8). Therefore, an estimate of the extreme value of any seismic effect extrR (denoted in EC8 as E_{Ed}) needs to be computed by combining the $\max R_{,X}$ and $\max R_{,Y}$ values which do not occur simultaneously (see also Sect. 2.4.3.2). According to EC8 provisions, this can be achieved by application of the SRSS spatial (directional) combination rule (clause §4.3.3.5.1(2) of EC8), that is,

$$E_{Ed} = \sqrt{E_{EdX}^2 + E_{EdY}^2} \quad \text{or} \quad \text{extrR} = \pm\sqrt{\max R_{,x} + \max R_{,y}}. \tag{3.8}$$

Alternatively, the following "percentage" combination rules can be considered for the purpose (clause §4.3.3.5.1(3) of EC8)

(a) $\pm E_{EdX} \pm 0.30 E_{EdY}$

(b) $\pm 0.30 E_{EdX} \pm E_{EdY}$

For the most involved case in which two or three seismic effects acting concurrently (vector of seismic effects) need to be considered under the simultaneous action of two horizontal orthogonal seismic components (see also Sect. 2.4.3.2), the expected (most probable) value of one seismic effect, $B_{,A}$, given that a different seismic effect, A, attains its expected extreme value, extrA, can be determined by the expression (Gupta and Singh 1977)

$$B_{,A} = \frac{P_{AB}}{extrA}. \tag{3.9}$$

In the above equation,

$$P_{AB} = P_{BA} = \sum_i \sum_j \varepsilon_{ij} \left(A_{i,X} B_{j,X} + A_{i,Y} B_{j,Y} \right); \quad i, j = 1, 2, \ldots N, \tag{3.10}$$

where N is the number of vibration modes considered in the analysis, $(A_{i,X}, B_{j,X})$ and $(A_{i,Y}, B_{j,Y})$ are the peak modal values of the seismic effects A and B for seismic action acting along the X and Y axes, respectively, and ε_{ij} is the correlation coefficient between modes i and j given by the expression (Der Kiureghian 1981):

$$\varepsilon_{ij} = \frac{8\sqrt{\zeta_i \zeta_j}\left(\zeta_i + \lambda \zeta_j\right)\lambda^{3/2}}{\left(1 - \lambda^2\right) + 4\zeta_i \zeta_j \lambda \left(1 + \lambda^2\right) + 4\left(\zeta_i^2 + \zeta_j^2\right)\lambda^2}, \tag{3.11}$$

in which ζ_i and ζ_j are the viscous damping ratios in modes i and j, respectively (commonly taken as equal to 0.05), and λ is the ratio of the natural periods of modes i and j ($\lambda = T_i/T_j$).

An important issue concerns the set of sectional forces required for the design of the r/c structural members. For example, the amount of longitudinal reinforcement of r/c frame columns is controlled by the concurrent action of three sectional stress resultants, i.e., axial force N and bending moments M_2 and M_3 as defined in Fig. 2.32 with respect to the member local axes. EC8 provides two options concerning the combinations of these three response quantities:

(i) The extreme values of the response parameters can be considered as simultaneous for design purposes (clause §4.3.3.5.1(2)c of EC8). According to this option, the following eight triads of internal stress resultants are obtained:

$$\begin{aligned} S_1 = -S_5 &= \begin{bmatrix} extrN \\ extrM_x \\ extrM_y \end{bmatrix}, S_2 = -S_6 = \begin{bmatrix} -extrN \\ extrM_2 \\ extrM_3 \end{bmatrix}, \\ S_3 = -S_7 &= \begin{bmatrix} extrN \\ -extrM_2 \\ extrM_3 \end{bmatrix}, S_4 = -S_8 = \begin{bmatrix} extrN \\ extrM_2 \\ -extrM_3 \end{bmatrix} \end{aligned} \tag{3.12}$$

(ii) The probable simultaneous values of response parameters can be used for design purposes (clause §4.3.3.5.1(2)c of EC8). A simplified approach is given in the Greek Seismic Code EAK2000 (Earthquake Planning and Protection Organization (EPPO) 2000) where the "unfavourable combinations" of internal forces are used for design purposes (Anastassiadis 1993; Gupta 1992). According to this approach, the following six triads of internal forces are obtained:

$$\mathrm{S}_1 = -\mathrm{S}_4 = \begin{bmatrix} extrN \\ M_{2,extrN} \\ M_{3,extrN} \end{bmatrix}, \mathrm{S}_2 = -\mathrm{S}_5 = \begin{bmatrix} N_{,extrM_2} \\ extrM_2 \\ M_{3,extrM_2} \end{bmatrix}, \quad \mathrm{S}_3 = -\mathrm{S}_6 = \begin{bmatrix} N_{,extrM_3} \\ M_{2,extrM_3} \\ extrM_3 \end{bmatrix} \tag{3.13}$$

where the comma after the first index denotes "corresponding to" (e.g., $M_{y,extrMx}$ is the simultaneous value of response parameter M_y corresponding to the extreme value of the response parameter M_x).

3.1.5.2 Lateral Force Method (LFM): FC-3.8

The single-mode response spectrum (lateral force) method (LFM) of analysis may be used for the seismic design of structures whose seismic response is predominantly translational along the principal axes X and Y, and is not significantly influenced by the higher-than-the-fundamental modes of vibration. According to clause §4.3.3.2.1(2) of EC8, the above conditions are satisfied for relative stiff and regular in elevation building structures based on the criteria shown in FC-3.6 (see also Sect. 2.4.3.1). Apart from its simplicity (it can be undertaken using solely software for linear static analysis), a significant advantage of the LFM over the MRSM is that it yields peak concurrent seismic effects equipped with a particular sign dependent on the considered direction (positive or negative) of the seismic action. Thus, the results of the LFM for each excitation direction can be readily interpreted and verified, as they satisfy global equilibrium equations. Note that this is not the case for MRSM, which yields expected *absolute* maximum values (both signs may apply) that do not occur concurrently and, therefore, do not satisfy global equilibrium equations. Compared to the MRSM, *the LFM yields conservative results provided that dynamic in-plan eccentricities are properly accounted for*. However, as discussed in Sect. 2.4.3.1, EC8 provisions do not consider any regularity in plan criterion (e.g., torsional sensitivity) for the application of the LFM to three-dimensional FE models and, thus, in some special cases of torsional sensitive buildings, LFM will not yield conservative results vis-à-vis the more general MRSM. In this regard, the use of LFM should not be preferred over MRSM. Instead, it is recommended to use the MRSM method, automated in the majority of commercial FE software for structural analysis, which yields more accurate and less conservative seismic effects and, thus, leads to more reliable and economical structural designs.

Further to the comments made on the range of applicability of the LFM in Sect. 2.4.3.1, some further notes on practical issues arising in the implementation of the method are given in the following paragraphs. The sequence of the required computational steps for the LFM are summarized in FC-3.8, as in clause §4.3.3.2 of EC8.

Points of lateral load application and accidental torsional effects

As in the case of the MRSM, the accidental torsional effects can be accounted for by additional static analyses for sets of torsional moments applied at every floor i given by Eq. (3.6). In this case, the lateral static forces F_i derived as prescribed in clause §4.3.3.2.3(2)P of EC8 (see also FC-3.8) are applied to the geometrical centers of gravity at each floor level i (M_i). The location of the M_i points can be determined by any of the three inertial discretization approaches discussed in the previous sub-section. Alternatively, the same lateral forces can be applied at four different locations displaced with respect to the nominal M_i by the accidental eccentricity, as shown in Fig. 3.3. In this case, there is no need to consider torsional moments. Finally, a third implicit approach to account for accidental torsional effects is offered in clause §4.3.3.2.4 of EC8, which circumvents the need to consider accidental eccentricities. The latter approach considers increasing the seismic effects in the individual structural members by means of a multiplication factor which depends on the location of structural members in-plan.

Estimation of seismic effects due to the two orthogonal horizontal seismic components acting simultaneously ("spatial combination")

The same comments made for the MRSM apply in the case of the LFM for seismic effects estimation accounting for the seismic action applied simultaneously along two orthogonal horizontal axes. The only difference is that, in the case of LFM, Eq. (3.8) used to derive the expected value of one seismic effect, $B_{,A}$, given that a different concurrent seismic effect, A, attains its expected extreme value, extrA, simplifies as (see Eqs. (3.9) and (3.10))

$$B_{,A} = \frac{A_{,X}}{\mathrm{extr}A} B_{,X} + \frac{A_{,Y}}{\mathrm{extr}A} B_{,Y}. \tag{3.14}$$

In the above equation, $(A_{,X}, B_{,X})$ and $(A_{,Y}, B_{,Y})$ are the values of the seismic effects A and B carrying the actual sign obtained from the application of the LFM for seismic action applied independently along two orthogonal horizontal axes X and Y, respectively, while $\mathrm{extr}A = \pm\sqrt{A_{,X}^2 + A_{,Y}^2}$.

Horizontal directions of lateral loads – "principal directions"

As discussed in Sect. 2.4.3.1, there is a level of ambiguity in defining the relevant horizontal directions along which lateral loads should be applied at each floor in the context of the LFM. Clause § 4.3.3.1(11)P of EC8, quoted below, provides the only recommendation within EC8 for addressing the above issue: "Whenever a spatial model is used, the design seismic action shall be applied along all relevant horizontal directions (with regard to the structural layout of the building) and their orthogonal horizontal directions. For buildings with resisting elements in two perpendicular directions these two directions shall be considered as the relevant directions." Clearly, the above clause is applicable only to a limited number of real-life r/c building structures. To this end, the use of the "virtual elastic axes for multistorey buildings" (Athanatopoulou and Doudoumis 2008; Makarios and Anastassiadis 1998) offers a viable solution to the definition of "principal horizontal axes" along which lateral forces can be applied in the case of r/c buildings with lateral load resisting structural systems not aligned along two orthogonal axes.

3.1.6 Accounting for the Vertical Component of the Seismic Action

According to clause §4.3.3.5.2 of EC8, the effect of the vertical component of the seismic ground motion should be considered in the analysis stage only if the design peak vertical ground acceleration α_{vg} is greater than 0.25 g, and only for certain structural components. These structural components are listed within FC-3.9, which includes the required verification checks and computational steps involved in accounting for the vertical component of the seismic action. Further comments and discussion on some of these steps are provided in the following paragraphs.

The minimum design peak vertical ground acceleration requirement

The condition of $\alpha_{vg} > 0.25$ g or, approximately, of $\alpha_{vg} > 2.5$ m/s^2 is rather stringent. It practically excludes considering the vertical component of the seismic action for the design of ordinary buildings (importance class II in clause §4.2.5 of EC8) in the vast majority of the "high seismicity" sites for which the "type 1" seismic response spectrum applies, according to clause §3.2.2.2(2)P of EC8 (i.e., sites where the earthquakes contributing most to its seismic hazard in the context of a probabilistic seismic hazard analysis have a surface-wave magnitude M_s greater than 5.5). Specifically, based on Table 3.4 of EC8 (§3.2.2.3), it is seen that the vertical component of the seismic action need only be considered for design horizontal ground acceleration $\alpha_g = \alpha_{vg}/0.9 > 0.25$ g/0.9 = 0.28 g for the above sites. This is rather high horizontal peak ground acceleration for a 10 % probability of being exceeded by a seismic event over a period of 50 years.

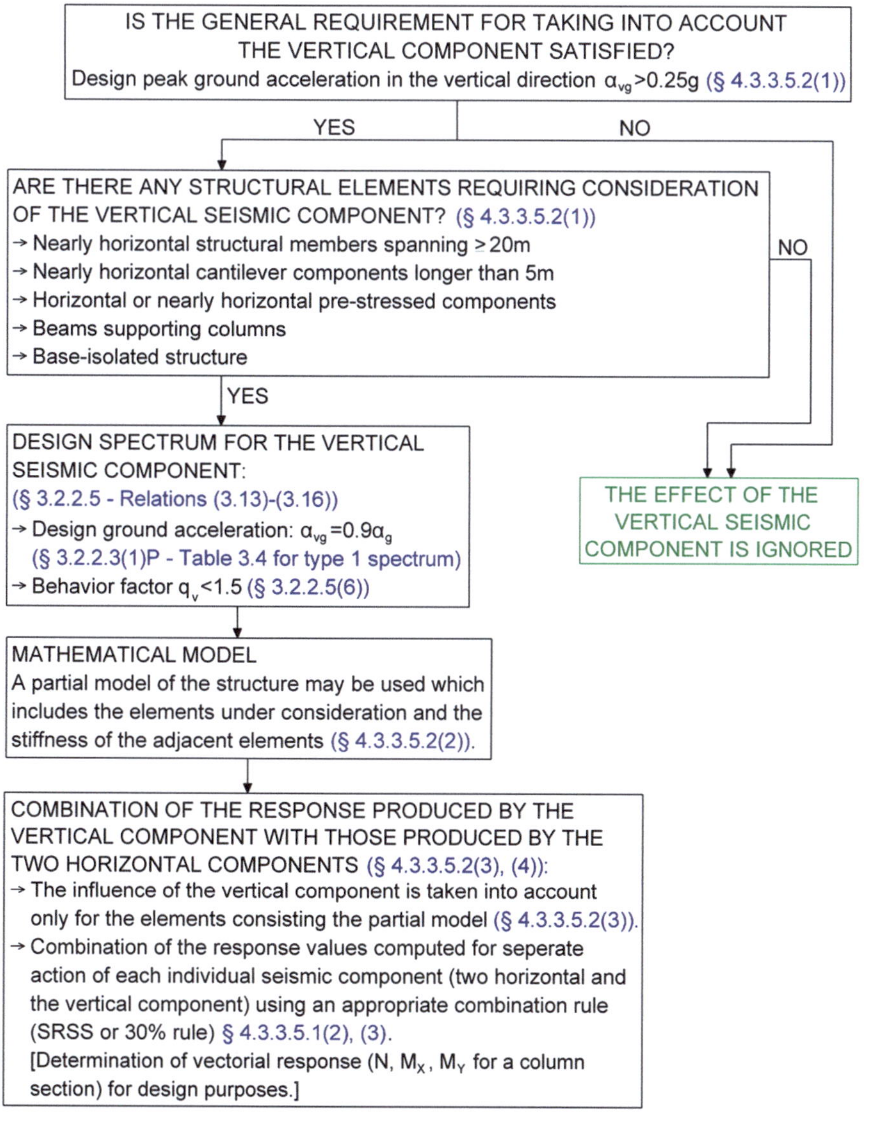

Flowchart 3.9 Accounting for the vertical component of seismic action

The EC8 Vertical Elastic Response Spectrum

The vertical component of the seismic action is represented by means of an elastic response spectrum which can be derived from the elastic response spectrum for the horizontal ground component (Eq. (2.1) for $q = 1$) by (§3.2.2.3)

- replacing α_g with the peak vertical ground acceleration α_{vg};
- setting the soil factor S = 1;
- replacing the multiplier 2.5 by 3.0; and
- using the following set of corner natural period values: $T_B = 0.05s$, $T_C = 0.15s$, and $T_D = 1.0s$.

Use of a partial model of the structure

It is interesting to note that, according to EC8, §4.3.3.5.2(2), "The analysis for determining the effects of the vertical component of the seismic action may be based on a partial model of the structure, which includes the elements on which the vertical component is considered to act §4.3.3.5.2(1) and takes into account the stiffness of the adjacent elements". In general, this simplified procedure gives divergence rates (in comparison to the analysis of the whole structure model by modal response spectrum method) which may or may not be conservative. However, it should be pointed out that the absolute values of the response parameters due to the vertical seismic component are, in general, small compared to the respective values due to the static vertical loads (Athanatopoulou et al. 1999). Therefore, this simplified procedure can be used in standard practice for conventional buildings, thus avoiding -as far as the vertical component of the seismic action is concerned- the analysis of the whole spatial structural model. However, such analysis is not a problem anymore given the availability of modern computers and professional analysis software.

Spatial combination of maximum response quantities due to simultaneous action of three orthogonal components of the seismic action

If, in addition to the concurrent action of two horizontal orthogonal components of the ground motion, the vertical component of the seismic action acts simultaneously, then extreme response quantities are determined either by the SRSS rule (Eq. 3.15) or by percentage combination rules (Eq. 3.16, §4.3.3.5.2(4)):

$$E_d = \sqrt{E_{Edx}^2 + E_{Edy}^2 + E_{Edz}^2}. \tag{3.15}$$

$$\begin{aligned} &(a) \quad \pm E_{Edx} \pm 0.30E_{Edy} \pm 0.30E_{Edz} \\ &(b) \quad \pm 0.30E_{Edx} \pm E_{Edy} \pm 0.30E_{Edz} \\ &(c) \quad \pm 0.30E_{Edx} \pm 0.30E_{Edy} \pm E_{Edz} \end{aligned} \tag{3.16}$$

3.2 Deformation-Based Verification Checks

EC8 prescribes two distinct verification checks involving deformation seismic effects derived from the analysis step results. The first check aims to verify the influence of second-order effects, an issue that has been critically discussed in Sect. 2.4.3.3. The check relies on Eq. (2.14), which requires the determination of the shear storey and of the interstorey drift (relative storey deformation) at each storey. Section 3.2.1 reviews all required calculations for implementation of the above verification check with the aid of two flowcharts (FC-3.10a and FC-3.10b).

ANALYSIS FOR (G+ψ_2Q) LOAD COMBINATION (vertical loads from the seismic design situation)

Determination of the total axial force at each storey k: $P^{(k)}_{tot,G+\psi_2 Q} = \sum_{j=1}^{M(k)} P^{(k)}_{j,G+\psi_2 Q}$

↓

4 ANALYSIS CASES FOR THE 4 LOCATIONS OF THE SEISMIC FORCES

↓

DETERMINATION FOR EACH STOREY k (k=1÷ L) THE SENSITIVITY COEFFICIENTS θ_X and θ_Y INDEPENDNTLY FOR EACH ANALYSIS CASE Lci (i=1÷4)

$$\theta^{(k)}_{X,Lci} = \frac{P^{(k)}_{tot,G+\psi_2 Q} \cdot d^{(k)}_{rX,Lci}}{V^{(k)}_{X(tot),Lci} \cdot h_k} \quad \text{and} \quad \theta^{(k)}_{Y,Lci} = \frac{P^{(k)}_{tot,G+\psi_2 Q} \cdot d^{(k)}_{rY,Lci}}{V^{(k)}_{Y(tot),Lci} \cdot h_k}$$

Notation: If an analysis case (1÷4) produces shear force $V_{X(tot)}$ or $V_{Y(tot)}$ equal to zero (e.g. due to symmetry), the corresponding value of θ_X or θ_Y will be also set to zero

↓

DIRECTIONAL COMBINATION OF $\theta^{(k)}_{X,Lci}$ and $\theta^{(k)}_{Y,Lci}$ (SRSS RULE):

→ Combination 1: Load case 1"+"3:

$$extr\theta^{(k)}_{X,1\text{-}3} = \sqrt{[\theta^{(k)}_{X,Lc1}]^2 + [\theta^{(k)}_{X,Lc3}]^2}$$
$$extr\theta^{(k)}_{Y,1\text{-}3} = \sqrt{[\theta^{(k)}_{Y,Lc1}]^2 + [\theta^{(k)}_{Y,Lc3}]^2}$$

→ Combination 2: Load case 1"+"4:

$$extr\theta^{(k)}_{X,1\text{-}4} = \sqrt{[\theta^{(k)}_{X,Lc1}]^2 + [\theta^{(k)}_{X,Lc4}]^2}$$
$$extr\theta^{(k)}_{Y,1\text{-}4} = \sqrt{[\theta^{(k)}_{Y,Lc1}]^2 + [\theta^{(k)}_{Y,Lc4}]^2}$$

→ Combination 3: Load case 2"+"3:

$$extr\theta^{(k)}_{X,2\text{-}3} = \sqrt{[\theta^{(k)}_{X,Lc2}]^2 + [\theta^{(k)}_{X,Lc3}]^2}$$
$$extr\theta^{(k)}_{Y,2\text{-}3} = \sqrt{[\theta^{(k)}_{Y,Lc2}]^2 + [\theta^{(k)}_{Y,Lc3}]^2}$$

→ Combination 4: Load case 2"+"4:

$$extr\theta^{(k)}_{X,2\text{-}4} = \sqrt{[\theta^{(k)}_{X,Lc2}]^2 + [\theta^{(k)}_{X,Lc4}]^2}$$
$$extr\theta^{(k)}_{Y,2\text{-}4} = \sqrt{[\theta^{(k)}_{Y,Lc2}]^2 + [\theta^{(k)}_{Y,Lc4}]^2}$$

↓

VERIFICATION CHECKS for $extr\theta^{(k)}_X$ and $extr\theta^{(k)}_Y$ at each storey k according to § 4.4.2.2(2)-(4):

→ If $extr\theta^{(k)}_X \le 0.1$ and $extr\theta^{(k)}_Y \le 0.1$: The second order effects are neglected.

→ If $0.1 < extr\theta^{(k)}_X \le 0.2$ or $0.1 < extr\theta^{(k)}_Y \le 0.2$: The second order effects are approximately taken into account by multiplying the seismic action effects by a factor equal to [1/(1-θ)] (where $\theta = \max\{extr\theta^{(k)}_X, extr\theta^{(k)}_Y\}$).

→ If $0.2 < extr\theta^{(k)}_X \le 0.3$ or $0.2 < extr\theta^{(k)}_Y \le 0.3$: Analysis taking into account second order effects should be undertaken.

→ If $extr\theta^{(k)}_X > 0.3$ or $extr\theta^{(k)}_Y > 0.3$: The structural system should be changed or modified.

where:

$V^{(k)}_{X(tot),Lci} = \sum_{j=1}^{M(k)} V^{(k)}_{Xj,Lci}$ and $V^{(k)}_{Y(tot),Lci} = \sum_{j=1}^{M(k)} V^{(k)}_{Yj,Lci}$ the total shear forces at the storey k in X and Y direction due to analysis for the load case Lci (i=1-4),

$d^{(k)}_{rX,Lci} = \frac{q_X}{M(k)} \cdot \sum_{j=1}^{M(k)} (u^{top}_{Xj,Lci} - u^{bot}_{Xj,Lci})$ and $d^{(k)}_{rY,Lci} = \frac{q_Y}{M(k)} \cdot \sum_{j=1}^{M(k)} (u^{top}_{Yj,Lci} - u^{bot}_{Yj,Lci})$

average relative displacements of storey k due to analysis for the load case Lci (i=1÷4).

h_k the storey's k height, q the behavior factor, and

M(k) is the total number of vertical elements at storey k.

Flowchart 3.10a Calculation of interstorey drift sensitivity coefficients using the LFM

ANALYSIS FOR $(G+\psi_2 Q)$ LOAD COMBINATION (vertical loads from the seismic design situation)

Determination of the total axial force at each storey k: $P^{(k)}_{tot,G+\psi_2 Q}=\sum_{j=1}^{M(k)} P^{(k)}_{j,G+\psi_2 Q}$

↓

MODAL ANALYSIS (N modes for each one of the four mass loacations)

↓

RESPONSE SPECTRUM ANALYSIS USING THE DESIGN SPECTRUM

(a) For excitation in X direction (N peak modal response values EXi)

(b) For excitation in Y direction (N peak modal response values EYi)

↓

DETERMINE SEPARATELY FOR EACH DIRECTION OF EXCITATION IN X AND Y AXES FOR EACH STOREY k (k=1÷L) AND FOR EACH MODE i (i=1÷L) THE SENSITIVITY COEFFICIENTS:

(a) Excitation X:

$$\theta^{(k)}_{X,EXi}=\frac{P^{(k)}_{tot,G+\psi_2 Q}\cdot d^{(k)}_{rX,EXi}}{V^{(k)}_{X(tot),EXi}\cdot h_k} \text{ and } \theta^{(k)}_{Y,EXi}=\frac{P^{(k)}_{tot,G+\psi_2 Q}\cdot d^{(k)}_{rY,EXi}}{V^{(k)}_{Y(tot),EXi}\cdot h_k}$$

(b) Excitation Y:

$$\theta^{(k)}_{X,EYi}=\frac{P^{(k)}_{tot,G+\psi_2 Q}\cdot d^{(k)}_{rX,EYi}}{V^{(k)}_{X(tot),EYi}\cdot h_k} \text{ and } \theta^{(k)}_{Y,EYi}=\frac{P^{(k)}_{tot,G+\psi_2 Q}\cdot d^{(k)}_{rY,EYi}}{V^{(k)}_{Y(tot),EYi}\cdot h_k}$$

Notation: If $V^{(k)}_{X(tot)}=0$ or $V^{(k)}_{Y(tot)}=0$ (e.g. due to symmetry), then the corresponding value of θ is zero.

↓

MODAL COMBINATION OF $\theta^{(k)}_{X,EXi}, \theta^{(k)}_{Y,EXi}, \theta^{(k)}_{X,EYi}, \theta^{(k)}_{Y,EYi}$ AT EACH STOREY k (CQC RULE):

→ Determination of the maximum values: $\max\theta^{(k)}_{X,EX}, \max\theta^{(k)}_{Y,EX}$ for excitation in X direction,

→ Determination of the maximum values: $\max\theta^{(k)}_{X,EY}, \max\theta^{(k)}_{Y,EY}$ for excitation in Y direction.

↓

DIRECTIONAL COMBINATION (SRSS RULE FOR $q_X=q_Y$):

(If $q_X \neq q_Y$, see remark (2) in § 3.2.5)

→ Determination of the extreme values: $\mathrm{extr}\theta^{(k)}_X=\sqrt{[\max\theta^{(k)}_{X,EX}]^2+[\max\theta^{(k)}_{X,EY}]^2}$

→ Determination of the extreme values: $\mathrm{extr}\theta^{(k)}_Y=\sqrt{[\max\theta^{(k)}_{Y,EX}]^2+[\max\theta^{(k)}_{Y,EY}]^2}$

↓

VERIFICATION CHECKS for $\mathrm{extr}\theta^{(k)}_X$ and $\mathrm{extr}\theta^{(k)}_Y$ at each storey k according to § 4.4.2.2(2)-(4):

→ If $\mathrm{extr}\theta^{(k)}_X \leq 0.1$ and $\mathrm{extr}\theta^{(k)}_Y \leq 0.1$: The second order effects are neglected.

→ If $0.1<\mathrm{extr}\theta^{(k)}_X \leq 0.2$ or $0.1<\mathrm{extr}\theta^{(k)}_Y \leq 0.2$: The second order effects are approximately taken into account by multiplying the seismic action effects by a factor equal to $[1/(1-\theta)]$ (where $\theta=\max\{\mathrm{extr}\theta^{(k)}_X, \mathrm{extr}\theta^{(k)}_Y\}$).

→ If $0.2<\mathrm{extr}\theta^{(k)}_X \leq 0.3$ or $0.2<\mathrm{extr}\theta^{(k)}_Y \leq 0.3$:
Analysis taking into account second order effects should be undertaken.

→ If $\mathrm{extr}\theta^{(k)}_X>0.3$ or $\mathrm{extr}\theta^{(k)}_Y>0.3$:
The structural system should be changed or modified.

where:

$V^{(k)}_{X(tot),EXi}=\sum_{j=1}^{M(k)} V^{(k)}_{Xj,EXi}$ and $V^{(k)}_{Y(tot),EXi}=\sum_{j=1}^{M(k)} V^{(k)}_{Yj,EXi}$ the total shear forces at the storey k in X and Y direction respectively for mode i under excitation along X axis.

$V^{(k)}_{X(tot),EYi}=\sum_{j=1}^{M(k)} V^{(k)}_{Xj,EYi}$ and $V^{(k)}_{Y(tot),EYi}=\sum_{j=1}^{M(k)} V^{(k)}_{Yj,EYi}$ the total shear forces at the storey k in X and Y direction respectively for mode i under excitation along Y axis.

$d^{(k)}_{rX,EXi}=\frac{q_X}{M(k)}\cdot\sum_{j=1}^{M(k)}(u^{top}_{Xj,EXi}-u^{bot}_{Xj,EXi})$ and $d^{(k)}_{rY,EXi}=\frac{q_Y}{M(k)}\cdot\sum_{j=1}^{M(k)}(u^{top}_{Yj,EXi}-u^{bot}_{Yj,EXi})$

average relative displacements of storey k for mode i (excitation in X direction).

$d^{(k)}_{rX,EYi}=\frac{q_X}{M(k)}\cdot\sum_{j=1}^{M(k)}(u^{top}_{Xj,EYi}-u^{bot}_{Xj,EYi})$ and $d^{(k)}_{rY,EYi}=\frac{q_Y}{M(k)}\cdot\sum_{j=1}^{M(k)}(u^{top}_{Yj,EYi}-u^{bot}_{Yj,EYi})$

average relative displacements of storey k for mode i (excitation in Y direction).

h_k the storey's k height, q the behavior factor, and

M(k) is the total number of vertical elements at storey k.

Flowchart 3.10b Calculation of interstorey drift sensitivity coefficients using the MRSM

The second check aims to ensure that the damage limitation design objective is satisfied by verifying that the interstorey drifts (relative storey deformations normalized by the storey height) remain below a certain threshold for a "frequent" seismic action which is less intense than the "design earthquake". The seismic effects for the considered frequent seismic action are obtained by reducing the deformation seismic effects computed in the analysis step for the design seismic action. In this regard, a second analysis (i.e., for reduced seismic input loads) is not required to be undertaken; hence, EC8 puts forth a two-tier seismic design with only a single equivalent linear analysis step for the design earthquake (see also Sect. 1.2). The required calculations for the implementation of this second deformation-based verification check are presented in Sect. 3.2.2 with the aid of two flowcharts (FC-3.11a and FC-3.11b).

From the practical implementation viewpoint, it is important to note that both the above verification checks require the computation of peak (design) interstorey drifts at each floor under seismic excitation along two principal directions. These peak values are straightforward to compute in case the lateral force method (LFM) is used in the analysis step. However, the computation of peak interstorey drifts becomes more involved in case the modal response spectrum method (MRSM) of analysis is employed which, by definition, provides non-concurrent peak displacement demands for each mode and for each direction of the seismic excitation. In this respect, sub-section 3.2.1.2 presents the theoretically rigorous method along with two other alternative approaches, commonly used by commercial software, to determine peak interstorey drifts from seismic effects derived by means of the MRSM.

3.2.1 Verification Check for Second Order (P- Δ) Effects (FC-3.10a and FC-3.10b)

According to EC8 ((§ 4.4.2.2(2)), second-order effects need not be accounted for in the design of r/c buildings if

$$\theta = \frac{P_{tot} d_r}{V_{tot} h} \leq 0.10, \tag{3.17}$$

where:

θ is the interstorey drift sensitivity coefficient
P_{tot} is the sum of the gravity loads of the seismic design combination of actions (G + ψ_2Q) as in clause §3.2.4 of EC8 (see also Sect. 2.3.1.4) carried by the vertical structural members of the considered storey;
V_{tot} is the total seismic storey shear;
h is the storey height; and

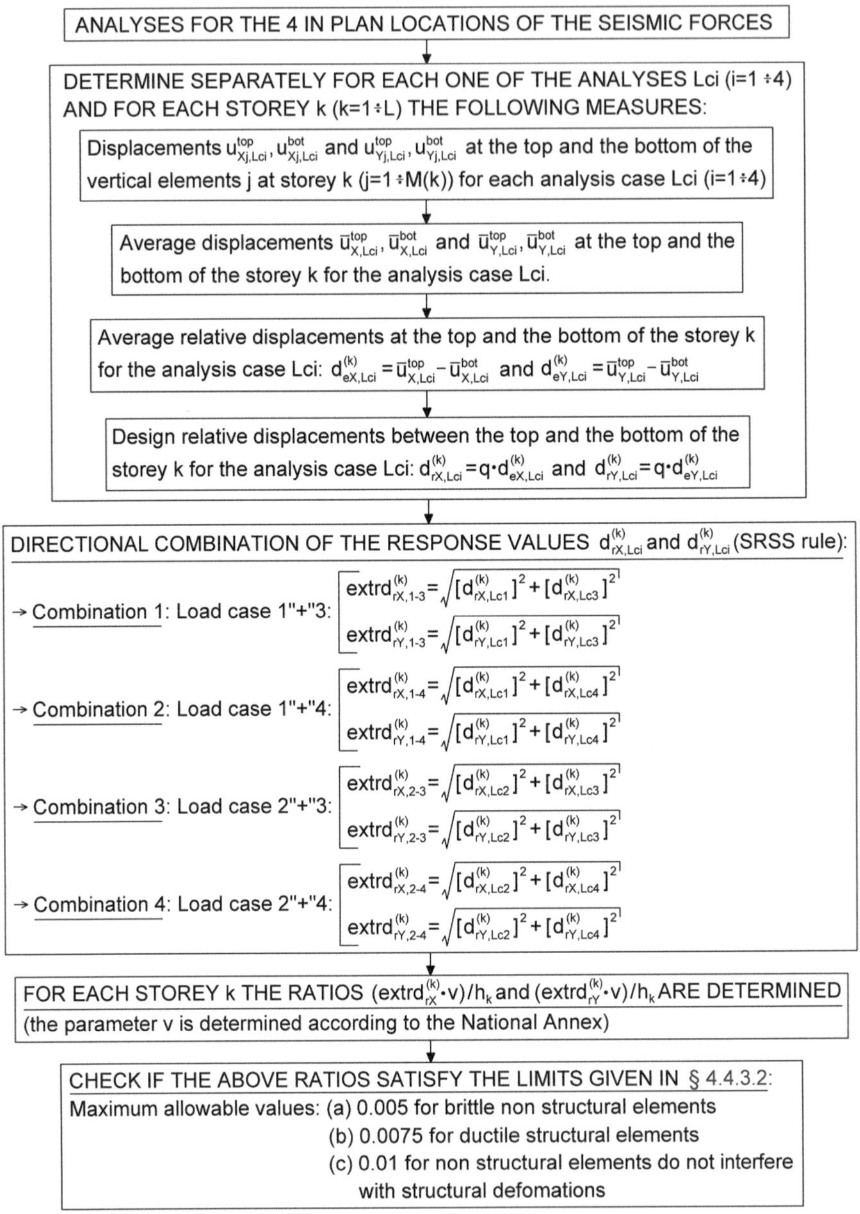

Flowchart 3.11a Damage limitation verification check for the LFM

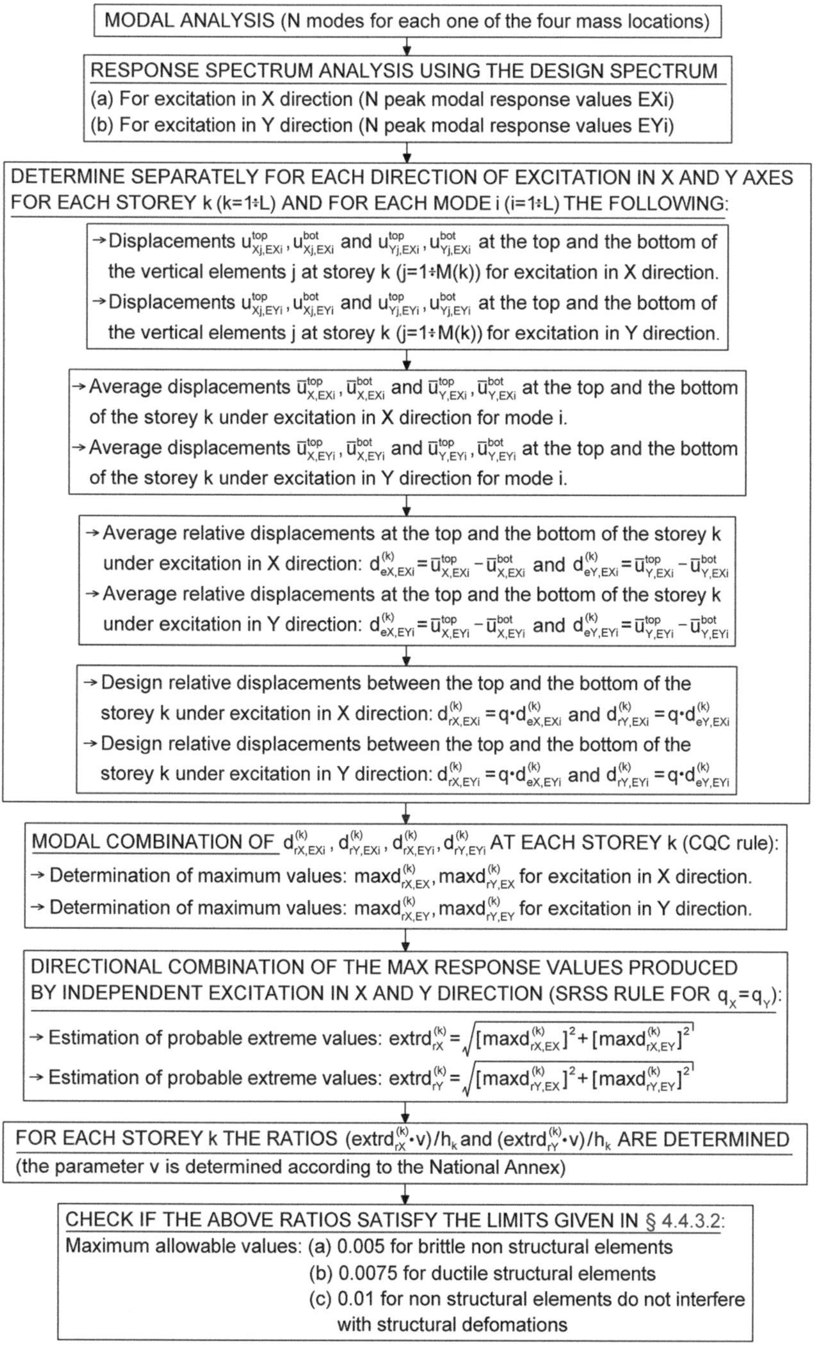

Flowchart 3.11b Damage limitation verification check for the MRSM

d_r is the design interstorey drift defined as the difference of the *average* (mean) lateral displacement at the top, d_s^t, and at the bottom, d_s^b, of the considered storey computed by the expression (clause §4.3.4 of EC8)

$$d_r = d_s^t - d_s^b = q\left(d_e^t - d_e^b\right). \tag{3.18}$$

In Eq. (3.18), d_s^t and d_e^b are the mean displacements of the top and bottom ends, respectively, from all vertical structural members of the considered storey computed by an equivalent linear analysis for the seismic design action defined by a design spectrum with an assumed behavior factor q. It is noted that, in the case of multistory r/c buildings, and especially in those buildings whose lateral load resisting system includes walls, the lateral storey displacements due to gravity loads can be assumed as negligible compared to the lateral storey displacements due to the design seismic action. It is further noted that, although not straightforwardly mentioned in EC8, the condition of Eq. (3.17) needs to be verified along both principal directions, X and Y, of a given building. Therefore, for each storey of the building, the interstorey drifts d_{rX} and d_{rY} and the storey shears $V_{X(tot)}$ and $V_{Y(tot)}$ are required to compute the interstorey drift sensitivity coefficients θ_X and θ_Y along the X and Y directions, respectively, as in

$$\theta_X = \frac{P_{tot} d_{rX}}{V_{X(tot)} h}; \quad \theta_Y = \frac{P_{tot} d_{rY}}{V_{Y(tot)} h}. \tag{3.19}$$

In this respect, special care needs to be exercised in case different behaviour factors, q_X and q_Y, are used to define the design seismic action (design spectrum) along directions X and Y, respectively, in the calculation of the average design interstorey drifts, d_{rX} and d_{rY}, using Eq. (3.18).

The following two sub-sections discuss the required computational steps for determining the peak "design" (extreme) interstorey drift sensitivity coefficients, extr$\theta_X^{(k)}$ and extr$\theta_Y^{(k)}$, at each storey k and along directions X and Y computed by means of the lateral force method (LFM) and of the modal response spectrum method (MRSM).

3.2.1.1 Calculation of Interstorey Drift Sensitivity Coefficients Using the Lateral Force Method (LFM)

In case the lateral force method (LFM) of analysis is adopted to compute seismic effects, the interstorey drift sensitivity coefficients can be calculated in a straightforward manner, since all structural response quantities appearing in Eq. (3.19) are derived from standard static analyses and, therefore, are concurrent and equipped with a specific sign. For example, the storey shears $V_{X(tot)}$ and $V_{Y(tot)}$ are determined as the algebraic sum of the shearing forces at the base of all vertical structural members of the considered storey along the principal axes X and Y, respectively.

Along these lines, flowchart FC-3.10a presents sequentially the required computations for determining the peak "design" (extreme) interstorey drift sensitivity coefficients, extr$\theta_X^{(k)}$ and extr$\theta_Y^{(k)}$, at storey k and along directions X and Y.

3.2.1.2 Calculation of Interstorey Drift Sensitivity Coefficients Using the Modal Response Spectrum Method (MRSM)

The application of the modal response spectrum method (MRSM) provides for the absolute (both signs may apply) expected (most probable) peak values of structural response quantities that do not occur concurrently. Therefore, in this case, the calculation of the interstorey drift sensitivity coefficient is less straightforward and requires a more involved treatment of the design seismic effects. For example, the average (mean) lateral displacements d_e^t and d_e^b cannot be simply determined as the algebraic mean value of the peak values of the top and bottom ends of the vertical structural members at a particular floor as obtained from the MRSM, since these peak values do not occur concurrently and do not carry any particular sign. Consequently, they cannot be subtracted, $\left(d_e^t - d_e^b\right)$, to compute the design interstorey drift as defined by Eq. (3.18). Similarly, the storey shears $V_{X(tot)}$ and $V_{Y(tot)}$ cannot be simply determined as the algebraic sum of the shearing forces at the base of all vertical structural members of the considered storey along the principal axes X and Y, respectively, since these forces do not occur concurrently and do not carry any particular sign. The use of such algebraic sums and differences may lead to non-conservative results not suited for design purposes. To this end, in the remainder of this subsection, the theoretically rigorous method along with two other alternative approaches for determining the peak interstorey drift sensitivity coefficients, extr$\theta_X^{(k)}$ and extr$\theta_Y^{(k)}$ from seismic effects derived using the MRSM are presented. Rigorous approach Flowchart FC-3.10b presents the required steps for determining the extreme (design) values of the interstorey drift sensitivity coefficients along the two principal directions X and Y using the MRSM in a theoretically rigorous manner. These steps are described below.

1. The application of standard modal analysis yields mode shapes (one for each i; i = 1,2,. . .,N, underlying mode of vibration), commonly normalized with respect to the generalized mass. These N mode shapes are proportional to the nodal displacement configuration of the structure assumed to vibrate according to each single, i, mode of vibration. Therefore, modal analysis involves the calculation of modal displacements of the nodes of the FE structural model, as well as modal shearing forces at the base of all vertical structural members. Using these modal quantities, the average interstorey drifts $d_{rXi}^{(k)}$ and $d_{rYi}^{(k)}$ and the sum of the shearing forces at the base of all vertical structural members $V_{Xi(tot)}^{(k)}$ and $V_{Yi(tot)}^{(k)}$ at each storey k and for each mode i are calculated. In this manner, N values for each of the above terms is computed at each storey k of the building. It is noted that the

above terms do not carry any information about the amplitude (intensity) of the design seismic action. Further, it is also noted that, due to the coupling of the degrees of freedom in space non-zero $d_{rXi}^{(k)}$ and $d_{rYi}^{(k)}$, and $V_{Xi(tot)}^{(k)}$ and $V_{Yi(tot)}^{(k)}$ terms appear concurrently in both principal directions X and Y.

2. The above modal terms are utilized to derive the seismic modal response quantities at each floor k and for each mode i by considering separately the design seismic action along the two principal directions X and Y. In particular, the following seismic modal structural response quantities are computed for seismic action along direction X

$$\begin{aligned} d_{rX,EXi}^{(k)} &= \left(v_{iX}\frac{S_{ai}}{\omega_i^2}\right)d_{rXi}^{(k)}; \quad & d_{rY,EXi}^{(k)} &= \left(v_{iX}\frac{S_{ai}}{\omega_i^2}\right)d_{rYi}^{(k)}; \\ V_{X(tot),EXi}^{(k)} &= \left(v_{iX}\frac{S_{ai}}{\omega_i^2}\right)V_{Xi(tot)}^{(k)}; \quad & V_{Y(tot),EXi}^{(k)} &= \left(v_{iX}\frac{S_{ai}}{\omega_i^2}\right)V_{Yi(tot)}^{(k)} \end{aligned} \tag{3.20}$$

and along direction Y

$$\begin{aligned} d_{rX,EYi}^{(k)} &= \left(v_{iY}\frac{S_{ai}}{\omega_i^2}\right)d_{rXi}^{(k)}; \quad & d_{rY,EYi}^{(k)} &= \left(v_{iY}\frac{S_{ai}}{\omega_i^2}\right)d_{rYi}^{(k)}; \\ V_{X(tot),EYi}^{(k)} &= \left(v_{iY}\frac{S_{ai}}{\omega_i^2}\right)V_{Xi(tot)}^{(k)}; \quad & V_{Y(tot),EYi}^{(k)} &= \left(v_{iY}\frac{S_{ai}}{\omega_i^2}\right)V_{Yi(tot)}^{(k)} \end{aligned}. \tag{3.21}$$

In Eqs. (3.20) and (3.21), v_{iX} and v_{iY} are the modal participation factors of mode i corresponding to the natural frequency ω_i for earthquake excitation along axes X and Y, respectively, expressed by the pseudo-spectral design ordinate $S_{\alpha i}$. Next, the seismic modal quantities of Eqs. (3.20) and (3.21) are used in conjunction with Eq. (3.19) to calculate the interstorey drift sensitivity coefficients $\theta_{X,EXi}^{(k)}$ and $\theta_{Y,EXi}^{(k)}$ for design seismic action along direction X, and $\theta_{X,EYi}^{(k)}$ and $\theta_{Y,EYi}^{(k)}$ for design seismic action along direction Y, respectively (see also FC-3.10b). Therefore, if all N mode shapes are taken into account, 2N seismic modal interstorey drift sensitivity coefficients $\theta_{Xi}^{(k)}$ and 2N seismic modal interstorey drift sensitivity coefficients $\theta_{Yi}^{(k)}$ are obtained at each k storey.

3. The third step involves modal combination (CQC rule) of the seismic modal interstorey drift sensitivity coefficients computed in the previous step at each k storey for seismic action along direction X (i.e., modal combination of $\theta_{X,EXi}^{(k)}$ and $\theta_{Y,EXi}^{(k)}$ values) and for seismic action along direction Y (i.e., modal combination of $\theta_{X,EYi}^{(k)}$ and $\theta_{Y,EYi}^{(k)}$ values). In this manner, the expected (most probable) peak values $\max\theta_{X,EX}^{(k)}$ for seismic action along direction X, and $\max\theta_{X,EY}^{(k)}$ for seismic action along direction Y of the interstorey drift sensitivity coefficient $\theta_X^{(k)}$, and the expected (most probable) peak values $\max\theta_{Y,EX}^{(k)}$ for seismic action along

direction X, and $\max \theta^{(k)}_{Y,EY}$ for seismic action along direction Y of the interstorey drift sensitivity coefficient $\theta^{(k)}_{Y}$ are derived for each storey k.

4. Finally, spatial combination (SRSS rule) is undertaken to compute the expected extreme value $\mathrm{extr}\,\theta^{(k)}_{X}$ of the interstorey drift sensitivity coefficient $\theta^{(k)}_{X}$ by combining the values $\max \theta^{(k)}_{X,EX}$ and $\max \theta^{(k)}_{X,EY}$ obtained in the previous step and the expected extreme value $\mathrm{extr}\theta^{(k)}_{Y}$ of the interstorey drift sensitivity coefficient $\theta^{(k)}_{Y}$ by combining the values $\max \theta^{(k)}_{Y,EX}$ and $\max \theta^{(k)}_{Y,EY}$ obtained in the previous step for all k storeys. These extreme values computed as

$$\begin{aligned} \mathrm{extr}\theta^{(k)}_{X} &= \sqrt{\left(\max\theta^{(k)}_{X,EX}\right)^2 + \left(\max\theta^{(k)}_{X,EY}\right)^2} \quad ; \quad \text{and} \\ \mathrm{extr}\theta^{(k)}_{X} &= \sqrt{\left(\max\theta^{(k)}_{X,EX}\right)^2 + \left(\max\theta^{(k)}_{X,EY}\right)^2} \end{aligned} \tag{3.22}$$

correspond to the simultaneous design seismic action along both principal axes X and Y. In this regard, the verification check of Eq. (3.17) for second-order effects needs to be undertaken by setting the interstorey drift sensitivity coefficient θ equal to the above extreme values.

3.2.2 *Verification Check for Maximum Interstorey Drifts (FC-3.11a and FC-3.11b)*

The verification check of clause §4.4.3.2(1) of EC8 aims to satisfy the damage (and, therefore, the economic loss) limitation requirement of clause §2.1(1)P of EC8 (see also Sect. 1.1.4) for a "frequent" seismic action which is less intense and has a higher probability of occurrence than the "design seismic action". This check verifies that the non-dimensional interstorey drift normalized by the storey height remains below a certain threshold (limiting value), as in (§4.4.3.2(1) of EC8)

$$\frac{d_r \cdot v}{h} = \frac{(q \cdot d_e) \cdot v}{h} \leq \text{limiting value}, \tag{3.23}$$

where:

d_r is the design interstorey drift defined in Sect. 3.2.1;
v is a reduction factor defined in the EC8 National Annexes which takes into account the lower return period (higher probability of occurrence) of the seismic action associated with the damage limitation requirement; and
h is the storey height.

In the above condition, the limiting value depends on the nature of the non-structural members, which should remain undamaged under the "frequent" seismic action. Although not straightforwardly mentioned in EC8, the verification check of Eq. (3.23) needs to be undertaken along both principal directions, X and Y, of a given building, that is, for both d_{rX} and d_{rY} intersotrey drifts normalized by the storey height as in

$$\begin{aligned} \frac{d_{rX}\cdot v}{h} &= \frac{(q_X\cdot d_{eX})\cdot v}{h} \leq \text{limiting value;} \quad \text{and} \\ \frac{d_{rY}\cdot v}{h} &= \frac{(q_Y\cdot d_{eY})\cdot v}{h} \leq \text{limiting value} \end{aligned} \tag{3.24}$$

Flowcharts FC-3.11a and FC-3.11b present the sequence of the required computational steps for undertaking the damage limitation verification check of Eq. (3.24) for the lateral force method (LFM) and for the modal response spectrum method (MRSM) of analysis, respectively. It is noted that these flowcharts assume that the same behaviour factor is adopted along both principal directions of the seismic action (i.e., $q = q_X = q_Y$). In case $q_X \neq q_Y$, the design interstorey drift d_r is computed from the elastic interstorey drift d_e as in Eq. (3.24). As a final note, it is highlighted that, in the case of the MRSM of analysis, the verification check starts by computing the seismic modal interstorey drifts separately for each mode of vibration. It then proceeds with the calculation of the expected peak and extreme values of the interstorey drifts for simultaneous seismic action along both principal directions by application of appropriate modal combination and spatial combination rules, as detailed in sub-section 3.2.1.2.

3.3 Special Requirements for Infill Walls in R/C Building Structures

Historically, the influence of non-engineered infill (or architectural) walls in single seismic action level of analysis for code-compliant seismic design of r/c buildings is usually neglected. This is mostly because, under the "rare" design, seismic action infill walls normally fail before the first plastic hinge takes place and, therefore, it is reasonable to be left out of the mathematical (FE) model used in the single seismic analysis step for the design seismic action. Still, infill walls do influence the *global* seismic response of r/c building structures. In particular, they contribute to the overall lateral stiffness of the structure for low to moderate seismic input action levels. Further, having a reduced yielding deformation compared to the r/c structural members, infill walls normally fail first under moderate-to-high levels of earthquake shaking. In this manner, they contribute to seismic energy dissipation, though not substantially, since common infill walls are rather brittle. Therefore, completely ignoring infill walls in seismic design introduces a non-negligible uncertainty to the actual seismic performance of the structure. In this regard, clause §4.3.6 of EC8 includes special provisions for characterizing the conditions under which infill walls are anticipated to influence significantly the seismic design in a non-conservative manner and to account for this influence in design.

It is noted that the level of influence of infill walls on the *global* seismic response of structures varies significantly and depends on the properties of both the infill walls (e.g., their distribution in plan and elevation and their degree of connectivity/interaction achieved during construction with the neighboring r/c structural members), and of the lateral load resisting system. To this end, according to clause §4.3.6.1(4) of EC8, infill walls may be completely neglected for the seismic design of r/c buildings with wall or wall-equivalent dual lateral load resisting systems. Importantly, given that in high seismic areas, the inclusion of strong r/c walls to resist seismic loads is commonly adopted in practical design, the need to consider the influence of infill walls in the seismic design of r/c buildings in such areas arises in few exceptional practical cases. Moreover, the special requirements of infill walls need to be considered only for frame or frame-equivalent dual systems and only for the high ductility class (DCH). Still, it is advisable to satisfy these requirements even for the medium ductility class (DCM).

In principle, EC8 considers infill walls with no special provisions for structural connection to the r/c lateral load resisting system as non-structural elements (clause §4.3.6.1(1)P(c) of EC8). Therefore, infill walls are not normally accounted for in the FE model of the load resisting structural system considered in the analysis stage of seismic design of r/c buildings. *However*, in the case of severe *uneven distribution of infill walls in-plan*, a three dimensional FE model that includes infill walls needs to be adopted in the analysis (clause §4.3.6.3.1(2) of EC8). To this end, a common modelling approach for infill walls is to consider equivalent X-bracings (axial only one-dimensional finite elements) with appropriate material and cross-sectional properties (e.g., axial rigidity EA) placed within the r/c frame openings occupied by infill panels (Fardis 2009). Furthermore, in case engineered masonry infill walls are included in the load resisting structural system, their analysis and design should be carried out by following the criteria and rules for confined masonry included in chapter 9 of EC8-part 1 (clause §4.3.6.1(5) of EC8).

Apart from potentially significant influence on the global seismic response of r/c buildings, EC8 recognizes that infill walls can have significant *local* adverse effects on the neighboring r/c columns (Penelis and Kappos 1997; Penelis and Penelis 2014) and prescribes special measures to be taken locally against these effects in clause §5.9. Specifically, clause §5.9.1 of EC8 warns of the increased vulnerability of ground floor infill walls. It further specifies that, if no detailed analysis is undertaken to account for potential failures of ground floor infill walls and their consequences, then ground floor r/c columns should be designed and detailed (i.e., confined) as critical members along their full height. Moreover, clause §5.9(2) of EC8 addresses the issue of short column formation in case one or more of the adjacent infill wall does not extend to the full free height of a column (see also Fig. 2.3). This issue is addressed by considering an effective capacity design shearing force according to clauses §5.4.2.3 and 5.5.2.2 to size and detail the r/c columns.

In summary, FC-3.12 provides an overview of the various EC8 requirements and additional measures involved in accounting for the potential local and global adverse influence and effects of infill walls in r/c building structures and points to the relevant clauses of EC8.

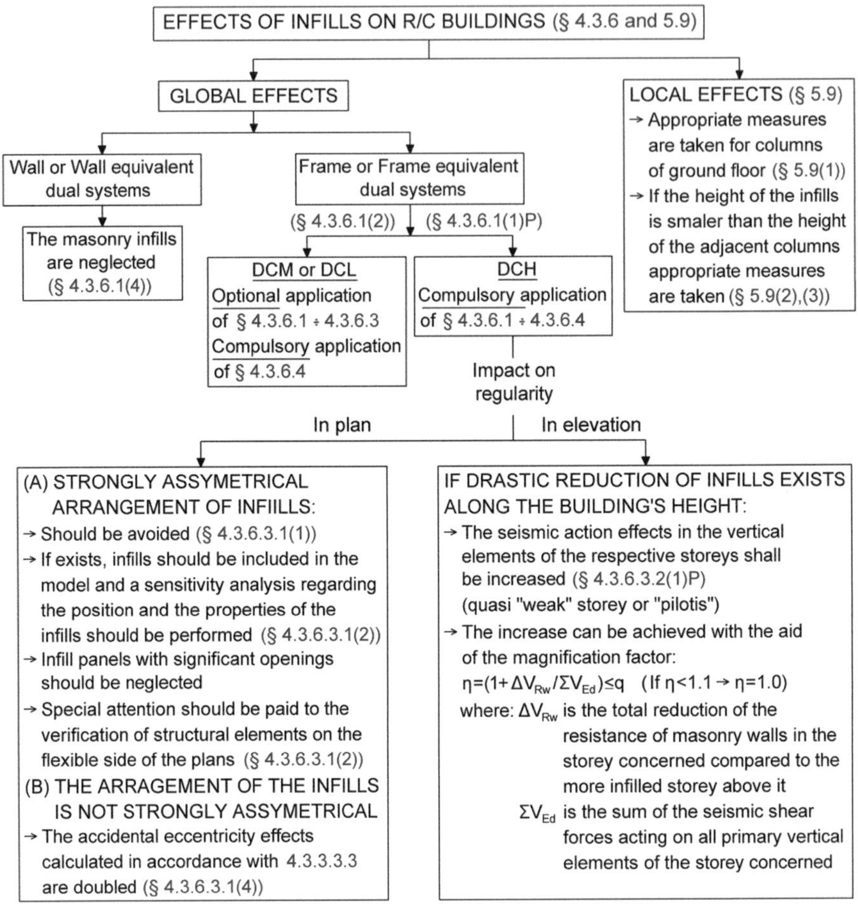

Flowchart 3.12 Influence of non-engineered infill walls in r/c buildings

3.4 Practical Recommendations for EC8 Compliant Seismic Analysis and Verification Checks

By considering collectively the discussions related

- to the selection of the desired seismic performance level at the onset of the seismic design process of r/c buildings (Sect. 1.4);
- to the expected behavior of r/c buildings designed for different seismic performance levels under the design seismic action (beginning of Chap. 3); and
- to particular choices made during the design process to reduce uncertainties in design and, therefore, to simplify the required EC8 verification checks (Sect. 3.1),

it is herein *recommended* by the authors to adopt the following practices in the EC8 compliant seismic design process for r/c buildings, whenever possible and/or applicable:

1. In case there is no written consent of the building owner with regards to the desired seismic performance level to be targeted at the design stage, the designer should choose to design for a relatively "high" performance and, therefore, to adopt a relatively low behavior factor q (e.g., q = 1.5–1.75) according to which the design seismic loads are reduced. In every case, the adopted behavior factor should be less than or equal to the maximum allowed value, $maxq_{allow}$, as prescribed by EC8 (see Sect. 3.1.4). It is further reminded that all EC8 regularity checks in plan and elevation and the computation of $maxq_{allow}$ can be circumvented by selecting $q \leq 1.5$, using a 3D structural model and applying the response spectrum analysis (see FC-3.1).
2. As a rule of thumb, the medium ductility class (DCM) should be preferred over the high ductility class (DCH) in areas of high seismicity. As a practical example in support of the above argument, it is noted that the shearing verification check for r/c walls in DCH is not easily satisfied in practice in areas of high seismicity mainly due to the enforced reduction by 40 % of the available shear capacity $V_{Rd,max}$ compared to the DCM (see FC-3.19b). In general, DCH should be adopted only in cases in which it is felt that the established local building practices, availability of technological means, and on-site supervision are sufficient to ensure that the prescribed stringent detailing requirements of DCH can actually be implemented during construction.
3. Structural modeling should involve a single spatial (three-dimensional) FE model, as opposed to two or more planar (two-dimensional) models.
4. The modal response spectrum method (MRSM) of analysis should be preferred over the lateral force method (LFM).
5. The deformation-based verification checks should be undertaken by following the rigorous procedures described in Sects. 3.2.1 and 3.2.2.
6. A final check should be made to ensure that the demand capacity ratio (DCR) at critical sections of r/c structural members observe a reasonably uniform distribution and preferably take values close to unity (≤ 1).

3.5 Determination of Design Seismic Effects for r/c Walls

Clauses §5.4.2.4 and §5.5.2.4 of EC8 list special design provisions for medium and high ductility class (DCM and DCH) r/c walls, respectively. The aim is to minimize the uncertainty of the post-yield seismic behavior of these structural members. At the core of these provisions lies the requirement to design r/c walls in flexure and shear by relying on design "envelope" bending moment and shearing force diagrams, respectively. These design diagrams are derived from the seismic effects computed in the analysis step for the seismic design load combinations. In what follows, a detailed discussion and comments are included on the derivation of the design envelope bending moment and shearing force diagrams for EC8-compliant design of r/c walls.

3.5.1 Envelope Bending Moment Diagram for Seismic Design of r/c Walls

The procedure for deriving the design envelope of the bending moment diagram for r/c walls is described in clause 5.4.2.4(5) of EC8 with the aid of Fig. 5.3 of EC8. This procedure concerns in-plane bending moments (i.e., bending moments about an axis normal to the vertical plane defined by the wall) developing along the height of the wall. According to clause 5.4.2.4(4)P of EC8, this procedure applies only in the case of slender walls with aspect ratio $H_w/L_w > 2$, where H_w is the height of the wall and L_w is the length of the long side of the wall in-plan. The detailing of an r/c wall for the thus derived envelope bending moment diagram ensures that the only region of the r/c wall where a plastic hinge can form (critical energy dissipation zone with high ductility demand) is at the base of the ground floor.

In other words, the seismic moment demand may only exceed the bending capacity of the r/c wall at its base; the r/c wall will behave elastically along the remainder of its height. In this manner, stringent detailing requirements for local ductile behavior apply only at the base of r/c walls and nowhere else along their height. Fig. 3.5 summarizes the procedure for determining the bending moment design envelope diagram for slender r/c walls according to EC8. It is noted that this procedure applies for buildings that do not observe any significant abrupt discontinuities in the mass, stiffness, and strength properties distribution in elevation. Additional explanatory comments on Fig. 3.5 are given below.

Explanatory comments related to Fig. 3.5

1. The procedure detailed in Fig. 3.5 needs to be applied twice (i.e., once by considering the bending moment diagram obtained from the $G+\psi_2Q+E$ seismic design combination and once by considering the bending moment diagram obtained from the $G+\psi_2Q$-E seismic design combination) for each of the (four) considered mass locations (see Fig. 3.4).
2. R/c walls are subject to biaxial bending with axial force. Therefore, the values of the in-plane bending moment, M_3, obtained from the design envelope need to be combined with appropriate design values of the out-of-plane bending moment, M_2, and the axial load, N, at each section along the height of the r/c wall to verify the adequacy of the reinforcement used. These design values are determined by considering triads of the expected peak values of the involved stress resultants (N, M_2, M_3) acting simultaneously at a particular section assuming that the "critical" in-plane bending moment attains its extreme value with both possible signs: $M_3 = \pm \text{extr}M_3$. As an example, the two triads (N, M_2, M_3) to be considered for the design/verification of an arbitrary section at a distance $x_N > \alpha_l$ from

Bending moment diagram of the wall

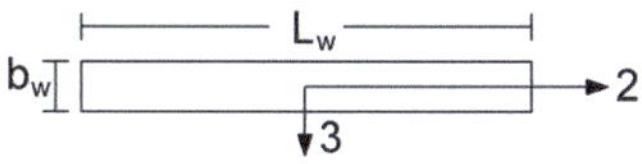

Action: Seismic design combination $G+\psi_2 Q \pm E$ (the probable extreme values produced by the seismic action are used in the above combination).

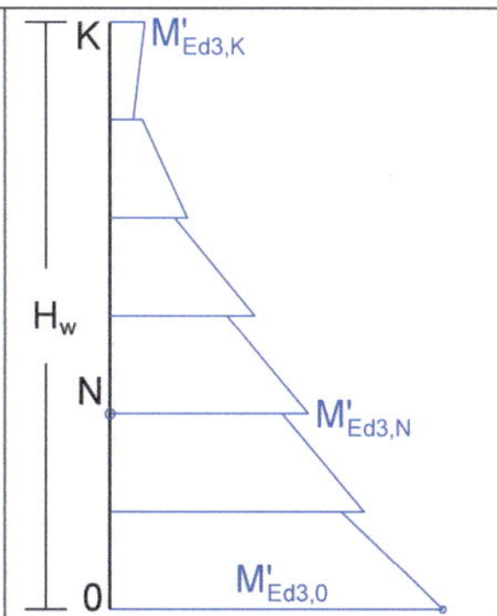

STEP 1: Join with a straight line points A and B:

- Point A corresponds to the moment value produced by the seismic design combination at the base of the wall (bending moments $M'_{Ed3,A}=M'_{Ed3,0}$).
- Point B corresponds to the moment value produced by the seismic design combination at such level that the straight line segment AB envelopes the moment diagram along the total height (In the figure beside the point B is at the top of the wall K: $M'_{Ed3,B}=M'_{Ed3,K}$).

Determine the angle φ: $\tan\varphi=\dfrac{M'_{Ed3,A}-M'_{Ed3,B}}{H_w}$

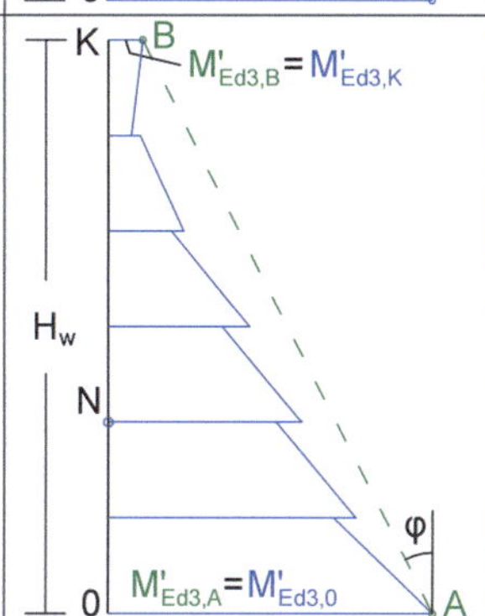

STEP 2:

- The line segment AB is shifted up by a distance α_l parallel to the wall axis.
- This shift results to the line segment CD:
  - The point C is defined by the length α_l in the vertical direction (line segment AC).
  - The point D is defined as the intersection of the line CD with the horizontal line passing through the point K.

The envelope for bending moments is the bilinear curve consisting of:

(a) The line segment AC of length α_l along which the bending moment is constant and equal to $M'_{Ed3,0}$.

(b) The line segment CD for the wall length $H_w-\alpha_l$ where the bending moment varies linearly and is given by the relation: $M_{Ed3}(x)=M_{Ed3,0}-(x-\alpha_l)\tan\varphi$

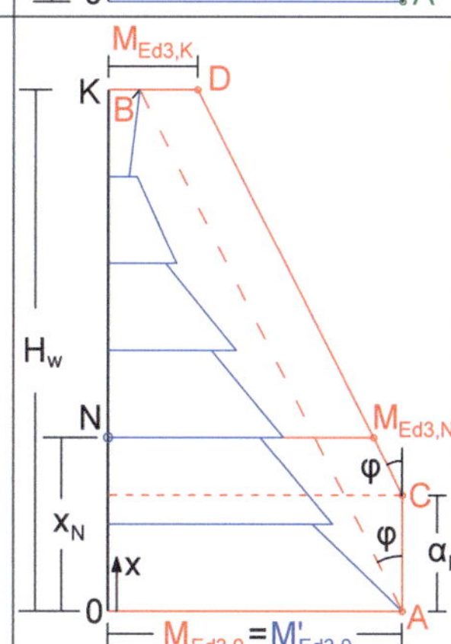

$\alpha_l=(z/2)(\cot\theta-\cot\alpha)$
[Eq. (9.2), EN1992-1-1]

Fig. 3.5 Calculation of envelope bending moment diagram for seismic design of r/c walls

Table 3.7 Final triads (N, M_2, M_3) for wall bending design

$N'_{Ed,N}$(kN)	$M'_{Ed2,N}$(kNm)	$M'_{Ed3,N}$(kNm)
$N_{,G}+\Psi_2 N_{,Q}+N_{,M3}$	$M_{2,G}+\Psi_2 M_{2,Q}+M_{2,M3}$	$M_{3,G}+\Psi_2 M_{3,Q}+exM_3$
$N_{,G}+\Psi_2 N_{,Q}-N_{,M3}$	$M_{2,G}+\Psi_2 M_{2,Q}-M_{2,M3}$	$M_{3,G}+\Psi_2 M_{3,Q}-exM_3$

⇩

$N'_{Ed,N}$(kN)	$M'_{Ed2,N}$(kNm)	$M_{Ed3,N}$(kNm)
$N_{,G}+\Psi_2 N_{,Q}+N_{,M3}$	$M_{2,G}+\Psi_2 M_{2,Q}+M_{2,M3}$	$M_{Ed3}(X_N)=M^{(+E)}_{Ed3,0}-(X_N-\alpha_l)\tan\varphi_{(+E)}$
$N_{,G}+\Psi_2 N_{,Q}-N_{,M3}$	$M_{2,G}+\Psi_2 M_{2,Q}-M_{2,M3}$	$M_{Ed3}(X_N)=M^{(-E)}_{Ed3,0}-(X_N-\alpha_l)\tan\varphi_{(-E)}$

the base of the wall are given in Table 3.7, where the symbols (+E) and (−E) are used to denote the values of the quantities $M_{Ed3,0}$ and $\tan\varphi$ derived from the design envelope bending moment corresponding to the design seismic combinations $G+\psi_2 Q+E$ and $G+\psi_2 Q-E$, respectively.

3. The displacement of the bending moment diagram by the "tension shift" distance α_l is determined by equation (9.2) of EC2-part 1 (clause §9.2.1.3 of EC2), repeated here for convenience

$$\alpha_l = \frac{z}{2}(\cos\theta - \cot\alpha). \tag{3.25}$$

In the above equation,

- z is the lever arm of the internal resisting forces which generates the resisting bending moment at the considered wall section. According to clause §5.5.3.4.2 (1) of the EC8, it can be taken as $z=0.8L_w$ for slender DCH walls. Since no special reference is made in EC8 regarding DCM walls, the same assumption can be made for the medium ductility class walls as well.
- α is the angle (inclination) between the transverse/shearing reinforcement and the longitudinal axis of the wall which is normally equal to 90°: the shearing reinforcement in walls forms an orthogonal grid with the longitudinal reinforcement. Therefore $\cot\alpha=0$.
- θ is the angle between the virtual "concrete compression strut" and the longitudinal axis of the wall (see clause §6.2.3 of EC2). According to clause §5.5.3.4.2 (1) of EC8 the values $\tan\theta=\cot\theta=1$ can be adopted for DCH walls. For the case of DCM walls, clause §5.4.3.4.1(1) of EC8 directs to EC2 for every issue related to the shearing capacity. Therefore, for DCM walls, the angle θ can assume any value such that $1\le\cot\theta\le 2.5$ according to clause §6.2.3(2) of EC2.

3.5.2 *Envelope Shear Force Diagram for Seismic Design of r/c Walls*

The procedure for deriving the design envelope of the shearing forces diagram for r/c walls from the shearing force diagrams obtained from the analysis step for the seismic design combination is described in clauses §5.4.2.4(7) and §5.4.2.4(8) of EC8 with the aid of Fig. 5.4 of EC8. This procedure is required to ensure accounting for the potential increase of the acting shearing forces close to the base of the wall upon yielding (clause §5.4.2.4(6)P of EC8). The values of the various parameters involved in the derivation of the shearing force design envelope to be adopted for DCM and for DCH walls are specified in clauses §5.4.2.4(7) and §5.5.2.4.1(7) of EC8, respectively. For the case of the DCH walls, a further distinction between slender walls (i.e., walls with aspect ratio $H_w/L_w > 2$) and squat walls (i.e., walls with aspect ratio $H_w/L_w < 2$) is made. The related specifications for DCH slender walls are given in clauses §5.4.2.4(7) of EC8, while clause §5.5.2.4.2(2) of EC8 is dedicated to squat walls. Fig. 3.6 summarizes the procedure for determining the shearing force design envelope diagram for slender r/c walls according to EC8. Additional explanatory comments on Fig. 3.6 are given below.

Explanatory comments related to Fig. 3.6

1. The procedure detailed in Fig. 3.6 needs to be applied twice (i.e., once by considering the bending moment diagram obtained from the $G+\psi_2Q+E$ seismic design combination and once by considering the bending moment diagram obtained from the $G+\psi_2Q-E$ seismic design combination) for each of the (four) considered mass locations (see Fig. 3.4).
2. The transverse reinforcement of r/c walls should be determined by considering the shearing forces of the design envelope instead of the shearing forces obtained from the analysis for the seismic design combination. In particular, consider an arbitrary section at a distance $x_N > H_w/3$ from the base of the wall, as indicated in Fig. 3.6. The value of the shearing force V_2 which the transverse reinforcement needs to accommodate should not come from the analysis step for the combination $G+\psi_2Q \pm E$, but from the design shearing force envelopes derived from the shearing force diagrams corresponding to the above combinations. That is,

$$V_{Ed2}^{(+E)}(x_N) = \left(\frac{3}{2} \times \frac{V_{Ed2,K}^{(+E)} - V_{Ed2,C}^{(+E)}}{H_w}\right) x_N - \frac{1}{2}\left(V_{Ed2,K}^{(+E)} - 3V_{Ed2,C}^{(+E)}\right) \tag{3.26}$$

corresponding to the combination $G+\psi_2Q+E$, and

$$V_{Ed2}^{(-E)}(x_N) = \left(\frac{3}{2} \times \frac{V_{Ed2,K}^{(-E)} - V_{Ed2,C}^{(-E)}}{H_w}\right) x_N - \frac{1}{2}\left(V_{Ed2,K}^{(-E)} - 3V_{Ed2,C}^{(-E)}\right) \tag{3.27}$$

corresponding to the combination $G+\psi_2Q-E$.

Shear force diagram of the wall (shear forces V'_{Ed2}):

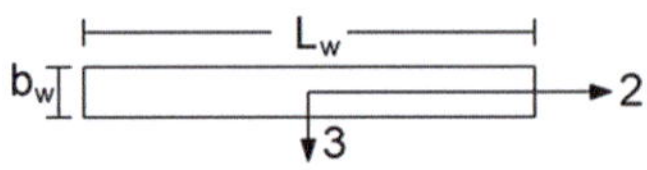

Action: Seismic design combination $G+\psi_2 Q\pm E$ (the probable extreme values produced by the seismic action are used in the above combination).

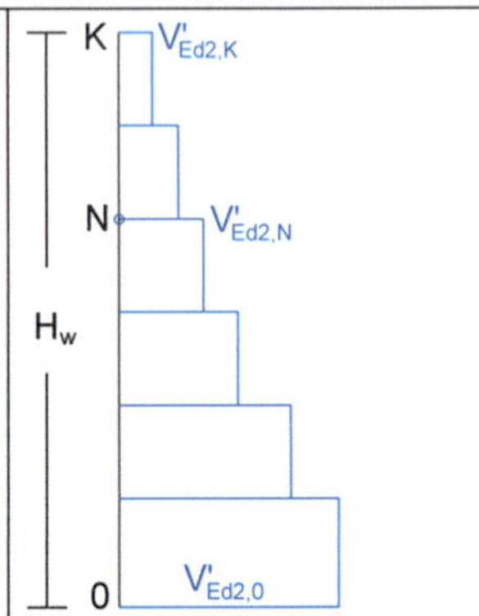

STEP 1: The values of shear forces produced by the seismic design situation are increased by an amplification factor ε:

- SHEAR FORCES $V_{Ed2}=\varepsilon\cdot V'_{Ed2}$
- AMPLIFICATION FACTOR ε

	"Squat" walls $H_w/L_w<2.0$	Slender walls $H_w/L_w>2.0$
DCM	1.5	1.5
DCH	$\gamma_{Rd}\cdot\left[\frac{M_{Rd}}{M_{Ed}}\right]\le q$	$\sqrt{\left[\frac{\gamma_{Rd}}{q}\cdot\frac{M_{Rd}}{M_{Ed}}\right]^2+0.1\left[\frac{S_e(T_c)}{S_e(T_1)}\right]^2}\le q$

At this stage the plot of the shear forces envelope for "squat" walls is completed.

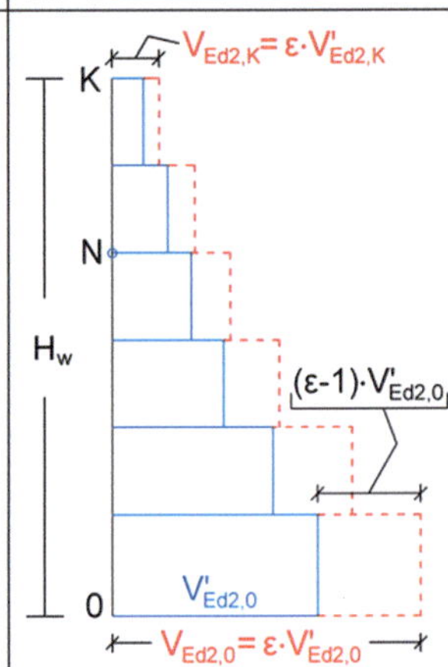

STEP 2: (only for slender walls)

- Up to the height $H_w/3$ the envelope for shear forces is that obtained in step 1. Hence the point A is defined.
- Determine the value of shear force at the top of the wall: $V_{Ed2,K}=\max\{(\varepsilon\cdot V'_{Ed2,K});(V_{Ed2,0}/2)\}$ Hence the point B is defined.
- Join points A and B with a straight line as we assume linear variation of the shear forces. The values of the shear forces are given by the realation: $V_{Ed2}(x)=\left[\frac{3}{2}\cdot\frac{V_{Ed2,K}-V_{Ed2,C}}{H_w}\right]\cdot x-\frac{1}{2}\cdot\left[V_{Ed2,K}-3\cdot V_{Ed2,C}\right]$

  Where: $V_{Ed2,C}=\varepsilon\cdot V'_{Ed2,C}=\varepsilon\cdot V'_{Ed2}(H_w/3)$

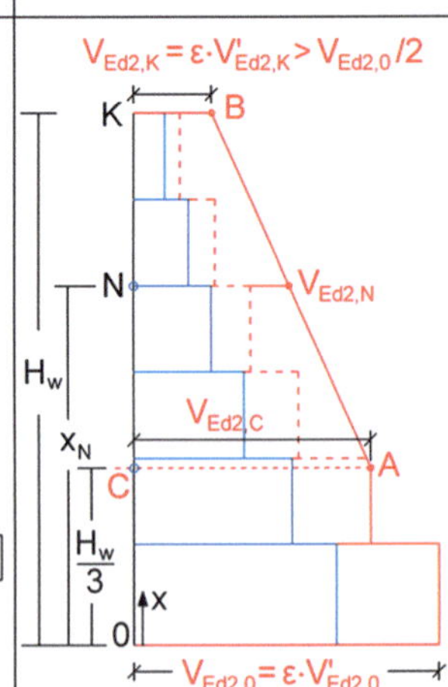

Fig. 3.6 Calculation of envelope shear force diagram for seismic design of r/c walls

3. The magnification factor ε required in step 1 of the procedure detailed in Fig. 3.6 is a function of the following parameters (clause §5.5.2.4.1(7) of EC8)

q	is the behavior factor used in the seismic analysis step;
M_{Ed}	is the design bending moment at the base of the wall;
V_{Ed}	is the design in-plane shearing force of the wall;
M_{Rd}	is the bending moment resistance at the base of the wall;
γ_{Rd}	is the overstrength coefficient due to the expected hardening of steel reinforcement. In the absence of case-specific material properties, γ_{Rd} can be taken equal to 1.2;
T_1	is the fundamental natural period of the structure along the direction of the V_{Ed} shearing force;
T_C	is the upper corner period of the plateau (constant spectral acceleration region) of the EC8 spectrum (Fig. 2.21); and
$S_e(T)$	is the spectral ordinate of the EC8 elastic response spectrum (Fig. 2.21)

4. If the building includes underground storeys acting as a box-type basement, as defined in clause §5.8.1(5) of EC8, then the design shearing force for the part of the r/c walls extending below the ground remains constant and corresponds to the value developed assuming that the wall attains its full flexural overstrength $\gamma_{Rd}M_{Rd}$, where $\gamma_{Rd} = 1.1$ for DCH and $\gamma_{Rd} = 1.2$ for DCM. Specifically, this design shearing force is determined as shown in the diagram provided in Fig. 3.7.

3.6 Detailing Requirements and Verification Checks for r/c Structural Members

In previous sections of this chapter, the rationale and computational steps involved in the seismic analysis stage of the seismic design process have been presented with the aid of 12 flowcharts along with relevant verification checks for adverse influences and deformation-based criteria. The next stage concerns the detailing of r/c structural members of the load resisting system based on cross-sectional stress resultants (i.e., moments, shear forces, and axial forces), or *design seismic effects*, computed in the seismic analysis stage. Specifically, this stage involves determining ratios of longitudinal and transverse steel reinforcement for all r/c structural members using the same member sizes adopted in the analysis stage such that all detailing requirements and verification checks of EC8-part 1 and of EC2-part 1 are met.

All the involved calculation steps and verification checks for EC2/EC8-compliant design/detailing of r/c beams, columns, and walls for medium and high

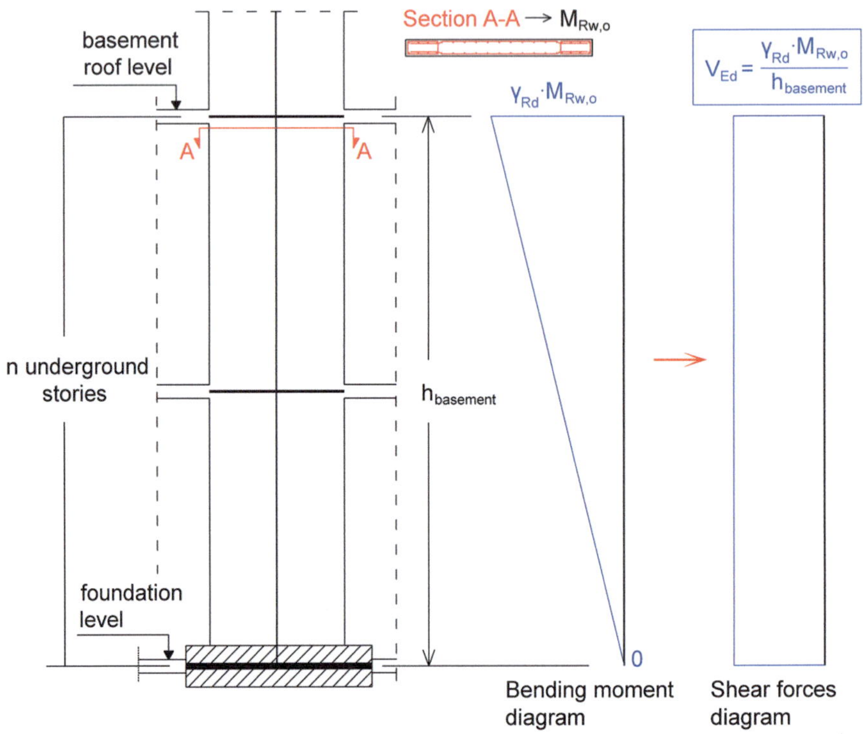

Fig. 3.7 Shear force envelope in underground storeys of r/c walls

Table 3.8 List of design and detailing flowcharts for r/c beams, columns, and walls

Structural member	Effect	Flowchart number
r/c beam	In flexure	13.a and 13.b
r/c column	In flexure	14a, 14b and 14c
r/c beam	In shear	15a, 15b and 15c
r/c column	In shear	16a, 16b, 16c and 16d
r/c beam-column joints	Miscellaneous effects	17
r/c wall	In flexure	18a and 18b
r/c wall	In shear	19a, 19b and 19c

ductility class r/c buildings are provided in a series of flowcharts. These are listed in Table 3.8 for ease of reference. The illustrative detailing calculations and verification checks included in Sect. 4.3 (example building C) follow these flowcharts (FC-3.13a, 3.13b, 3.14a, 3.14b, 3.14c, 3.15a, 3.15b, 3.15c, 3.16a, 3.16b, 3.16c, 3.16d, 3.17, 3.18a, 3.18b, 3.19a, 3.19b and 3.19c).

INPUT DATA

Geometry and building properties

Preliminary beam design: width (b_w), height (h_w)
Preliminary column design at the joints: width (b_c), height (h_c)
Behavior factor in the direction of the frame: q_0
Uncoupled translational eigenperiod T & corner period T_C

Material properties

Characteristic strength: f_{yk}, f_{ck}

DCM → f_{ck}>16MPa

DCH → f_{ck}>20MPa

Design actions for each loading combination (rectangular grid)

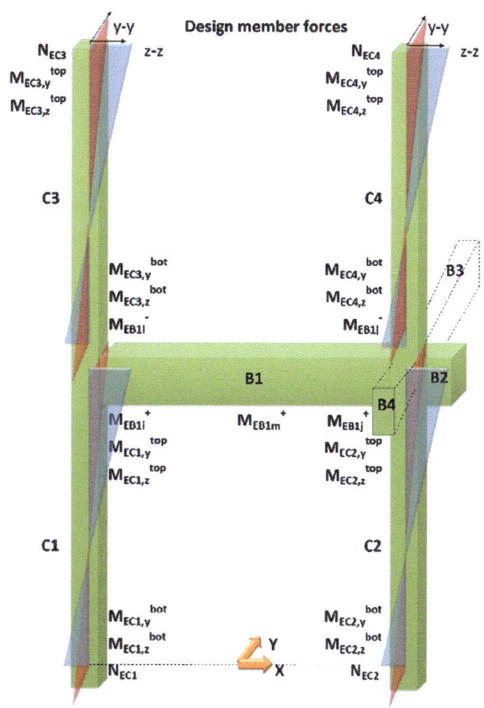

Flowchart 3.13a Design of beam for flexure (1/2)

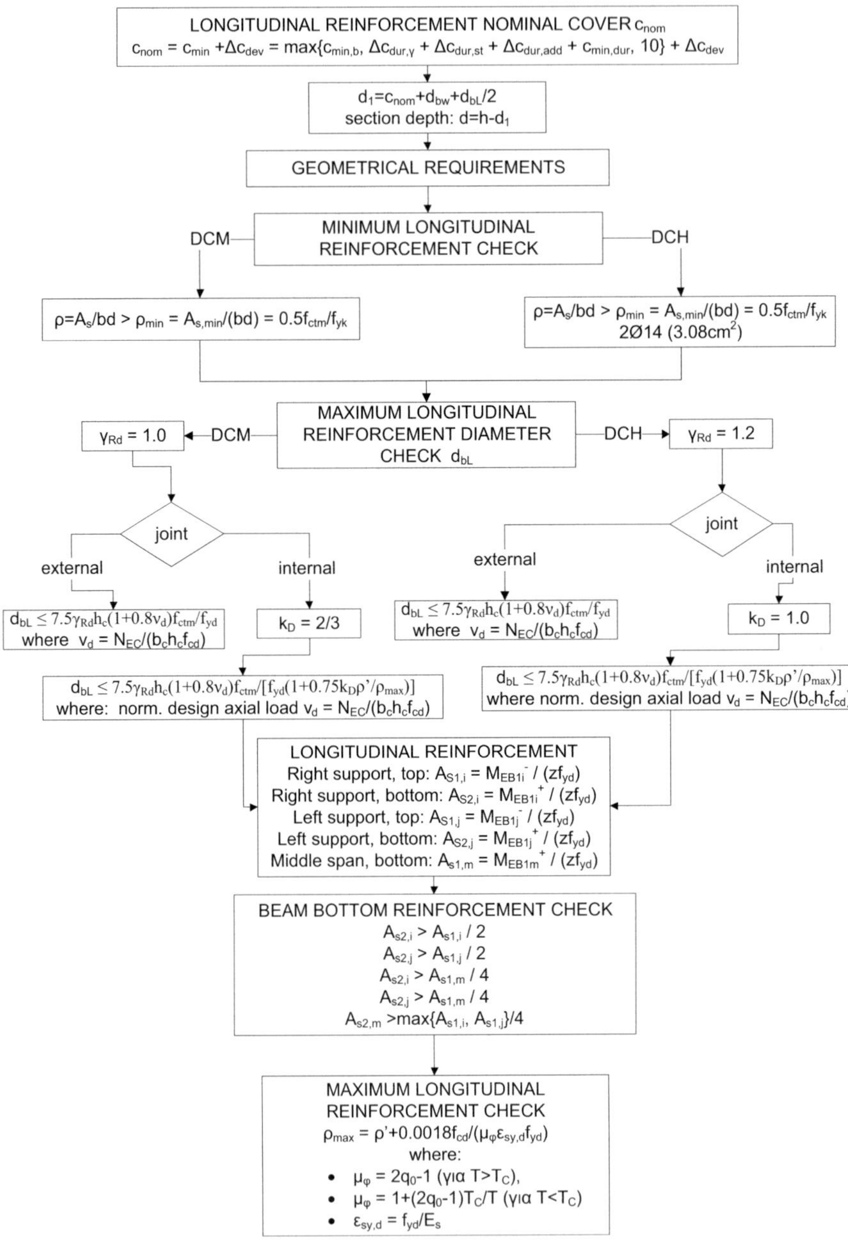

Flowchart 3.13b Design of beam for flexure (2/2)

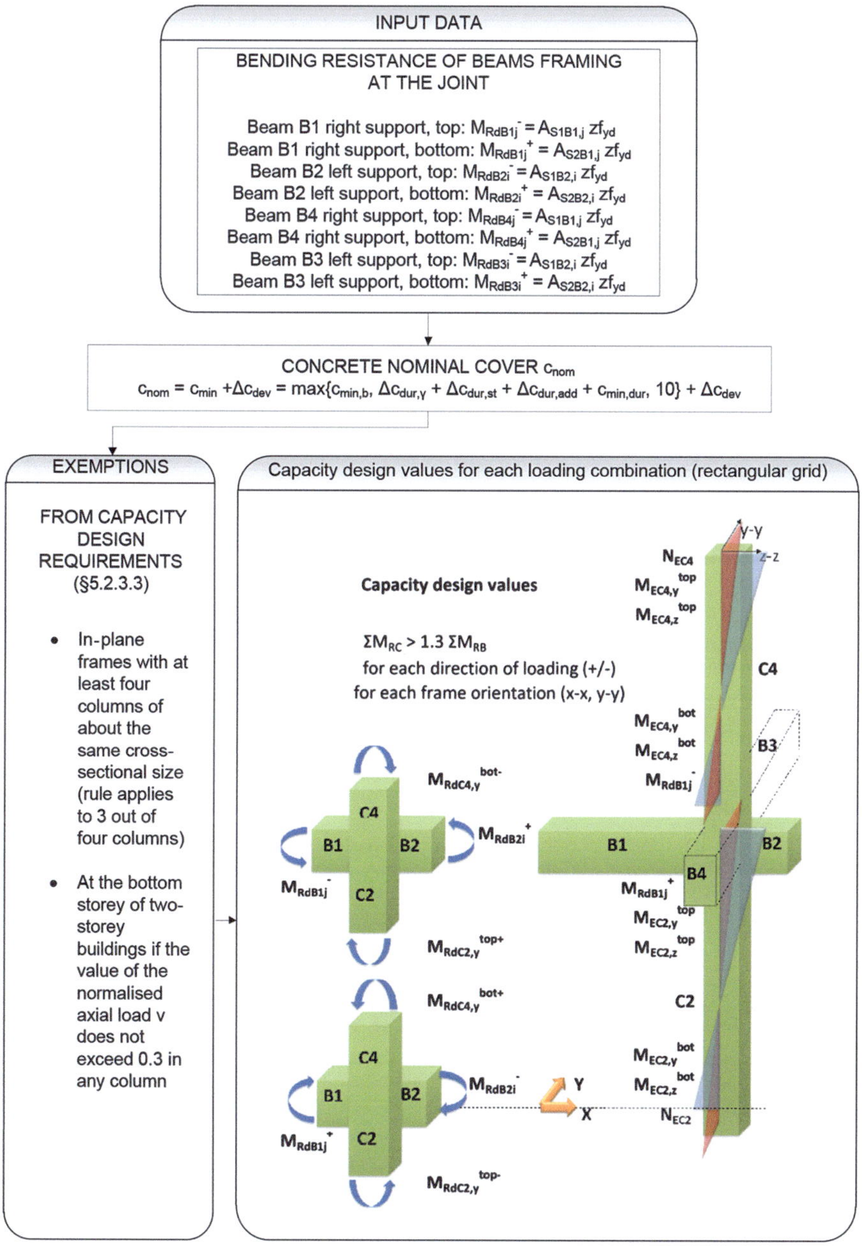

Flowchart 3.14a Design of column for flexure (1/3)

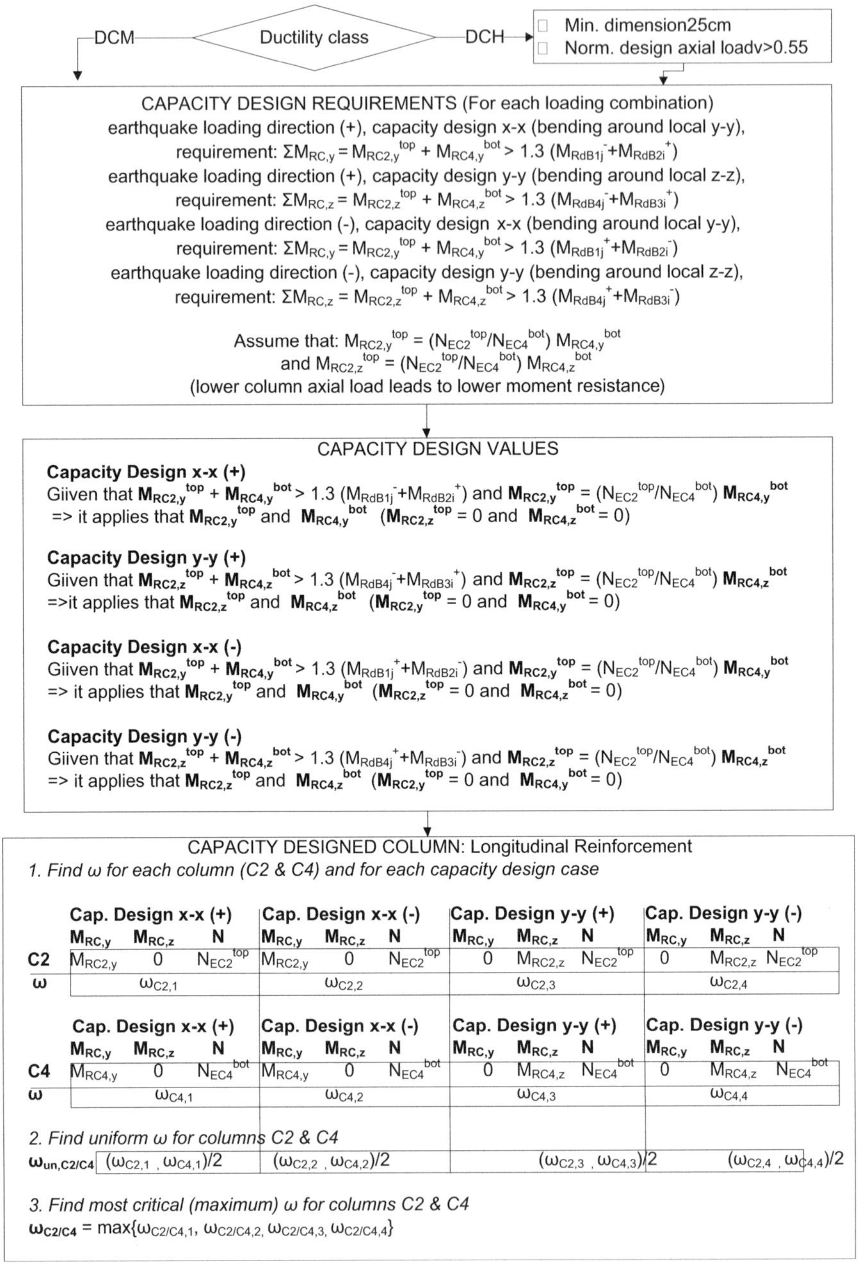

Flowchart 3.14b Design of column for flexure (2/3)

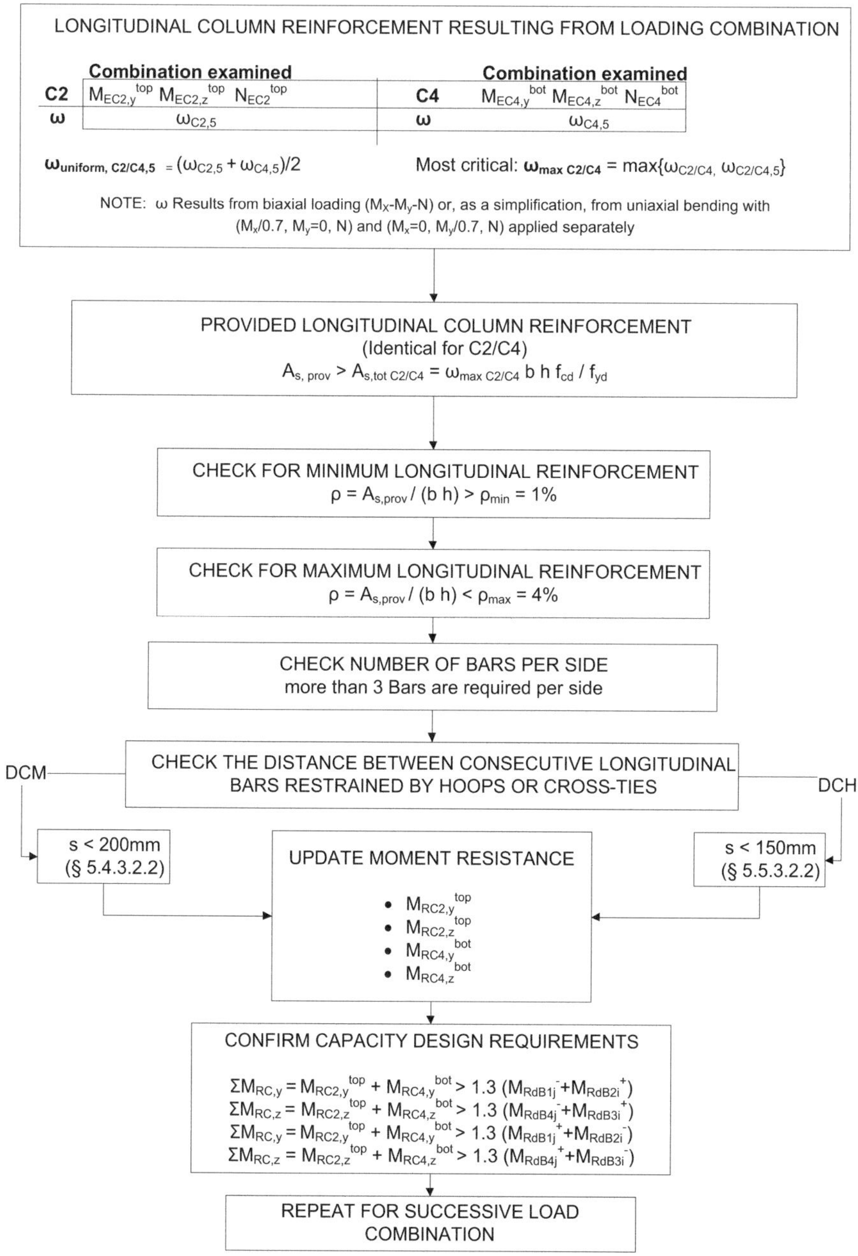

Flowchart 3.14c Design of column for flexure (3/3)

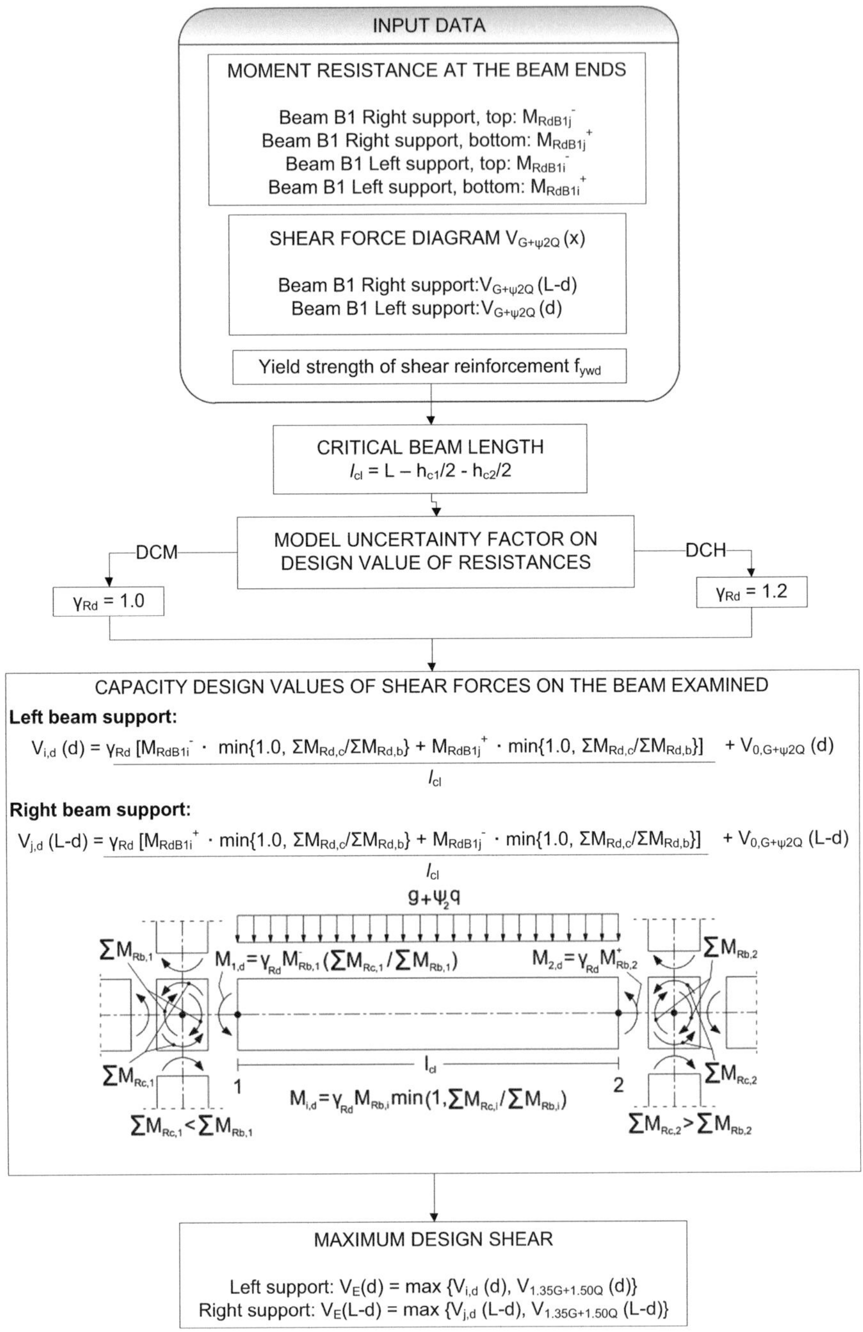

Flowchart 3.15a Design of beam for shear (1/3)

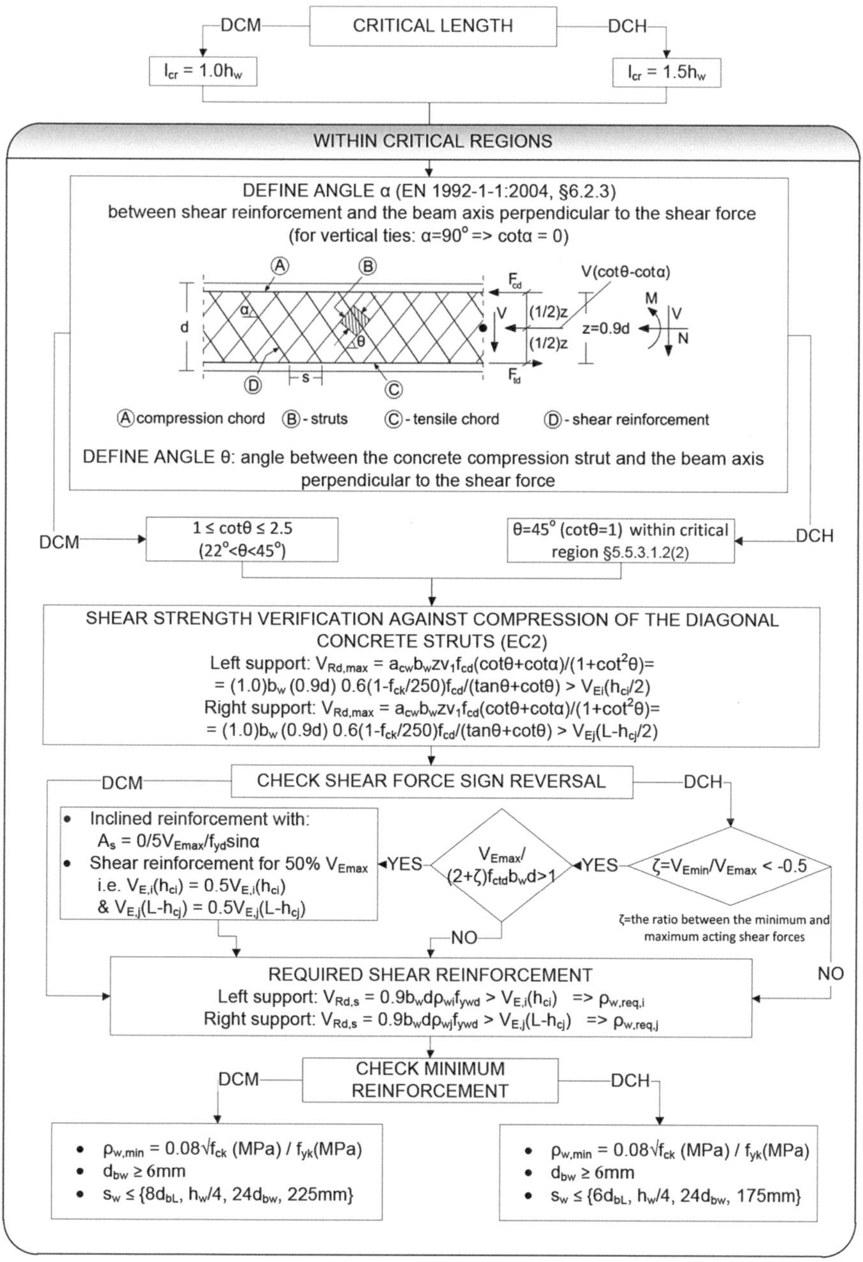

Flowchart 3.15b Design of beam for shear (2/3)

OUTSIDE CRITICAL REGION

DEFINE ANGLE α (EN 1992-1-1:2004, §6.2.3)
between shear reinforcement and the beam axis perpendicular to the shear force
(for vertical ties: $\alpha=90^{o} \Rightarrow \cot\alpha = 0$)

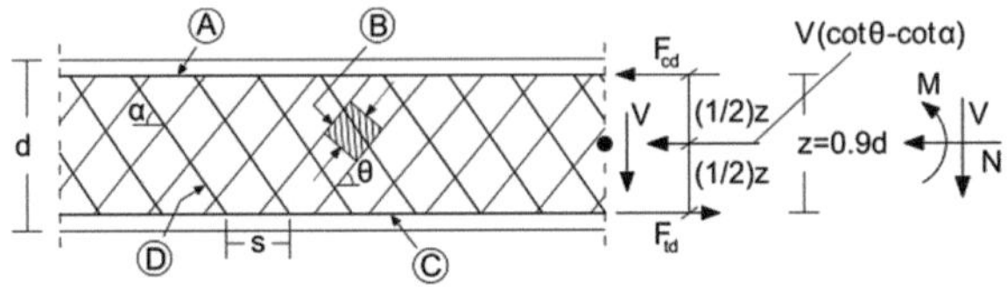

Ⓐ compression chord Ⓑ - struts Ⓒ - tensile chord Ⓓ - shear reinforcement

DEFINE ANGLE θ: angle between the concrete compression strut and the beam axis perpendicular to the shear force ($1 \leq \cot\theta \leq 2.5$ outside critical region)

↓

REQUIRED SHEAR REINFORCEMENT

Left support: $V_{Rd,s} = 0.9 b_w d \rho_{wi} f_{ywd} > V_{E,i}(l_{cr}) \Rightarrow \rho_{w,req,i}$
Right support: $V_{Rd,s} = 0.9 b_w d \rho_{wj} f_{ywd} > V_{E,j}(L-l_{cr}) \Rightarrow \rho_{w,req,j}$

↓

CHECK MINIMUM REINFORCEMENT

- $\rho_{w,min} = 0.08\sqrt{f_{ck}}\,(MPa) / f_{yk}(MPa)$
- $s_w \leq 0.75d$

Flowchart 3.15c Design of beam for shear (3/3)

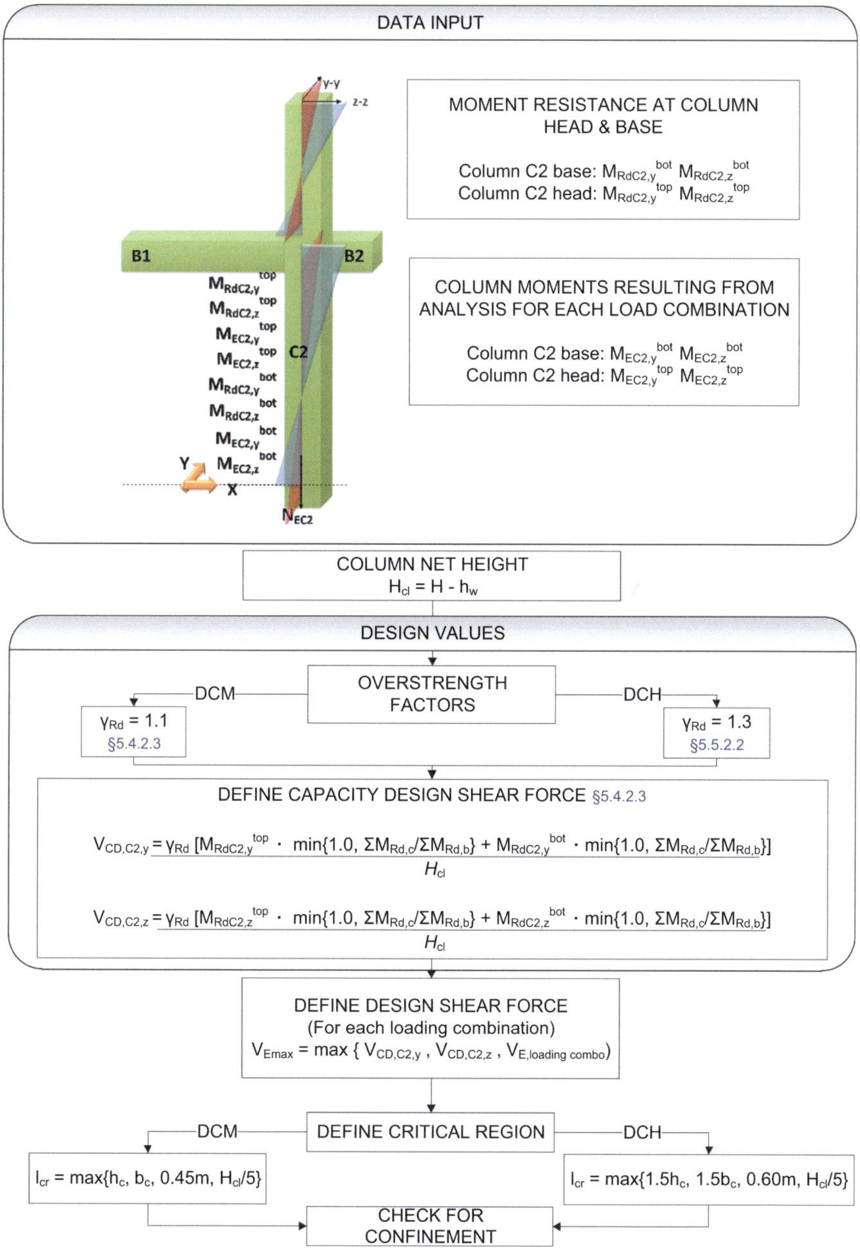

Flowchart 3.16a Design of column for shear (1/4)

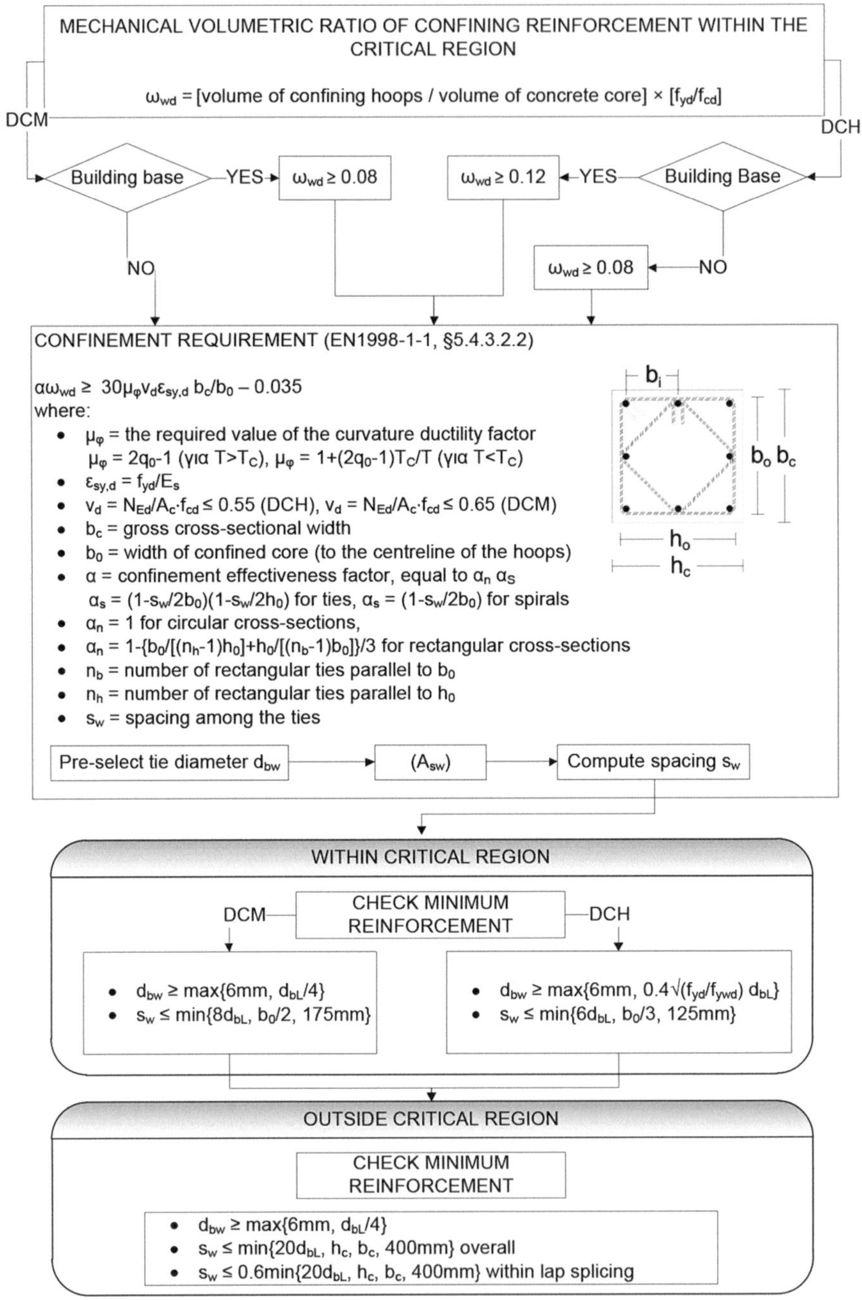

Flowchart 3.16b Design of column for shear (2/4)

DEFINE ANGLE α (EN 1992-1-1:2004, §6.2.3)
between shear reinforcement and the beam axis perpendicular to the shear force
(for vertical ties: $\alpha = 90^\circ \Rightarrow \cot\alpha = 0$)

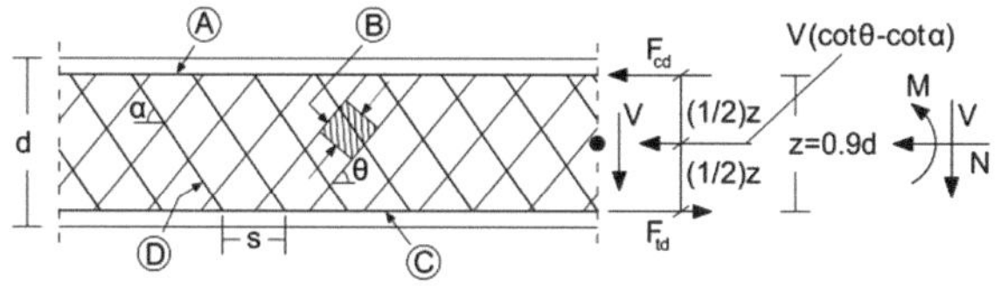

Ⓐ compression chord Ⓑ - struts Ⓒ - tensile chord Ⓓ - shear reinforcement

DEFINE ANGLE θ: angle between the concrete compression strut and the beam axis perpendicular to the shear force ($1 \le \cot\theta \le 2.5$ outside critical region)

↓

SHEAR STRENGTH VERIFICATION AGAINST COMPRESSION OF THE DIAGONAL CONCRETE STRUTS (EC2)

$$V_{Rd,max} = a_{cw} b_w z v_1 f_{cd} (\cot\theta + \cot\alpha)/(1+\cot^2\theta) = (1.0) b_w (0.9d)\, 0.6(1-f_{ck}/250) f_{cd}/(\tan\theta + \cot\theta) > V_{Emax}$$

↓

REQUIRED shear reinforcement

$$V_{Rd,s} = 0.9 b_w d \rho_{wi} f_{ywd} > V_{Emax} \Rightarrow \rho_{w,req}$$

↓

WITHIN CRITICAL REGION

VERIFY MINIMUM REINFORCEMENT

DCM:

- $s_w \le \min\{8d_{bL}, b_0/2, 175\text{mm}\}$

DCH:

- $s_w \le \min\{6d_{bL}, b_0/3, 125\text{mm}\}$

↓

OUTSIDE CRITICAL REGION

VERIFY MINIMUM REINFORCEMENT

- $s_w \le \min\{20d_{bL}, h_c, b_c, 400\text{mm}\}$ overall
- $s_w \le 0.6\min\{20d_{bL}, h_c, b_c, 400\text{mm}\}$ within lap splicing

Flowchart 3.16c Design of column for shear (3/4)

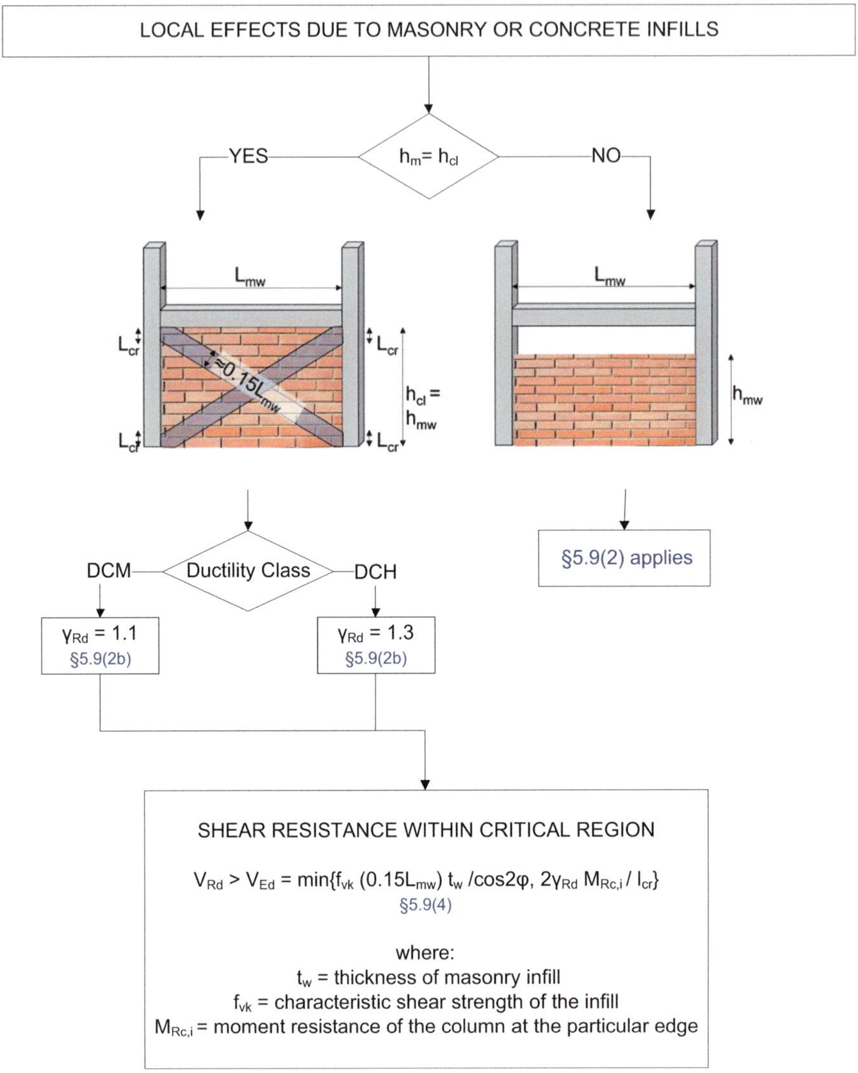

Flowchart 3.16d Design of column for shear (4/4)

VERIFICATION AT THE BEAM-COLUMN JOINT

Ductility Class

- DCM → BEAM-COLUMN JOINT VERIFICATION NOT REQUIRED
- DCH → BEAM-COLUMN JOINT VERIFICATION REQUIRED

Joint: interior / exterior

HORIZONTAL SHEAR FORCE ACTING ON THE CONCRETE CORE OF THE JOINT §5.5.2.3

interior	exterior
$V_{jhd} = \gamma_{Rd}\,(A_{s1}+A_{s2})f_{yd}-V_c$	$V_{jhd} = \gamma_{Rd}\,(A_{s1})f_{yd}-V_c$

where:
$\gamma_{Rd} > 1.2$
V_c = shear force in the column above the joint, from the analysis in the seismic design situation

SHEAR RESISTANCE AGAINST DIAGONAL COMPRESSION §5.5.3.3(2)

interior	exterior
$V_{jhd} \leq n f_{cd}\sqrt{1-v_d b_j h_{jc}/\eta}$	$V_{jhd} \leq 0.8\, n\, f_{cd}\sqrt{1-v_d b_j h_{jc}/\eta}$

where:
$\eta = 0.6(1-f_{ck}/250)$
v_d = norm . design axial force of the upper column $= N_{Ed}/A_c \cdot f_{cd}$

SHEAR RESISTANCE AGAINST DIAGONAL TENSION §5.5.3.3(3)

$$A_{sh}\, f_{wyd} / b_j\, h_{jw} \geq (V_{jhd}/b_j\, h_{jc})^2/(f_{ctd}+v_d f_{cd}) - f_{ctd}$$

where:
A_{sh} = total area of the horizontal hoops
V_{jhd} = as above
b_j = as follows:
if $b_c > b_w$: $b_j = \min\{b_c; (b_w + 0{,}5.h_c)\}$
if $b_c < b_w$: $b_j = \min\{b_w; (b_c + 0{,}5.h_c)\}$
f_{ctd} = design value of the tensile strength of concrete, in accordance with EN1992-1-1:2004

* alternatively, the following provision §5.5.3.3(4) applies

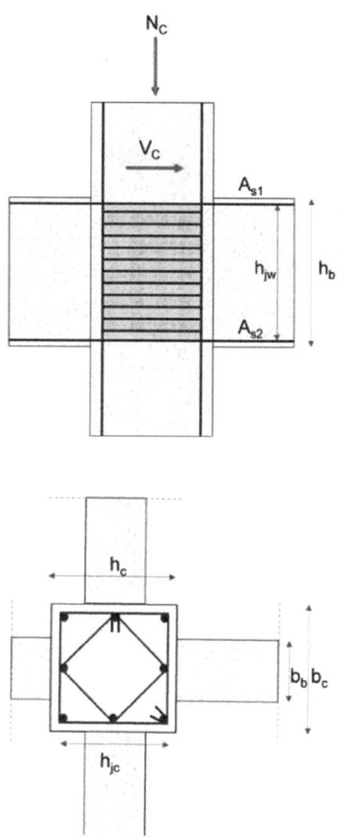

Flowchart 3.17 Verification at the beam-column joint

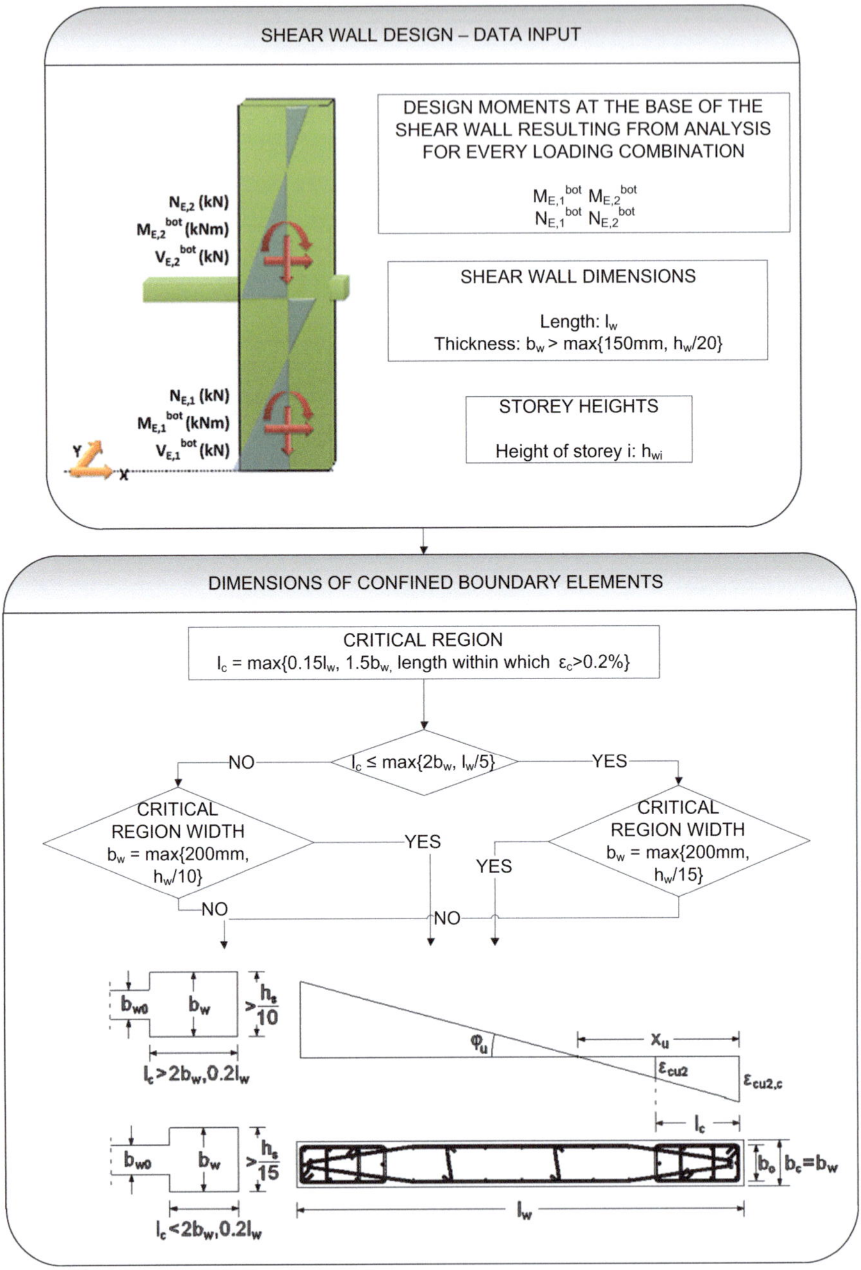

Flowchart 3.18a Design of walls to flexure (1/2)

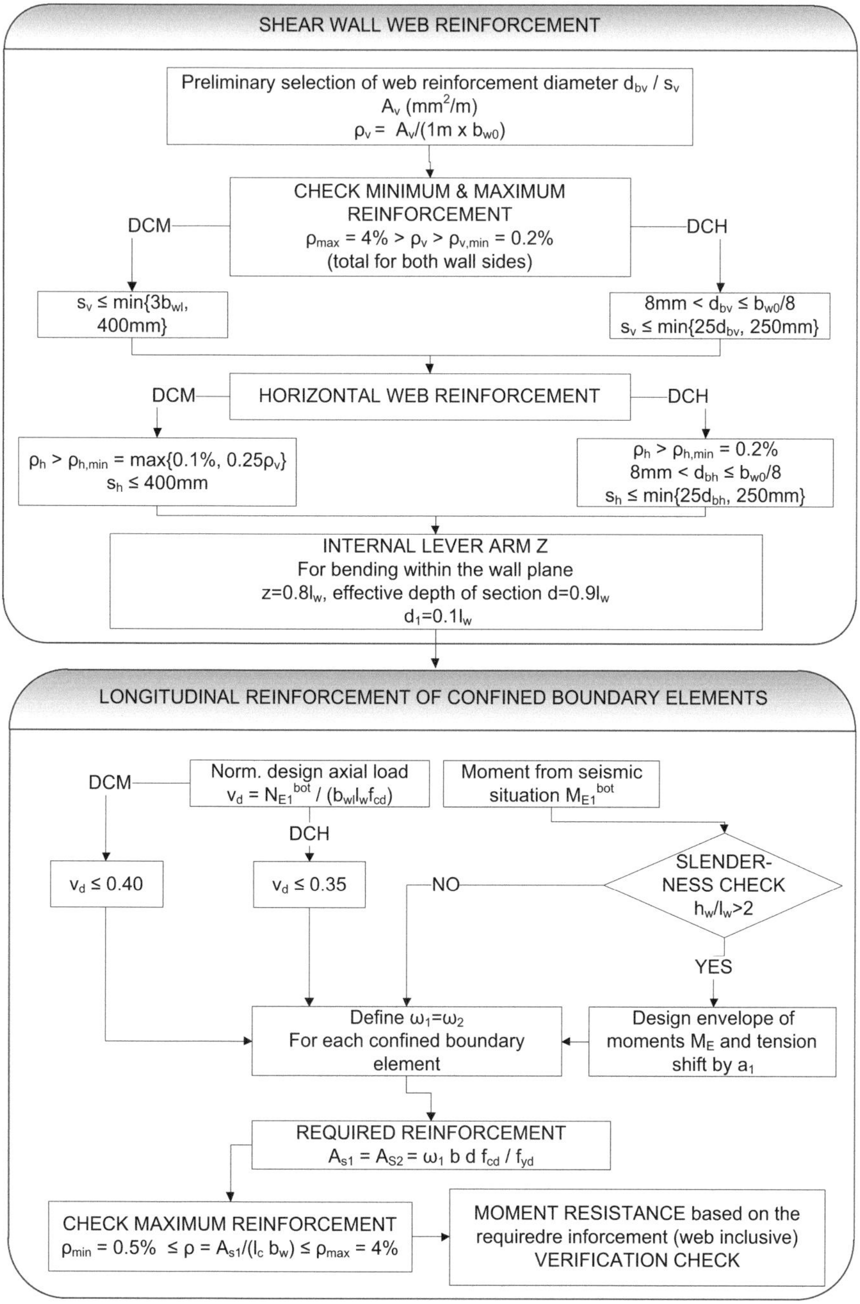

Flowchart 3.18b Design of walls to flexure (2/2)

WALL DESIGN TO SHEAR FORCES - DATA INPUT

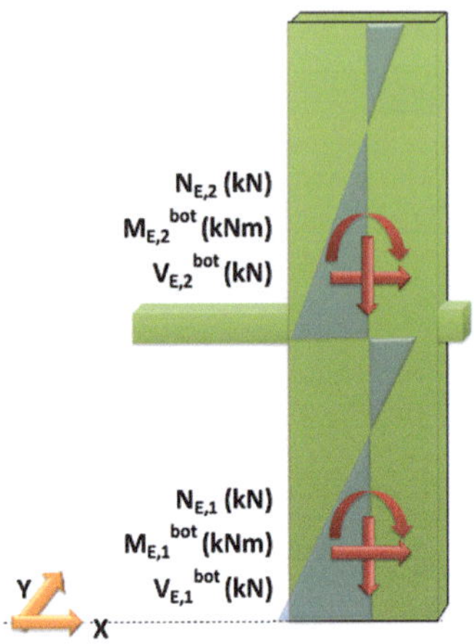

DESIGN MOMENTS AT THE BASE OF THE SHEAR WALL RESULTING FROM ANALYSIS FOR EVERY LOADING COMBINATION

$V_{E,1}^{bot}$ $V_{E,2}^{bot}$ N_{Ed}

MOMENT RESISTANCE AT THE BASE OF THE SHEAR WALL

$M_{Rd,1}^{bot}$

SHEAR WALL DIMENSIONS

Length: l_w
Thickness: $b_w > \max\{150\text{mm}, h_{st}/20\}$

BUILDING PROPERTIES

Total number of storeys: n
Storey height i: h_{wi}
Behaviour factor: q_0
Fundamental period: T_1
Spectral acceleration: $S_e(T)$

DESIGN SHEAR FORCE

Ductility Class

DCM → Multiply V_{E1}^{bot} by $\varepsilon=1.5$

DCH → SLENDERNESS CHECK $H_w/l_w>2$

YES → Multiply V_{E1}^{bot} by
$\varepsilon=\sqrt{[(1.2M_{Rd}/M_{Ed})^2+0.1(qS_e(T_c)/S_e(T_1))^2]} \le q$
§5.5.2.4.1

NO → Multiply V_{E1}^{bot} by
$\varepsilon=1.2M_{Rd}/M_{Ed} \le q$

Critical wall height

$h_{cr} \ge \max\{l_w, H_w/6\}$
$h_{cr} \le \min\{2l_w, h_{storey}\}$ for n≤6 storeys
$h_{cr} \le \min\{2l_w, 2h_{storey}\}$ for n>6 storeys

Flowchart 3.19a Design of walls to shear (1/3)

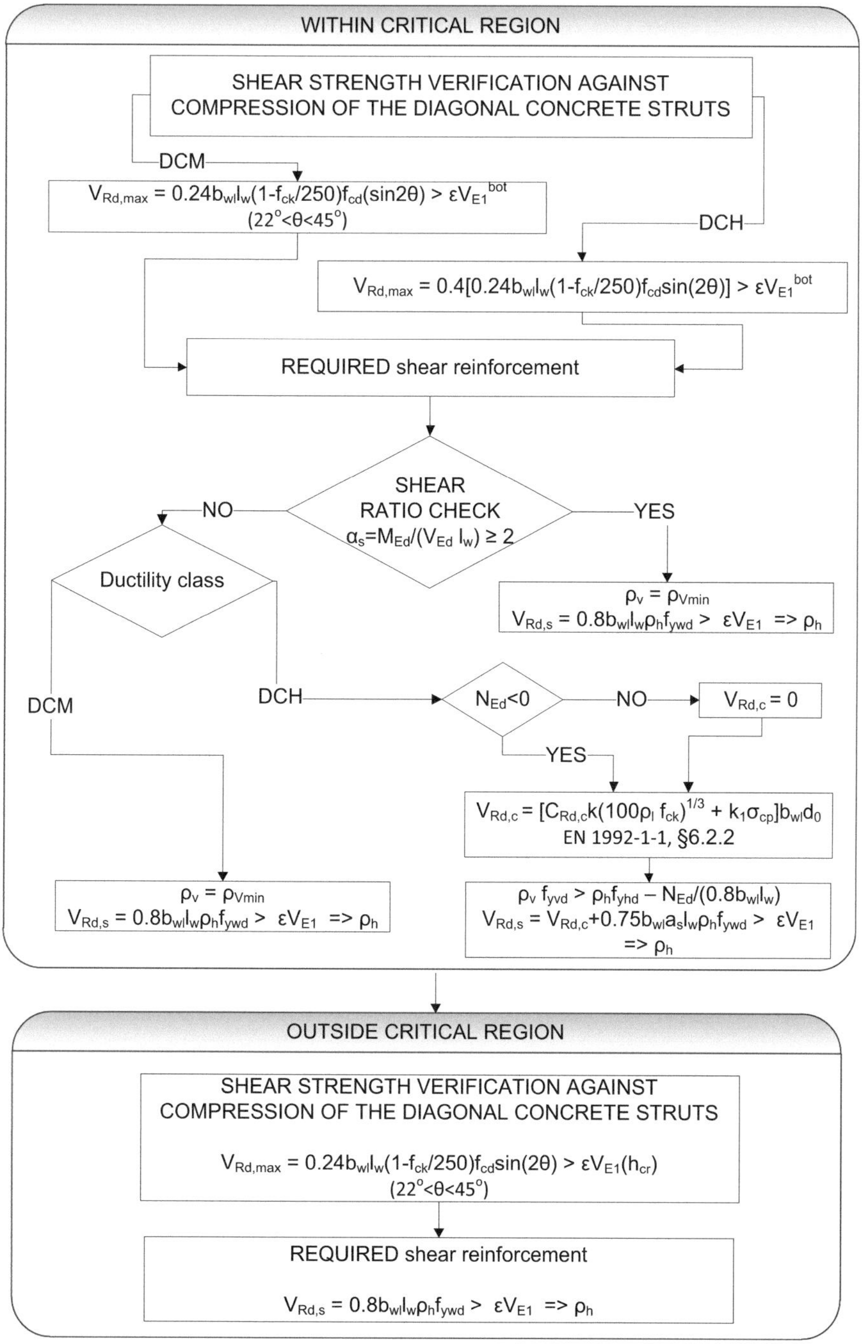

Flowchart 3.19b Design of walls to shear (2/3)

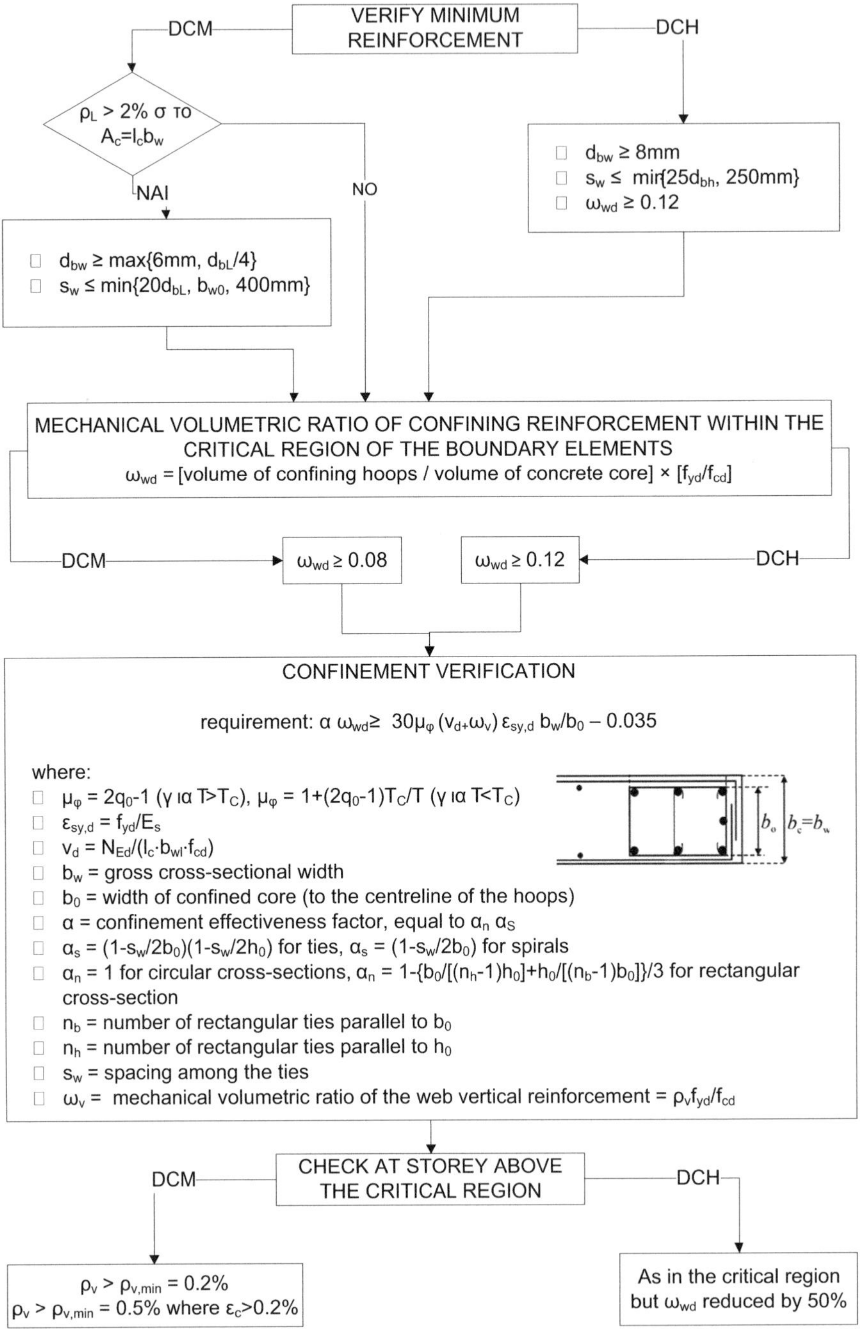

Flowchart 3.19c Design of walls to shear (3/3)

References

Anastassiadis K (1993) Directions seismic defavorables et combinaisons defavorables des efforts. Ann l' I.T.B.T.P 512:82–99

Anastassiadis K, Avramidis IE, Panetsos P (2002) Concurrent design forces in structures under three-component orthotropic seismic excitation. Earthq Spectra 18:1–17

Athanatopoulou AM, Doudoumis IN (2008) Principal directions under lateral loading in multistorey asymmetric buildings. Struct Des Tall Spec Build 17:773–794. doi:10.1002/tal

Athanatopoulou AM, Avramidis IE, Xenidis H (1999) Vertical component of seismic action – partial model and whole structure model. In: EURODYN'99, Prague, pp 7–10

Athanatopoulou AM, Makarios TK, Anastassiadis K (2006) Elastic earthquake analysis of isotropic asymmetric multistory buildings. Struct Des Tall Spec Build 15:417–443

CEN (2004a) European Standard EN 1998-1. Eurocode 8: design of structures for earthquake resistance, Part 1: general rules, seismic actions and rules for buildings. Committee for Standarization, Design, European Committee for Standardization, Brussels, Belgium

CEN (2004b) European Standard EN 1992-1-1. Eurocode 2: design of concrete structures, Part 1-1: general rules and rules for buildings, Committee for Standarization. Brussels, Belgium

Chopra AK (2007) Dynamics of structures—theory and applications to earthquake engineering, 3rd edn. Pearson, Prentice-Hall, Upper Saddle River

Der Kiureghian A (1981) A response spectrum method for random vibration analysis of MDOF systems. Earthq Eng Struct Dyn 9:419–435

Earthquake Planning and Protection Organization (EPPO) (2000) Greek seismic code EAK2000 (amended in 2003), Athens, Greece (in Greek)

Fardis MN (2006) Vision for the development of seismic codes (in Greek). In: 15th Hellenic concrete conference, pp 191–198

Fardis MN (2009) Seismic design, assessment and retrofit of concrete buildings, based on Eurocode 8. Springer, Dordrecht

Gupta AK (1992) Response spectrum method in seismic analysis and design of structures. CRC Press, New York. Weiler, WA

Gupta AK, Singh MP (1977) Design of column sections subjected to three components of earthquake. Nucl Eng Des 41:129–133

Hejal R, Chopra AK (1989) Earthquake analysis of a class of torsionally coupled buildings. Earthq Eng Struct Dyn 18:305–323

Hellenic Organization for Standarization (2009) Standard 1498-1. Greek National Annex to Eurocode 8 – Part 1: general rules, seismic actions and rules for buildings (in Greek). Athens

Kan CL, Chopra AK (1977) Elastic earthquake analysis of a class of torsionally coupled buildings. J Struct Div ASCE 103:821–838

Makarios TK, Anastassiadis K (1998) Real and fictitious elastic axis of multi-storey buildings: theory. Struct Des Tall Spec Build 7:33–55

Penelis GG, Kappos AJ (1997) Earthquake-resistant concrete structures. E & FN Spon, London

Penelis GG, Penelis GG (2014) Concrete buildings in seismic regions. CRC Press/Taylor & Francis Group, London

Rosman R (1984) Seismische Lasten mehr- und vielgeschossiger Bauwerke. Beton-und Stahlbetonbau 79(8):201–208

Chapter 4
EC8-Compliant Seismic Analysis and Design Examples

Abstract This chapter presents three numerical benchmark example problems considering three different structures to illustrate the implementation of various clauses of Eurocode 8 (EC8) for the seismic design of ordinary reinforced concrete (r/c) buildings. Pertinent numerical input data and selected output data are reported in tabular form to facilitate the use of these example problems as tutorials for commercial seismic structural analysis and design software packages according to EC8. In particular, the first example problem involves undertaking EC8-compliant seismic analysis and deformation-based verification checks using both the lateral force and the modal response spectrum methods for a 5-storey dual r/c building with a single in-plan symmetry and with fixed support conditions. The outcomes from each analysis method are juxtaposed and compared. Further, the second example problem considers a 5-storey dual torsionally sensitive r/c building with fixed support conditions to perform seismic analysis using the modal response spectrum method along with post-analysis deformation-based verification checks. Finally, the third example considers a 4-storey dual r/c building with an r/c core and one underground storey (basement) resting on compliant soil. Emphasis is placed on the modelling of the box-type foundation, including pad and strip footings, and of the supporting ground using linear translational and rotational springs. The modal response spectrum method of analysis is used for the EC8-compliant design of the structure. Complete detailing and design verification checks according to Eurocode 2 and Eurocode 8, including capacity design rules, are presented in detail for selected r/c structural members, namely, for two beams, one column and a wall.

Keywords Benchmark EC8 design problems • Structural regularity • Torsional sensitivity • Behavior factor • Lateral force method • Modal response spectrum method • Design seismic effects • Second-order effects • Interstorey drift sensitivity coefficient • Interstorey drift • Seismic detailing or r/c structural members • Compliant soil

In this final chapter, three numerical examples, A, B, and C, of EC8-compliant seismic analysis and design are provided pertaining to three different r/c building structures. Each example is presented in a distinct self-contained section of the

I. Avramidis et al., *Eurocode-Compliant Seismic Analysis and Design of R/C Buildings*, Geotechnical, Geological and Earthquake Engineering 38,
DOI 10.1007/978-3-319-25270-4_4

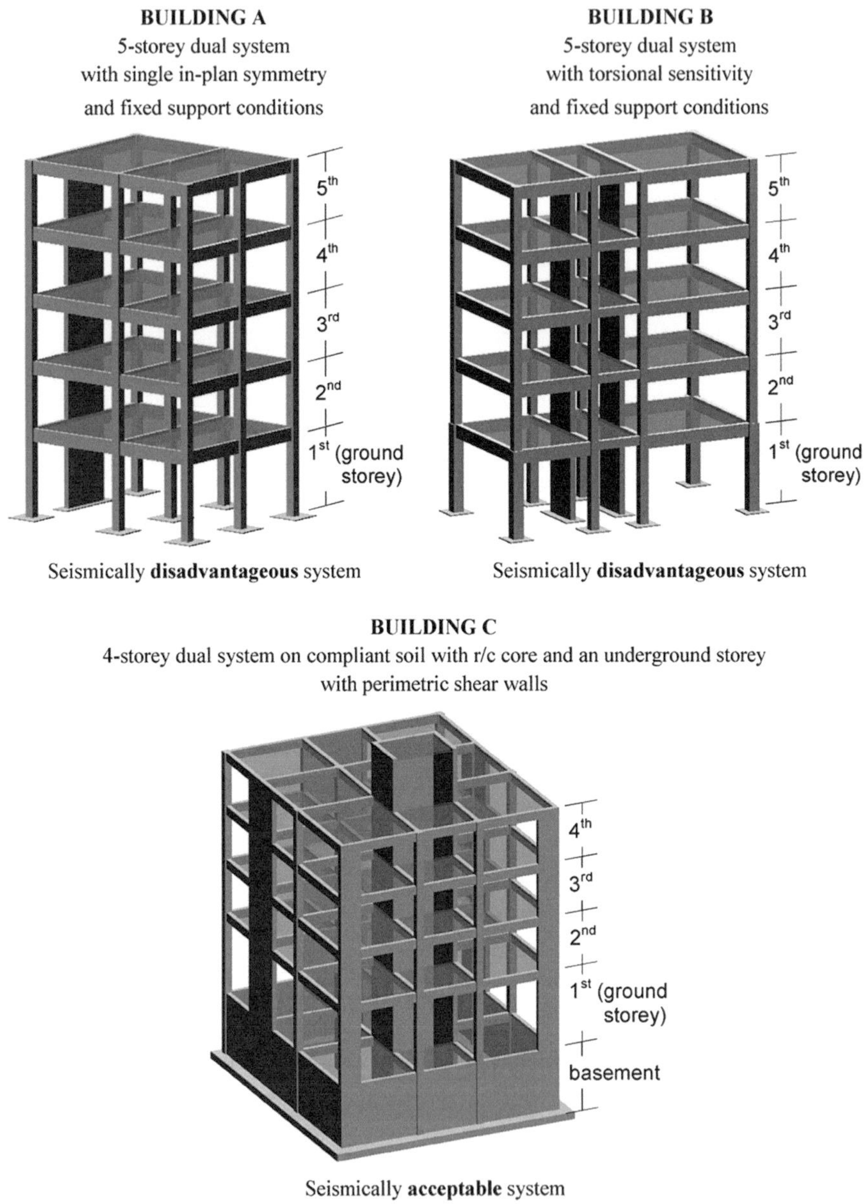

Fig. 4.1 Load-resisting structural systems of the considered example buildings

chapter. The load-resisting structural systems of the considered buildings are shown in Fig. 4.1. The first two examples, A and B, aim to illustrate the implementation of different clauses of EC8 related to the analysis and deformation-based verification checks reviewed in Sects. 3.1 and 3.2. To this aim, they involve buildings with simple architectural plan-views of limited dimensions that do not qualify for real-life structures. More importantly, it is emphasized that both structures violate certain fundamental conceptual design rules for earthquake resistant buildings, such as the absence of shear walls along one "principal direction" of a structure (building A), and the eccentrically in-plan arrangement of a single r/c core (building B). In this respect, buildings A and B should be taken as counter-examples of proper seismic design to be avoided in practice. Still, due to the simplicity of the involved structures, examples A and B allow for:

- reporting, in a concise and instructive manner, all the required seismic analysis input data and pertinent numerical results derived from the various EC8-compliant analysis steps for illustration and pedagogic purposes;
- serving as benchmark analysis example problems which can be readily input and run to gain familiarity with and to verify any commercial structural analysis software package (see also Sect. 2.5).

Specifically, building A (studied in Sect. 4.1) comprises a dual lateral load resisting structural system with fixed support conditions having a single in-plan axis of symmetry. The structural system is characterized by *high eccentricity* (distance between the center of mass and the fictitious elastic axis; see Sect. 3.1.1.1 and FC-3.2a), since it includes a single shear wall placed at the perimeter of the structure. The shear wall is modelled using an equivalent frame model, as detailed in Sect. 2.3.3.3. *Both* the lateral force method (LFM) and the modal response spectrum method (MRSM) of EC8 are employed in the analysis of the structure to make possible the comparison of analysis results from the two equivalent linear analysis methods of EC8.

Building B (studied in Sect. 4.2) comprises an "almost" double symmetric in-plan dual system (the asymmetry along principal axis Y is relatively small) with fixed support conditions. This structure is classified as *torsionally sensitive* according to EC8, providing for the opportunity to exemplify the implementation of EC8 clauses pertinent to torsionally sensitive structures. The MRSM of EC8 is used in the analysis.

Contrary to buildings A and B, building C (studied in Sect. 4.3) can be construed as a realistic r/c structure without being overly complex or of large dimensions. It has a lateral load resisting structural system conforming to all practical conceptual design rules for earthquake resistance. In example C, both analysis and detailing/design verification results are reported for selected structural members, namely for a beam, a column, and a shear wall. To this end, example C serves the purpose of a benchmark analysis *and design* example problem. It is noted that building C includes a Π-shaped r/c core modeled using an equivalent frame model with three equivalent columns (Fig. 2.43). Further, the building includes an underground storey (basement) with perimetric r/c shear walls which is included in the finite element (FE) model and rests on compliant soil modelled via linear vertical translational and rotational springs.

Overview of EC8 implementation steps in the provided numerical examples

As discussed in Chap. 3 (Fig. 3.2), the initial phase (phase A) of the seismic design procedure involves

- specifying the topology of the load resisting structural system based on the architectural plans and conceptual design rules for earthquake resistance;
- selecting the desirable/targeted level of seismic performance (Fig. 3.1) in agreement with the building owner; and
- pre-sizing all structural members.

In all three numerical examples of this chapter, it is assumed that the above three steps (i.e., phase A of the seismic design procedure) have already been undertaken and, therefore, the geometry of the load resisting structural system and the dimensions of all r/c sections are known. In this regard, each of the three sub-sections of this chapter begins with

- listing the geometrical and material properties of the structural system and the EC8-compatible seismic loading data, and proceeds with
- reporting the assumptions, simplifications, and idealizations considered in constructing the structural FE and loading numerical model used in the EC8-compliant seismic analysis;
- undertaking all required steps involved in the EC8-compliant seismic analysis stage; and
- detailing and verification of selected r/c structural members (for building C-Sect. 4.3 only).

In particular, the EC8-compliant seismic analysis stage follows the general FC-3.1 (see also Sect. 3.1) which involves:

- Verification check for structural regularity in plan and elevation (Sect. 3.1.1; FC-3.2, 3.2a, and 3.3);
- Classification of the lateral load resisting system (Sect. 3.1.2; FC-3.4);
- Selection of the ductility capacity class (Sect. 3.1.3);
- Determination of the maximum allowed behaviour factor q value (Sect. 3.1.4; FC-3.5);
- Choosing an equivalent linear seismic analysis method (Sect. 3.1.5; FC-3.6);
- Undertaking structural static analysis for the calculation of response quantities (effects) due to the permanent gravity load combination;
- Undertaking seismic analysis for the calculation of the seismic response quantities/effects (Sect. 3.1.5; FC-3.7, 3.8 and 3.9); and
- Deformation-based verification checks for second-order phenomena and interstorey drift demands (Sect. 3.2; FC-3.10a, 3.10b 3.11a, and 3.11b).

The full list of the above tasks is undertaken in the first two examples, buildings A and B, for which the maximum allowed behaviour factor q according to EC8 is chosen and, therefore, the maximum possible (allowed) utilization of the ductility capacity of the structure is considered to resist the nominal "design earthquake".

However, the proposed design recommendations of Sect. 3.4 are adopted in the third example (building C), namely,

- a relatively low value for the behaviour factor, $q = 2 < maxq_{allow} = 3.0$, is taken, implying a relatively high desired seismic performance level with limited utilization of the ductility capacity of the structure for the nominal "design earthquake";
- the medium ductility class (DCM) is assumed in design;
- a three-dimensional FE model (default choice) is considered; and
- the MRSM of analysis is used.

All required (linear) structural analyses have been conducted using the SAP2000 commercial structural analysis software package (CSI 2012), while the detailing and design verifications of structural members for building C (Sect. 4.3.10) have been performed through custom-made MS Excel spreadsheets.

For user's convenience and to facilitate the readability of tables containing intermediate and final analysis results, all notations and symbols used in the following Sects. 4.1, 4.2 and 4.3 are summarized in Table 4.141 at the end of this chapter.

4.1 Example A: Five-Storey Single Symmetric In-Plan Building with Dual Lateral Load-Resisting Structural System

4.1.1 Geometric, Material, and Seismic Action Data

Units

Length: m; Force: kN; Mass t ($1\ t = 10^3$ kg), Time: s.

Material properties for reinforced concrete

Modulus of Elasticity: $E = 2.9 \cdot 10^7\ kN/m^2$; Poisson ratio: $v = 0.2$; weight per unit volume: $\gamma = 25\ kN/m^3$.

In-plan description and geometry of building A

The example building A is a five storey structure with rectangular in-plan geometry having a single horizontal axis of symmetry along direction X as defined in Fig. 4.2. The lateral-load resisting system of the building includes a single r/c wall oriented along direction Y and positioned on the perimeter of the building. The in-plan dimensions of building A, along with the storey heights and the cross-sectional dimensions (in cm) of its vertical r/c members (columns and wall), which remain constant along the full height of the building, are shown in Fig. 4.2. Structural members for which analysis results are reported in detail are marked in red. For the beams, l and r denote their left and right ends, respectively.

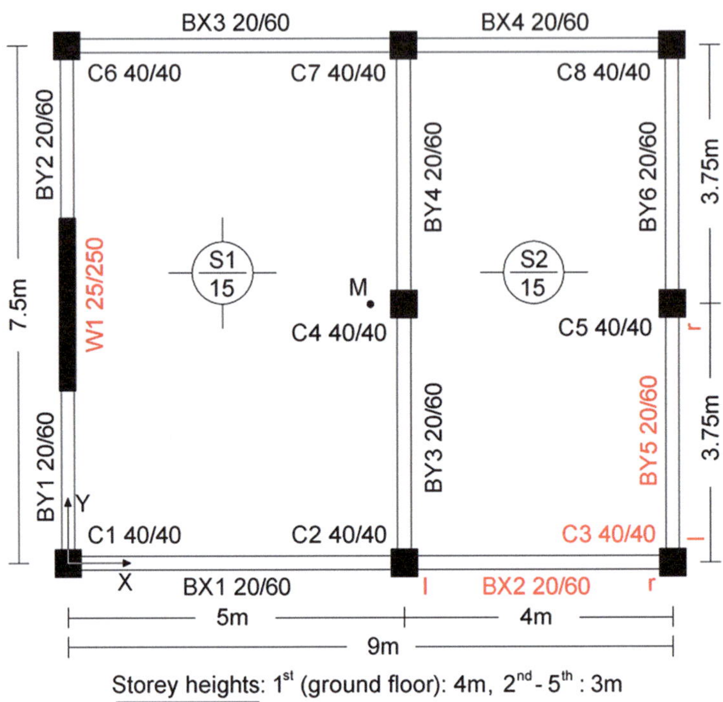

Fig. 4.2 Typical floor plan of building A

Gravity loads imposed on beams and slabs

- Double-layered masonry walls occupy the full storey height at all storeys along the perimeter of the building. These infill walls impose 3.6 kN/m^2 of "permanent" weight on all exterior beams except the beams of the top storey. The exterior beams of the top storey accommodate a "permanent" uniform distributed load of 3.6 kN/m, corresponding to a double-layered masonry roof parapet 1 m in height.
- Single-layered masonry walls occupy the full height at all storeys along the full length of the interior beams BY3 and BY4. These infill walls impose 2.1 kN/m^2 of "permanent" weight on all interior beams except the beams of the top storey.
- Permanent floor finishings of 1.3 kN/m^2 weight evenly distributed in plan is assumed.
- The assumed "live" gravity loads (variable action) applied to the floor slabs are taken as $Q = 2$ kN/m^2 evenly distributed in plan.

Directions of the horizontal seismic action (clause §4.3.3.1(11)P of EC8)

Axes X and Y shown in Fig. 4.2 can be unambiguously identified as the "principal" orthogonal axes along which the input seismic action is assumed to act for design purposes.

Design spectrum data (clause §3.2.2.5 of EC8, Type1)

- Peak ground acceleration: $a_{gR} = 0.24$ g
- Ground type: C ($S = 1.15$, $T_B = 0.20$ s, $T_C = 0.6$ s)
- Importance category: II (residential building) → $\gamma_I = 1$
- Damping coefficient: $\zeta = 5$ %
- Behaviour factor q: to be determined in Sect. 4.1.6 below

4.1.2 Modeling Assumptions

4.1.2.1 Structural Modeling Assumptions

A spatial (three-dimensional) FE model is considered which accounts for flexural, shear, axial, and torsional deformations of r/c structural members. The infill walls are not included in the FE model, assuming their influence on the lateral stiffness and strength of the building structure to be negligible (see clause §4.3.1(8) of EC8). The beam-column joints are modeled as perfectly rigid (see clause §4.3.1(2) of EC8) using rigid offsets (arms) at the end of FE members, as shown in Fig. 2.37a (see Sect. 2.3.3.2).

Modeling of floor slabs

Floor slabs are assumed to act as perfectly rigid diaphragms in their plane (see Sect. 2.3.3.1). The actual height level of these diaphragms is defined in the considered FE model, as shown in Fig. 2.37a.

Effective flange width of beams

The effective width b_{eff} of the upper flange of the beams included in the FE model is given in Table 4.1. It is computed according to clause §5.3.2.1 of EC2 as

- $b_{eff} = b_w + 0.2 \cdot l_o$, for T-shaped beams, and
- $b_{eff} = b_w + 0.1 \cdot l_o$, for L-shaped beams,

where b_w is the width of the beam web and $l_0 = 0.85 \cdot L$, in which L is the length of the beam excluding its rigid offsets.

Effective rigidities of structural members (see §2.3.2.1)

The flexural rigidity (EI) and the shear rigidity (GA_s) are assumed equal to 50 % of the values corresponding to uncracked gross section properties for all r/c members (clause §4.3.1(7) of EC8). The torsional rigidity (GJ) is taken to be equal to 10 % of the value corresponding to uncracked gross section properties for all r/c members, as discussed in Sect. 2.3.2.1 (Fardis 2009). However, the axial rigidity (EA) of structural members *is not reduced* compared to the value corresponding to uncracked gross section properties given that the vertical structural members are under compression due to the gravity loads and all beam

Table 4.1 Assumed effective width of the beams of the FE model for building A

Beam members	BX1,BX3	BX2,BX4	BY1,BY2	BY3,BY4	BY5,BY6
b_{eff}	0.625 m	0.54 m	0.42 m	0.84 m	0.52 m

members are considered to be part of the perfectly rigid diaphragms within the plane of the floor slabs. It is noted that clause §5.4(2) of EC2 allows for considering the uncracked gross section properties to compute the rigidity of r/c structural members under gravity loads. Nevertheless, given that the reduction is considered generally towards safety, it also preserved for the vertical loads, exactly as in the analyses for the seismic loads. Therefore, in all analyses, static and seismic, the same model of the structure is employed.

Modeling of the r/c wall W1

The r/c wall W1 is modeled by means of an equivalent frame model, as discussed in Sect. 2.3.3.3 and shown in Fig. 2.38d. The model comprises an equivalent column positioned at the center of gravity of the actual shear wall, which is connected to the beams at each floor level by means of "virtual" perfectly rigid arms of length 2.5/2 = 1.25 m each. The uncracked gross section properties of the equivalent column are computed as

- Area: $A = 0.25 \cdot 2.5$ (in m^2);
- Second moment of area about the X axis: $I_{XX} = (0.25 \cdot 2.5^3)/12$ (in m^4);
- Second moment of area about the Y axis: $I_{YY} = (0.25^3 \cdot 2.5)/12$ (in m^4);
- Effective shearing area along the X axis: $A_{sX} = (5/6)A$ (in m^2);
- Effective shearing area along the Y axis: $A_{sY} = (5/6)A$ (in m^2); and
- Polar moment of inertia: $J = a \cdot 0.25^3 \cdot 2.5$ (in m^4).

The effective rigidities of the equivalent column computed from the above properties are reduced to account for concrete cracking according to the previously mentioned assumptions. Note that the constant involved in determining the polar moment of inertia can be computed for rectangular cross-sections with dimensions (d × t) as (Oden 1967)

$$a = \frac{1}{3}\left[1 - \left(\frac{192t}{\pi^5 d}\right)\tanh\left(\frac{\pi d}{2t}\right)\right],$$

though other expressions are also applicable (see Fig. 2.43). Herein, for $d = 2.5$ m and $t = 0.25$ m, one obtains $a = 0.312$.

4.1.2.2 Vertical Load Modeling Assumptions

- The permanent (self weight and finishings) and variable evenly distributed area loads carried by the slabs are transferred to the beams using triangular and trapezoidal tributary areas (rule of 45° or 60°). In this manner, they are distributed along the length of the beams following triangular or trapezoidal distributions.
- The masonry infill walls' self weight (permanent loads) is transferred directly onto the beams and is computed without accounting for any of the existing architectural openings in the infill walls.
- The self weight of r/c beams and the infill walls carried by the beams is considered to be uniformly distributed along the length of the beams.
- The self weight of r/c columns is modeled as a uniformly distributed axial load along the height of the columns.

4.1.2.3 Mass/Inertial Modeling Assumptions

- The total mass of each storey is lumped at the center of gravity M (geometrical center) of the corresponding floor rigid diaphragm.
- The total mass of each storey comprises:
  - The own mass of the storey slab and beams, including all finishings;
  - The own mass of the masonry infill walls resting on the storey beams (ignoring any openings);
  - The own mass of the vertical r/c members (columns and walls) extending above and below the considered storey slab up to the middle of their total storey height; and
  - The mass that corresponds to the variable gravity load of the seismic design load combination, as defined in clauses §3.2.4(2)P, 4.2.4(2)P and 4.3.1(10)P of EC8.

The mass of each storey of building A is computed from the gravity loads of the seismic design load combination, as detailed in Table 4.2

4.1.3 Verification Checks for Regularity for Building A

The rationale of the regularity verification checks in plan and elevation has been discussed in detail in Sect. 3.1.1. Herein, it is reminded that structural regularity may influence

- the choice of the structural FE model to be used in analysis (planar or spatial), which depends only on the regularity in plan check;

Table 4.2 Storey mass and gravity loads of the seismic design load combination for building A

Storey	Permanent load G_k [kN]	Variable action Q_k [kN]	ψ_2[a]	φ[b]	$\psi_E = \varphi \cdot \psi_2$	Combination of actions $G_k + \psi_E \cdot Q_k$[c] [kN]	Mass [t]
1st	858.06	135	0.3	0.8	0.24	890.46	90.77
2nd	834.25	135	0.3	0.8	0.24	866.65	88.34
3rd	834.25	135	0.3	0.8	0.24	866.65	88.34
4th	834.25	135	0.3	0.8	0.24	866.65	88.34
5th	607.07	135	0.3	1.0	0.3	647.57	66.00
Total sum	3967.88	675				4137.98	421.80

[a]The combination coefficient ψ_2 of the quasi-permanent value of the variable action Q_k ("live" gravity loads) is given in Eurocode 0 (CEN 2002) Annex A1 and is taken to be equal to 0.3 assuming that building A is an ordinary occupancy residential or office building

[b]The reduction factor φ is given in Table 4.2 of EC8 (clause §4.2.4 of EC8). It is herein assumed to be equal to 1.0 for the top storey and equal to 0.8 for the rest of the storeys, which are assumed to have correlated occupancies. This assumption is particularly valid for residential buildings at night time and for office buildings during day hours

[c]The storey masses are computed from this gravity load combination, as specified in clause §3.2.4 (2)P of EC8

- the choice of the seismic structural analysis method LFM, which depends only on the regularity in elevation check; and
- the value of the maximum allowable behaviour factor q, which depends on both the regularity in plan and elevation checks.

It is further reminded that, by adopting a spatial FE structural model, applying the MRSM of analysis, and choosing a value ≤ 1.5 for the behaviour factor q_{dem}, the need to undertake the regularity verification checks is circumvented (see FC-3.1). In fact, these modeling and analysis choices are generally recommended (see Sect. 3.4). However, in what follows, the EC8 regularity checks are carried out for building A to illustrate the application of the relevant provisions.

4.1.3.1 Verification Checks for Regularity in Elevation

The verification check procedure for regularity in elevation follows FC-3.3 and the pertinent outcomes for building A are summarized in Table 4.3 (see also Sect. 3.1.1.2). Based on the conditions (a) to (d) of the latter table, building A is classified as regular in elevation according to EC8. This classification entails that:

- the LFM of analysis can be used provided that condition (a) of clause §4.3.3.2.1 (2) of EC8 (see also FC-3.6) is satisfied as well. This condition is verified in Sect. 4.1.10.1 below.
- the basic value of the behaviour factor, q_0, can be used without reducing it by 20 % (clause §5.2.2.2(3) of EC8).

Table 4.3 Verification check procedure for regularity in elevation for building A

	Condition/Criterion	Is the condition/criterion satisfied?
(a)	Clause §4.2.3.3(2) of EC8: All lateral load resisting systems (cores, walls, frames) run without interruption from the foundation to the top of the building.	**YES**
(b)	Clause §4.2.3.3(3) of EC8: The lateral stiffness and the mass of the individual storeys remain constant or reduce gradually, without abrupt changes, from the base to the top of the building.	**YES** The sections of all structural members remain the same along the height and no abrupt mass changes occur.
(c)	Clause §4.2.3.3(5) of EC8: When setbacks are present, the additional conditions given in this clause shall be met.	**DOES NOT APPLY** No setbacks are present in the example building.
(d)	Clause §4.2.3.3(4) of EC8: In framed buildings, the ratio of the actual storey resistance to the resistance required by the analysis should not vary disproportionately between adjacent storeys.	**DOES NOT APPLY** Building A includes an r/c wall, thus it is not a frame structural system.
(e)	Clause §4.3.6.3.2 of EC8: If there are considerable irregularities in elevation (e.g., drastic reduction of infills in one or more storeys compared to the others), the seismic action effects in the vertical elements of the respective storeys shall be increased.	**DOES NOT APPLY** The masonry infills of building A are uniformly distributed along its height. Hence, no increase in seismic action effects is required.

4.1.3.2 Verification Checks for Regularity in Plan

The verification check procedure for regularity in plan follows FC-3.2 and FC-3.2a, and the pertinent outcomes for building A are summarized in Table 4.4 (see also Sect. 3.1.1.1).

Notably, the condition/criterion (e) of Table 4.4 involves checking the validity of the inequalities of Eqs. (3.1) and (3.2) and requires the computation of the structural eccentricities, e_{oX} and e_{oY}, and of the torsional radii, r_X and $r_{Y,}$ along the principal axes X and Y, respectively, for each storey. To this aim, the location of the center of stiffness (shear center) needs to be defined at each floor (see Sect. 3.1.1.1 and Appendix B). However, as discussed in Sect. 3.1.1.1, the shear center at each floor and the associated elastic vertical axis cannot be uniquely defined in closed-form for the general, and quite common, case of asymmetric multi-storey buildings with dual lateral load resisting systems comprising frames and r/c walls, as is the case with building A along axis Y. In this regard, the so-called "fictitious elastic vertical axis", whose traces at each floor level observe, approximately, the property of the shear center, can be used as a surrogate for the elastic vertical axis, as proposed in (Makarios and Anastassiadis 1998; Hellenic Organization for Standarization 2009). The required procedure for determining the structural eccentricities and the torsional radii by considering the fictitious elastic vertical axis is summarized in FC-3.2a.

Table 4.4 Verification check procedure for regularity in plan for building A

Condition/Criterion		Is the condition/criterion satisfied?
(a)	Clause §4.2.3.2(2) of EC8: With respect to the lateral stiffness and mass distribution, the building structure shall be approximately symmetrical in plan with respect to two orthogonal axes.	**NO** The r/c wall introduces asymmetry with respect to the axis Y.
(b)	Clause §4.2.3.2(3) of EC8: The plan configuration is compact, i.e., each floor is delimited by a polygonal convex line.	**YES** The plan configuration of the example building is orthogonal without setbacks.
(c)	Clause §4.2.3.2(4) of EC8: The floors have a large in-plane stiffness, i.e., they behave as rigid diaphragms (Floor plans with L, Π, H, I, X shape should be carefully examined).	**YES** The floor slabs of the example building are compact in shape without setbacks or openings.
(d)	Clause §4.2.3.2(5) of EC8: The slenderness of the building in plan shall be not higher than 4: $\lambda = L_{max}/L_{min} \leq 4$	**YES** ($\lambda = 9/7.5 = 1.2 < 4$)
(e)	Clause §4.2.3.2(6) of EC8: Check the inequalities (Eqs. (3.1) and (3.2)): $e_{oX} \leq 0.30 \bullet r_X$; $e_{oY} \leq 0.30 \bullet r_Y$; $r_X \geq l_s$; $r_Y \geq l_s$	See Tables 4.6 and 4.7
(f)	Clause §4.3.6.3.1 of EC8: Strongly irregular, unsymmetrical or non-uniform arrangements of masonry infills in plan should be avoided.	**YES** The masonry infill walls uniformly distributed in plan.

Table 4.5 Triangular distribution of torsional moments and lateral forces along the height of building A for an arbitrary base shear equal to 5000 kN (Equation (4.11) of EC8)

Storey k	m [t]	J_m [t·m^2]	z_k [m]	$m_k \cdot z_k$	F_{Xk} [kN] or F_{Yk} [kN] or M_{Zk} [kNm]
1st	90.77	1038.19	4.00	363.08	446.09
2nd	88.34	1010.39	7.00	618.40	759.79
3rd	88.34	1010.39	10.00	883.43	1085.41
4th	88.34	1010.39	13.00	1148.46	1411.04
5th	66.01	754.87	16.00	1056.18	1297.66
Total sum	421.80			4069.56	5000.00

Specifically, the determination of the structural eccentricities requires that a linear static analysis be performed first for torsional moments, M_{Zk}, about the gravitational axis Z at each floor diaphragm k (loading case "M"). These torsional moments may follow a "triangular" distribution along the height of the building by application of equation (4.11) of EC8. Table 4.5 reports the values of the torsional moments along with the pertinent steps for computing these values for building A, assuming an arbitrarily taken total base shear equal to 5000 kN. Evidently, this arbitrarily chosen value has no influence on the values of the subsequently calculated structural eccentricities and torsional stiffness radii.

Next, the coordinates X_{Po} and Y_{Po} of the "center of twist" P_o at the storey k lying closer to the 80 % level of the total height of the building (4th storey for building A), through which the fictitious elastic vertical axis passes, are determined by the relationships $X_{Po} = X_{Mk} - (u_{Y(Mk)}/\theta_{Z(Mk)})$ and $Y_{Po} = Y_{Mk} - (u_{X(Mk)}/\theta_{Z(Mk)})$ included in FC-3.2. In the above relationships, X_{Mk} and Y_{Mk} are the coordinates of the center of mass of storey k ($X_{Mk} = 4.5$ m and $Y_{Mk} = 3.75$ m for building A), and $u_{X(Mk)}$, $u_{Y(Mk)}$, and $\theta_{Z(Mk)}$ are the translations along axes X, Y, and the rotation abouz axis Z, respectively, of point M (Fig. 4.2) due to the loading case "M". For building A and for the torsional moments of Table 4.5, the aforementioned nodal displacements are: $u_{X(Mk)} = 0$, $u_{Y(Mk)} = 0.01542$ m, and $\theta_{Z(Mk)} = 0.00578$. Therefore, *the coordinates of point P_o for building A, through which the fictitious elastic vertical axis passes, are computed as (with regard to α coordinate system X-Y having its origin at the geometric center of column C1* (Fig. 4.2):

- $X_{Po} = X_{Mk} - (u_{Y(Mk)}/\theta_{Z(Mk)}) = 4.5 - 0.01542/0.00578 = 1.832$ m; and
- $Y_{Po} = Y_{Mk} - (u_{X(Mk)}/\theta_{Z(Mk)}) = 3.75 - 0/0.00578 = 3.75$ m

and *the structural eccentricities common to all storeys of building A are*:

- $e_{oX} = X_{Mk} - X_{Po} = 4.5 - 1.832 = 2.668$ m; and
- $e_{oY} = X_{Mk} - X_{Po} = 3.75 - 3.75 = 0$ m.

Furthermore, the determination of the torsional stiffness radii requires two additional static linear analyses to be performed considering horizontal (lateral) forces applied at the traces of the previously defined fictitious elastic axis at each floor diaphragm of the building along the assumed directions of the seismic action, that is, axes X and Y (loading cases "F_X" and "F_Y", respectively). These lateral forces may follow a "triangular" distribution along the height of the building by application of equation (4.11) of EC8, as in the case of the torsional moments previously considered. Table 4.5 reports the values of the lateral forces along with the pertinent steps for computing these values for building A, assuming a total base shear equal to 5000 kN, i.e., equal to the previous arbitrarily chosen base shear.

Subsequently, the torsional stiffness radii r_X and r_Y corresponding to the fictitious elastic vertical axis are determined by the relationships (Makarios and Anastassiadis, 1998) $r_X = (u_{Y(FY)}/\theta_{Z(Mk)})^{1/2}$ and $r_Y = (u_{X(FX)}/\theta_{Z(Mk)})^{1/2}$ included in FC-3.2. In the above relationships, $u_{Y(FY)}$ and $u_{X(FY)}$ are the translations of point P_o along axis Y due to the loading case F_Y and along axis X due to the loading case F_X, respectively. For building A and for the lateral forces of Table 4.5, the aforementioned nodal displacements are: $u_{Y(FY)} = 0.0842$ m and $u_{X(FX)} = 0.306$ m. Therefore, *the torsional stiffness radii common to all storeys of building A are*:

- $r_X = (u_{Y(FY)}/\theta_{Z(Mk)})^{1/2} = (0.0842/0.00578)^{1/2} = 3.817$; and
- $r_Y = (u_{X(FX)}/\theta_{Z(Mk)})^{1/2} = (0.306/0.00578)^{1/2} = 7.276$.

Having computed the structural eccentricities and torsional radii for the fictitious elastic axis, the inequalities of Eqs. (3.1) and (3.2) are checked in Tables 4.6 and 4.7, respectively. As shown in the penultimate column of Table 4.6, building A does not satisfy the inequality of Eq. (3.1) (i.e., equation (4.1a) of EC8) along principal axis X. Therefore, the building is not regular in plan.

Table 4.6 Verification check of inequality in Eq. (3.1), Sect. 3.1.1.1

Storey	e_{oX} [m]	e_{oY} [m]	r_X [m]	r_Y [m]	$e_{oX} < 0.3 \cdot r_X$	$e_{oY} < 0.3 \cdot r_Y$
1st	2.67	0.00	3.82	7.28	NO	YES
2nd	2.67	0.00	3.82	7.28	NO	YES
3rd	2.67	0.00	3.82	7.28	NO	YES
4th	2.67	0.00	3.82	7.28	NO	YES
5th	2.67	0.00	3.82	7.28	NO	YES

Table 4.7 Verification check of inequality in Eq. (3.2), Sect. 3.1.1.1

Storey	l_s[a] [m]	r_X [m]	r_Y [m]	$r_X \geq l_s$	$r_Y \geq l_s$
1st	3.382	3.82	7.28	YES	YES
2nd	3.382	3.82	7.28	YES	YES
3rd	3.382	3.82	7.28	YES	YES
4th	3.382	3.82	7.28	YES	YES
5th	3.382	3.82	7.28	YES	YES

[a]The radius of gyration l_s for rectangular floor plans such as that of building A is obtained using the following relationship: $l_s = [(L_x^2 + L_Y^2)/12]^{1/2} = [(9.0^2 + 7.5^2)/12]^{1/2} = 3.382$ m

4.1.4 *Classification of the Lateral Load-Resisting Structural System of Building A*

The rationale of classifying building structures according to the properties of their lateral load-resisting structural system and the implications of this classification in the seismic design process have been discussed in detail in Sect. 3.1.2. The classification procedure follows the steps delineated in FC-3.4, which include:

(1) Verification check for torsional sensitivity based on the criterion of Eq. (3.2): $r_x \geq l_s$ and $r_y \geq l_s$, that is, torsional stiffness radius ≥ radius of gyration along principal axes X and Y (Equation (4.1b) of clause §4.2.3.2(6) of EC8); and
(2) Verification check of the fraction of the total base shear resisted by the frames (i.e., columns) and the r/c walls at the assumed foundation level along the considered directions of the seismic action, that is, principal axes X and Y.

Verification (1) (torsional sensitivity check) is performed according to the procedure of FC-3.2a, which has already been undertaken in the previous section as part of the regularity in plan verification check. The results reported in Table 4.7 suggest that the inequalities $r_x \geq l_s$ and $r_y \geq l_s$ are satisfied for all storeys and, therefore, *building A is not torsionally sensitive*.

Verification (2) (check of the percentage of the total base shear resisted by r/c walls) is performed according to the procedure detailed in Sect. 3.1.2. Specifically,

(a) two independent static linear analyses should normally be carried out considering horizontal (lateral) forces applied at each floor diaphragm of the building along the assumed directions of the seismic action, that is, axes X and Y. However, since no wall is oriented along axis X, only the case of lateral

Table 4.8 Verification check of the percentage of the total base shear resisted by r/c wall W1 for lateral forces along axis Y applied at the traces of the fictitious elastic axis

Member	Analysis in direction of axis Y		
	Wall	$V_{Y,walls}$ [kN]	$V_{Y,columns}$ [kN]
C1	NO	–	132.37
C2	NO	–	152.09
C3	NO	–	196.26
C4	NO	–	187.89
C5	NO	–	236.64
C6	NO	–	132.37
C7	NO	–	152.09
C8	NO	–	196.26
W1	YES	3614.02	–
Total sum		3614.02	1385.97
Percentage		**72.28 %**	**27.72 %**

forces applied along axis Y are required for the classification of building A, as detailed further below. The distribution of the lateral forces along the building height is taken to be "triangular" according to equation (4.11) of EC8. The value of the applied total base shear can be chosen arbitrarily, since only the distribution of the base shear among walls and columns as a percentage of its total value is of interest. As discussed in Sect. 3.1.2, EC8 does not specify the point of action of the lateral forces at each floor. Herein, the two most obvious options are considered for the sake of comparison, namely, lateral forces are applied *at the traces of the fictitious elastic axis* determined in Sect. 4.1.3.1 at each floor diaphragm, and lateral forces are applied *at the center of mass of each floor diaphragm*. In what follows, numerical results from both cases are examined and discussed vis-à-vis.

(b) for each of the loading cases involving lateral forces along the X and Y axes, the percentage of the base shear V_X and V_Y resisted by walls and columns along axes X and Y, respectively, are evaluated and verified separately against the classification criteria included in FC-3.4. For building A, only the parts of the base shear V_Y resisted by the walls, $V_{Y,walls}$, and by the columns, $V_{Y,columns}$, are of interest.

Classification of building A along principal axis X

There are no r/c walls oriented along axis X. Therefore, *building A is classified as a frame structural system along axis X*, without the need to take any computational step.

Classification of building A along principal axis Y

There exists a single wall, W1, oriented along axis Y and, therefore, the classification of building A along this axis necessitates undertaking verification check (2), as previously detailed. Table 4.8 summarizes the pertinent numerical results obtained by application of linear static analysis for the lateral forces along direction Y reported in Table 4.5 applied *at the traces of the fictitious elastic axis*

Table 4.9 Verification check of the percentage of the total base shear resisted by r/c wall W1 for lateral forces along axis Y applied at the center of mass

Member	Analysis in direction of axis Y		
	Wall	$V_{Y,walls}$ [kN]	$V_{Y,columns}$ [kN]
C1	NO	–	85.98
C2	NO	–	294.84
C3	NO	–	472.45
C4	NO	–	356.01
C5	NO	–	568.24
C6	NO	–	85.98
C7	NO	–	294.84
C8	NO	–	472.45
W1	YES	2369.22	–
Sum		2369.22	2630.79
Percentage		**47.38 %**	**52.62 %**

determined in Sect. 4.1.3.1. Based on these results, *building A is classified as a wall structural system along axis Y*, since the wall W1 resists more than 65 % of the total base shear applied to the structure along the considered axis.

Similarly, Table 4.9 summarizes some pertinent numerical results obtained by application of linear static analysis for the lateral forces along direction Y reported in Table 4.5 applied *at the center of mass of each floor diaphragm*. Based on these results, *building A is classified as a dual frame-equivalent structural system along axis Y*, since the columns resist less than 65 % but more than 50 % of the total base shear applied to the structure along the considered axis.

Therefore, it is seen that the same building may be classified into a different type of structural system depending on the assumed in-plan location of the points through which the lateral statically applied forces are acting in performing the pertinent classification verification check of the percentage of the total base shear resisted by walls. The difference in classification influences the value of the maximum allowable value of the behaviour factor q which can be adopted in the analysis stage, as will be further seen in Sect. 4.1.6.

4.1.5 Selection of Ductility (Capacity) Class of Building A

The rationale of deciding upon the desirable ductility capacity class according to EC8 has been discussed in Sect. 3.1.3. Herein, the following relevant points are highlighted:

- the "low" ductility class (DCL) should be avoided by all means in regions of high seismicity;
- the choice of ductility class influences the maximum allowable value $\max q_{allow}$ of the behaviour factor q which may be used to reduce the input seismic action in the seismic analysis. Higher $\max q_{allow}$ values apply for the "high" ductility class

(DCH) compared to those corresponding to the "medium" ductility class (DCM); and
- selecting the DCH should be considered only for sites where stringent DCH detailing requirements for r/c buildings can be practically guaranteed during construction (see also Sect. 3.4).

For building A, it is decided to adopt the DCH, though the case of selecting the DCM is also considered in the next section to quantify the influence that the selection of different ductility class has on the $\max q_{allow}$ value.

4.1.6 *Determination of the Maximum Allowed Behaviour Factor for Building A*

A detailed presentation of the procedure for determining the maximum allowed value, $\max q_{allow}$, of the behaviour factor q has been provided in Sect. 3.1.4. Herein, the following relevant points are highlighted:

- the values of the behaviour factor determined by the relevant EC8 provisions are the maximum allowable values, $\max q_{allow}$, which may be considered to reduce the level of the seismic input action (i.e., the spectral ordinates of the EC8 elastic spectrum) in the context of an equivalent linear seismic analysis;
- adopting the maximum allowable value for the behaviour factor, $q = \max q_{allow}$, which is not mandatory, entails full utilization of the assumed ductility capacity of the considered building to resist the nominal design seismic action; and
- adopting a lower than the $\max q_{allow}$ value for the behaviour factor, $1 \leq q < \max q_{allow}$, entails that the nominal design seismic action is resisted without utilization of the full assumed ductility capacity of the structure. This further implies that less structural damage is induced by the design earthquake compared to the expected damage for $q = \max q_{allow}$ while the reserved (unutilized) ductility capacity is used to resist more severe levels of seismic action than the nominal design level.

The procedure of determining the $\max q_{allow}$ follows FC-3.5. The latter flowchart suggests that the value of $\max q_{allow}$ depends, among other factors, on the classification of the lateral load resisting system of a building into one of the types of structural systems defined by EC8. However, building A has been classified differently along the Y axis, namely as a wall system or as a dual frame-equivalent system, depending on the assumed points of application of the horizontal forces considered in the pertinent verification check (2) of Sect. 4.1.4. In this regard, both the above alternative structural systems are considered in this section to exemplify the influence of structural system classification on the value of the $\max q_{allow}$. It is further instructive to determine the value of $\max q_{allow}$ for both viable alternatives with regards to the ductility class, that is, high (DCH) and medium (DCM), for the two aforementioned structural system types. Therefore, in the remainder of this section, $maxq_{allow}$ values are obtained and compared for the following four cases for building A

- Case (I): DCH- frame system along the X axis; wall system along the Y axis
- Case (II): DCM- frame system along the X axis; wall system along the Y axis
- Case (III): DCH- frame system along the X axis; dual frame-equivalent system along the Y axis
- Case (IV): DCM- frame system along the X axis; dual frame-equivalent system along the Y axis

In this manner, the potential range of values of the maximum allowable behaviour factor is quantified for different ductility classes and structural system type classification.

Case (I): DCH- frame system along the X axis; wall system along the Y axis

According to Table 3.4 (table 5.1 of EC8), the basic value of the behaviour factor q_o is a function of the ratio a_u/a_1 and is given by

- Along X axis: $q_{0,X} = 4.5(a_u/a_1)$
- Along Y axis: $q_{0,Y} = 4.0(a_u/a_1)$

The ratio a_u/a_1 is estimated from Table 3.5 (clause §5.2.2.2(5a) of EC8) as

- Along the X axis: $a_u/a_1 = 1.3$
- Along the Y axis: $a_u/a_1 = 1.0$

The building is not regular in plan. Therefore, according to clause §5.2.2.2(6) of EC8, the value of a_u/a_1 should be taken equal to the mean value between 1.0 and the value calculated on the basis of clause §5.2.2.2(5) of EC8, as above. Thus, the following updated values for the ratio a_u/a_1 apply

- Along the X axis: $a_u/a_1 = 1.3 \times a_u/a_1 = (1.0 + 1.3)/2 = 1.15$
- Along the Y axis: $a_u/a_1 = 1.0 \times a_u/a_1 = (1.0 + 1.0)\ 2 = 1.0$

Hence, the following values of the behaviour factor q_o are reached

- Along the X axis: $q_{0,X} = 4.5 \times 1.15 = 5.2$
- Along the Y axis: $q_{0,Y} = 4.0 \times 1.00 = 4.0$

Further, building A is regular in elevation. Therefore, it is not required to reduce (by 20 %) the above values of the basic behaviour factor.

The reduction factor k_w is computed according to Table 3.6 (clause §5.2.2.2(11) P of EC8)

- Along the X axis: $k_w = 1.0$
- Along the Y axis: $k_w = (1 + a_0) / 3 \leq 1$;

where $a_0 =$ (shear wall height h_w) / (length of shear wall section L_w).
Thus, $a_0 = 16.0/2.5 = 6.4$ and $k_w = (1 + 6.4)/3 = 2.46 > 1.0 \rightarrow k_w = 1.0$

Finally, the maximum allowable behaviour factors $maxq_{allow}$ are computed from Eq. (3.4) (equation (5.1) of EC8)

- Along the X axis: $maxq_{allow} = q_X = q_{0,X} \cdot k_w = 5.2 \cdot 1.0 = 5.2$
- Along the Y axis: $maxq_{allow} = q_Y = q_{0,Y} \cdot k_w = 4.0 \cdot 1.0 = 4.0 \neq q_X$

It is observed that, for the considered case, the maximum allowable value of the behaviour factor is not equal along the X and Y axes, which is, in general, allowed (see clause §4.3.3.5.1 (4) of EC8). However, adopting different values of the behaviour factor along the two orthogonal directions of the seismic action, $q_X \neq q_Y$, requires caution in applying the modal response spectrum method of analysis, as discussed in Sect. 3.1.4.4.

Case (II): DCM- frame system along the X axis; wall system along the Y axis

According to Table 3.4 (table 5.1 of EC8), the basic value of the behaviour factor q_o is a function of the ratio a_u/a_1 and is given by

- Along the X axis: $q_{0,X} = 3.0(a_u/a_1)$
- Along the Y axis: $q_{0,Y} = 3.0$

The ratio a_u/a_1 is estimated from Table 3.5 (clause §5.2.2.2(5a) of EC8) as

- Along the X axis: $a_u/a_1 = 1.3$

The building is not regular in plan. Therefore, according to clause §5.2.2.2(6) of EC8, the value of a_u/a_1 should be taken equal to the mean value between 1.0 and the value calculated on the basis of clause §5.2.2.2(5) of EC8, as above. Thus, the following updated values for the ratio a_u/a_1 apply

- Along the X axis: $a_u/a_1 = 1.3 \rightarrow a_u/a_1 = (1.0 + 1.3) / 2 = 1.15$

Hence, the following values of the behaviour factor q_o are reached

- Along the X axis: $q_{0,X} = 3.0 \cdot 1.15 = 3.45$
- Along the Y axis: $q_{0,Y} = 3.0 = 3.0$

Further, building A is regular in elevation. Therefore, it is not required to reduce (by 20 %) the above values of the basic behaviour factor.

The reduction factor k_w is computed according to Table 3.6 (clause §5.2.2.2(11) P of EC8)

- Along the X axis: $k_w = 1.0$
- Along the Y axis: $k_w = (1 + a_0) / 3 \leq 1$;

where $a_0 =$ (shear wall height h_w) / (length of shear wall section L_w).
Thus, $a_0 = 16.0/2.5 = 6.4$ and $k_w = (1 + 6.4) / 3 = 2.46 > 1.0 \times k_w = 1.0$

Finally, the maximum allowable behaviour factors $maxq_{allow}$ are computed from Eq. (3.4) (equation (5.1) of EC8)

- Along the X axis: $maxq_{allow} = q_X = q_{0,X} x k_w = 3.45 \times 1.0 = 3.45$
- Along the Y axis: $maxq_{allow} = q_Y = q_{0,Y} x k_w = 3.0 \times 1.0 = 3.0 \neq q_X$

It is observed that, as in case (I), the maximum allowable value of the behaviour factor is not equal along the X and Y axes, which is, in general, allowed (see clause §4.3.3.5.1 (4) of EC8). However, adopting different values of the behaviour factor along the two orthogonal directions of the seismic action, $q_X \neq q_Y$, requires caution in applying the modal response spectrum method of analysis, as discussed in Sect. 3.1.4.4.

Case (III): DCH- frame system along the X axis; dual frame-equivalent system along the Y axis

According to Table 3.4 (table 5.1 of EC8), the basic value of the behaviour factor q_o is a function of the ratio a_u/a_1 and is given by

- Along the X axis: $q_{0,X} = 4.5(a_u/a_1)$
- Along the Y axis: $q_{0,Y} = 4.5(a_u/a_1)$

The ratio a_u/a_1 is estimated from Table 3.5 (clause §5.2.2.2(5a) of EC8) as

- Along the X axis: $a_u/a_1 = 1.3$
- Along the Y axis: $a_u/a_1 = 1.3$

The building is not regular in plan. Therefore, according to clause §5.2.2.2(6) of EC8, the value of a_u/a_1 should be taken equal to the mean value between 1.0 and the value calculated on the basis of clause §5.2.2.2(5) of EC8, as above. Thus, the following updated values for the ratio a_u/a_1 apply

- Along the X axis: $a_u/a_1 = 1.3 \times a_u/a_1 = (1.0 + 1.3) / 2 = 1.15$
- Along the Y axis: $a_u/a_1 = 1.3 \times a_u/a_1 = (1.0 + 1.3) / 2 = 1.15$

Hence, the following values of the behaviour factor q_o are reached

- Along the X axis: $q_{0,X} = 4.5 \times 1.15 = 5.2$
- Along the Y axis: $q_{0,Y} = 4.5 \times 1.15 = 5.2$

Further, building A is regular in elevation. Therefore, it is not required to reduce (by 20 %) the above values of the basic behaviour factor.

The reduction factor k_w is computed according to Table 3.6 (clause §5.2.2.2(11) P of EC8)

- Along the X axis: $k_w = 1.0$
- Along the Y axis: $k_w = 1.0$

Finally, the maximum allowable behaviour factors $maxq_{allow}$ are computed from Eq. (3.4) (equation (5.1) of EC8)

- Along the X axis: $maxq_{allow} = q_X = q_{0,X} x\ k_w = 5.2 \times 1.0 = 5.2$
- Along the Y axis: $maxq_{allow} = q_Y = q_{0,Y} x\ k_w = 5.2 \times 1.0 = 5.2 = q_X$

It is observed that, contrary to the previous two cases (I) and (II) considered, the maximum allowable values of the behaviour factor are equal along axes X and Y.

Case (IV): DCM- frame system along the X axis; dual frame-equivalent system along the Y axis

According to Table 3.4 (table 5.1 of EC8), the basic value of the behaviour factor q_o is a function of the ratio a_u/a_1 and is given by

- Along the X axis: $q_{0,X} = 3.0(a_u/a_1)$
- Along the Y axis: $q_{0,Y} = 3.0(a_u/a_1)$

The ratio a_u/a_1 is estimated from Table 3.5 (clause §5.2.2.2(5a) of EC8) as

- Along the X axis: $a_u/a_1 = 1.3$
- Along the Y axis: $a_u/a_1 = 1.3$

The building is not regular in plan. Therefore, according to clause §5.2.2.2(6) of EC8, the value of a_u/a_1 should be taken equal to the mean value between 1.0 and the value calculated on the basis of clause §5.2.2.2(5) of EC8, as above. Thus, the following updated values for the ratio a_u/a_1 apply

- Along the X axis: $a_u/a_1 = 1.3 \times a_u/a_1 = (1.0 + 1.3) / 2 = 1.15$
- Along the Y axis: $a_u/a_1 = 1.3 \times a_u/a_1 = (1.0 + 1.3) / 2 = 1.15$

Hence, the following values of the behaviour factor q_o are reached

- Along the X axis: $q_{0,X} = 3.0 \times 1.15 = 3.45$
- Along the Y axis: $q_{0,Y} = 3.0 \times 1.15 = 3.45$

Further, building A is regular in elevation. Therefore, it is not required to reduce (by 20 %) the above values of the basic behaviour factor.

The reduction factor k_w is computed according to Table 3.6 (clause §5.2.2.2(11) P of EC8)

- Along the X axis: $k_w = 1.0$
- Along the Y axis: $k_w = 1.0$

Finally, the maximum allowable behaviour factors $maxq_{allow}$ are computed from Eq. (3.4) (equation (5.1) of EC8)

- Along the X axis: $maxq_{allow} = q_X = q_{0,X}xkw = 3.45 \times 1.0 = 3.45$
- Along the Y axis: $maxq_{allow} = q_Y = q_{0,Y}xkw = 3.45 \times 1.0 = 3.45 = q_X$

It is observed that, as in the previously considered case (III), the maximum allowable values of the behaviour factor are equal along axes X and Y.

The results for the four cases, (I), (II), (III), and (IV) presented are summarized in Table 4.10 to facilitate comparison and to demonstrate the wide range of $maxq_{allow}$ values obtained from different assumptions and choices made.

Building A has been decided to be designed as a DCH structure (see Sect. 4.1.5) and, therefore, the values 5.2 and 4.0 apply for the behaviour factor. It is further decided to *adopt a common behaviour factor along directions X and Y equal to the smallest of the above maximum allowable values:* $q = q_X = q_Y = 4.0$.

Table 4.10 Maximum allowable behaviour factors for building A for the 4 different cases considered

	(I)	(II)	(III)	(IV)
	DCH	DCM	DCH	DCM
$maxq_{allow}$	Wall system in the Y direction	Wall system in the Y direction	Dual frame-equivalent system in the Y direction	Dual frame-equivalent system in the Y direction
q_X	5.20	3.45	5.20	3.45
q_Y	4.00	3.00	5.20	3.45

4.1.7 *Selection of an Equivalent Linear Method of Seismic Analysis for Building A*

The modal response spectrum method (MRSM) is of general use for the seismic analysis of any building structure regardless of regularity conditions in plan and elevation. As such, it is applicable to building A. Moreover, the MRSM is recommended to be used even in the seismic analysis of buildings for which the simpler lateral force method (LFM) can be used. This is due to the challenges and practical difficulties in the application of LFM discussed in Sect. 2.4.3.1 and because the MRSM provides for more accurate numerical results.

According to Table 3.1 (table 4.1 of EC8), the LFM can be used for the seismic analysis of non-regular in plan but regular in elevation buildings, such as building A, provided that the uncoupled fundamental natural periods along directions X and Y, T_{1X} and T_{1Y}, respectively, are smaller than min{$4T_c$, 2.0 s} (see FC-3.6). In general, the verification of the above criterion requires undertaking two pertinent modal analyses. For the building under consideration, the uncoupled fundamental natural period along axis X results from a modal analysis with mass position 1, as the building is symmetric with regard to axis X. Hence only one additional modal analysis is performed with mass positioned at the center of mass and considering all degrees of freedom along axis X restraint. The obtained values for the uncoupled fundamental natural periods are $T_{1X} = 0.88$ s and $T_{1Y} = 0.443$ s, which satisfy the applicability criterion of the LFM. Consequently, both the MRSM and the LFM of analysis can be applied to building A and both are considered. Indicative numerical results (seismic effects) from the application of the MRSM are reported in Sect. 4.1.9, while seismic effects obtained from the LFM are included in Sect. 4.1.10. Finally, Sect. 4.1.11 compares and discusses seismic effects obtained by both EC8 equivalent linear methods of seismic analysis.

4.1.8 *Static Analysis for Gravity Loads of the Design Seismic Loading Combination (G "+" ψ_2Q) for Building A*

The design seismic loading combination involves gravitational (statically applied) permanent and quasi-permanent variable actions (see Eq. (2.12)). Effects due to these actions can be derived separately by means of standard static analysis and

Table 4.11 Sectional stress resultants of wall W1, column C3, and beams BX2 and BY5 at the ground storey for the permanent and quasi-permanent variable actions of the design seismic loading combination (G "+" ψ_2Q) [Sign convention follows Fig. 2.32]

Member	Position	N [kN]	V2 [kN]	V3 [kN]	T [kNm]	M2 [kNm]	M3 [kNm]
W1	Bottom	−684.76	0.03	0.0	0.0	0.0	0.278
	Top	−622.26	0.03	0.0	0.0	0.0	0.158
C3	Bottom	−327.14	1.90	−1.54	0.0	−2.01	2.63
	Top	−312.34	1.90	−1.54	0.0	3.71	−4.41
BX2	Left end	0.0	−29.18	0.0	0.0431	0.0	−20.47
	Mid-span	0.0	−3.12	0.0	0.0431	0.0	10.18
	Right end	0.0	26.39	0.0	0.0431	0.0	−12.58
BY5	Left end	0.0	−24.89	0.0	0.0695	0.0	−9.31
	Mid-span	0.0	2.69	0.0	0.0695	0.0	11.16
	Right end	0.0	34.78	0.0	0.0695	0.0	−20.22

superposed to the effects due to the seismic (accidental) action. Table 4.11 reports stress resultants (effects) due to the gravity loads of the design seismic loading combination at critical cross-sections of selected r/c structural members, namely, the wall W1, the column C1, and the beams BX2 and BY5 at the ground (1st) storey.

4.1.9 *Seismic Analysis of Building A Using the Modal Response Spectrum Method and Deformation-Based Verification Checks*

The implementation of the MRSM of seismic analysis follows FC-3.7. Four different spatial FE models are considered in the analysis corresponding to the positioning of the center of mass of each floor diaphragm at four different sets of locations (positions 1 to 4, as shown in Table 4.12) to account for accidental mass eccentricity (see also Sect. 3.1.5.1 and Fig. 3.4). The accidental eccentricities, $\pm e_{aX}$ and $\pm e_{aY}$, along the principal axes X and Y, respectively, define the four displaced locations of the mass center measured from the geometric center of each slab. These eccentricities are computed in Table 4.12 taken to eb equal to 5 % of the length of building A, $L_X = 9.0$ m and $L_Y = 7.5$ m, along axes X and Y, respectively, assuming that the masonry infill walls are evenly distributed in plan. Further, the polar moment of inertia about the gravitational axis of each floor diaphragm is computed with respect to the displaced position as $J_{mi} = J_m + m e_{ai}^2$, $i = X,Y$, where $J_m = m(L_X^2 + L_Y^2)/12$ is the polar moment of inertia with respect to the geometric center of each slab, as reported in Table 4.5.

Table 4.12 Accidental eccentricities and polar moment of inertia for seismic excitation along X and Y axes for building A

		Accidental eccentricities [m]		Polar moment of inertia [tm^2]		
Storey	Mass [t]	e_{aX}[a]	e_{aY}[b]	J_{mX}[c]	J_{mY}[d]	
1st	90.77	0.45	0.375	1056.56	1050.946	
2nd–4th	88.34	0.45	0.375	1028.28	1022.811	
5th	66.00	0.45	0.375	768.24	764.156	

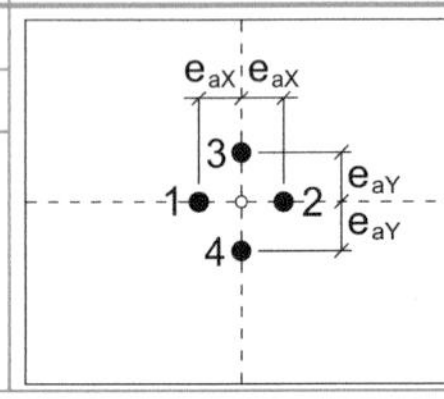

[a]$e_{aX} = 0.05L_X$
[b]$e_{aY} = 0.05L_Y$
[c]$J_{mX} = J_m + me_{aX}^2$
[d]$J_{mY} = J_m + me_{aY}^2$

Table 4.13 Natural periods of the four considered FE structural models (center of mass displaced by $\pm e_{aX}$ and $\pm e_{aY}$ as shown in Table 4.12) for building A

	Natural period [s]			
Mode shape	Mass position 1	Mass position 2	Mass position 3	Mass position 4
1	0.881	0.881	0.883	0.883
2	0.598	0.674	0.634	0.634
3	0.316	0.281	0.297	0.297
4	0.273	0.273	0.274	0.274
5	0.183	0.207	0.194	0.194
6	0.148	0.148	0.148	0.148
7	0.103	0.115	0.108	0.108
8	0.097	0.097	0.097	0.097
9	0.091	0.0815	0.086	0.086

4.1.9.1 Modal Analysis Results

Table 4.13 lists the natural periods corresponding to the first 9 mode shapes of vibration for the four considered FE structural models (position of center of mass 1 to 4, as shown in Table 4.12) of building A derived from standard modal analysis.

Further, Table 4.14 reports the modal participation mass ratios for each mode (i.e., ratios of effective modal mass over the total mass of building A) along axes X and Y, as well as the corresponding cumulative modal participation mass ratios for the four considered FE models. The latter results suggest that at least the first 5 mode shapes need to be considered to satisfy the criterion of clause 4.3.3.3.1(3) of EC8 for all the four FE models, that is, a sufficient number of modes are considered in the MRSM such that 90 % or more of the total oscillatory mass along both principal axes is activated. Therefore, in all ensuing numerical results reported, only the first 5 mode shapes are utilized in implementing the MRSM of analysis.

Table 4.14 Modal participation mass ratios and cumulative participation mass ratios as percentages of the total mass of building A

	Mass position 1				Mass position 2			
	Individual mode (%)		Cumulative sum (%)		Individual mode (%)		Cumulative sum (%)	
Mode shape	X	Y	X	Y	X	Y	X	Y
1	**91.93**	**0.00**	**91.93**	**0.00**	**91.93**	**0.00**	**91.93**	**0.00**
2	**0.00**	**66.22**	**91.93**	**66.21**	**0.00**	**69.84**	**91.93**	**69.84**
3	**0.00**	**21.16**	**91.93**	**87.38**	**0.00**	**17.5**	**91.93**	**87.34**
4	**6.405**	**0.00**	**98.33**	**87.38**	**6.404**	**0.00**	**98.33**	**87.34**
5	**0.00**	**6.204**	**98.33**	**93.58**	**0.00**	**7.19**	**98.33**	**94.53**
6	1.24	0.00	99.57	93.58	1.24	0.00	99.57	94.53
7	0.000	0.004	99.57	93.58	0.00	0.35	99.57	94.88
8	0.35	0.00	99.92	93.58	0.35	0.00	99.92	94.88
9	0.00	4.705	99.92	98.29	0.00	3.76	99.92	98.64
	Mass position 3				Mass position 4			
	Individual mode		Cumulative sum		Individual mode		Cumulative sum	
Mode shape	X	Y	X	Y	X	Y	X	Y
1	**91.57**	**0.21**	**91.57**	**0.21**	**91.57**	**0.21**	**91.57**	**0.21**
2	**0.345**	**68.00**	**91.92**	**68.21**	**0.345**	**68.00**	**91.92**	**68.21**
3	**0.034**	**19.27**	**91.95**	**87.48**	**0.034**	**19.27**	**91.95**	**87.48**
4	**6.37**	**0.00**	**98.32**	**87.48**	**6.37**	**0.00**	**98.32**	**87.48**
5	**0.02**	**6.62**	**98.34**	**94.10**	**0.02**	**6.62**	**98.34**	**94.10**
6	1.23	0.00	99.57	94.10	1.23	0.00	99.57	94.10
7	0.00	0.20	99.57	94.30	0.00	0.20	99.57	94.30
8	0.35	0.00	99.92	94.30	0.35	0.00	99.92	94.30
9	0.00	4.14	99.92	98.44	0.00	4.14	99.92	98.44

4.1.9.2 Selected Design Seismic Effects (Sectional Stress Resultants)

In this section, the design seismic effects at critical sections (sectional stress resultants) for the column C3, the wall W1, and the beams BX2 and BY5 of the ground (1st) storey (see Fig. 4.2) obtained by means of the MRSM are presented in tabular form.

Vertical structural members C3 and W1 (bi-axial bending with axial force)

In the case of the vertical structural members C3 and W1, which need to be designed for bi-axial bending with axial force, the design values of the three concurrent pertinent stress resultants ("design triads"), namely moments M_2, M_3 and axial force N, as defined in Fig. 2.32, are reported. Following the conservative approach of clause §4.3.3.5.1(2)c of EC8, these design triads may comprise the extreme values of the M_2, M_3, and N, as defined in Eq. (3.8), for simultaneous seismic action along the two principal axes X and Y. According to this approach, the 8 design triads of Eq. (3.12) need to be considered for each position of floor

Table 4.15 Extreme values of stress resultants in column C3 (ground storey) of building A

	N [kN]		M_2 [kNm]		M_3 [kNm]	
Mass position	Bottom	Top	Bottom	Top	Bottom	Top
1	±281.76	±281.76	±126.70	±102.44	±124.43	±96.15
2	±282.73	±282.73	±132.82	±107.45	±125.76	±97.24
3	±283.99	±283.99	±130.60	±105.61	±122.94	±95.04
4	±280.70	±280.70	±130.63	±105.66	±127.00	±98.18

Table 4.16 Extreme values of stress resultants in wall W1 (ground storey) of building A

	N [kN]		M_2 [kNm]		M_3 [kNm]	
Mass position	Bottom	Top	Bottom	Top	Bottom	Top
1	±61.16	±61.16	±899.22	±48.26	±120.96	±46.01
2	±61.16	±61.16	±703.78	±39.60	±120.96	±46.01
3	±61.02	±61.02	±803.96	±43.90	±120.68	±45.90
4	±61.02	±61.02	±803.96	±43.90	±120.68	±45.90

mass. Alternatively, the design triads may be compiled by considering the extreme value of a single stress resultant together with the expected (most probable) values of the other two stress resultants attained concurrently. In the latter case, the 6 design triads of Eq. (3.13) for each position of floor mass can be considered derived by means of a simplified approach detailed in the Greek Seismic Code EAK2000 (Earthquake Planning and Protection Organization (EPPO) 2000) and assuming simultaneous seismic action along the two principal axes X and Y (see last paragraph in Sect. 3.1.5.1).

In particular, the following three computational steps are taken to determine the design triads for the considered vertical members (see also Sect. 3.1.5.1):

(1) The peak (seismic) modal values of the considered stress resultants are obtained separately by application of the MRSM for each of the directions of the seismic excitation X and Y. Next, these modal values are superposed by means of the CQC rule for modal combination (clause §4.3.3.3.2(3)P of EC8) to derive the (non-concurrent) maximum values of stress resultants for seismic excitation along axes X and Y, independently.
(2) The SRSS rule for spatial combination (clause §4.3.3.5.1(2)b of EC8) is employed to obtain the extreme values of the considered stress resultants from the maximum values derived in the previous step for simultaneous seismic action along the X and Y horizontal directions. Tables 4.15 and 4.16 report the thus obtained extreme values of the M_2, M_3, and N stress resultants developing at the bottom and the top of the structural members C3 and W1, respectively, at the ground (1st) storey of building A for all four different FE models used in the analysis. As previously discussed, EC8 allows for compiling 8 design triads for each of the four eccentrically positioned mass centers comprising these extreme M_2, M_3, and N values with alternating signs according to Eq. (3.12) assumed to act concurrently in each section. However, the above design triads may lead to

Table 4.17 Design triads (expected -most probable- concurrent values of N, M_2, and M_3 stress resultants for simultaneous seismic action along axes X and Y) for column C3 (ground storey) of building A [Extreme values in bold]

Mass position		N [kN]		M_2 [kNm]		M_3 [kNm]	
		Bottom	Top	Bottom	Top	Bottom	Top
1	extrN	**281.76**	**281.76**	69.08	−55.32	−66.00	48.84
	extrM_2	153.62	−152.17	**126.7**	**102.44**	50.04	41.25
	extrM_3	−149.44	143.11	50.95	43.95	**124.43**	**96.15**
	−extrN	**−281.76**	**−281.76**	−69.08	55.32	66.00	−48.84
	−extrM_2	−153.62	152.17	**−126.7**	**−102.44**	−50.04	−41.25
	−extrM_3	149.44	−143.11	−50.95	−43.95	**−124.43**	**−96.15**
2	extrN	**282.73**	**282.73**	73.04	−58.55	−63.50	46.92
	extrM_2	155.49	−154.05	**132.82**	**107.45**	53.31	43.78
	extrM_3	−142.76	136.41	56.30	48.37	**125.76**	**97.24**
	−extrN	**−282.73**	**−282.73**	−73.04	58.55	63.50	−46.92
	−extrM_2	−155.49	154.05	**−132.82**	**−107.45**	−53.31	−43.78
	−extrM_3	142.76	−136.41	−56.30	−48.37	**−125.76**	**−97.24**
3	extrN	**283.99**	**283.99**	74.80	−59.96	−62.77	46.35
	extrM_2	162.65	−161.26	**130.60**	**105.61**	48.26	39.88
	extrM_3	−145.01	138.51	51.27	44.32	**122.94**	**95.04**
	−extrN	**−283.99**	**−283.99**	−74.80	59.96	62.77	−46.35
	−extrM_2	−162.65	161.26	**−130.60**	**−105.61**	−48.26	−39.88
	−extrM_3	145.01	−138.51	−51.27	−44.32	**−122.94**	**−95.04**
4	extrN	**280.70**	**280.70**	69.12	−55.36	−65.08	48.10
	extrM_2	148.52	−147.06	**130.63**	**105.66**	55.76	45.68
	extrM_3	−143.84	137.52	57.35	49.16	**127.00**	**98.18**
	−extrN	**−280.70**	**−280.70**	−69.12	55.36	65.08	−48.10
	−extrM_2	−148.52	147.06	**−130.63**	**−105.66**	−55.76	−45.68
	−extrM_3	143.84	−137.52	−57.35	−49.16	**−127.00**	**−98.18**

overly conservative detailing of cross-sections (see also Fardis 2009). In this regard, since EC8 allows for the use of more accurate methods to estimate the probable concurrent values of more than one seismic effect due to simultaneous seismic action along two horizontal axes without, nevertheless, specifying any, the simplified approach of EAK2000 is herein considered, as detailed in Sect. 3.1.5.1. Tables 4.17 and 4.18 present the 6 design triads at the bottom and the top of the structural members C3 and W1, respectively, at the ground (1st) storey of building A for all four different FE models used in the analysis obtained by application of the aforementioned simplified approach. The single extreme value attained by a certain stress resultant in each design triad is noted by bold faced fonts.

(3) Finally, the seismic design triads derived in the previous step are superposed to the corresponding stress resultants of the considered structural members due to the gravitational permanent and quasi-permanent variable actions summarized in Table 4.11 (Sect. 4.1.8) to obtain the design triads for the EC8 design seismic

Table 4.18 Design triads (expected -most probable- concurrent values of N, M_2, and M_3 stress resultants for simultaneous seismic action along axes X and Y) for wall W1 (ground storey) of building A [Extreme values in bold]

Mass position		N [kN]		M_2 [kNm]		M_3 [kNm]	
		Bottom	Top	Bottom	Top	Bottom	Top
1	extrN	**61.16**	**61.16**	0.00	0.00	117.95	−44.12
	extrM_2	0.00	0.00	**899.22**	**48.27**	0.00	0.00
	extrM_3	59.63	−58.65	0.00	0.00	**120.96**	**46.01**
	−extrN	**−61.16**	**−61.16**	0.00	0.00	−117.95	44.12
	−extrM_2	0.00	0.00	**−899.22**	**−48.27**	0.00	0.00
	−extrM_3	−59.63	58.65	0.00	0.00	**−120.96**	**−46.01**
2	extrN	**61.16**	**61.16**	0.00	0.00	117.95	−44.12
	extrM_2	0.00	0.00	**703.77**	**39.60**	0.00	0.00
	extrM_3	59.63	−58.65	0.00	0.00	**120.96**	**46.01**
	−extrN	**−61.16**	**−61.16**	0.00	0.00	−117.95	44.12
	−extrM_2	0.00	0.00	**−703.77**	**−39.60**	0.00	0.00
	−extrM_3	−59.63	58.65	0.00	0.00	**−120.96**	**−46.01**
3	extrN	**61.02**	**61.02**	85.72	−0.40	117.66	−44.02
	extrM_2	6.51	−0.55	**803.96**	**43.90**	11.84	1.68
	extrM_3	59.49	−58.51	78.89	1.61	**120.68**	**45.90**
	−extrN	**−61.02**	**−61.02**	−85.72	0.40	−117.66	44.02
	−extrM_2	−6.51	0.55	**−803.96**	**−43.90**	−11.84	−1.68
	−extrM_3	−59.49	58.51	−78.89	−1.61	**−120.68**	**−45.90**
4	extrN	**61.02**	**61.02**	−85.72	0.40	117.66	−44.02
	extrM_2	−6.51	0.55	**803.96**	**43.90**	−11.84	−1.68
	extrM_3	59.49	−58.51	−78.89	−1.61	**120.68**	**45.90**
	−extrN	**−61.02**	**−61.02**	85.72	−0.40	−117.66	44.02
	−extrM_2	6.51	−0.55	**−803.96**	**−43.90**	11.84	1.68
	−extrM_3	−59.49	58.51	78.89	1.61	**−120.68**	**−45.90**

loading combination G "+" Ψ_2Q "±" E. The thus obtained triads are reported in Tables 4.19 and 4.20.

Beams BX2 and BY5 (uni-axial bending)

The previously described 3 steps are applied to obtain the extreme values of the moment M_3 and of the shearing force V_2 at critical cross-sections (left end, right end, and at midspan) of the beams BX2 and BY5 of the ground (1st) storey of building A (Fig. 4.2), which need to be designed for uni-axial bending. However, in this case, the procedure of obtaining the design seismic effects for simultaneous seismic action along the two principal axes X and Y is significantly simplified by the fact that only a single seismic effect (i.e., stress resultant M_3 and corresponding shearing force V_2) is required in the detailing of beam sections, as opposed to the vector of the three concurrently acting seismic effects (triads) N, M_2, M_3 considered for the case of vertical structural members.

Table 4.19 Design triads for column C3 (ground storey) of building A for the seismic design load combination G "+" ψ_2Q "±" E [Extreme values in bold]

Mass position		N [kN]		M_2 [kNm]		M_3 [kNm]	
		Bottom	Top	Bottom	Top	Bottom	Top
1	extrN	**−45.38**	**−30.58**	67.06	−51.62	−63.36	44.43
	extrM_2	−173.52	−464.51	**124.69**	**106.14**	52.67	36.84
	extrM_3	−476.58	−169.23	48.94	47.66	**127.07**	**91.74**
	−extrN	**−608.90**	**−594.10**	−71.09	59.03	68.63	−53.25
	−extrM_2	−480.76	−160.17	**−128.71**	**−98.73**	−47.40	−45.66
	−extrM_3	−177.70	−455.45	−52.96	−40.25	**−121.80**	**−100.56**
2	extrN	**−44.41**	**−29.61**	71.03	−54.84	−60.87	42.51
	extrM_2	−171.65	−466.39	**130.81**	**111.16**	55.94	39.37
	extrM_3	−469.90	−175.93	54.29	52.08	**128.39**	**92.83**
	−extrN	**−609.87**	**−595.07**	−75.05	62.25	66.14	−51.33
	−extrM_2	−482.63	−158.29	**−134.83**	**−103.75**	−50.68	−48.19
	−extrM_3	−184.37	−448.75	−58.31	−44.67	**−123.13**	**−101.65**
3	extrN	**−43.15**	**−28.35**	72.79	−56.26	−60.14	41.94
	extrM_2	−164.49	−473.60	**128.59**	**109.31**	50.90	35.47
	extrM_3	−472.15	−173.83	49.26	48.02	**125.57**	**90.63**
	−extrN	**−611.13**	**−596.33**	−76.81	63.67	65.41	−50.76
	−extrM_2	−489.79	−151.08	**−132.61**	**−101.90**	−45.63	−44.29
	−extrM_3	−182.13	−450.85	−53.28	−40.61	**−120.30**	**−99.45**
4	extrN	**−46.44**	**−31.64**	67.11	−51.65	−62.44	43.69
	extrM_2	−178.62	−459.40	**128.62**	**109.37**	58.39	41.27
	extrM_3	−470.98	−174.82	55.34	52.87	**129.63**	**93.77**
	−extrN	**−607.84**	**−593.04**	−71.13	59.06	67.71	−52.51
	−extrM_2	−475.66	−165.28	**−132.64**	**−101.96**	−53.13	−50.09
	−extrM_3	−183.30	−449.86	−59.36	−45.46	**−124.36**	**−102.59**

Specifically, Tables 4.21 and 4.22 report the extreme values of M_3 and V_2 for the beams BX2 and BY5, respectively, for all four FE models considered in the analysis. These are obtained by first computing the maximum values of M_3 and V_2 by modal combining the peak (seismic) modal values of these stress resultants along the directions X and Y of the seismic action, separately, using the CQC modal combination rule and, then, by application of the SRSS rule for spatial combination to the previously computed maximum values. Further, Tables 4.23 and 4.24 report the values of M_3 and V_2 for the beams BX2 and BY5, respectively, for the seismic design loading combination G "+" Ψ_2Q "±" E for which the sections of BX2 and BY5 need to be detailed. The latter values have been obtained by superposing the extreme values of the seismic effects of Tables 4.21 and 4.22 to the corresponding stress resultants of the considered structural members due to the gravitational permanent and quasi-permanent variable actions summarized in Table 4.11 (Sect. 4.1.8).

Table 4.20 Design triads for wall W1 (ground storey) of building A for the seismic design load combination G "+" ψ_2Q "±" E [Extreme values in bold]

Mass position		N [kN]		M_2 [kNm]		M_3 [kNm]	
		Bottom	Top	Bottom	Top	Bottom	Top
1	extrN	**−623.60**	**−561.10**	0.00	0.00	118.23	−43.97
	extrM_2	−684.76	−622.26	**899.22**	**48.27**	0.28	0.16
	extrM_3	−625.13	−680.91	0.00	0.00	**121.24**	**46.17**
	−extrN	**−745.92**	**−683.42**	0.00	0.00	−117.67	44.28
	−extrM_2	−684.76	−622.26	**−899.22**	**−48.27**	0.28	0.16
	−extrM_3	−744.39	−563.61	0.00	0.00	**−120.69**	**−45.85**
2	extrN	**−623.60**	**−561.10**	0.00	0.00	118.23	−43.97
	extrM_2	−684.76	−622.26	**703.77**	**39.60**	0.28	0.16
	extrM_3	−625.13	−680.91	0.00	0.00	**121.24**	**46.17**
	−extrN	**−745.92**	**−683.42**	0.00	0.00	−117.67	44.28
	−extrM_2	−684.76	−622.26	**−703.77**	**−39.60**	0.28	0.16
	−extrM_3	−744.39	−563.61	0.00	0.00	**−120.69**	**−45.85**
3	extrN	**−623.74**	**−561.24**	85.72	−0.40	117.94	−43.86
	extrM_2	−678.25	−622.81	**803.96**	**43.90**	12.12	1.84
	extrM_3	−625.27	−680.77	78.89	1.61	**120.96**	**46.06**
	−extrN	**−745.78**	**−683.28**	−85.72	0.40	−117.39	44.17
	−extrM_2	−691.27	−621.71	**−803.96**	**−43.90**	−11.57	−1.52
	−extrM_3	−744.25	−563.75	−78.89	−1.61	**−120.40**	**−45.75**
4	extrN	**−623.74**	**−561.24**	−85.72	0.40	117.94	−43.86
	extrM_2	−691.27	−621.71	**803.96**	**43.90**	−11.57	−1.52
	extrM_3	−625.27	−680.77	−78.89	−1.61	**120.96**	**46.06**
	−extrN	**−745.78**	**−683.28**	85.72	−0.40	−117.39	44.17
	−extrM_2	−678.25	−622.81	**−803.96**	**−43.90**	12.12	1.84
	−extrM_3	−744.25	−563.75	78.89	1.61	**−120.40**	**−45.75**

Table 4.21 Extreme values of stress resultants in beam BX2 (ground storey) of building A

	V2 [kN]			M3 [kNm]		
Mass position	Left end	Midspan	Right end	Left end	Midspan	Right end
1	±91.76	±91.76	±91.76	±146.75	±18.42	±183.58
2	±92.54	±92.54	±92.54	±147.99	±18.59	±185.16
3	±90.49	±90.49	±90.49	±144.73	±18.16	±181.04
4	±93.53	±93.53	±93.53	±149.56	±18.79	±187.14

4.1.9.3 Verification Check of the Influence of Second Order Effects

The rationale of the verification check for second-order effects and its implications in the seismic design process have been discussed in detail in Sects. 2.4.3.3 and 3.2.1. This deformation-based verification check involves determination of the interstorey drift sensitivity coefficients θ_X and θ_Y along the principal directions X and Y, respectively, for all storeys defined in Eq. (3.17). For the case of the MRSM

Table 4.22 Extreme values of stress resultants in beam BY5 (ground storey) of building A

	V_2 [kN]			M_3 [kNm]		
Mass position	Left end	Midspan	Right end	Left end	Midspan	Right end
1	±85.67	±85.67	±85.67	±162.47	±19.00	±124.54
2	±89.65	±89.65	±89.65	±170.03	±19.89	±130.31
3	±88.16	±88.16	±88.16	±167.24	±19.59	±128.12
4	±88.31	±88.31	±88.31	±167.45	±19.56	±128.41

Table 4.23 Design effects beam BX2 (ground storey) of building A for the seismic design load combination G "+" ψ_2Q "±" E

		V_2 [kN]			M_3 [kNm]		
Mass position	Loading combination	Left end	Midspan	Right end	Left end	Midspan	Right end
1	$G+\psi_2Q+E$	62.58	88.64	118.14	126.28	28.6	170.99
	$G+\psi_2Q-E$	−120.93	−94.87	−65.37	−167.21	−8.24	−196.16
2	$G+\psi_2Q+E$	63.36	89.42	118.93	127.52	28.77	172.57
	$G+\psi_2Q-E$	−121.72	−95.66	−66.15	−168.45	−8.41	−197.74
3	$G+\psi_2Q+E$	61.31	87.37	116.88	124.26	28.34	168.45
	$G+\psi_2Q-E$	−119.67	−93.61	−64.1	−165.2	−7.98	−193.62
4	$G+\psi_2Q+E$	64.35	90.41	119.92	129.09	28.98	174.56
	$G+\psi_2Q-E$	−122.7	−96.65	−67.14	−170.03	−8.61	−199.73

Table 4.24 Design effects beam BY5 (ground storey) of building A for the seismic design load combination G "+" ψ_2Q "±" E

		V_2 [kN]			M_3 [kNm]		
Mass position	Loading combination	Left end	Midspan	Right end	Left end	Midspan	Right end
1	$G+\psi_2Q+E$	60.78	88.36	120.45	153.15	30.16	104.32
	$G+\psi_2Q-E$	−110.56	−82.98	−50.89	−171.78	−7.84	−144.76
2	$G+\psi_2Q+E$	64.76	92.34	124.43	160.71	31.05	110.09
	$G+\psi_2Q-E$	−114.54	−86.96	−54.87	−179.34	−8.73	−150.53
3	$G+\psi_2Q+E$	63.27	90.85	122.94	157.92	30.75	107.9
	$G+\psi_2Q-E$	−113.05	−85.48	−53.38	−176.55	−8.43	−148.34
4	$G+\psi_2Q+E$	63.42	91	123.09	158.14	30.72	108.19
	$G+\psi_2Q-E$	−113.2	−85.63	−53.53	−176.76	−8.4	−148.63

of analysis, a recommended "rigorous approach" of estimating the coefficients θ_X and θ_Y via Eq. (3.20) has been presented in Sect. 3.2.1.2 In this section, the interstorey drift sensitivity coefficients θ_X and θ_Y obtained by this approach are presented for the FE structural model of building A with centers of mass positioned in location 1 as defined in Table 4.12. The lateral absolute and relative floor translations due to the gravitational loads of the design seismic loading combination (G "+" ψ_2Q) are negligible compared with the corresponding translations due to the design seismic action and, therefore, are ignored.

Calculation of the coefficients θ_X and θ_Y

The procedure for calculating the coefficients θ_X and θ_Y according to the rigorous approach described in Sect. 3.2.1.2 follows FC-3.10b. The first step of this procedure involves calculating the values of the interstorey drift sensitivity coefficients along directions X and Y for each mode i considered in the analysis and at each building storey k, for design seismic action along direction X, that is, $\theta^{(k)}_{X,EXi}$ and $\theta^{(k)}_{Y,EXi}$, and for design seismic action along direction Y, that is, $\theta^{(k)}_{X,EYi}$ and $\theta^{(k)}_{Y,EYi}$. Tables 4.25, 4.26, 4.27, and 4.28 present in tabular form the required calculations for determining the four peak (seismic) "modal" interstorey drifts for the second mode shape (i = 2) and for the ground storey (k = 1). Note that this particular mode shape has been selected since it is not a purely translational one (as is the case of the 1st mode shape, which is purely translational along the axis of symmetry X of building A; see table 4.14) and, thus, involves translations along axis X for excitation along axis Y.

Specifically, the last row of Tables 4.25, 4.26, 4.27, and 4.28 reports:

- the sum of the gravity loads of the seismic design combination of actions (G + ψ_2Q) at the considered storey, P_{tot};
- the total seismic storey shears $V^{(1)}_{X(tot),EX2}$, $V^{(1)}_{Y(tot),EX2}$, $V^{(1)}_{X(tot),EY2}$, $V^{(1)}_{Y(tot),EY2}$, respectively (see also Eqs. (3.18) and (3.19)); and
- the average (mean) lateral design interstorey drift $d^{(1)}_{rX,EX2}$, $d^{(1)}_{rY,EX2}$, $d^{(1)}_{rX,EY2}$, $d^{(1)}_{rY,EY2}$, respectively (see also Eqs. (3.18) and (3.19)).

From Table 4.25, the following result for the peak (seismic) "modal" interstorey drift sensitivity coefficient $\theta^{(2)}_{X,EX2}$ is reached:

$$\theta^{(1)}_{X,EX2} = \left|\frac{P^{(1)}_{tot} \cdot d^{(1)}_{rX,EX2}}{V^{(1)}_{X(tot),EX2} \cdot h^{(1)}}\right| = \left|\frac{-4267.05 \cdot 0.00}{0.00 \cdot 4.00}\right| = N/D,$$

indicating that this particular coefficient is non-definable (N/D) since the $d^{(1)}_{rX,EX2}$ and $V^{(1)}_{X(tot),EX2}$ terms in the numerator and the denominator of the above ratio, respectively, are zero. This occurs for two reasons. Firstly, the modal participation factor for the 2nd mode shape and for excitation along axis X, ν_{2X}, is zero (see also Tables 4.14 and 4.31), and, thus, all entries of the 4th and 8th column of Table 4.25 are zero. Secondly, the non-zero translational displacements along X, u^t_{X2j}, and the corresponding non-zero shearing forces, $V^{(1)}_{X2j}$, at the bottom of columns C1 ~ C3 and C6 ~ C8, are anti-symmetric with respect to the X axis and, therefore, the mean value $d^{(1)}_{rX2}$ and the total shear storey $V^{(1)}_{X2(tot)}$ are equal to zero, as can be readily seen from the 3rd and 5th columns of Table 4.25. In fact, these non-zero translations and base shears along direction X will always be anti-symmetric for the 2nd mode

Table 4.25 Determination of the peak (seismic) modal interstorey drift $d_{rX,EX2}^{(1)}$ along direction X (ground storey of building A, 2nd mode shape; seismic excitation along X)

Vertical member j	Axial load $(G+\psi_2 Q)$ [kN]	Shear force $V_{X2j}^{(1)}$ [kN/m]	$V_{Xj,EX2}^{(1)} = \left(v_{2X}\frac{S_{a2}}{\omega_2^2}\right) \cdot V_{X2j}^{(1)}$ [kN]	u_{X2j}^{t} [−]	u_{X2j}^{b} [−]	$d_{rX2,j}^{(1)} = q(u_{X2j}^{t} - u_{X2j}^{b})$ [−]	$\left(v_{2X}\frac{S_{a2}}{\omega_2^2}\right) \cdot d_{rX,j}^{(1)}$ [cm]
C1	−300.20	−90.82	0.00	−1.78	0.00	−7.12	0.00
C2	−579.45	−110.44	0.00	−1.78	0.00	−7.12	0.00
C3	−327.14	−94.10	0.00	−1.78	0.00	−7.12	0.00
C4	−770.00	0.00	0.00	0.00	0.00	0.00	0.00
C5	−398.70	0.00	0.00	0.00	0.00	0.00	0.00
C6	−300.20	90.82	0.00	1.78	0.00	7.12	0.00
C7	−579.45	110.44	0.00	1.78	0.00	7.12	0.00
C8	−327.14	94.10	0.00	1.78	0.00	7.12	0.00
W1	−684.76	0.00	0.00	0.00	0.00	0.00	0.00
	$P_{tot}^{(1)}$ [a]	$V_{X2(tot)}^{(1)}$ [b]	$V_{X(tot),EX2}^{(1)}$			$d_{rX2}^{(1)}$ [c]	$d_{rX,EX2}^{(1)}$
	−4267.05	0.00	0.00			0.00	0.00

The third column reports the modal values of shear forces corresponding to the second mode shape if the normalized mode shape components are considered as real displacements

v_{2X} is the modal participation factor for the second mode shape and excitation along axis X

The values of columns 5–7 are multiplied by 100. The mode shapes are normalized so that the generalized mass has unit value

[a]The value of the modal participation factor for the 2nd mode shape and for excitation along axis X, v_{2X}, is equal to zero and, thus, all contributions of the j members to the modal seismic shear storey, $V_{X(tot),EX2}^{(1)}$, and to the modal design interstorey drift, $d_{rX,EX2}^{(1)}$, are also zero

[b]The sum of the modal shearing forces along axis X from all j elements for the 2nd mode shape, $V_{X2(tot)}^{(1)}$, is equal to zero

[c]The average (mean) value of the modal design (inelastic) interstorey drifts along axis X from all j elements for the 2nd mode shape, $d_{rX2}^{(1)}$, is equal to zero

Table 4.26 Determining the peak (seismic) modal interstorey drift $d^{(1)}_{rY,EX2}$ along direction Y (ground storey of building A, 2nd mode shape; seismic excitation along X)

Vertical member j	Axial load (G + ψ_2Q) [kN]	Shear force $V^{(1)}_{Y2j}$ [kN/m]	$V^{(1)}_{Yj,EX2} = \left(v_{2X}\frac{S_{a2}}{\omega_2^2}\right) \cdot V^{(1)}_{Y2j}$ [kN]	u^t_{Y2j} [–]	u^b_{Y2j} [–]	$d^{(1)}_{rY2,j} = q(u^t_{Y2j} - u^b_{Y2j})$ [–]	$\left(v_{2X}\frac{S_{a2}}{\omega_2^2}\right) \cdot d^{(1)}_{rY,j}$ [cm]
C1	−300.20	−20.40	0.00	−0.33	0.00	−1.34	0.00
C2	−579.45	−142.87	0.00	−2.71	0.00	−10.83	0.00
C3	−327.14	−240.41	0.00	−4.61	0.00	−18.43	0.00
C4	−770.00	−171.11	0.00	−2.71	0.00	−10.83	0.00
C5	−398.70	−288.37	0.00	−4.61	0.00	−18.43	0.00
C6	−300.20	−20.40	0.00	−0.33	0.00	−1.34	0.00
C7	−579.45	−142.87	0.00	−2.71	0.00	−10.83	0.00
C8	−327.14	−240.41	0.00	−4.61	0.00	−18.43	0.00
W1	−684.76	−576.31	0.00	−0.33	0.00	−1.34	0.00
	$P^{(1)}_{tot}$		$V^{(1)}_{Y(tot),EX2}$				$d^{(1)}_{rY,EX2}$
	−4267.05		0.00				0.00

The values of columns 5–7 are multiplied by 100. The mode shapes are normalized so that the generalized mass has unit value
The value of the modal participation factor for the 2nd mode shape and for excitation along axis X, v_{2X}, is equal to zero. Hence, all contributions of the j members to the peak (seismic) modal storey shear, $V^{(1)}_{Y(tot),EX2}$, and to the peak modal design interstorey drift, $d^{(1)}_{rY,EX2}$, are also zero

Table 4.27 Determination of the peak (seismic) modal interstorey drift $d_{rX,EY2}^{(1)}$ along direction X (ground storey of building A, 2nd mode shape; seismic excitation along Y)

Vertical member j	Axial load (G + ψ_2Q) [kN]	Shear force $V_{X2j}^{(1)}$ [kN/m]	$V_{Xj,EY2}^{(1)} = \left(v_{2Y}\frac{S_{a2}}{\omega_2^2}\right) \cdot V_{X2j}^{(1)}$ [kN]	u_{X2j}^{t} [−]	u_{X2j}^{b} [−]	$d_{rX2,j}^{(1)} = q(u_{X2j}^{t} - u_{X2j}^{b})$ [−]	$\left(v_{2Y}\frac{S_{a2}}{\omega_2^2}\right) \cdot d_{rX,j}^{(1)}$ [cm]
C1	−300.20	−90.82	23.28	−1.78	0.00	−7.12	1.83
C2	−579.45	−110.44	28.32	−1.78	0.00	−7.12	1.83
C3	−327.14	−94.10	24.13	−1.78	0.00	−7.12	1.83
C4	−770.00	0.00	0.00	0.00	0.00	0.00	0.00
C5	−398.70	0.00	0.00	0.00	0.00	0.00	0.00
C6	−300.20	90.82	−23.28	1.78	0.00	7.12	−1.83
C7	−579.45	110.44	−28.32	1.78	0.00	7.12	−1.83
C8	−327.14	94.10	−24.13	1.78	0.00	7.12	−1.83
W1	−684.76	0.00	0.00	0.00	0.00	0.00	0.00
	$P_{tot}^{(1)}$ [a]	$V_{X2(tot)}^{(1)}$ [b]	$V_{X(tot),EY2}^{(1)}$			$d_{rX2}^{(1)}$ [c]	$d_{rX,EY2}^{(1)}$
	−4267.05	0.00	0.00			0.00	0.00

The values of columns 5–7 are multiplied by 100. The mode shapes are normalized so that the generalized mass has unit value

[a]The term $v_{2Y} \cdot S_{a2}/\omega_2^2$ (2nd mode shape; excitation along axis Y) is equal to -0.25638 (see also Table 4.31) for mode shapes normalized so that the generalized mass has unit value

[b]The sum of the modal shearing forces along axis X from all j elements for the 2nd mode shape, $V_{X2(tot)}^{(1)}$, is equal to zero

[c]The average (mean) value of the modal design (inelastic) interstorey drifts from all j elements for the 2nd mode shape and for excitation along axis X, $d_{rX2}^{(1)}$, is equal to zero

Table 4.28 Determination of the peak (seismic) modal interstorey drift $d^{(1)}_{rY,EY2}$ along direction Y (ground storey of building A, 2nd mode shape; seismic excitation along Y)

Vertical member j	Axial load $(G+\psi_2 Q)$ [kN]	Shear force $V^{(1)}_{Y2j}$ [kN/m]	$V^{(1)}_{Yj,EY2} = \left(v_{2Y}\frac{S_{a2}}{\omega_2^2}\right) \cdot V^{(1)}_{Y2j}$ [kN]	u^t_{Y2j} [−]	u^b_{Y2j} [−]	$d^{(1)}_{rY2,j} = q(u^t_{Y2j} - u^b_{Y2j})$ [−]	$\left(v_{2Y}\frac{S_{a2}}{\omega_2^2}\right) \cdot d^{(1)}_{rY,j}$ [cm]
C1	−300.20	−20.40	5.23	−0.33	0.00	−1.34	0.34
C2	−579.45	−142.87	36.63	−2.71	0.00	−10.83	2.78
C3	−327.14	−240.41	61.64	−4.61	0.00	−18.43	4.72
C4	−770.00	−171.11	43.87	−2.71	0.00	−10.83	2.78
C5	−398.70	−288.37	73.93	−4.61	0.00	−18.43	4.72
C6	−300.20	−20.40	5.23	−0.33	0.00	−1.34	0.34
C7	−579.45	−142.87	36.63	−2.71	0.00	−10.83	2.78
C8	−327.14	−240.41	61.64	−4.61	0.00	−18.43	4.72
W1	−684.76	−576.31	147.76	−0.33	0.00	−1.34	0.34
	$P^{(1)}_{tot}$		$V^{(1)}_{Y(tot),EY2}$				$d^{(1)}_{rY,EY2}$
	−4267.05		472.56				2.62

The values of columns 5–7 are multiplied by 100. The mode shapes are normalized so that the generalized mass has unit value
The product $v_{2Y} \cdot S_{a2}/\omega_2^2$ (2nd mode shape; excitation along axis Y) is equal to -0.25638 (see also Table 4.31)

shape, no matter what the considered direction of the seismic action, since the 2nd mode shape of building A does not include any translational component along the X axis; it involves only a translational component along axis Y and a rotational component about the vertical (gravitational) axis Z passing through the assumed center of mass. The latter rotational component results in anti-symmetric non-zero deformations and stresses along the X direction for structural members that do not lie on the axis of symmetry X.

From a practical viewpoint, it is important to note that the fact that $\theta^{(1)}_{X,EX2}$ is not defined does not have any effect in the subsequent modal superposition step, since the 2nd mode is not activated for seismic excitation along the X axis. The same holds for the 3rd and the 5th mode of building A, as will be further discussed below (see also Tables 4.29 and 4.31).

Similarly to $\theta^{(2)}_{X,EX2}$, the peak (seismic) modal interstorey drift sensitivity coefficient $\theta^{(2)}_{Y,EX2}$ cannot be defined, since both the $d^{(1)}_{rY,EX2}$ and $V^{(1)}_{Y(tot),EX2}$ terms in the numerator and the denominator of the ratio

$$\theta^{(1)}_{Y,EX2} = \left| \frac{P^{(1)}_{tot} \cdot d^{(1)}_{rY,EX2}}{V^{(1)}_{Y(tot),EX2} \cdot h^{(1)}} \right| = \left| \frac{-4267.05 \cdot 0.00}{0.00 \cdot 4.00} \right| = N/D,$$

respectively, attain zero values (see Table 4.26). In this case, this occurs because the modal participation factor for the 2nd mode shape and for excitation along the X axis, ν_{2X}, is zero because X is the axis of symmetry. Therefore, all entries in the 4th and 8th columns of Table 4.26 are zero, despite the fact that the sum of the shear storey and the average translation along the Y axis of the 1st floor slab are not zero. Nevertheless, the fact that $\theta^{(2)}_{Y,EX2}$ cannot be defined does not have any effect in the subsequent modal superposition step, as the 2nd mode is not activated for seismic excitation along the X axis.

As in the case of the two previous peak modal interstorey drift sensitivity coefficients examined, coefficient $\theta^{(2)}_{X,EY2}$ expressed as (see Table 4.26),

$$\theta^{(1)}_{X,EY2} = \left| \frac{P^{(1)}_{tot} \cdot d^{(1)}_{rX,EY2}}{V^{(1)}_{X(tot),EY2} \cdot h^{(1)}} \right| = \left| \frac{-4267.05 \cdot 0.00}{0.00 \cdot 4.00} \right| = N/D$$

is non-definable. This is because the average of peak modal interstorey drifts from all vertical elements along direction X, $d^{(1)}_{rX,EY2}$, and the peak modal storey shearing force along direction X, $V^{(1)}_{X(tot),EY2}$, are zero due to the translations along X and the shearing forces at the base of the vertical structural members being anti-symmetric with respect to the X axis of symmetry. However, the peak modal interstorey drift sensitivity coefficient $\theta^{(1)}_{Y,EY2}$ attains a real value, (see Table 4.28)

Table 4.29 Peak modal (seismic) interstorey drift sensitivity coefficients for seismic excitation along the X axis for building A (mass position 1 in Table 4.12)

Mode	1st storey		2nd storey		3rd storey		4th storey		5th storey	
	$\theta_{X,EX}$	$\theta_{Y,EX}$	$\theta_{X,EX}$	$\theta_{Y,EX}$	$\theta_{X,EX}$	$\theta_{Y,EX}$	$\theta_{X,EX}$	$\theta_{Y,EX}$	$\theta_{X,EX}$	$\theta_{Y,EX}$
1	0.106	N/D	0.076	N/D	0.052	N/D	0.035	N/D	0.019	N/D
2	N/D	N/D	N/D	N/D	N/D	N/D	N/D	N/D	N/D	N/D
3	N/D	N/D	N/D	N/D	N/D	N/D	N/D	N/D	N/D	N/D
4	0.087	N/D	0.22	N/D	0.045	N/D	0.029	N/D	0.015	N/D
5	N/D	N/D	N/D	N/D	N/D	N/D	N/D	N/D	N/D	N/D

Table 4.30 Peak modal (seismic) interstorey drift sensitivity coefficients for seismic excitation along the Y axis for building A (mass position 1 in Table 4.12)

Mode	1st storey		2nd storey		3rd storey		4th storey		5th storey	
	$\theta_{X,EY}$	$\theta_{Y,EY}$	$\theta_{X,EY}$	$\theta_{Y,EY}$	$\theta_{X,EY}$	$\theta_{Y,EY}$	$\theta_{X,EY}$	$\theta_{Y,EY}$	$\theta_{X,EY}$	$\theta_{Y,EY}$
1	N/D	N/D	N/D	N/D	N/D	N/D	N/D	N/D	N/D	N/D
2	N/D	0.059	N/D	0.040	N/D	0.025	N/D	0.017	N/D	0.010
3	N/D	0.002	N/D	0.010	N/D	0.010	N/D	0.008	N/D	0.007
4	N/D	N/D	N/D	N/D	N/D	N/D	N/D	N/D	N/D	N/D
5	N/D	0.040	N/D	0.014	N/D	0.026	N/D	0.016	N/D	0.008

Table 4.31 Values of the product $v_{ik} \cdot S_{ai}/\omega_i^2$ for each mode shape i and for the two considered directions of the seismic action k = X or Y, assuming mode shape normalization with respect to the generalized mass of each mode (building A; mass position 1 in Table 4.12)

Mode shape i	Seismic excitation along the X axis $v_{iX} \cdot S_{ai}/\omega_i^2$	Seismic excitation along the Y axis $v_{iY} \cdot S_{ai}/\omega_i^2$
1	−0.44650	0.0
2	0.0	−0.25638
3	0.0	−0.04046
4	0.01661	0.0
5	0.0	−0.00742

$$\theta^{(1)}_{Y,EY2} = \left| \frac{P^{(1)}_{tot} \cdot d^{(1)}_{rY,EY2}}{V^{(1)}_{Y(tot),EY2} \cdot h^{(1)}} \right| = \left| \frac{-4267.05 \cdot 0.0262}{472.56 \cdot 4.00} \right| = 0.059,$$

and, therefore, contributes to the subsequent modal supersposition step.

Tables 4.29 and 4.30 report the peak (seismic) modal interstorey drift sensitivity coefficients at all storeys for the first 5 mode shapes of building A for seismic excitation along the X and Y directions, respectively. These results are obtained by performing similar calculations as those presented in Tables 4.25, 4.26, 4.27, and 4.28 to determine the ground storey interstorey drift sensitivity coefficients for the 2nd mode shape. Further, Table 4.31 collects the values of the products $v_{iX} \cdot S_{ai}/\omega_i^2$ and $v_{iY} \cdot S_{ai}/\omega_i^2$ used for the calculation of the peak (seismic) modal interstorey drift sensitivity coefficients of Tables 4.29 and 4.30 for all the 5 modes considered ($i = 1,2,\ldots,5$).

Notably, the 2nd, 3rd, and 5th mode shapes for seismic excitation along the X direction are not excited, since they do not include a translational component along the X axis (i.e., $v_{2X} = v_{3X} = v_{5X} = 0$ in Table 4.31). Consequently, none of the modal coefficients θ_X and θ_Y are defined for the above mode shapes for seismic excitation along the X direction in Table 4.29. Additionally, no modal coefficients θ_Y are defined for the 1st and 4th mode shapes for seismic excitation along the Y direction, since, for the case, the centers of mass are located in position 1 (see figure in Table 4.12) the 1st and 4th mode shapes are purely translational along the X axis (see also Table 4.14). Consequently, for the particular FE model of building A considered, no modal contributions for the θ_Y coefficients exist for seismic excitation along the X direction, while modal contributions for the θ_X coefficients for the same direction of excitation need to be considered only for the 1st and 4th mode shapes.

For seismic excitation along the Y direction, the modal coefficients θ_X are not defined for the 2nd, 3rd and 5th mode shapes because the lateral displacements and the base shearing forces along the X axis are anti-symmetric. This is because the above three mode shapes do not involve any translation component along the axis of symmetry X (see also Table 4.14). Moreover, none of the modal coefficients θ_X

Table 4.32 Expected extreme values of the interstorey drift sensitivity coefficients θ_X and θ_Y for building A (mass position 1 in Table 4.12)

Storey	$\text{extr}\theta_X^{(k)}$	$\text{extr}\theta^{(k)}$
1	**0.137**	0.071
2	**0.233**	0.039
3	0.069	0.038
4	0.045	0.025
5	0.024	0.014

and θ_Y are defined for the 1st and 4th mode shapes for seismic excitation along the Y direction in Table 4.30. This is due to the fact that the latter two modes are not excited, since they are purely translational along axis X (i.e., $v_{1Y} = v_{4Y} = 0$ in Table 4.31). Overall, for the particular FE model of building A considered, no modal contributions for the θ_X coefficients exist for seismic excitation along the Y direction, while modal contributions for the θ_Y coefficients for the same direction of excitation need to be considered only for the 2nd, 3rd, and 5th modes.

Having determined the modal interstorey drift sensitivity coefficients, the maximum value of the interstorey drift sensitivity coefficients is found at each storey and for each direction of seismic excitation by modal combination using the CQC rule. Next, spatial combination using the SRSS rule is considered to evaluate the expected extreme values of the interstorey drift sensitivity coefficients θ_X and θ_Y for all storeys of building A due to simultaneous design seismic action along the principal directions X and Y. These extreme values are given in Table 4.32.

As expected, the extreme values of the interstorey drift sensitivity coefficients θ_Y are smaller than the θ_X for all storeys due to the wall W1, which renders the structure stiffer along the Y principal direction. Further, in all storeys, the criterion of clause §4.4.2.2 (2) of EC8 is satisfied along direction Y, that is, $\text{extr}\theta_Y \leq 0.1$, and, therefore, second-order effects need not be accounted for. However, this is not the case for direction X. For the ground storey, it is found that $0.1 \leq \text{extr}\theta_X = 0.137 \leq 0.2$ and, therefore, for this particular storey, all seismic effects derived from the analysis step (i.e., stress resultants and deformations) must be increased by a factor of $1/(1\text{-extr}\theta_X) = 1.159$. Further, for the 2nd storey, it is found that $0.2 \leq \text{extr}\theta_X = 0.233 \leq 0.3$. Thus, according to clause §4.4.2.2 (2) of EC8, a more accurate (geometrically non-linear) analysis needs to be undertaken to account for second-order effects in a direct manner. Such an analysis option may not be available in commercial structural analysis and design software. Further, these large values of the interstorey drift sensitivity coefficients θ_X are due to the fact that building A does not adhere to the standard conceptual design rules for earthquake resistant buildings: there are no walls along the X-direction, while the wall T1 induces a significant static eccentricity (distance between shear center and center of gravity) leading to excessive floor slab rotations about the vertical axis and, therefore, to excessive horizontal displacements under the design seismic action. In this regard, the recommended practical approach to follow is to re-consider the conceptual design of the lateral load-resisting system of building A by making appropriate modifications along the lines of Sect. 2.1.1 to reduce the static eccentricity and to increase stiffness along the X axis using properly oriented walls.

4.1.9.4 Verification Check for Maximum Interstorey Drift Demands

The aims and rationale of the verification check for the maximum allowed interstorey drifts (or damage limitation verification check) and its implications in the seismic design process have been discussed in detail in Sect. 3.2.2. This deformation-based verification check relies on Eq. (3.23) and involves determination of the design interstorey drifts d_{rX} and d_{rY} along the principal axes X and Y, respectively, for all building storeys and for simultaneous design seismic action along the X and Y directions. The computational steps that need to be taken to estimate the expected extreme values of d_{rX} and d_{rY} from displacements (seismic effects) derived from the MRSM of analysis are provided in FC-3.11b. Firstly, the seismic modal average values of d_{rX} and d_{rY} are computed separately for each storey and for each mode shape from modal analysis data (i.e., normalized mode shape ordinates). Next, the expected maximum interstorey drifts are computed for each direction of the seismic action by combining the seismic modal values using the CQC rule. Finally, the expected extreme values of interstorey drifts for simultaneous seismic action along both principal directions X and Y are computed by means of spatial combination using the SRSS rule.

In this section, the computational steps for obtaining the interstorey drifts d_{rX} and d_{rY} are presented and the verification check of Eq. (3.23) is undertaken for the FE structural model of building A with centers of mass positioned at location 1, as defined in Table 4.12. As in the deformation check performed in the previous section, the absolute and relative horizontal translations due to the gravitational loads of the design seismic loading combination (G "+" ψ_2Q) are assumed to be negligible compared with the corresponding translations due to the design seismic action and, therefore, are ignored.

Table 4.33 reports the modal ordinates (normalized mode shape displacements), u_X and u_Y along principal directions X and Y, respectively, of the upper end of all vertical structural members at all storeys for the 2nd mode shape obtained from standard modal analysis. The average (mean) values of the above quantities, u_{Xm} and u_{Ym}, corresponding to each floor slab are computed in the last row of Table 4.33. Next, Table 4.34 provides the average elastic relative floor slab displacements (interstorey drifts), $d_{eX}^{(k)}$ and $d_{eY}^{(k)}$, and the corresponding inelastic (design) interstorey drifts $d_{rX}^{(k)}$ and $d_{rY}^{(k)}$ for the 2nd mode shape and for all k storeys. By performing similar operations, the design interstorey drifts $d_{rX}^{(k)}$ and $d_{rY}^{(k)}$ for the 5 first mode shapes considered in the analysis of building A are obtained and reported in Table 4.35.

The seismic modal values of the interstorey drifts for each direction of the seismic excitation X and Y are obtained by multiplying the modal interstorey drift values of Table 4.35 with the products $v_{iX} \cdot S_{ai}/\omega_i^2$ and $v_{iY} \cdot S_{ai}/\omega_i^2$, respectively, given in Table 4.31. The thus obtained seismic modal values are reported in Tables 4.36 and 4.37 for seismic excitation along directions X and Y, respectively.

Finally, the expected extreme values of interstorey drifts for simultaneous seismic action along both principal directions X and Y are computed (Table 4.38)

Table 4.33 Modal ordinates [values*100] at the upper end (top) of vertical members for the 2nd mode shape of building A (mass position 1 in Table 4.12)

	1st storey		2nd storey		3rd storey		4th storey		5th storey	
Vertical member	u^t_{X2j}	u^t_{Y2j}	u^t_{X2j}	u^t_{Y2j}	u^t_{X2j}	u^t_{Y2j}	u^t_{X2j}	u^t_{Y2j}	u^t_{X2j}	u^t_{Y2j}
C1	−1.78	−0.33	−2.50	−0.69	−3.03	−1.03	−3.36	−1.32	−3.51	−1.55
C2	−1.78	−2.71	−2.50	−4.03	−3.03	−5.07	−3.36	−5.81	−3.51	−6.22
C3	−1.78	−4.61	−2.50	−6.70	−3.03	−8.29	−3.36	−9.39	−3.51	−9.96
W1	0.00	−0.33	0.00	−0.69	0.00	−1.03	0.00	−1.32	0.00	−1.55
C4	0.00	−2.71	0.00	−4.03	0.00	−5.07	0.00	−5.81	0.00	−6.22
C5	0.00	−4.61	0.00	−6.70	0.00	−8.29	0.00	−9.39	0.00	−9.96
C6	1.78	−0.33	2.50	−0.69	3.03	−1.03	3.36	−1.32	3.51	−1.55
C7	1.78	−2.71	2.50	−4.03	3.03	−5.07	3.36	−5.81	3.51	−6.22
C8	1.78	−4.61	2.50	−6.70	3.03	−8.29	3.36	−9.39	3.51	−9.96
Average values	0.00	−2.55	0.00	−3.81	0.00	−4.80	0.00	−5.51	0.00	−5.91

Table 4.34 Modal mean interstorey drifts [values *100] for the 2nd mode shape of building A (mass position 1 in Table 4.12)

Storey k	u^t_{Xm2}	u^b_{Xm2}	$d^{(k)}_{eX2} = u^t_{Xm2} - u^b_{Xm2}$	$d^{(k)}_{rX2} = q \cdot d^{(k)}_{eX2}$	u^t_{Ym2}	u^b_{Ym2}	$d^{(k)}_{eY2} = u^t_{Ym2} - u^b_{Ym2}$	$d^{(k)}_{rY2} = q \cdot d^{(k)}_{eY2}$
1	0.00	0.00	0.00	0.00	−2.55	0.00	−2.55	−10.20
2	0.00	0.00	0.00	0.00	−3.81	−2.55	−1.26	−5.03
3	0.00	0.00	0.00	0.00	−4.80	−3.81	−0.99	−3.97
4	0.00	0.00	0.00	0.00	−5.51	−4.80	−0.71	−2.84
5	0.00	0.00	0.00	0.00	−5.91	−5.51	−0.40	−1.60

Table 4.35 Modal mean interstorey drifts [values *100] of building A (mass position 1 in Table 4.12)

	1st storey		2nd storey		3rd storey		4th storey		5th storey	
Mode i	$d_{rXi}^{(1)}$	$d_{rYi}^{(1)}$	$d_{rXi}^{(2)}$	$d_{rYi}^{(2)}$	$d_{rXi}^{(3)}$	$d_{rYi}^{(3)}$	$d_{rXi}^{(4)}$	$d_{rYi}^{(4)}$	$d_{rXi}^{(5)}$	$d_{rYi}^{(5)}$
1	−9.93	0.00	−6.05	0.00	−4.50	0.00	−3.13	0.00	−1.68	0.00
2	0.00	−10.2	0.00	−5.03	0.00	−3.97	0.00	−2.84	0.00	−1.60
3	0.00	0.62	0.00	−3.24	0.00	−4.03	0.00	−3.84	0.00	−3.20
4	22.50	0.00	0.96	0.00	−14.9	0.00	−20.3	0.00	−14.8	0.00
5	0.00	−22.3	0.00	0.90	0.00	13.89	0.00	18.19	0.00	12.44

The values in Tables 4.33, 4.34, and 4.35 are obtained by mode shape normalization so that the generalized mass has unit value

Table 4.36 Seismic modal interstorey drifts [in cm] for seismic excitation along the X direction of building A (mass position 1 in Table 4.12)

Mode i	1st storey		2nd storey		3rd storey		4th storey		5th storey	
	$d_{rXi,EX}^{(1)}$	$d_{rYi,EX}^{(1)}$	$d_{rXi,EX}^{(2)}$	$d_{rYi,EX}^{(2)}$	$d_{rXi,EX}^{(3)}$	$d_{rYi,EX}^{(3)}$	$d_{rXi,EX}^{(4)}$	$d_{rYi,EX}^{(4)}$	$d_{rXi,EX}^{(5)}$	$d_{rYi,EX}^{(5)}$
1	4.43	0.00	2.70	0.00	2.01	0.00	1.40	0.00	0.75	0.00
2	0.00	0.00	0.00	0.00	0.00	0.00	0.00	0.00	0.00	0.00
3	0.00	0.00	0.00	0.00	0.00	0.00	0.00	0.00	0.00	0.00
4	0.37	0.00	0.02	0.00	−0.25	0.00	−0.34	0.00	−0.25	0.00
5	0.00	0.00	0.00	0.00	0.00	0.00	0.00	0.00	0.00	0.00

Table 4.37 Seismic modal interstorey drifts [in cm] for seismic excitation along the Y direction of building A (mass position 1 in Table 4.12)

	1st storey		2nd storey		3rd storey		4th storey		5th storey	
Mode i	$d_{rXi,EY}^{(1)}$	$d_{rYi,EY}^{(1)}$	$d_{rXi,EY}^{(2)}$	$d_{rYi,EY}^{(2)}$	$d_{rXi,EY}^{(3)}$	$d_{rYi,EY}^{(3)}$	$d_{rXi,EY}^{(4)}$	$d_{rYi,EY}^{(4)}$	$d_{rXi,EY}^{(5)}$	$d_{rYi,EY}^{(5)}$
1	0.00	0.00	0.00	0.00	0.00	0.00	0.00	0.00	0.00	0.00
2	0.00	2.62	0.00	1.29	0.00	1.02	0.00	0.73	0.00	0.41
3	0.00	−0.03	0.00	0.13	0.00	0.16	0.00	0.16	0.00	0.13
4	0.00	0.00	0.00	0.00	0.00	0.00	0.00	0.00	0.00	0.00
5	0.00	0.17	0.00	−0.01	0.00	−0.10	0.00	−0.13	0.00	−0.09

Table 4.38 Expected (most probable) extreme values of the relative floor displacements [in cm] of building A (mass position 1 in Table 4.12)

Storey	$\text{extr}d_{rX}^{(k)}$	$\text{extr}d_{r}^{(k)}$
1	4.45	2.62
2	2.70	1.30
3	2.03	1.04
4	1.44	0.76
5	0.79	0.44

Table 4.39 Expected extreme values of ratios ($v\cdot d_{rX}$)/h and ($v\cdot d_{rY}$)/h of Eq. (3.23) for building A (mass position 1 in Table 4.12)

Storey k	$\text{extr}d_{rX}^{(k)}$ [cm]	$\text{extr}d_{r}^{(k)}$ [cm]	v	h_k [cm]	$\frac{v\cdot \text{extr}d_{rX}^{(k)}}{h_k}$	$\frac{v\cdot \text{extr}d_{r}^{(k)}}{h_k}$
1	4.45	2.62	0.5	400	**0.0056**	0.0033
2	2.70	1.30	0.5	300	0.0045	0.0022
3	2.03	1.04	0.5	300	0.0034	0.0017
4	1.44	0.76	0.5	300	0.0024	0.0013
5	0.79	0.44	0.5	300	0.0013	0.0007

by first applying the CQC rule for modal combination to the seismic modal interstorey drift values of Tables 4.36 and 4.37 and, subsequently, by applying the SRSS rule for spatial combination.

Having determined the expected extreme values of the interstorey drifts, the damage limitation verification check of clause §4.4.3.2 of EC8 can be performed by comparing the ratios of Eq. (3.23) reported in Table 4.39 with pertinent limiting values. Assuming that building A has brittle non-structural infill walls, the maximum allowed interstorey drift according to clause 4.4.3.2(1) of EC8 is equal to 0.5 % of the storey height, i.e., ($v\cdot$ $\text{extr}d_{rX}^{(k)}$)/h <0.005 and ($v\cdot$ $\text{extr}d_{rY}^{(k)}$)/h <0.005 for all k storeys. The above condition is not met at the ground storey along the X axis. This is expected, since building A has no r/c walls oriented along the X axis.

4.1.10 *Seismic Analysis of Building A Using the Lateral Force Method and Deformation-Based Verification Checks*

In this section, the application of the lateral force method (LFM) of seismic analysis for building A is considered. First, the EC8 prescribed applicability conditions of the LFM are verified. Next, the LFM is implemented for building A following the FC-3.8 (see also Sect. 3.1.5.2). Finally, representative numerical data (design seismic effects for selected structural members) from application of the LFM are presented and the required deformation-based verification checks of EC8 are undertaken. It is noted that critical comments pertaining to the appropriateness of the EC8 prescribed applicability conditions of the LFM and on the problems arising in the practical implementation of the LFM for the seismic analysis of building structures have been provided in Sect. 2.4.3.1. This section focuses on the numerical application of the LFM for the seismic analysis of building A adopting the use of a spatial (three-dimensional) FE structural model and considering four different sets of points of application for the lateral forces to account for accidental torsional effects.

4.1.10.1 Check of Lateral Force Method Applicability Conditions

According to clause §4.3.3.2.1(2) of EC8, the LFM of analysis is applicable to building structures satisfying the following two conditions/criteria:

- the building is regular in height according to clause §4.2.3.3 of EC8; and
- the fundamental natural periods of vibration T_1 of the building in the two main directions are smaller than min{$4T_c$, 2.0 s}, where T_C is the corner period signifying the end of the constant pseudo-acceleration region (plateau) of the EC8 elastic spectrum (see Sect. 2.3.1.2 and Fig. 2.21).

It is reminded that building A was found to be regular in elevation upon performing the pertinent regularity verification check detailed in Sect. 4.1.3.1. Therefore, the above regularity condition is met. The verification of the second condition involves, firstly, the definition of the two "main" directions of building A, and, secondly, the determination of the fundamental natural period of vibration of building A in these two directions.

With regards to the two "main" directions, it is noted that EC8 does not provide any immediate suggestions or guidance to facilitate their specification. However, in clause §4.3.3.1(11)P of EC8, it is stated that: "Whenever a spatial model is used, the design seismic action shall be applied along all relevant horizontal directions (with regard to the structural layout of the building) and their orthogonal horizontal directions. For buildings with resisting elements in two perpendicular directions these two directions shall be considered as the relevant directions." Clearly, building A belongs to the latter class of buildings, since its vertical structural elements are arranged along the two perpendicular horizontal axes X and Y defined in Fig. 4.2. In this regard, the main directions of building A are assumed to coincide with the directions of the X and Y axes in the context of the LFM.

Furthermore, the "fundamental natural periods of vibration T_1 in the two main directions" (clause §4.3.3.2.1(2)a) of EC8) are herein interpreted as the first natural periods corresponding to the uncoupled mode shapes along directions X and Y. In general, the determination of the uncoupled natural period along any horizontal direction of a spatial FE model is not a straightforward task, since the dynamic degrees of freedom are generally coupled. Therefore, this task can only be accomplished by considering appropriately modified spatial FE models with artificially restrained degrees of freedom (see also Sect. 2.4.3.1). Specifically, the fundamental uncoupled natural periods along the main axes X and Y (i.e., T_{1X} and T_{1Y}, respectively) can be obtained by modal analysis of spatial FE models in which:

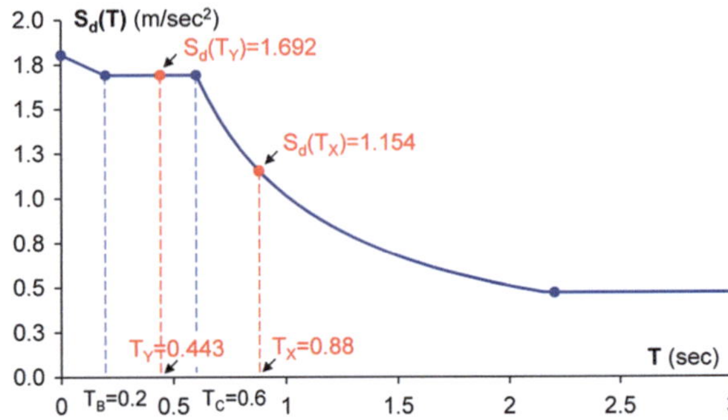

Fig. 4.3 EC8 design spectrum ordinates for application of LFM to building A

(i) all the translational nodal dynamic degrees of freedom along the directions Y and X, respectively, are restrained, and, simultaneously, (ii) all the rotational nodal dynamic degrees of freedom about the vertical axis Z are restrained. In this manner, T_{1X} is derived from a version of the spatial FE model of the building allowed to oscillate only along the direction X and, similarly, T_{1Y} is derived from a different version of the same spatial FE model allowed to oscillate only along the direction Y.

In the special case of building A, T_{1X} can be obtained directly from the standard unrestrained FE model, since the axis X is an axis of symmetry and, therefore, a set of uncoupled mode shapes along X is found by application of modal analysis. Thus, only one additional FE model needs to be considered to compute T_{1Y}. In particular, the two uncoupled fundamental natural periods of building A are computed as $T_{1X} = 0.880$ s and $T_{1Y} = 0.443$ s. For a corner period T_C equal to 0.6 s (see Sect. 4.1.1 and Fig. 4.3), it is found that the second applicability condition for LFM is satisfied for building A, since

- $T_{1X} = 0.880\ \text{s} < \min\{4 \times 0.6 = 2.4\ \text{s},\ 2\ \text{s}\} = 2\ \text{s}$; and
- $T_{1Y} = 0.443\ \text{s} < \min\{4 \times 0.6 = 2.4\ \text{s},\ 2\ \text{s}\} = 2\ \text{s}$.

Therefore, the LFM method of analysis can be applied for the seismic design of building A.

4.1.10.2 Determination of the Lateral Seismic Forces and Points of Action

The seismic base shear (total sum of the lateral seismic forces) is computed by means of equation (4.5) of EC8 (clause §4.3.3.2.2(1)P of EC8), which reads as

$$F_b = S_d(T_1) \cdot m \cdot \lambda,$$

Table 4.40 Distribution of the lateral seismic forces at the storey levels of building A

Storey k	m [t]	z [m]	$S_d(T_{1X})$ [m/s^2]	F_{bX} [kN]	F_{kX} [kN]	$S_d(T_{1Y})$ [m/s^2]	F_{bY} [kN]	F_{kY} [kN]
1	90.77	4.0	1.154	413.7	36.91	1.692	606.7	54.13
2	88.34	7.0			62.86			92.19
3	88.34	10.0			89.8			131.7
4	88.34	13.0			116.75			171.22
5	66.0	16.0			107.4			157.46
Total sum	421.80				413.7			606.7

where

- m is the total oscillatory mass of building A computed as m = 421.80 t (Table 4.2);
- λ is a correction factor, which for building A, in this case, is equal to:
  - X-direction: $T_{1X} = 0.880$ s $< 2xT_C = 2 \times 0.6 = 1.2$ s, therefore, $\lambda_X = 0.85$ because building A has more than two floors, and
  - Y-direction: $T_{1Y} = 0.443$ s $< 2xT_C = 2 \times 0.6 = 1.2$ s, therefore, $\lambda_Y = 0.85$ because the building has more than two floors;
- $S_d(T_1)$ is the EC8 design spectrum ordinate corresponding to the uncoupled fundamental natural period of building A along each of the considered directions of the seismic action, as shown in Fig. 4.3.

Application of the previous equation along the X and Y directions of seismic excitation using the design spectral ordinates of Fig. 4.3 yields the following base shear values:

- $F_{bX} = S_d(T_{1X})m\lambda = 1.154\ \ 421.80\ \ 0.85 = 413.7kN$, and
- $F_{bY} = S_d(T_{1Y})m\lambda = 1.692\ \ 421.80\ \ 0.85 = 606.7kN$.

The above base shear forces are distributed at the floor levels z_k, k = 1,2,..,5, along the height of building A according to equation (4.11) of EC8 (clause §4.3.3.2.3 (3) of EC8), which reads as

$$F_k = F_b \frac{z_k m_k}{\sum_j z_j m_j}.$$

The underlying computations involved in the application of the latter equation along the directions X and Y are summarized in Table 4.40. In this manner, the set of lateral seismic forces F_{kX} and F_{kY} acting at each floor k along directions X and Y, respectively, are reached. Lastly, the positions of action of these forces at each floor diaphragm of building A are determined according to clause §4.3.2(1)P of EC8 to take into account the accidental torsional effects due to uncertainties in

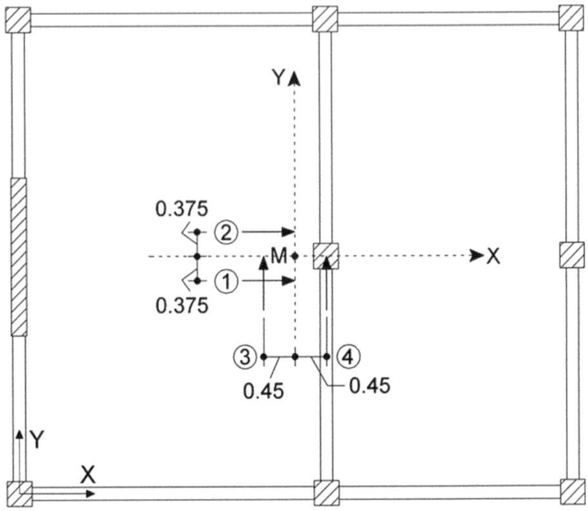

Fig. 4.4 Four positions of application of the horizontal seismic forces at each floor of building A

the floor mass distribution and to the torsional seismic excitation component caused by a potential spatial variation of the ground seismic motion. Specifically, the statically applied lateral seismic forces F_{kX} and F_{kY} are displaced with respect to the nominal center of mass (assuming even mass distribution) by a distance equal to the accidental eccentricity (computed in Table 4.12) perpendicular to each considered direction of the seismic action. Consequently, the seismic forces F_{kX} and F_{kY} are acting at points located $\pm$ 0.375 m and $\pm$ 0.45 m away from the center of gravity of each floor along directions Y and X, respectively. Therefore, four different static analyses need to be carried out in total: two for lateral seismic forces F_{kX} acting along the X direction and located at positions 1 and 2, as shown in Fig. 4.4, and two for lateral seismic forces F_{kY} acting along the Y direction and located at positions 3 and 4, as shown in Fig. 4.4.

4.1.10.3 Selected Design Seismic Effects (Sectional Stress Resultants)

In this section, the design seismic effects at critical sections (sectional stress resultants) for the column C3, the wall W1, and the beams BX2 and BY5 at the ground (1st) storey obtained by means of the LFM are presented in tabular form. Note that these are the same structural members for which the design seismic effects for the MRSM are previously given in Sect. 4.1.9.2.

Vertical structural members C3 and W1 (bi-axial bending with axial force)

The vertical structural members W1 and C3 need to be designed for bi-axial bending with axial force. Therefore, the design values of the moments M_2, M_3 and axial force N, as defined in Fig. 2.32, acting concurrently ("design triads") are herein reported. In particular, Tables 4.41 and 4.42 summarize design triads for the

Table 4.41 Stress resultants in column C3 (ground storey) for the 4 individual LFM analyses

Force position	Loading case	N [kN]		M_2 [kNm]		M_3 [kNm]	
		Bottom	Top	Bottom	Top	Bottom	Top
$F_x(-e_y)$	1	−222.57	−222.57	7.24	−6.36	108.34	−82.43
$F_x(+e_y)$	2	−234.93	−234.93	−6.26	4.54	102.06	−77.48
$F_y(-e_x)$	3	166.23	166.23	105.79	−84.94	36.93	−30.70
$F_y(+e_x)$	4	187.89	187.89	129.45	−104.04	47.93	−39.36

Table 4.42 Stress resultants in wall W1 (ground storey) for the 4 individual LFM analyses

		N [kN]		M_2 [kNm]		M_3 [kNm]	
Force position	Loading case	Bottom	Top	Bottom	Top	Bottom	Top
$F_x(-e_y)$	1	60.65	60.65	−59.44	0.02	111.38	−41.23
$F_x(+e_y)$	2	60.65	60.65	59.44	−0.02	111.38	−41.23
$F_y(-e_x)$	3	0.00	0.00	1268.02	13.92	0.00	0.00
$F_y(+e_x)$	4	0.00	0.00	1059.74	13.99	0.00	0.00

column C3 and for the wall W1, respectively, obtained from the four individual linear static analyses defined in Fig. 4.4 involving the lateral seismic forces of Table 4.40.

Next, the expected extreme values of the above stress resultants (seismic effects) for simultaneous design seismic action along directions X and Y are computed by spatial combination using the SRSS rule of the seismic effects corresponding to independent seismic action along directions X and Y (clause §4.3.3.5.1(2) of EC8). The pertinent combinations that need to be considered from the previous 4 static analyses corresponding to different loading cases are (see also Fig. 4.4):

- Loading case 1 "+" Loading case 3 (Combination 1–3)
- Loading case 1 "+" Loading case 4 (Combination 1–4)
- Loading case 2 "+" Loading case 3 (Combination 2–3)
- Loading case 2 "+" Loading case 4 (Combination 2–4)

As in the case of the MRSM, EC8 allows for compiling design triads comprising the thus computed extreme values of the seismic effects (i.e., extrM_2, extrM_3, and extrN) assuming that they act concurrently. However, this assumption leads to quite conservative detailing. In this regard, since EC8 allows for the use of more accurate methods to estimate the expected concurrent values of more than one seismic effect due to simultaneous seismic action along two horizontal axes without, nevertheless, specifying any (clause §4.3.3.5.1(2)c) of EC8), the simplified approach of the Greek Seismic Code EAK2000 (Earthquake Planning and Protection Organization (EPPO) 2000) is herein considered as in the case of the MRSM in Sect. 4.1.9.2. Tables 4.43 and 4.44 report the thus obtained 6 design triads at the bottom and the top of the structural members C3 and W1, respectively, at the ground (1st) storey of building A for the above four combinations of the four static LFM analysis cases.

Table 4.43 Design triads (expected -most probable- concurrent values of N, M_2, and M_3 stress resultants for simultaneous seismic action along axes X and Y) for column C3 (ground storey) [Extreme values in bold]

Combination		N [kN]		M_2 [kNm]		M_3 [kNm]	
		Bottom	Top	Bottom	Top	Bottom	Top
1–3	extrN	**277.79**	**277.79**	57.50	−45.73	−64.70	47.67
	extrM_2	150.63	−149.15	**106.04**	**85.18**	44.25	36.77
	extrM_3	−157.02	150.56	40.99	35.60	**114.46**	**87.96**
	−extrN	**−277.79**	**−277.79**	−57.50	45.73	64.70	−47.67
	−extrM_2	−150.63	149.15	**−106.04**	**−85.18**	−44.25	−36.77
	−extrM_3	157.02	−150.56	−40.99	−35.60	**−114.46**	**−87.96**
1–4	extrN	**291.27**	**291.27**	77.97	−62.26	−51.87	37.59
	extrM_2	175.16	−173.96	**129.65**	**104.24**	53.91	44.32
	extrM_3	−127.52	119.87	59.00	50.57	**118.47**	**91.35**
	−extrN	**−291.27**	**−291.27**	−77.97	62.26	51.87	−37.59
	−extrM_2	−175.16	173.96	**−129.65**	**−104.24**	−53.91	−44.32
	−extrM_3	127.52	−119.87	−59.00	−50.57	**−118.47**	**−91.35**
2–3	extrN	**287.79**	**287.79**	66.21	−52.77	−61.98	45.52
	extrM_2	179.81	−178.54	**105.97**	**85.06**	30.84	26.52
	extrM_3	−164.35	157.18	30.11	27.06	**108.54**	**83.34**
	−extrN	**−287.79**	**−287.79**	−66.21	52.77	61.98	−45.52
	−extrM_2	−179.81	178.54	**−105.97**	**−85.06**	−30.84	−26.52
	−extrM_3	164.35	−157.18	−30.11	−27.06	**−108.54**	**−83.34**
2–4	extrN	**300.82**	**300.82**	85.74	−68.53	−49.77	35.92
	extrM_2	199.02	−197.96	**129.60**	**104.14**	42.94	35.95
	extrM_3	−132.78	124.35	49.36	43.08	**112.76**	**86.91**
	−extrN	**−300.82**	**−300.82**	−85.74	68.53	49.77	−35.92
	−extrM_2	−199.02	197.96	**−129.60**	**−104.14**	−42.94	−35.95
	−extrM_3	132.78	−124.35	−49.36	−43.08	**−112.76**	**−86.91**

Table 4.44 Design triads (expected -most probable- concurrent values of N, M_2, and M_3 stress resultants for simultaneous seismic action along axes X and Y) for wall W1 (ground storey) [Extreme values in bold]

		N [kN]		M_2 [kNm]		M_3 [kNm]	
Combination		Bottom	Top	Bottom	Top	Bottom	Top
1–3	extrN	**60.65**	**60.65**	−59.44	0.02	111.38	−41.23
	extrM$_2$	−2.84	0.09	**1269.41**	**13.92**	−5.22	−0.06
	extrM$_3$	60.65	−60.65	−59.44	−0.02	**111.38**	**41.23**
	−extrN	**−60.65**	**−60.65**	59.44	−0.02	−111.38	41.23
	−extrM$_2$	2.84	−0.09	**−1269.41**	**−13.92**	5.22	0.06
	−extrM$_3$	−60.65	60.65	59.44	0.02	**−111.38**	**−41.23**
1–4	extrN	**60.65**	**60.65**	−59.44	0.02	111.38	−41.23
	extrM$_2$	−3.40	0.09	**1061.41**	**13.99**	−6.24	−0.06
	extrM$_3$	60.65	−60.65	−59.44	−0.02	**111.38**	**41.23**
	−extrN	**−60.65**	**−60.65**	59.44	−0.02	−111.38	41.23
	−extrM$_2$	3.40	−0.09	**−1061.41**	**−13.99**	6.24	0.06
	−extrM$_3$	−60.65	60.65	59.44	0.02	**−111.38**	**−41.23**
2–3	extrN	**60.65**	**60.65**	59.44	−0.02	111.38	−41.23
	extrM$_2$	2.84	−0.09	**1269.41**	**13.92**	5.22	0.06
	extrM$_3$	60.65	−60.65	59.44	0.02	**111.38**	**41.23**
	−extrN	**−60.65**	**−60.65**	−59.44	0.02	−111.38	41.23
	−extrM$_2$	−2.84	0.09	**−1269.41**	**−13.92**	−5.22	−0.06
	−extrM$_3$	−60.65	60.65	−59.44	−0.02	**−111.38**	**−41.23**
2–4	extrN	**60.65**	**60.65**	59.44	−0.02	111.38	−41.23
	extrM$_2$	3.40	−0.09	**1061.41**	**13.99**	6.24	0.06
	extrM$_3$	60.65	−60.65	59.44	0.02	**111.38**	**41.23**
	−extrN	**−60.65**	**−60.65**	−59.44	0.02	−111.38	41.23
	−extrM$_2$	−3.40	0.09	**−1061.41**	**−13.99**	−6.24	−0.06
	−extrM$_3$	−60.65	60.65	−59.44	−0.02	**−111.38**	**−41.23**

– Lastly, the seismic design triads derived previously are superposed to the corresponding stress resultants of the considered structural members due to the gravitational permanent and quasi-permanent variable actions summarized in Table 4.11 (Sect. 4.1.8) to obtain the design triads for the EC8 design seismic loading combination G "+" Ψ_2Q "±" E. The thus obtained triads are reported in Tables 4.45 and 4.46.

Beams BX2 and BY5 (uni-axial bending)

A similar procedure as in the case of the vertical members C3 and W1 is applied to obtain the extreme values of the moment M_3 and the shearing force V_2 at critical cross-sections (left end and right end) of the beams BX2 and BY5 of the ground (1st) storey of building A, which need to be designed for uni-axial bending. However, in this case, the procedure of obtaining the design seismic effects for simultaneous seismic action along the two principal axes X and Y is significantly

Table 4.45 Design triads for column C3 (ground storey) for the seismic design load combination G "+" ψ_2Q "±" E [Extreme values in bold]

Combination		N [kN]		M_2 [kNm]		M_3 [kNm]	
		Bottom	Top	Bottom	Top	Bottom	Top
1–3	extrN	**−49.35**	**−34.55**	55.49	−42.02	−62.07	43.26
	extrM_2	−176.51	−461.49	**104.03**	**88.89**	46.88	32.36
	extrM_3	−484.16	−161.78	38.98	39.31	**117.09**	**83.55**
	−extrN	**−604.93**	**−590.13**	−59.51	49.44	67.33	−52.08
	−extrM_2	−477.77	−163.19	**−108.05**	**−81.47**	−41.62	−41.18
	−extrM_3	−170.12	−462.90	−43.00	−31.89	**−111.83**	**−92.37**
1–4	extrN	**−35.87**	**−21.07**	75.96	−58.55	−49.24	33.18
	extrM_2	−151.98	−486.30	**127.64**	**107.95**	56.54	39.91
	extrM_3	−454.66	−192.47	56.99	54.28	**121.10**	**86.94**
	−extrN	**−618.41**	**−603.61**	−79.98	65.97	54.50	−42.00
	−extrM_2	−502.30	−138.38	**−131.66**	**−100.53**	−51.28	−48.73
	−extrM_3	−199.62	−432.21	−61.01	−46.86	**−115.84**	**−95.76**
2–3	extrN	**−39.35**	**−24.55**	64.20	−49.06	−59.35	41.11
	extrM_2	−147.33	−490.88	**103.96**	**88.77**	33.47	22.11
	extrM_3	−491.49	−155.16	28.10	30.77	**111.17**	**78.93**
	−extrN	**−614.93**	**−600.13**	−68.22	56.48	64.61	−49.93
	−extrM_2	−506.95	−133.80	**−107.98**	**−81.35**	−28.21	−30.93
	−extrM_3	−162.79	−469.52	−32.12	−23.35	**−105.91**	**−87.75**
2–4	extrN	**−26.32**	**−11.52**	83.73	−64.82	−47.14	31.51
	extrM_2	−128.12	−510.30	**127.59**	**107.85**	45.57	31.54
	extrM_3	−459.92	−187.99	47.35	46.79	**115.39**	**82.50**
	−extrN	**−627.96**	**−613.16**	−87.75	72.24	52.40	−40.33
	−extrM_2	−526.16	−114.38	**−131.61**	**−100.43**	−40.31	−40.36
	−extrM_3	−194.36	−436.69	−51.37	−39.37	**−110.13**	**−91.32**

simplified by the fact that only a single seismic effect (i.e., stress resultant M_3 and corresponding shearing force V_2) is required in the detailing of beam sections, as opposed to the vector of the three concurrently acting seismic effects (triads) N, M_2, M_3 considered for the case of vertical structural members.

Tables 4.47 and 4.48 report the values of M_3 and V_2 for the beams BX2 and BY5, respectively, obtained from the four individual linear static analyses defined in Fig. 4.4. Next, these values are combined spatially using the SRSS rule for the four combinations of the four static loading cases identified previously and added to the corresponding stress resultants of the considered structural members due to the gravitational permanent and quasi-permanent variable actions summarized in Table 4.11 (Sect. 4.1.8). The thus obtained design effects are given in Tables 4.49 and 4.50 for the beams BX2 and BY5, respectively.

Table 4.46 Design triads for wall W1 (ground storey) for the seismic design load combination G "+" ψ_2Q "±" E [Extreme values in bold]

Combi-nation		N [kN]		M_2 [kNm]		M_3 [kNm]	
		Bottom	Top	Bottom	Top	Bottom	Top
1–3	extrN	**−624.11**	**−561.61**	−59.44	0.02	111.66	−41.07
	extrM$_2$	−687.60	−622.17	**1269.41**	**13.92**	−4.94	0.09
	extrM$_3$	−624.11	−682.91	−59.44	−0.02	**111.66**	**41.39**
	−extrN	**−745.41**	**−682.91**	59.44	−0.02	−111.11	41.39
	−extrM$_2$	−681.92	−622.35	**−1269.41**	**−13.92**	5.49	0.22
	−extrM$_3$	−745.41	−561.61	59.44	0.02	**−111.11**	**−41.07**
1–4	extrN	**−624.11**	**−561.61**	−59.44	0.02	111.66	−41.07
	extrM$_2$	−688.16	−622.17	**1061.41**	**13.99**	−5.96	0.09
	extrM$_3$	−624.11	−682.91	−59.44	−0.02	**111.66**	**41.39**
	−extrN	**−745.41**	**−682.91**	59.44	−0.02	−111.11	41.39
	−extrM$_2$	−681.36	−622.35	**−1061.41**	**−13.99**	6.52	0.22
	-extrM$_3$	−745.41	−561.61	59.44	0.02	**−111.11**	**−41.07**
2–3	extrN	**−624.11**	**−561.61**	59.44	−0.02	111.66	−41.07
	extrM$_2$	−681.92	−622.35	**1269.41**	**13.92**	5.49	0.22
	extrM$_3$	−624.11	−682.91	59.44	0.02	**111.66**	**41.39**
	−extrN	**−745.41**	**−682.91**	−59.44	0.02	−111.11	41.39
	−extrM$_2$	−687.60	−622.17	**−1269.41**	**−13.92**	−4.94	0.09
	−extrM$_3$	−745.41	−561.61	−59.44	−0.02	**−111.11**	**−41.07**
2–4	extrN	**−624.11**	**−561.61**	59.44	−0.02	111.66	−41.07
	extrM$_2$	−681.36	−622.35	**1061.41**	**13.99**	6.52	0.22
	extrM$_3$	−624.11	−682.91	59.44	0.02	**111.66**	**41.39**
	−extrN	**−745.41**	**−682.91**	−59.44	0.02	−111.11	41.39
	−extrM$_2$	−688.16	−622.17	**−1061.41**	**−13.99**	−5.96	0.09
	−extrM$_3$	−745.41	−561.61	−59.44	−0.02	**−111.11**	**−41.07**

Table 4.47 Stress resultants in beam BX2 (ground storey) for the 4 individual LFM analyses

		V_2 [kN]		M_3 [kNm]	
Force position	Loading case	Left end	Right end	Left end	Right end
$F_x(-e_y)$	1	83.31	83.31	133.35	−166.58
$F_x(+e_y)$	2	78.93	78.93	126.37	−157.78
$F_y(-e_x)$	3	20.54	20.54	32.85	−41.09
$F_y(+e_x)$	4	28.22	28.22	45.07	−56.50

4.1.10.4 Verification Check of the Influence of Second Order Effects

The rationale for the verification check for second-order effects and its implications in the seismic design process have been discussed in detail in Sects. 2.4.3.3 and 3.2.1. This deformation-based verification check involves determination of the interstorey drift sensitivity coefficients θ_X and θ_Y along the principal directions X and Y, respectively, for all storeys defined in Eq. (3.17). This is achieved in the

Table 4.48 Stress resultants in beam BY5 (ground storey) for the 4 individual LFM analyses

Load position	Loading case	V_2 [kN]		M_3 [kNm]	
		Left end	Right end	Left end	Right end
$F_x(-e_y)$	1	7.03	7.03	12.21	−11.33
$F_x(+e_y)$	2	−2.22	−2.22	−5.32	2.11
$F_y(-e_x)$	3	73.24	73.24	138.80	−106.56
$F_y(+e_x)$	4	89.44	89.44	169.51	−130.12

Table 4.49 Design effects beam BX2 (ground storey) for the seismic design load combination G"+"ψ_2Q "±" E

Combination		V_2 [kN]		M_3 [kNm]	
		Left end	Right end	Left end	Right end
1–3	$G+\psi_2Q+E$	56.63	112.20	116.86	158.99
	$G+\psi_2Q-E$	−114.99	−59.42	−157.80	−184.15
1–4	$G+\psi_2Q+E$	58.78	114.35	120.29	163.32
	$G+\psi_2Q-E$	−117.14	−61.57	−161.23	−188.48
2–3	$G+\psi_2Q+E$	52.38	107.95	110.10	150.46
	$G+\psi_2Q-E$	−110.74	−55.17	−151.04	−175.62
2–4	$G+\psi_2Q+E$	54.64	110.21	113.70	155.01
	$G+\psi_2Q-E$	−113.00	−57.43	−154.64	−180.17

Table 4.50 Design effects beam BY5 (ground storey) for the seismic design load combination G"+"ψ_2Q "±" E

Combination		V_2 [kN]		M_3 [kNm]	
		Left end	Right end	Left end	Right end
1−3	$G+\psi_2Q+E$	48.69	108.36	130.03	86.94
	$G+\psi_2Q-E$	−98.47	−38.80	−148.65	−127.38
1−4	$G+\psi_2Q+E$	64.83	124.50	160.64	110.39
	$G+\psi_2Q-E$	−114.61	−54.94	−179.26	−150.83
2−3	$G+\psi_2Q+E$	48.39	108.06	129.60	86.36
	$G+\psi_2Q-E$	−98.17	−38.50	−148.22	−126.80
2−4	$G+\psi_2Q+E$	64.58	124.25	160.28	109.91
	$G+\psi_2Q-E$	−114.36	−54.69	−178.90	−150.35

context of the LFM by following the steps delineated in FC-3.10a. The first step involves calculating the θ_X and θ_Y coefficients independently for each of the four LFM loading cases (Lci, i = 1,2,3,4) defined in Fig. 4.4, that is,

- Loading case 1 (Lc1, seismic loads along X): $\theta^{(k)}_{X,Lc1}$ and $\theta^{(k)}_{Y,Lc1}$
- Loading case 2 (Lc2, seismic loads along X): $\theta^{(k)}_{X,Lc2}$ and $\theta^{(k)}_{Y,Lc2}$
- Loading case 3 (Lc3, seismic loads along Y): $\theta^{(k)}_{X,Lc3}$ and $\theta^{(k)}_{Y,Lc3}$
- Loading case 4 (Lc4, seismic loads along Y): $\theta^{(k)}_{X,Lc4}$ and $\theta^{(k)}_{Y,Lc4}$

Next, the above interstorey drift sensitivity coefficients are spatially superposed by means of the SRSS rule such that seismic effects due to seismic excitation along the X direction are combined with seismic effects due to seismic excitation along the Y direction. Hence, the following four combinations need to be considered

- Loading case 1 "+" Loading case 3 (Combination 1–3)
- Loading case 1 "+" Loading case 4 (Combination 1–4)
- Loading case 2 "+" Loading case 3 (Combination 2–3)
- Loading case 2 "+" Loading case 4 (Combination 2–4)

In this manner, the expected extreme values of the interstorey drift sensitivity coefficients extrθ_X and extrθ_Y are obtained due to simultaneous design seismic along the horizontal directions X and Y. It is noted that the lateral absolute and relative floor translations due to the gravitational loads of the design seismic loading combination (G "+" ψ_2Q) are negligible compared with the corresponding translations due to the design seismic action and, therefore, are ignored.

To illustrate the first step of the above procedure, Table 4.51 presents all required calculations for computing the interstorey drift $d^{(1)}_{rX,Lc1}$ at the ground storey corresponding to the first loading case shown in Fig. 4.4. Then, the interstorey drift sensitivity coefficient is determined by

$$\theta^{(1)}_{X,Lc1} = \left| \frac{P^{(1)}_{tot} \cdot d^{(1)}_{rX,Lc1}}{V^{(1)}_{X(tot),Lc1} \cdot h^{(1)}} \right| = \left| \frac{-4267.1 \cdot 0.0412}{413.72 \cdot 4.00} \right| = 0.106.$$

The interstorey drift sensitivity coefficients $\theta^{(k)}_{X,Lc1}$ for all k storeys are calculated in a similar manner and reported in Table 4.52.

Further, Table 4.53 summarizes the required calculations for computing the interstorey drift sensitivity coefficients at all storeys along axis Y for loading case 1, that is, $\theta^{(k)}_{Y,Lc1}$.

It is seen that the sum of the shearing forces of the vertical elements along the direction Y is equal to zero for loading case 1, since the structure is subject to lateral forces applied along the X direction. Therefore, the interstorey drift sensitivity coefficients along axis Y computed by Eq. (3.17) cannot be defined. The same holds for loading case 2, while for loading cases 3 and 4, only the coefficients θ_Y are defined, since for these cases, the structure is subject to lateral forces applied along the Y direction. Hence, spatial supersposition simplifies for the four combinations as follows:

- Combination 1–3: $\text{extr}\theta^{(k)}_{X,1-3} = \sqrt{\left(\theta^{(k)}_{X,Lc1}\right)^2} = \theta^{(k)}_{X,Lc1}$; $\text{extr}\theta^{(k)}_{Y,1-3} = \sqrt{\left(\theta^{(k)}_{Y,Lc3}\right)^2} = \theta^{(k)}_{Y,Lc3}$
- Combination 1–4: $\text{extr}\theta^{(k)}_{X,1-4} = \sqrt{\left(\theta^{(k)}_{X,Lc1}\right)^2} = \theta^{(k)}_{X,Lc1}$; $\text{extr}\theta^{(k)}_{Y,1-4} = \sqrt{\left(\theta^{(k)}_{Y,Lc4}\right)^2} = \theta^{(k)}_{Y,Lc4}$
- Combination 2–3: $\text{extr}\theta^{(k)}_{X,2-3} = \sqrt{\left(\theta^{(k)}_{X,Lc2}\right)^2} = \theta^{(k)}_{X,Lc2}$; $\text{extr}\theta^{(k)}_{Y,2-3} = \sqrt{\left(\theta^{(k)}_{Y,Lc3}\right)^2} = \theta^{(k)}_{Y,Lc3}$
- Combination 2–4: $\text{extr}\theta^{(k)}_{X,2-4} = \sqrt{\left(\theta^{(k)}_{X,Lc2}\right)^2} = \theta^{(k)}_{X,Lc2}$; $\text{extr}\theta^{(k)}_{Y,2-4} = \sqrt{\left(\theta^{(k)}_{Y,Lc4}\right)^2} = \theta^{(k)}_{Y,Lc4}$

Table 4.51 Procedure for determining the interstorey drift $d^{(1)}_{rX,Lc1}$ at the ground storey of the building based on results produced by analysis case 1

Vertical member j	Axial load (G + ψ_2Q) [kN]	Shear force $V^{(1)}_{X,Lc1,j}$ [kN]	$u^t_{X,Lc1,j}$ [cm]	$u^b_{X,Lc1,j}$ [cm]	$d^{(1)}_{eX,Lc1,j} = u^t_{X,Lc1,j} - u^b_{X,Lc1,j}$ [cm]	$d^{(1)}_{rX,Lc1,j} = q \cdot d^{(1)}_{eX,Lc1,j}$ [cm]
C1	−300.20	49.53	1.06	0.00	1.06	4.24
C2	−579.45	62.98	1.06	0.00	1.06	4.24
C3	−327.14	51.56	1.06	0.00	1.06	4.24
C4	−770.00	28.64	1.03	0.00	1.03	4.12
C5	−398.70	28.43	1.03	0.00	1.03	4.12
C6	−300.20	46.58	1.00	0.00	1.00	4.00
C7	−579.45	59.32	1.00	0.00	1.00	4.00
C8	−327.14	48.53	1.00	0.00	1.00	4.00
W1	−684.76	38.15	1.03	0.00	1.03	4.12
	$P^{(1)}_{tot}$	$V^{(1)}_{X(tot),Lc1}$ [a]				$d^{(1)}_{rX,Lc1}$ [b]
	−4267.05	413.72				4.12

[a] Sum of the shearing forces along axis X from all j elements for loading case 1

[b] Average (mean) value of the design (inelastic) interstorey drifts along axis X from all j elements for loading case 1

Table 4.52 Interstorey drift sensitivity coefficients θ_X for each building storey based on the results produced by analysis case 1

Storey k	$u_{Xm,Lc1}^{t(k)}$ [a] [cm]	$u_{Xm,Lc1}^{b(k)}$ [b] [cm]	$d_{eX,Lc1}^{(k)} = u_{Xm,Lc1}^{t(k)} - u_{Xm,Lc1}^{b(k)}$ [cm]	$d_{rX,Lc1}^{(k)} = qd_{eX,Lc1}^{(k)}$ [cm]	$\lvert P_{tot}^{(k)} \rvert$ [kN]	$\lvert V_{X(tot),Lc1}^{(k)} \rvert$ [kN]	$h^{(k)}$ [cm]	$\theta_{X,Lc1}^{(k)}$
1	1.03	0.00	1.03	4.12	4267.05	413.72	400	0.106
2	1.67	1.03	0.64	2.57	3335.08	376.81	300	0.076
3	2.17	1.67	0.50	1.99	2460.34	313.95	300	0.052
4	2.53	2.17	0.36	1.45	1585.59	224.15	300	0.034
5	2.74	2.53	0.20	0.82	710.85	107.40	300	0.018

[a] Average (mean) horizontal displacement (interstorey drift) along the X axis of the floor slab of the k storey for loading case 1
[b] Average (mean) horizontal displacement along the X axis of the floor slab of the (k-1) storey for loading case 1

Table 4.53 Interstorey drift sensitivity coefficients θ_Y for each building storey based on the results produced by analysis case 1

Storey k	$u_{Ym,Lc1}^{t(k)}$ [a] [cm]	$u_{Ym,Lc1}^{b(k)}$ [b] [cm]	$d_{e,Lc1}^{(k)} = u_{Ym,Lc1}^{t(k)} - u_{Ym,Lc1}^{b(k)}$ [cm]	$d_{rY,Lc1}^{(k)} = q \cdot d_{eY,Lc1}^{(k)}$ [cm]	$\lvert P_{tot}^{(k)} \rvert$ [kN]	$\lvert V_{Y(tot),Lc1}^{(k)} \rvert$ [kN]	$h^{(k)}$ [cm]	$\theta_{Y,Lc1}^{(k)}$
1	0.029	0.000	0.029	0.115	4267.05	0.00	400	N/D[c]
2	0.039	0.029	0.011	0.043	3335.08	0.00	300	N/D
3	0.047	0.039	0.008	0.031	2460.34	0.00	300	N/D
4	0.052	0.047	0.005	0.020	1585.59	0.00	300	N/D
5	0.054	0.052	0.002	0.007	710.85	0.00	300	N/D

[a]Average (mean) horizontal displacement along the Y axis of the floor slab of the k storey for loading case 1
[b]Average (mean) horizontal displacement along the Y axis of the floor slab of the (k-1) storey for loading case 1
[c]Non definable value (division by zero)

Table 4.54 Expected extreme interstorey drift sensitivity coefficient values

	$\text{extr}\theta_X^{(k)}$			$\text{extr}\theta_Y^{(k)}$		
Storey	Loading case 1	Loading case 2	Max value	Loading case 3	Loading case 4	Max value
1	**0.106**	**0.106**	**0.106**	0.043	0.050	0.050
2	0.076	0.076	0.076	0.027	0.031	0.031
3	0.052	0.052	0.052	0.021	0.023	0.023
4	0.034	0.034	0.034	0.014	0.016	0.016
5	0.018	0.018	0.018	0.008	0.009	0.009

Finally, Table 4.54 collects the expected extreme values of the interstorey drift sensitivity coefficients for all storeys which coincide with the coefficients θ_X obtained from loading cases 1 and 2 and with the coefficients θ_Y obtained from loading cases 3 and 4. As expected, the loading cases 1 and 2 yield equal interstorey drift sensitivity coefficients θ_X, since X is an axis of symmetry for building A (see Fig. 4.4). Further, coefficients θ_Y attain lower values than coefficients θ_X at all storeys due to the W1 wall, which renders building A significantly stiffer along the Y direction. Lastly, for the ground storey, it is found that $0.1 \leq \text{extr}\theta_X = 0.106 \leq 0.2$ and, therefore, for this particular storey, the design seismic effects derived from the analysis step must be increased by a factor of $1/(1\text{-extr}\theta_X) = 1.12$. In this respect, it is recommended to apply the above multiplication factor not only to the bending moments within the X-Z vertical plane due to the lateral seismic forces, but to all stress resultants for biaxial bending with axial force (design triads N, M_2, M_3) computed from the design seismic action combination $G + \psi_2 Q \pm E$.

As a final note, it is observed by comparing the numerical data of Table 4.54 with those reported in Table 4.32 that the extreme values of the interstorey drift sensitivity coefficients due to simultaneous seismic action along directions X and Y obtained by means of the LFM differ significantly from those determined using the rigorous approach presented in Sects. 3.2.1.2 and 4.1.9.3 in the context of the MRSM.

4.1.10.5 Verification Check for Maximum Interstorey Drift Demands

The verification check for the maximum allowed interstorey drifts (or damage limitation verification check) involves determination of the design interstorey drifts d_{rX} and d_{rY} along the principal axes X and Y, respectively, for all building storeys and for simultaneous design seismic action along the X and Y directions (expected "extreme" values $\text{extr}d_{rX}^{(k)}$ and $\text{extr}d_{rY}^{(k)}$) and relies on Eq. (3.23) (see Sect. 3.2.2). The computational steps for estimating the expected $\text{extr}d_{rX}^{(k)}$ and $\text{extr}d_{rY}^{(k)}$ from displacements (seismic effects) derived from the LFM of analysis are given in FC-3.11a. Specifically, the design interstorey drifts d_{rX} and d_{rY} are computed separately for each loading case indicated in Fig. 4.4 as follows:

- Loading case 1 (Lc1, seismic loads along X): $d^{(k)}_{rX,Lc1}$ and $d^{(k)}_{rY,Lc1}$
- Loading case 2 (Lc2, seismic loads along X): $d^{(k)}_{rX,Lc2}$ and $d^{(k)}_{rY,Lc2}$
- Loading case 3 (Lc3, seismic loads along Y): $d^{(k)}_{rX,Lc3}$ and $d^{(k)}_{rY,Lc3}$
- Loading case 4 (Lc4, seismic loads along Y): $d^{(k)}_{rX,Lc4}$ and $d^{(k)}_{rY,Lc4}$

Next, the above interstorey drifts are spatially superposed by means of the SRSS rule such that seismic effects due to seismic excitation along the X direction are combined with seismic effects due to seismic excitation along the Y direction. Therefore, the same four combinations used in the previous section for the second-order effects verification check need to be considered here as well. The resulting expected extreme values of the design interstorey drifts at each storey k are used to form the ratios (extr$d^{(k)}_{rX}$·v)/h and (extr$d^{(k)}_{rY}$·v)/h for which the damage limitation verification check is performed in accordance with clause §4.3.2(1) of EC8. As in all previous cases, the lateral absolute and relative floor translations due to the gravitational loads of the design seismic loading combination (G "+" ψ_2Q) are negligible compared with the corresponding translations due to the design seismic action and, therefore, are ignored.

For illustration, Table 4.55 summarizes the end displacements of vertical members at all storeys obtained from the first loading case of Fig. 4.4 and reports their average (mean) values. Next, Table 4.56 details the required calculations for determining the interstorey drift sensitivity coefficients for the first loading case, $d^{(k)}_{rX,Lc1}$ and $d^{(k)}_{rY,Lc1}$ from the previous average values.

Following the same procedure, the design interstorey drift results are computed for the remaining 3 loading cases and collected in Table 4.57. It is noted that, due to the symmetry of building A with respect to axis X, the horizontal displacements along the X axis of vertical structural members due to lateral forces along the Y direction (loading cases 3 and 4) are anti-symmetric. Therefore, the average floor diaphragm displacements along X are zero for loading cases 3 and 4.

Lastly, the expected extreme values of the design (inelastic) interstorey drifts at each storey k due to simultaneous seismic action along the X and Y directions are reported in Table 4.58. These are obtained by application of the SRSS rule for combining the 4 loading cases of Fig. 4.4, as indicated in the previous table. Further, Table 4.58 also presents the ratios (extr$d^{(k)}_{rX}$·v)/h and (extr$d^{(k)}_{rY}$·v)/h required for the damage limitation verification check in accordance with clause §4.3.2(1) of EC8 (see also Eq. (3.23)).

Assuming that building A has brittle non-structural infill walls, the maximum allowed interstorey drift ratios defined in clause §4.4.3.2(1) of EC8 is equal to 0.5 % of the storey height, i.e., (v· extr$d^{(k)}_{rX}$)/h $<$0.005 and (v· extr$d^{(k)}_{rY}$)/h $<$0.005 for all k storeys. The above condition is not met at the ground storey along the X axis due to

Table 4.55 Translational displacements [cm] at the upper end (top) of the vertical members along the X and Y axes obtained by loading case 1

Vertical member	1st storey		2nd storey		3rd storey		4th storey		5th storey	
	$u_{X,Lc1}^{t(1)}$	$u_{Y,Lc1}^{t(1)}$	$u_{X,Lc1}^{t(2)}$	$u_{Y,Lc1}^{t(2)}$	$u_{X,Lc1}^{t(3)}$	$u_{Y,Lc1}^{t(3)}$	$u_{X,Lc1}^{t(4)}$	$u_{Y,Lc1}^{t(4)}$	$u_{X,Lc1}^{t(5)}$	$u_{Y,Lc1}^{t(5)}$
C1	1.06	−0.01	1.72	−0.02	2.23	−0.03	2.60	−0.03	2.81	−0.04
C2	1.06	0.03	1.72	0.04	2.23	0.05	2.60	0.06	2.81	0.06
C3	1.06	0.06	1.72	0.09	2.23	0.12	2.60	0.13	2.81	0.14
W1	1.03	−0.01	1.67	−0.02	2.17	−0.03	2.53	−0.03	2.74	−0.04
C4	1.03	0.03	1.67	0.04	2.17	0.05	2.53	0.06	2.74	0.06
C5	1.03	0.06	1.67	0.09	2.17	0.12	2.53	0.13	2.74	0.14
C6	1.00	−0.01	1.63	−0.02	2.11	−0.03	2.46	−0.03	2.66	−0.04
C7	1.00	0.03	1.63	0.04	2.11	0.05	2.46	0.06	2.66	0.06
C8	1.00	0.06	1.63	0.09	2.11	0.12	2.46	0.13	2.66	0.14
Average values	1.03	0.03	1.67	0.04	2.17	0.05	2.53	0.05	2.74	0.05

Table 4.56 Design (inelastic) interstorey drifts for loading case 1

Storey k	$u_{Xm,Lc1}^{t(k)}$ [a] [cm]	$u_{Xm,Lc1}^{b(k)}$ [b] [cm]	$d_{eX,Lc1}^{(k)} = u_{Xm,Lc1}^{t(k)} - u_{Xm,Lc1}^{b(k)}$ [cm]	$d_{rX,Lc1}^{(k)} = q \cdot d_{eX,Lc1}^{(k)}$ [cm]	$u_{Ym,Lc1}^{t(k)}$ [c] [cm]	$u_{Ym,Lc1}^{b(k)}$ [d] [cm]	$d_{eY,Lc1}^{(k)} = u_{Ym,Lc1}^{t(k)} - u_{Ym,Lc1}^{b(k)}$ [cm]	$d_{rY,Lc1}^{(k)} = q \cdot d_{eY,Lc1}^{(k)}$ [cm]
1	1.03	0.00	1.03	4.12	0.03	0.00	0.03	0.11
2	1.67	1.03	0.64	2.57	0.04	0.03	0.01	0.04
3	2.17	1.67	0.50	1.99	0.05	0.04	0.01	0.03
4	2.53	2.17	0.36	1.45	0.05	0.05	0.01	0.02
5	2.74	2.53	0.20	0.82	0.05	0.05	0.00	0.01

$d_{eX,Lc1}^{(k)}$ relative elastic displacement in the X direction for loading case Lc1

$d_{eY,Lc1}^{(k)}$ relative elastic displacement in the Y direction for loading case Lc2

[a] Average (mean) horizontal displacement along the X axis of the k storey floor slab for loading case 1

[b] Average (mean) horizontal displacement along the X axis of the (k-1) storey floor slab for loading case 1

[c] Average (mean) horizontal displacement along the Y axis of the k storey floor slab for loading case 1

[d] Average (mean) horizontal displacement along the Y axis of the (k-1) storey floor slab for loading case 1

Table 4.57 Design (inelastic) interstorey drifts [cm] for all 4 loading cases shown in Fig. 4.4

	Loading case 1		Loading case 2		Loading case 3		Loading case 4	
Storey	$d^{(k)}_{rX,Lc1}$	$d^{(k)}_{rY,Lc1}$	$d^{(k)}_{rX,Lc2}$	$d^{(k)}_{rY,Lc2}$	$d^{(k)}_{rX,Lc3}$	$d^{(k)}_{rY,Lc3}$	$d^{(k)}_{rX,Lc4}$	$d^{(k)}_{rY,Lc4}$
1	4.12	0.11	4.12	−0.11	0.00	2.42	0.00	2.82
2	2.57	0.04	2.57	−0.04	0.00	1.37	0.00	1.52
3	1.99	0.03	1.99	−0.03	0.00	1.16	0.00	1.27
4	1.45	0.02	1.45	−0.02	0.00	0.90	0.00	0.97
5	0.82	0.01	0.82	−0.01	0.00	0.56	0.00	0.59

Table 4.58 Expected extreme values of ratios ($v \cdot d_{rX}$)/h and ($v \cdot d_{rY}$)/h of Eq. (3.23)

Storey k	extr$d^{(k)}_{rX}$ [cm]	extr$d^{(k)}_{rY}$ [cm]	v	h [cm]	(v· extr$d^{(k)}_{rX}$)/h	(v· extr$d^{(k)}_{rY}$)/h
Combination 1–3						
1	4.12	2.42	0.5	400	**0.0052**	0.0030
2	2.57	1.37	0.5	300	0.0043	0.0023
3	1.99	1.16	0.5	300	0.0033	0.0019
4	1.45	0.90	0.5	300	0.0024	0.0015
5	0.82	0.56	0.5	300	0.0014	0.0009
Combination 1–4						
1	4.12	2.82	0.5	400	**0.0052**	0.0035
2	2.57	1.52	0.5	300	0.0043	0.0025
3	1.99	1.27	0.5	300	0.0033	0.0021
4	1.45	0.97	0.5	300	0.0024	0.0016
5	0.82	0.59	0.5	300	0.0014	0.0010
Combination 2–3						
1	4.12	2.42	0.5	400	**0.0052**	0.0030
2	2.57	1.37	0.5	300	0.0043	0.0023
3	1.99	1.16	0.5	300	0.0033	0.0019
4	1.45	0.90	0.5	300	0.0024	0.0015
5	0.82	0.56	0.5	300	0.0014	0.0009
Combination 2–4						
1	4.12	2.82	0.5	400	**0.0052**	0.0035
2	2.57	1.52	0.5	300	0.0043	0.0025
3	1.99	1.27	0.5	300	0.0033	0.0021
4	1.45	0.97	0.5	300	0.0024	0.0016
5	0.82	0.59	0.5	300	0.0014	0.0010

the relatively flexible lateral load resisting system along axis X which does not include any r/c wall.

As a final comment, it is observed by comparing the ratios (v· extr$d^{(k)}_{rX}$)/h and (v· extr$d^{(k)}_{rY}$)/h reported in Table 4.58 with those of Table 4.39 that the two different EC8 analysis methods considered, that is, the LFM and the MRSM, yield very similar results with regards to extreme interstorey drift demands. However, this conclusion is not general and applies only to the herein examined example building.

4.1.11 Comparison of Design Seismic Effects for Building A Obtained from the MRSM and the LFM

This section compares vis-à-vis the following numerical data (design seismic effects) obtained by application of the modal response spectrum method (MRSM) and of the lateral force method (LFM) of seismic analysis for building A:

- displacements of the center of gravity of all floor slabs;
- design bending moments and shearing forces for the beams BX2 and BY5 at the ground (1st) storey;
- expected extreme values of N, M_2, M_3 stress resultants for the column C3 at the ground (1st) storey; and
- expected extreme values of N, M_2, M_3 stress resultants for the wall W1 at the ground (1st) storey.

Comparison of displacements of floor slab centers of gravity

Table 4.59 collects the horizontal translations along the X and Y axes, u_X and u_Y, respectively, and the rotations about the gravitational Z axis, φ_z, of the nominal centers of gravity (assuming even mass distribution in plan) of all floor slabs of building A for the seismic combination of design actions $G + \Psi_2 Q \pm E$ obtained by application of the MRSM and the LFM of analysis. The percentage differences are also reported normalized by the values obtained from the MRSM. The considered displacements indicate that the MRSM is slightly more conservative than the LFM in the case of translations along axis X and rotations about axis Z, while the LFM is more conservative (i.e., yields larger values) for translations along axis Y. It is emphasized that the above observations apply for the considered building example and cannot be generalized for other structures.

Comparison of design seismic effects for beams BX2 and BY5 at the 1st storey

Tables 4.60 and 4.61 report the extreme bending moments and shearing forces at the ends of beams BX2 and BY5 (see Fig. 4.2) at the ground (1st) storey for the seismic combination of design actions $G + \Psi_2 Q \pm E$ obtained by application of the MRSM and the LFM of analysis. The percentage differences are also reported normalized by the values obtained from the MRSM. The considered numerical data indicate that the MRSM is more conservative than the LFM with regard to the design seismic effects of beam BX2, while the two methods yield practically the same design seismic effects for the BY5 beam member.

Table 4.59 Displacements at the floor centers of mass obtained from the MRSM and LFM

	Translation u_X [cm]			Translation u_Y [cm]			Rotation φ_Z [mrad]		
Storey	MRSM	LFM	Difference (%)	MRSM	LFM	Difference (%)	MRSM	LFM	Difference (%)
Combination $G+\Psi_2Q+E$									
1	1.12	1.03	7.38	0.65	0.69	−5.06	1.32	1.18	10.04
2	1.80	1.68	6.35	0.97	1.06	−9.31	1.87	1.64	12.31
3	2.30	2.18	5.05	1.21	1.37	−12.79	2.28	1.98	13.28
4	2.65	2.55	3.89	1.39	1.61	−15.31	2.55	2.20	13.74
5	2.84	2.75	3.10	1.50	1.75	−17.15	2.68	2.29	14.49
Combination $G+\Psi_2Q$-E									
1	−1.11	−1.03	7.44	−0.65	−0.69	−5.06	−1.32	−1.18	10.04
2	−1.78	−1.66	6.41	−0.97	−1.06	−9.31	−1.87	−1.64	12.31
3	−2.27	−2.16	5.10	−1.21	−1.37	−12.8	−2.28	−1.98	13.28
4	−2.62	−2.52	3.93	−1.39	−1.61	−15.3	−2.55	−2.20	13.74
5	−2.81	−2.72	3.13	−1.50	−1.75	−17.1	−2.68	−2.29	14.49

Table 4.60 Design values of bending moments and shearing forces for beam BX2 at the ground storey for the seismic combination of design actions $G + \Psi_2 Q \pm E$ obtained through MRSM and LFM

		Moment M_3 [kNm]			Shearing force V_2 [kN]		
Position		MRSM	LFM	Difference (%)	RSM	LFM	Difference (%)
Left end	max	129.09	120.29	6.82	64.35	58.78	8.66
	min	−170.03	−161.23	5.18	−122.7	−117.14	4.53
Right end	max	174.56	163.32	6.44	119.92	114.35	4.64
	min	−199.73	−188.48	5.63	−67.14	−61.57	8.30

Table 4.61 Design values of bending moments and shearing forces for beam BY5 at the ground storey for the seismic combination of design actions $G + \Psi_2 Q \pm E$ obtained from MRSM and LFM

		Moment M_3 [kNm]			Shear force V_2 [kN]		
Position		MRSM	LFM	Difference (%)	MRSM	LFM	Difference (%)
Left end	max	160.71	160.64	0.04	64.76	64.83	−0.11
	min	−179.34	−179.26	0.04	−114.54	−114.61	−0.06
Right end	max	110.09	110.39	−0.27	124.43	124.5	−0.06
	min	−150.53	−150.83	−0.20	−54.87	−54.94	−0.13

Comparison of extreme stress resultant values of column C3 at ground storey

Table 4.62 summarizes design triads (moments M_2, M_3 and axial force N, as defined in Fig. 2.32) at the ends of the column C3 comprising the extreme (non-concurrent) values of the effects for the design seismic combination of actions $G + \Psi_2 Q \pm E$ obtained by application of the MRSM (for all different positions of the center of mass defined in Table 4.12) and the LFM (for all different combinations of the loading cases defined in Fig. 4.4). A safe conclusion as to which of the two analysis methods is more conservative in this case cannot be drawn by direct comparison of the data included in Table 4.62. To facilitate a comparison, Table 4.63 collects only the extreme values of each stress resultant considered in the previous table. It is seen that the MRSM yields slightly higher absolute values of the extreme design seismic effects, with the exception of the axial force N. This conclusion holds only for this particular building example. It is further noted that a fairer comparison in terms of conservatism can be accomplished in this case in terms of the required reinforcement at the critical cross-sections.

Comparison of extreme stress resultant values of wall W1 at ground storey

Table 4.64 summarizes design triads (moments M_2, M_3 and axial force N, as defined in Fig. 2.32) at the ends of the wall W1 comprising the extreme (non-concurrent) values of the effects for the design seismic combination of actions $G + \Psi_2 Q \pm E$ obtained by application of the MRSM (for all different positions of the

Table 4.62 Extreme design values of N, M_2, M_3 stress resultants for column C3 (ground storey) for the seismic design load combination G "+" ψ_2Q "±" E obtained from the MRSM (for all mass positions of Table 4.12) and the LFM (for all combinations of the loading cases of Fig. 4.4)

MRSM mass position	Seismic design combination	N [kN]		M_2 [kNm]		M_3 [kNm]	
		Bottom	Top	Bottom	Top	Bottom	Top
1	$G + \psi_2Q + extrE$	−45.38	−30.58	124.69	106.14	127.07	91.74
	$G + \psi_2Q - extrE$	−608.90	−594.10	−128.71	−98.73	−121.80	−100.56
2	$G + \psi_2Q + extrE$	−44.41	−29.61	130.81	111.16	128.39	92.83
	$G + \psi_2Q - extrE$	−609.87	−595.07	−134.83	−103.75	−123.13	−101.65
3	$G + \psi_2Q + extrE$	−43.15	−28.35	128.59	109.31	125.57	90.63
	$G + \psi_2Q - extrE$	−611.13	−596.33	−132.61	−101.90	−120.30	−99.45
4	$G + \psi_2Q + extrE$	−46.44	−31.64	128.62	109.37	129.63	93.77
	$G + \psi_2Q - extrE$	−607.84	−593.04	−132.64	−101.96	−124.36	−102.59
LFM combination	Seismic design situation	N [kN]		M_2 [kNm]		M_3 [kNm]	
		Bottom	Top	Bottom	Top	Bottom	Top
1–3	$G + \psi_2Q + extrE$	−49.35	−34.55	104.03	88.89	117.09	83.55
	$G + \psi_2Q - extrE$	−604.93	−590.13	−108.05	−81.47	−111.83	−92.37
1–4	$G + \psi_2Q + extrE$	−35.87	−21.07	127.64	107.95	121.1	86.94
	$G + \psi_2Q - extrE$	−618.41	−603.61	−131.66	−100.53	−115.84	−95.76
2–3	$G + \psi_2Q + extrE$	−39.35	−24.55	103.96	88.77	111.17	78.93
	$G + \psi_2Q - extrE$	−614.93	−600.13	−107.98	−81.35	−105.91	−87.75
2–4	$G + \psi_2Q + extrE$	−26.32	−11.52	127.59	107.85	115.39	82.5
	$G + \psi_2Q - extrE$	−627.96	−613.16	−131.61	−100.43	−110.13	−91.32

Table 4.63 Extreme values of N, M_2, M_3 stress resultants for column C3 (ground storey) for the seismic design load combination G "+" ψ_2Q "±" E obtained from the MRSM (for all mass positions of Table 4.12) and the LFM (for all combinations of the loading cases of Fig. 4.4)

	Top		Bottom	
Peak extreme design values	MRSM	LFM	MRSM	LFM
max exN [kN]	−28.35	−11.52	−43.15	−26.32
min exN [kN]	−596.33	−613.16	−611.13	−627.96
max exM_2 [kNm]	111.16	107.95	130.81	127.64
min exM_2 [kNm]	−103.75	−100.53	−134.83	−131.66
max exM_3 [kNm]	93.77	86.94	129.63	121.10
min exM_3 [kNm]	−102.59	−95.76	−124.36	−115.84

center of mass defined in Table 4.12) and the LFM (for all different combinations of the loading cases defined in Fig. 4.4). A safe conclusion as to which of the two analysis methods is more conservative in this case cannot be drawn by direct comparison of the data included in Table 4.64. To facilitate a comparison, Table 4.65 collects only the extreme values of each stress resultant considered in the previous table. It is seen that the two methods yield similar results, with the exception of moment M_2, for which a large discrepancy is observed. As before, this conclusion holds only for this particular building example while a fairer comparison in terms between the two methods of analysis can be accomplished in terms of the required reinforcement at the bottom of wall W1.

4.2 Example B: Five-Storey Torsionally Sensitive Building with Dual Lateral Load-Resisting Structural System

4.2.1 Geometric, Material, and Seismic Action Data

Units

Length: m; Force: kN; Mass t (1 t = 10^3 kg), Time: s.

Material properties for reinforced concrete

Modulus of Elasticity: $E = 2.9 \cdot 10^7$ kN/m^2; Poisson ratio: $v = 0.2$; weight per unit volume: $\gamma = 25$ kN/m^3.

In-plan description and geometry of building B

The example building B is a five storey structure with rectangular in-plan geometry having a single horizontal axis of symmetry along direction X as defined in Fig. 4.5. The lateral-load resisting system of the building includes two r/c walls oriented along direction Y and positioned close to the geometric center of the plan. The in-plan dimensions of building A along with the cross-sectional dimensions

Table 4.64 Extreme design values of N, M_2, M_3 stress resultants for wall W1 (ground storey) for the seismic design load combination G "+" ψ_2Q "±" E obtained from the MRSM (for all mass positions of Table 4.12) and the LFM (for all combinations of the loading cases of Fig. 4.4)

RSM Mass position	Seismic design situation	N [kN]		M_2 [kNm]		M_3 [kNm]	
		Bottom	Top	Bottom	Top	Bottom	Top
1	$G+\psi_2Q$ +extrE	−623.6	−561.1	899.22	48.27	121.24	46.17
	G $+\psi_2Q-$ extrE	−745.92	−683.42	−899.22	−48.27	−120.69	−45.85
2	$G+\psi_2Q$ +extrE	−623.6	−561.1	703.77	39.6	121.24	46.17
	G $+\psi_2Q-$ extrE	−745.92	−683.42	−703.77	−39.60	−120.69	−45.85
3	$G+\psi_2Q$ +extrE	−623.74	−561.24	803.96	43.9	120.96	46.06
	G $+\psi_2Q-$ extrE	−745.78	−683.28	−803.96	−43.90	−120.40	−45.75
4	$G+\psi_2Q$ +extrE	−623.74	−561.24	803.96	43.9	120.96	46.06
	G $+\psi_2Q-$ extrE	−745.78	−683.28	−803.96	−43.90	−120.40	−45.75
LFM combination	Seismic design situation	N [kN]		M_2 [kNm]		M_3 [kNm]	
		Bottom	Top	Bottom	Top	Bottom	Top
1–3	$G+\psi_2Q$ +extrE	−624.11	−561.61	1269.41	13.92	111.66	41.39
	G $+\psi_2Q-$ extrE	−745.41	−682.91	−1269.41	−13.92	−111.11	−41.07
1–4	$G+\psi_2Q$ +extrE	−624.11	−561.61	1061.41	13.99	111.66	41.39
	G $+\psi_2Q-$ extrE	−745.41	−682.91	−1061.41	−13.99	−111.11	−41.07
2–3	$G+\psi_2Q$ +extrE	−624.11	−561.61	1269.41	13.92	111.66	41.39
	G $+\psi_2Q-$ extrE	−745.41	−682.91	−1269.41	−13.92	−111.11	−41.07
2–4	$G+\psi_2Q$ +extrE	−624.11	−561.61	1061.41	13.99	111.66	41.39
	G $+\psi_2Q-$ extrE	−745.41	−682.91	−1061.41	−13.99	−111.11	−41.07

(in cm) of its vertical r/c members (columns and walls) are shown in Fig. 4.5. Note that the columns of the 1st (ground) storey have larger cross-sectional dimensions compared to the columns of the 2nd–5th storeys. Storey heights are also reported for the previous figure. Structural members for which analysis results are reported in detail are marked in red. For the beams, l and r denote their left and right ends, respectively.

Table 4.65 Extreme values of N, M_2, M_3 stress resultants for wall W1 (ground storey) for the seismic design load combination G "+" ψ_2Q "±" E obtained from the MRSM (for all mass positions of Table 4.12) and the LFM (for all combinations of the loading cases of Fig. 4.4)

	Top		Bottom	
Extreme design values	RSM	LFM	RSM	LFM
max exN [kN]	−561.10	−561.61	−623.60	−624.11
min exN [kN]	−683.42	−682.91	−745.92	−745.41
max exM_2 [kNm]	48.27	13.99	899.22	1269.41
min exM_2 [kNm]	−48.27	−13.99	−899.22	−1269.40
max exM_3 [kNm]	46.17	41.39	121.24	111.66
min exM_3 [kNm]	−45.85	−41.07	−120.69	−111.11

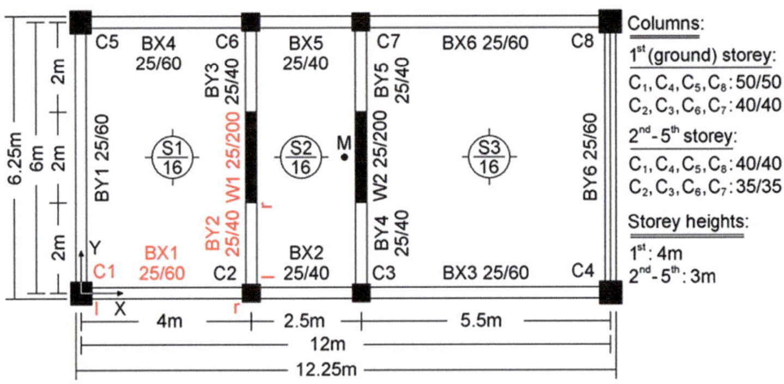

Fig. 4.5 Typical floor plan of building B

Gravity loads imposed on beams and slabs

- Double-layered masonry walls occupy the full storey height at all storeys along the perimeter of the building. These infill walls impose 3.6 kN/m^2 of "permanent" weight on all exterior beams except the beams of the top storey. The exterior beams of the top storey accommodate a "permanent" uniform distributed load of 3.6 kN/m, corresponding to a double-layered masonry roof parapet 1 m in height. There are no infill walls in the interior of the building.
- Permanent floor finishings of 1.3 kN/m^2 weight evenly distributed in plan is assumed.
- The assumed "live" gravity loads (variable action) applied to the floor slabs is taken as Q = 2 kN/m^2 evenly distributed in plan.

Directions of the horizontal seismic action (clause §4.3.3.1(11)P of EC8)

Axes X and Y shown in Fig. 4.5 can be unambiguously identified as the "principal" orthogonal axes along which the input seismic action is assumed to act for design purposes.

Design spectrum data (clause §3.2.2.5 of EC8, Type1)

- Peak ground acceleration: $a_{gR} = 0.20$ g
- Ground type: B ($S = 1.2$, $T_B = 0.15$ s, $T_C = 0.5$ s)
- Importance category: II (residential building) $\rightarrow \gamma_I = 1$
- Damping coefficient: $\zeta = 5$ %
- Behaviour factor q: to be determined in Sect. 4.2.6 below

4.2.2 *Modeling Assumptions*

4.2.2.1 Structural Modeling Assumptions

A spatial (three-dimensional) finite element (FE) model is considered which accounts for flexural, shear, axial, and torsional deformations of r/c structural members. The infill walls are not included in the FE model, assuming their influence on the lateral stiffness and strength of the building structure to be negligible (see clause §4.3.1(8) of EC8). The beam-column joints are modeled as perfectly rigid (see clause §4.3.1(2) of EC8) using rigid offsets (arms) at the end of FE members, as shown in Fig. 2.37a (see Sect. 2.3.3.2).

Modeling of floor slabs

Floor slabs are assumed to act as perfectly rigid diaphragms in their plane (see Sect. 2.3.3.1). The actual height level of these diaphragms is defined in the considered FE model, as shown in Fig. 2.37a.

Effective flange width of beams

The effective width b_{eff} of the upper flange of the beams included in the FE model is given in Table 4.66. It is computed according to clause §5.3.2.1 of EC2 as

- $b_{eff} = b_w + 0.2 \cdot l_o$, for T-shaped beams, and
- $b_{eff} = b_w + 0.1 \cdot l_o$, for L-shaped beams,

where b_w is the width of the beam web, $l_0 = 0.85 \cdot L$ for beams occupying external frame bays, and $l_0 = 0.70 \cdot L$ for beams occupying internal frame bays in which L is the length of the beam excluding its rigid offsets. Note that the beams of the ground storey are slightly shorter than the beams of the upper storeys, since the ground floor

Table 4.66 Assumed effective width of the beams of the FE model for building B

Beam members	Length [m]	b_w [m]	Bay	Shape	b_{eff} [m]
BX1,BX4	3.625	0.25	external	L	0.558
BX2,BX5	2.15	0.25	internal	L	0.401
BX3,BX6	5.125	0.25	external	L	0.686
BY1,BY6	5.60	0.25	external	L	0.726
BY2,BY3,BY4,BY5	1.825	0.25	external	T	0.560

columns have larger dimensions. However, for the sake of simplicity, all beams are assumed to have the same length L at all storeys, as shown in Table 4.66.

Effective rigidities of structural members (see §2.3.2.1)

The flexural rigidity (EI) and the shear rigidity (GA_s) are assumed equal to 50 % of the values corresponding to uncracked gross section properties for all r/c members (clause §4.3.1(7) of EC8). The torsional rigidity (GJ) is taken to be equal to 10 % of the value corresponding to uncracked gross section properties for all r/c members, as discussed in Sect. 2.3.2.1 (Fardis 2009). However, the axial rigidity (EA) of structural members is not reduced compared to the value corresponding to uncracked gross section properties given that the vertical structural members are under compression due to the gravity loads and all beam members are considered to be part of the perfectly rigid diaphragms within the plane of the floor slabs. It is noted that clause §5.4(2) of EC2 allows for considering the uncracked gross section properties to compute the rigidity of r/c structural members under gravity loads. Nevertheless, given that the reduction is generally considered towards safety, it also preserved for the vertical loads, exactly as in the analyses for the seismic loads. Therefore, in all analyses, static and seismic, the same model of the structure is employed.

Modeling of the r/c walls W1 and W2

The r/c walls of building B are modeled by means of an equivalent frame model, as discussed in Sect. 2.3.3.3 and shown in Fig. 2.38d. The model comprises an equivalent column positioned at the center of gravity of the actual shear wall, which is connected to the beams at each floor level by means of "virtual" perfectly rigid arms of length 2.0/2 = 1.00 m each. The uncracked gross section properties of the equivalent column are computed as

- Area: $A = 0.25 \cdot 2.0$ (in m^2);
- Second moment of area about the X axis: $I_{XX} = (0.25 \cdot 2.0^3)/12$ (in m^4);
- Second moment of area about the Y axis: $I_{YY} = (0.25^3 \cdot 2.0)/12$ (in m^4);
- Effective shearing area along the X axis: $A_{sX} = (5/6)A$ (in m^2);
- Effective shearing area along the Y axis: $A_{sY} = (5/6)A$ (in m^2); and
- Polar moment of inertia: $J = a \cdot 0.25^3 \cdot 2.0$ (in m^4).

The effective rigidities of the equivalent column computed from the above properties are reduced to account for concrete cracking according to the previously mentioned assumptions. Note that the constant involved in determining the polar moment of inertia can be computed for rectangular cross-sections with dimensions ($d \times t$) as (Oden 1967)

$$a = \frac{1}{3}\left[1 - \left(\frac{192t}{\pi^5 d}\right)\tanh\left(\frac{\pi d}{2t}\right)\right],$$

though other expressions are also applicable (see also Fig. 2.43). Herein, for d = 2.0 m and t = 0.25 m, one obtains a = 0.307.

4.2.2.2 Vertical Load Modeling Assumptions

- The permanent (self weight and finishings) and variable evenly distributed area loads carried by the slabs are transferred to the beams using triangular and trapezoidal tributary areas (rule of 45° or 60°). In this manner, they are distributed along the length of the beams following triangular or trapezoidal distributions.
- The masonry infill walls' self weight (permanent loads) is transferred directly onto the beams and is computed without accounting for any of the existing architectural openings in the infill walls.
- The self weight of the r/c beams and the infill walls carried by the beams is considered to be uniformly distributed along the length of the beams.
- The self weight of the r/c columns is modeled as a uniformly distributed axial load along the height of the columns.

4.2.2.3 Mass/Inertial Modeling Assumptions

- The total mass of each storey is lumped at the center of gravity M (geometrical center) of the corresponding floor rigid diaphragm.
- The total mass of each storey comprises:
  - The self mass of the storey slab and beams, including all finishings;
  - The self mass of the masonry infill walls resting on the storey beams (ignoring any openings);
  - The own mass of the vertical r/c members (columns and walls) extending above and below the considered storey slab up to the middle of their total storey height; and
  - The mass that corresponds to the variable gravity load of the seismic design load combination, as defined in clauses §3.2.4(2)P, 4.2.4(2)P and 4.3.1(10)P of EC8.

The mass of each storey of building B is computed from the gravity loads of the seismic design load combination, as detailed in Table 4.67.

4.2.3 Verification Checks for Regularity for Building B

The rationale for the regularity verification checks has been discussed in detail in Sect. 3.1.1. Herein, the EC8 regularity checks in plan and elevation are carried out for building B following the FC-3.2 and FC-3.3, respectively.

Table 4.67 Storey mass and gravity loads of the seismic design load combination for building A

Storey	Permanent load G_k [kN]	Variable action Q_k [kN]	ψ_2[a]	φ[b]	$\psi_E = \varphi \cdot \psi_2$	Combination of actions $G_k + \psi_E \cdot Q_k$[c] [kN]	Mass [t]
1st	995.27	153.25	0.3	0.8	0.24	1032.02	105.20
2nd	950.07	153.25	0.3	0.8	0.24	986.82	100.59
3rd	950.07	153.25	0.3	0.8	0.24	986.82	100.59
4th	950.07	153.25	0.3	0.8	0.24	986.82	100.59
5th	711.58	153.25	0.3	1.0	0.3	757.52	77.22
Total sum	4557.06	766.25				4750.00	484.19

[a]The combination coefficient ψ_2 of the quasi-permanent value of the variable action Q_k ("live" gravity loads) is given in Eurocode 0 (CEN 2002) Annex A1 and is taken to be equal to 0.3 assuming that building A is an ordinary occupancy residential or office building
[b]The reduction factor φ is given in Table 4.2 of EC8 (clause §4.2.4 of EC8). It is herein assumed to be equal to 1.0 for the top storey and equal to 0.8 for the rest of the storeys, which are assumed to have correlated occupancies. This assumption is particularly valid for residential buildings at night time and for office buildings during day hours
[c]The storey masses are computed from this gravity load combination, as specified in clause §3.2.4 (2)P of EC8

4.2.3.1 Verification Check for Regularity in Elevation

Following the required steps of FC-3.3 for checking for regularity in elevation for building B, it is readily verified that all vertical r/c structural members extend uninterrupted to the full height of the building and that there are no setbacks in elevation. Further, it is shown in Sect. 4.2.4 that building B has a wall lateral load resisting system along the Y direction and, therefore, there is no requirement to check for strength distribution in elevation. However, the stiffness and mass distribution along the height of building B change. Therefore, pertinent checks are undertaken to quantify the severity of this change, as described in the following paragraphs.

Check for storey stiffness variation along the height of building B

Both the storey height and the section size of columns are reduced above the first (ground) storey and remain constant among the higher storeys of building B. Thus, the change in the storey lateral stiffness along the X and Y directions between the first and subsequent storeys needs to be quantified to verify the criterion of clause §4.2.3.3(3) of EC8 on constant or gradually varying stiffness in elevation. To this aim, the following measure of the lateral stiffness of the storey k is considered

$$K_k = \sum_{j=1}^{J} \left(\frac{EI_{kj}}{h_k} + \frac{EI_{(k+1)j}}{h_{(k+1)}} \right),$$

where the index j refers to the vertical structural members. Building B has J = 10 vertical structural members (8 columns and 2 walls). Application of the above expression along the two principal directions yields the following numerical results for the material and geometric properties given in Sect. 4.2.1:

- $K_{1X} = \sum_{j=1}^{10} \left(\frac{EI_{1jX}}{h_1} + \frac{EI_{2jX}}{h_2} \right) = 250668.75 + 181189.58 = 431858.33 \ kN/m$

- $K_{2X} = \sum_{j=1}^{10} \left(\frac{EI_{2jX}}{h_2} + \frac{EI_{3jX}}{h_3} \right) = 181189.60 + 181189.60 = 362379.20 \ kN/m$

- $K_{2X} = K_{3X} = K_{4X} = 362379.20 \ kN/m$

- $K_{1Y} = \sum_{j=1}^{10} \left(\frac{EI_{1jY}}{h_1} + \frac{EI_{2jY}}{h_2} \right) = 2629575.00 + 3353064.60 = 5982639.60 \ kN/m$

- $K_{2Y} = \sum_{j=1}^{10} \left(\frac{EI_{2jY}}{h_2} + \frac{EI_{3jY}}{h_3} \right) = 3353064.60 + 3353064.60 = 6706129.20 \ kN/m$

- $K_{2Y} = K_{3Y} = K_{4Y} = 6706129.20 \ kN/m.$

Therefore, the stiffness reduction in between the ground storey and subsequent storeys along the X and Y directions are

- $\Delta K_X = K_{2X}\text{-}K_{1X} = 362{,}379.20 - 431{,}858.33 = -69{,}479.13 \approx -0.16K_{1X}$ or 16 % reduction, and
- $\Delta K_Y = K_{2Y}\text{-}K_{1Y} = 6{,}706{,}129.20 - 5{,}982{,}639.60 = 723{,}489.60 \approx 0.12K_{1Y}$ or 12 % increase,

respectively. Even though EC8 does not specify limiting values for the variation of storey stiffness above which buildings are not classified as regular in elevation, the above computed changes of stiffness can be safely regarded as being "gradual" and should not cause any adverse effects to the dynamic response of the building.

Check for mass distribution along the height of building B

Table 4.68 reports the variation of storey mass along the height of building B. As in the case of storey stiffness variation, EC8 does not specify limiting values for the mass variation in elevation above which buildings are not classified as regular in elevation. Still, the mass variations computed in Table 4.68 are not significant and decrease along the height of the building. Therefore, they can be treated as being "gradual" without causing any adverse effects to the dynamic response of the building.

Table 4.68 Storey mass variation along the height of building B

Storey k	m_k	$\Delta m_k = m_k - m_{k-1}$	$\lvert \Delta m_k / m_{k-1} \rvert$	Variation
1st	105.20			
2nd	100.59	−4.61	$\lvert -0.044 \rvert$	4.4 % reduction
3rd	100.59	0.00	0.00	No variation
4th	100.59	0.00	0.00	No variation
5th	77.22	−23.37	$\lvert -0.232 \rvert$	23 % reduction

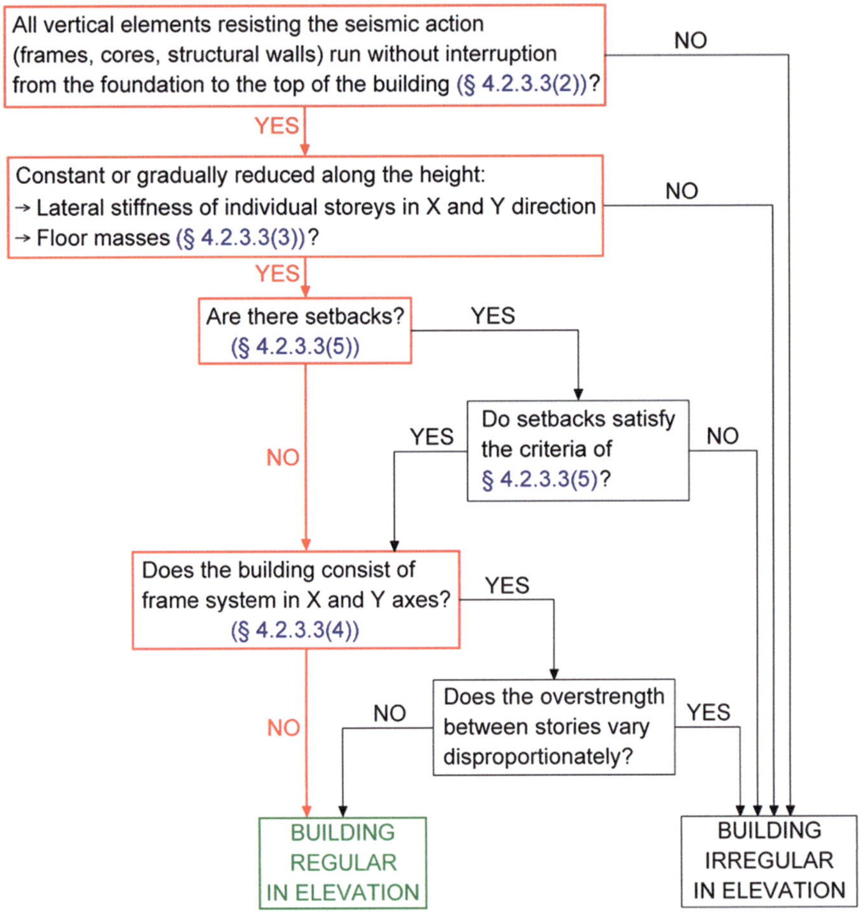

Fig. 4.6 Regularity in elevation verification check for building B following FC-3.3

Taking into account all previous checks and comments, building B is classified as being regular in elevation. The path of the relevant logical and computational steps following FC-3.3 closely is indicated in Fig. 4.6 in red. The above classification allows for

- using the LFM of analysis provided that condition (a) of clause §4.3.3.2.1(2) of EC8 is satisfied as well. Still, only the MRSM of analysis will be employed for the seismic analysis of building B in this example.
- a reduction of 20 % to the basic (reference) value of the behaviour factor, q_0, (see clause §5.2.2.2(3) of the EC8).

4.2.3.2 Verification Check for Regularity in Plan

The verification check procedure for regularity in plan follows FC-3.2 and FC-3.2a. The outcomes from the pertinent logical and computational steps for building B are summarized in Fig. 4.7, which closely follows FC-3.2 (the path taken is marked in

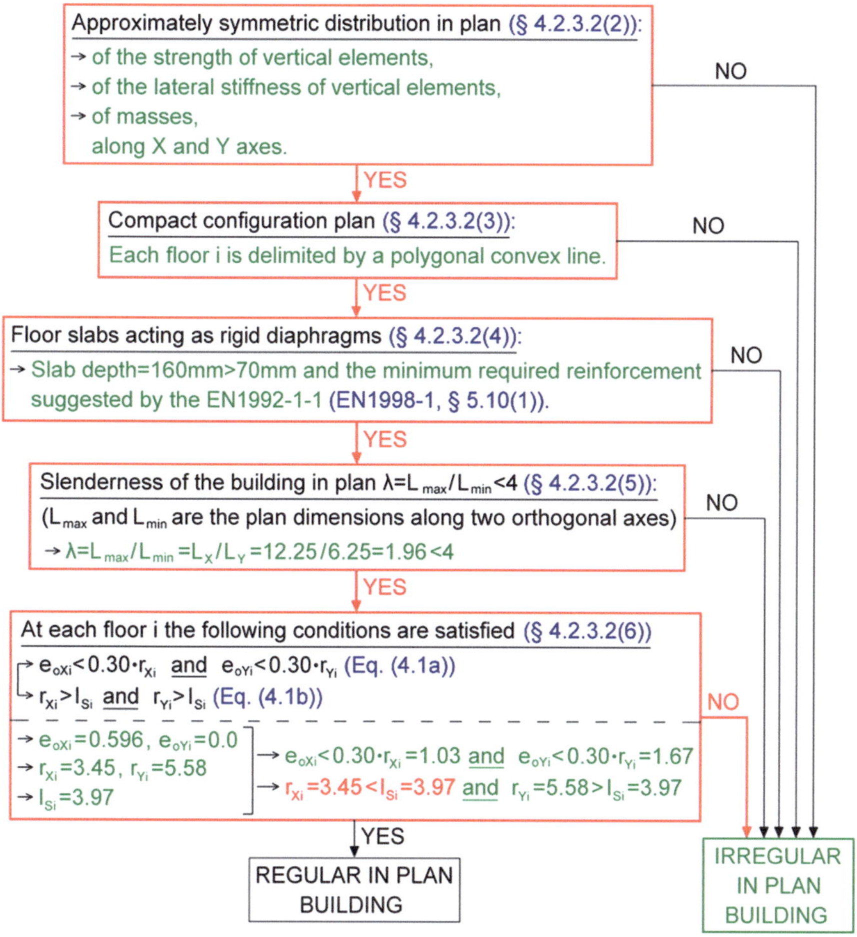

Fig. 4.7 Regularity in plan verification check for building B following FC-3.2

Table 4.69 Triangular distribution of torsional moments and lateral forces along the height of building B for an arbitrary base shear equal to 5000 kN (Equation (4.11) of EC8)

Storey k	m [t]	J_m [t · m^2]	z_k [m]	$m_k \cdot z_k$	F_{Xk} [kN] / F_{Yk} [kN] / M_{Zk} [kNm]
1st	105.20	1658.00	4.00	420.80	450.14
2nd	100.59	1585.40	7.00	704.15	753.25
3rd	100.59	1585.40	10.00	1005.94	1076.07
4th	100.59	1585.40	13.00	1307.72	1398.89
5th	77.22	1217.00	16.00	1235.50	1321.65
Total Sum	484.17			4674.11	5000.00

red). Focusing on checking the validity of the inequalities of Eqs. (3.1) and (3.2) (clause §4.2.3.2(6) of EC8), the same procedure is herein followed as in Sect. 4.1.3.2 for building A, to compute the structural eccentricities, e_{oX} and e_{oY}, and the torsional radii, r_X and r_Y, along the principal axes X and Y, respectively, for each floor diaphragm of building B with the aid of the fictitious elastic vertical axis.

Specifically, the determination of the structural eccentricities requires that a linear static analysis be performed first for torsional moments, M_{Zk}, about the gravitational axis Z at each floor diaphragm k (loading case "M"). These torsional moments can follow a "triangular" distribution along the height of the building by application of equation (4.11) of EC8. Table 4.69 reports the values of the torsional moments along with the pertinent steps for computing these values for building B, assuming an arbitrarily taken total base shear equal to 5000 kN. As mentioned in Sect. 4.1.3.2, this arbitrarily chosen value has no influence on the values of the subsequently calculated structural eccentricities and torsional stiffness radii.

Next, the coordinates X_{Po} and Y_{Po} of the "center of twist" P_o at the storey k lying closer to the 80 % level of the total height of the building (4th storey for building B), through which the fictitious elastic vertical axis passes, are determined by the relationships $X_{Po} = X_{Mk} - (u_{Y(Mk)}/\theta_{Z(Mk)})$ and $Y_{Po} = Y_{Mk} - (u_{X(Mk)}/\theta_{Z(Mk)})$ included in FC-3.2. In the above relationships, X_{Mk} and Y_{Mk} are the coordinates of the center of mass of storey k ($X_{Mk} = 6.0$ m and $Y_{Mk} = 3.00$ m for building B with regard to the coordinate system having its origin at the geometric center of column C1), and $u_{X(Mk)}$, $u_{Y(Mk)}$, and $\theta_{Z(Mk)}$ are the translations along axes X, Y, and the rotation about axis Z, respectively, of point M (Fig. 4.5) due to the loading case "M". For building B and the torsional moments of Table 4.69, the aforementioned nodal displacements are: $u_{X(Mk)} = 0$, $u_{Y(Mk)} = 0.0043$ m, and $\theta_{Z(Mk)} = 0.00721$. Therefore, the coordinates of point P_o for building B, through which the fictitious elastic vertical axis passes, are computed as:

- $X_{Po} = X_{Mk} - (u_{Y(Mk)}/\theta_{Z(Mk)}) = 6.0 - 0.0043/0.00721 = 5.404$ m; and
- $Y_{Po} = Y_{Mk} - (u_{X(Mk)}/\theta_{Z(Mk)}) = 3.0 - 0/0.00721 = 3.00$ m

and the structural eccentricities common to all storeys of building B are:

- $e_{oX} = X_{Mk} - X_{Po} = 6.0 - 5.404 = 0.596$ m; and
- $e_{oY} = Y_{Mk} - Y_{Po} = 3.0 - 3.0 = 0$ m.

Table 4.70 Verification check of inequality in Eq. (3.1), Sect. 3.1.1.1, for building B

Storey	e_{oX} [m]	e_{oY} [m]	r_X [m]	r_Y [m]	$e_{oX} < 0.3 \cdot r_X$	$e_{oY} < 0.3 \cdot r_Y$
1st	0.596	0.00	3.45	5.58	YES	YES
2nd	0.596	0.00	3.45	5.58	YES	YES
3rd	0.596	0.00	3.45	5.58	YES	YES
4th	0.596	0.00	3.45	5.58	YES	YES
5th	0.596	0.00	3.45	5.58	YES	YES

Table 4.71 Verification check of inequality in Eq. (3.2), Sect. 3.1.1.1, for building B

Storey	l_s[a] [m]	r_X [m]	r_Y [m]	$r_X \geq l_s$	$r_Y \geq l_s$
1st	3.97	3.45	5.58	NO	YES
2nd	3.97	3.45	5.58	NO	YES
3rd	3.97	3.45	5.58	NO	YES
4th	3.97	3.45	5.58	NO	YES
5th	3.97	3.45	5.58	NO	YES

[a]The radius of gyration l_s for a rectangular floor plan such as that of building B is obtained using the following relationship: $l_s = [(L_X^2 + L_Y^2)/12]^{1/2} = [(12.25^2 + 6.25^2)/12]^{1/2} = 3.97$ m

Furthermore, the torsional stiffness radii r_X and r_Y corresponding to the fictitious elastic vertical axis are determined by the relationships $r_X = (u_{Y(FY)}/\theta_{Z(Mk)})^{1/2}$ and $r_Y = (u_{X(FX)}/\theta_{Z(Mk)})^{1/2}$ included in FC-3.2. In the above relationships, $u_{Y(FY)}$ and $u_{X(FX)}$ are the translations of point P_o along axis Y due to the loading case "F_Y"and along axis X due to the loading case "F_X", respectively. Loading cases "F_X" and "F_Y" involve lateral forces applied at the traces of the previously defined fictitious elastic axis at each floor diaphragm of the building along the assumed directions of the seismic action, that is, axes X and Y, respectively. The considered lateral forces may follow a "triangular" distribution along the height of the building by application of equation (4.11) of EC8, as calculated in Table 4.69. For building B and the lateral forces of Table 4.69, the aforementioned nodal displacements are: $u_{Y(FY)} = 0.0857$ m and $u_{X(FX)} = 0.2245$ m. Therefore, the torsional stiffness radii common to all storeys of building B are:

- $r_X = (u_{Y(FY)}/\theta_{Z(Mk)})^{1/2} = (0.0857/0.00721)^{1/2} = 3.45$; and
- $r_Y = (u_{X(FX)}/\theta_{Z(Mk)})^{1/2} = (0.2245/0.00721)^{1/2} = 5.58$.

Having computed the structural eccentricities and torsional radii for the fictitious elastic axis, the inequalities of Eqs. (3.1) and (3.2) are checked in Tables 4.70 and 4.71, respectively. As shown in the penultimate column of Table 4.70, building B does not satisfy the inequality of Eq. (3.2) (i.e., equation (4.1b) of EC8) along principal axis X. Therefore, the building is not regular in plan.

4.2.4 *Classification of the Lateral Load-Resisting Structural System of Building B*

The rationale of classifying building structures according to the properties of their lateral load-resisting structural system and the implications of this classification for the seismic design process have been discussed in detail in Sect. 3.1.2. The classification procedure follows the steps delineated in FC-3.4. The first step of this procedure is the torsional sensitivity verification check, which follows FC-3.2a. This check has already been undertaken in the previous section during the regularity in plan verification check, and building B was found to be torsionally sensitive (see Table 4.71). Therefore, the lateral load-resisting system of building B is classified as torsionally sensitive, and this classification suffices for the purpose of determining the maximum allowable value of the behaviour factor q to be discussed in Sect. 4.2.6.

However, according to clause §5.1.2(1) of EC8, a building that is classified as torsionally sensitive may belong, on the basis of the percentage of the total base shear resisted by walls, to any one type of structural system (i.e., frame, wall or dual system). For instance, in the context of the regularity in elevation verification check, there is a requirement to distinguish between lateral load resisting systems of the frame and of the non-frame type (see FC-3.4), which is independent from the classification of the building in terms of torsional sensitivity. Further, the capacity design rule to ensure "weak beams-strong columns" for frame and dual frame-equivalent structural systems expressed by Equation (4.29) of EC8 can be omitted for wall systems irrespective of whether these are torsionally sensitive or not (see clause §4.4.2.3 of EC8). Therefore, an additional check along the direction Y of building B is undertaken to quantify the percentage of the total base shear resisted by the two walls W1 and W2. In particular, the loading case F_{Yk} comprising the lateral forces reported in Table 4.69 applied along the Y axis at the traces of the fictitious elastic axis determined in Sect. 4.2.3.2 is considered (see also Sect. 4.1.4). Table 4.72 summarizes pertinent numerical results obtained by linear static analysis

Table 4.72 Verification check of the percentage of the total base shear resisted by r/c walls for the loading case F_{Yk} defined in Table 4.69

Member	Analysis in direction of axis Y		
	Wall	$V_{Y,walls}$ [kN]	$V_{Y,columns}$ [kN]
C1	NO	–	102.63
C2	NO	–	105.18
C3	NO	–	125.54
C4	NO	–	257.36
C5	NO	–	102.63
C6	NO	–	105.18
C7	NO	–	125.54
C8	NO	–	257.36
W1	YES	1708.33	–
W2	YES	2110.25	–
Total sum		3818.58	1181.42
Percentage		**76.37 %**	**23.63 %**

for the above loading case. Based on these results, building B is classified as a wall structural system along axis Y since the walls W1 and W2 resist more than 65 % of the total base shear applied to the structure along the considered axis. Note that a similar check is not required to be undertaken along axis X as there are no walls oriented along direction X.

4.2.5 Selection of Ductility (Capacity) Class of Building B

The rationale of deciding upon the desirable ductility capacity class according to EC8 has been discussed in Sect. 3.1.3. Furthermore, some relevant points have been highlighted in Sect. 4.1.5. For building B, it is decided to adopt the medium ductility class (DCH)..

4.2.6 Determination of the Maximum Allowed Behaviour Factor for Building B

A detailed presentation of the procedure for determining the maximum allowed value, $maxq_{allow}$, of the behaviour factor q has been provided in Sect. 3.1.4. Moreover, some relevant points have been highlighted in Sect. 4.1.6. The procedure of determining the $maxq_{allow}$ follows FC-3.5. The path taken (marked in red) and the outcomes of the pertinent logic and computational steps (in green fonts) for building B are given in Fig. 4.8, which follows the above flowchart closely. It is seen that the $maxq_{allow}$ value of the behaviour factor is equal to 2.0: this value is adopted in applying the MRSM of analysis.

4.2.7 Selection of an Equivalent Linear Method of Seismic Analysis for Building B

The modal response spectrum method (MRSM) is of general use for the seismic analysis of any building structure regardless of regularity conditions in plan and elevation. As such, it is applicable to building B, and this is the analysis method that is exclusively considered in this example to obtain the seismic effects reported in Sect. 4.2.9.

Still, it is noted in passing that, since building B is regular in elevation and its uncoupled fundamental natural periods along directions X and Y, $T_{1X} = 0.798$ s and $T_{1Y} = 0.486$ s, respectively, are smaller than min{$4T_c$, 2.0 s}, then the lateral force method (LFM) of analysis could also be applied (see Table 3.1, FC-3.2 and Sect. 4.1.7).

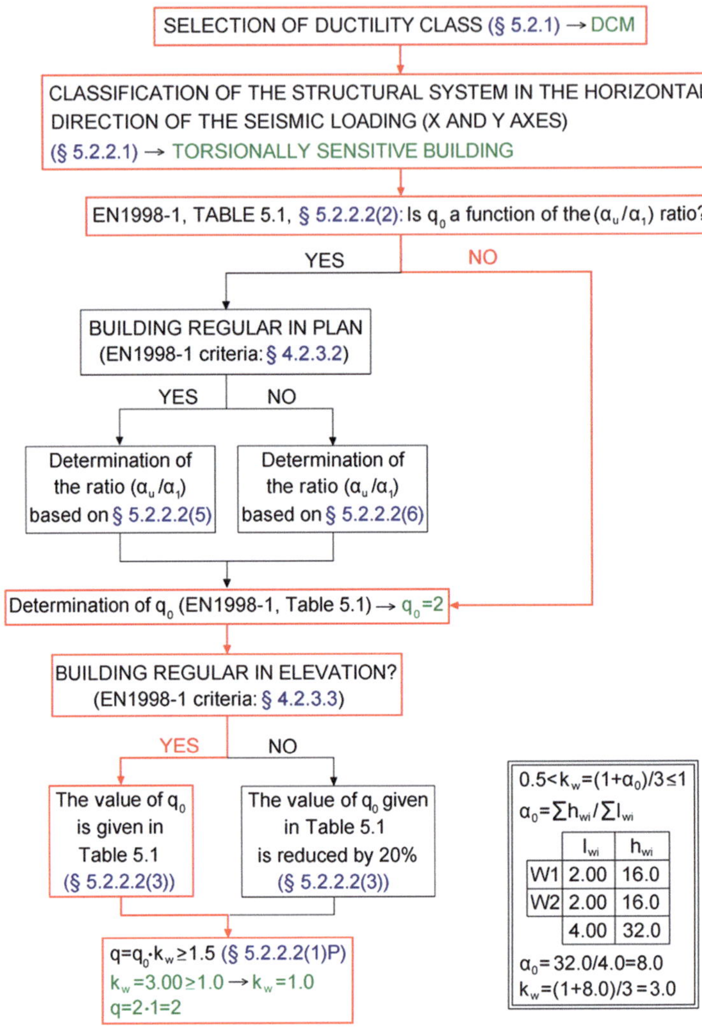

Fig. 4.8 Determination of the maximum allowable value of the behaviour factor q for building B

4.2.8 Static Analysis for Gravity Loads of the Design Seismic Loading Combination (G "+" ψ_2Q) for Building B

The design seismic loading combination involves gravitational (statically applied) permanent and quasi-permanent variable actions (see Eq. (2.12)). Effects due to these actions can be derived separately by means of standard static analysis and superposed to the effects due to the seismic (accidental) action. Table 4.73 reports stress resultants (effects) due to the gravity loads of the design seismic loading combination at critical cross-sections of selected r/c structural members, namely, the wall W1 and the column C1 at the ground storey and beams BX1 and BY2 at the ground (1st) storey (Fig. 4.5).

Table 4.73 Sectional stress resultants of wall W1, column C1 and beams BX1 and BY2 at the ground storey for the permanent and quasi-permanent variable actions of the design seismic loading combination (G "+" ψ_2Q)

Member	Position	N [kN]	V_2 [kN]	V_3 [kN]	T [kNm]	M_2 [kNm]	M_3 [kNm]
W1	bottom	−670.84	−0.13	0.00	0.00	0.00	−0.54
	top	−620.84	−0.13	0.00	0.00	0.00	−0.014
C1	bottom	−476.19	−4.21	−9.00	0.00	−11.63	−6.17
	top	−453.07	−4.21	−9.00	0.00	21.68	9.41
BX1	left end	0.00	−29.35	0.00	0.199	0.00	−14.11
	midspan	0.00	−0.033	0.00	0.199	0.00	13.59
	right end	0.00	25.66	0.00	0.199	0.00	−10.64
BY2	left end	0.00	−8.75	0.00	0.088	0.00	−0.476
	midspan	0.00	3.19	0.00	0.088	0.00	2.95
	right end	0.00	22.59	0.00	0.088	0.00	−8.47

Sign convention follows Fig. 2.32

4.2.9 Seismic Analysis of Building B Using the Modal Response Spectrum Method and Deformation-Based Verification Checks

The implementation of the modal response spectrum method (MRSM) of seismic analysis follows FC-3.7. Four different spatial FE models are considered in the analysis corresponding to the positioning of the center of mass of each floor diaphragm at four different sets of locations (positions 1 to 4, as shown in Table 4.74) to account for accidental mass eccentricity (see also Sect. 3.1.5.1 and Fig. 3.4). The accidental eccentricities, $\pm e_{aX}$ and $\pm e_{aY}$, along the principal axes X and Y, respectively, define the four displaced locations of the mass center measured from the geometric center of each slab. These eccentricities are computed in Table 4.74 taken to be equal to 5 % of the length of building A, $L_X = 12.25$ m and $L_Y = 6.25$ m (Fig. 4.5), along axes X and Y, respectively, assuming that the masonry infill walls are evenly distributed in plan. Further, the polar moment of inertia about the gravitational axis of each floor diaphragm is computed with respect to the displaced position as $J_{mi} = J_m + m \cdot e_{ai}^2$, $i = X,Y$, where $J_m = m \cdot (L_X^2 + L_Y^2)/12$ is the polar moment of inertia with respect to the geometric center of each slab, as reported in Table 4.74.

4.2.9.1 Modal Analysis Results

Table 4.75 lists the natural periods corresponding to the first 9 mode shapes of vibration for the four considered FE structural models (position of center of mass 1 to 4, as shown in Table 4.74) of building B derived from standard modal analysis.

Further, Table 4.76 reports the modal participation mass ratios for each mode (i.e., ratios of effective modal mass over the total mass of building B) for excitation

Table 4.74 Accidental eccentricities and polar moment of inertia for seismic excitation along X and Y axes for building B

		Accidental eccentricities [m]		Polar moment of inertia [tm^2]		
Storey	Mass [t]	e_{aX}[a]	e_{aY}[b]	J_{mX}[c]	J_{mY}[d]	
1st	105.20	0.6125	0.3125	1697.46	1668.27	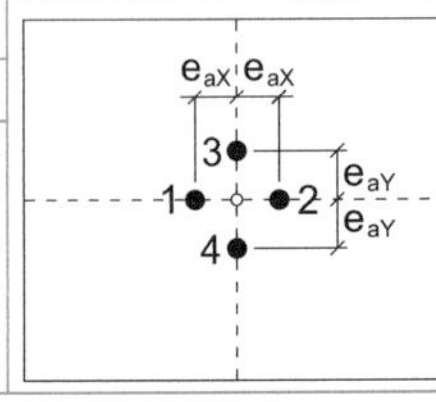
2nd–4th	100.59	0.6125	0.3125	1623.08	1595.16	
5th	77.22	0.6125	0.3125	1245.99	1224.56	

[a] $e_{aX} = 0.05L_X$
[b] $e_{aY} = 0.05L_Y$
[c] $J_{mX} = J_m + me_{aX}^2$
[d] $J_{mY} = J_m + me_{aY}^2$

Table 4.75 Natural periods of the four considered FE structural models (center of mass displaced by $\pm e_{aX}$ and $\pm e_{aY}$ as shown in Table 4.74) for building B

	Natural period [s]			
Mode shape	Mass position 1	Mass position 2	Mass position 3	Mass position 4
1	0.798	0.798	0.800	0.800
2	0.571	0.632	0.585	0.585
3	0.487	0.440	0.470	0.470
4	0.258	0.258	0.259	0.259
5	0.182	0.199	0.186	0.186
6	0.147	0.147	0.147	0.147
7	0.142	0.130	0.138	0.138
8	0.102	0.110	0.103	0.103
9	0.798	0.798	0.800	0.800

along axes X and Y, as well as the corresponding cumulative modal participation mass ratios for the four considered FE models. The latter results suggest that at least the first 7 mode shapes need to be considered to satisfy the criterion of clause 4.3.3.3.1(3) of EC8 for all the four FE models, that is, a sufficient number of modes are considered in the MRSM such that 90 % or more of the total oscillatory mass along both principal axes is activated. Therefore, in all ensuing numerical results reported, the first 7 mode shapes are utilized in implementing the MRSM of analysis.

4.2.9.2 Selected Design Seismic Effects (Sectional Stress Resultants)

In this section, the design seismic effects at critical sections (sectional stress resultants) for the column C1, the wall W1 and the beams BX1 and BY2 of the

Table 4.76 Modal participation mass ratios and cumulative participation mass ratios as percentages of the total mass of building B

	Mass position 1				Mass position 2			
	Individual mode (%)		Cumulative sum (%)		Individual mode (%)		Cumulative sum (%)	
Mode shape	X	Y	X	Y	X	Y	X	Y
1	89.02	0.00	**89.02**	**0.00**	89.02	0.00	**89.02**	**0.00**
2	0.00	0.01	**89.02**	**0.01**	0.00	30.03	**89.02**	**30.03**
3	0.00	82.35	**89.02**	**82.36**	0.00	52.30	**89.02**	**82.33**
4	8.22	0.00	**97.24**	**82.36**	8.22	0.00	**97.24**	**82.33**
5	0.00	0.00	97.24	**82.36**	0.00	3.79	**97.24**	**86.12**
6	1.98	0.00	99.22	**82.36**	1.98	0.00	99.22	**86.12**
7	0.00	12.77	99.22	**95.13**	0.00	8.57	99.22	**94.69**
8	0.00	0.00	99.22	95.13	0.00	1.29	99.22	95.98
	Mass position 3				Mass position 4			
	Individual mode		Cumulative sum		Individual mode		Cumulative sum	
Mode shape	X	Y	X	Y	X	Y	X	Y
1	88.39	0.033	**88.39**	**0.033**	88.39	0.033	**88.39**	**0.033**
2	0.60	16.14	**88.99**	**16.17**	0.60	16.14	**88.99**	**16.17**
3	0.03	66.21	**89.02**	**82.38**	0.03	66.21	**89.02**	**82.38**
4	8.16	0.003	**97.18**	**82.39**	8.16	0.003	**97.18**	**82.39**
5	0.057	1.613	97.24	**84.00**	0.057	1.613	97.24	**84.00**
6	1.97	0.003	99.21	**84.00**	1.97	0.003	99.21	**84.00**
7	0.001	11.07	99.21	**95.07**	0.001	11.07	99.21	**95.07**
8	0.025	0.382	99.23	95.45	0.025	0.382	99.23	95.45

ground (1st) storey (Fig. 4.5) obtained by means of the MRSM are presented in tabular form.

Vertical structural members C1 and W1 (Bi-axial Bending with Axial Force)

In the case of the vertical structural members C1 and W1, which need to be designed for bi-axial bending with axial force, the design values of the three concurrent pertinent stress resultants ("design triads"), namely moments M_2, M_3 and axial force N, as defined in Fig. 2.32, are reported. Following the conservative approach of clause §4.3.3.5.1(2)c of EC8, these design triads may comprise the extreme values of the M_2, M_3, and N, as defined in Eq. (3.8) for simultaneous seismic action along the two principal axes X and Y. According to this approach, the 8 design triads of Eq. (3.12) need to be considered for each position of floor mass. Alternatively, the design triads may be compiled by considering the extreme value of a single stress resultant together with the expected (most probable) values of the other two stress resultants attained concurrently. In the latter case, the 6 design triads of Eq. (3.13) for each position of floor mass can be considered derived by means of a simplified approach detailed in the Greek Seismic Code EAK2000 (Gupta 1992; Anastassiadis 1993; Earthquake Planning and Protection

Table 4.77 Extreme values of stress resultants in column C1 (ground storey) of building B

	N [kN]		M_2 [kNm]		M_3 [kNm]	
Mass position	Bottom	Top	Bottom	Top	Bottom	Top
1	±442.93	±442.93	±94.47	±49.43	±267.32	±169.48
2	±420.46	±420.46	±152.51	±88.04	±277.21	±175.14
3	±434.56	±434.56	±151.25	±85.51	±260.69	±165.04
4	±429.16	±429.16	±151.25	±85.51	±281.79	±177.94

Table 4.78 Extreme values of stress resultants in wall W1 (ground storey) of building B

	N [kN]		M_2 [kNm]		M_3 [kNm]	
Mass position	Bottom	Top	Bottom	Top	Bottom	Top
1	±39.88	±39.88	±1832.56	±94.82	±98.36	±26.53
2	±39.88	±39.88	±1502.86	±101.72	±98.36	±26.53
3	±39.56	±39.56	±1781.29	±102.12	±97.57	±26.32
4	±39.56	±39.56	±1781.28	±102.12	±97.57	±26.32

Organization (EPPO) 2000) and assuming simultaneous seismic action along the two principal axes X and Y.

In particular, the following three computational steps are taken to determine the design triads for the considered vertical members (see also Sect. 3.1.5.1):

(1) The peak (seismic) modal values of the considered stress resultants are obtained separately by application of the MRSM for each of the directions of the seismic excitation X and Y. Next, these modal values are superposed by means of the CQC rule for modal combination (clause §4.3.3.3.2(3)P of EC8) to derive the (non-concurrent) maximum values of stress resultants for seismic excitation along axes X and Y, independently.

(2) The SRSS rule for spatial combination (clause §4.3.3.5.1(2)b of EC8) is employed to obtain the extreme values of the considered stress resultants from the maximum values derived in the previous step for simultaneous seismic action along the X and Y horizontal directions. Tables 4.77 and 4.78 report the thus obtained extreme values of the M_2, M_3, and N stress resultants developing at the bottom and the top of the structural members C1 and W1, respectively, at the ground storey of building B for all four different FE models used in the analysis. As previously discussed, EC8 allows for compiling 8 design triads for each of the four eccentrically positioned mass centers comprising these extreme M_2, M_3, and N values with alternating signs according to Eq. (3.12) assumed to act concurrently in each section. However, the above design triads may lead to overly conservative detailing of cross-sections (see also Fardis 2009). In this regard, since EC8 allows for the use of more accurate methods to estimate the probable concurrent values of more than one seismic effect due to simultaneous seismic action along two horizontal axes without, nevertheless, specifying any, the simplified approach presented in Sect. 3.1.5.1 is herein considered.

Table 4.79 Design triads (expected -most probable- concurrent values of N, M_2, and M_3 stress resultants for simultaneous seismic action along axes X and Y) for column C1 (ground storey) of building B [Extreme values in bold]

Mass position		N [kN]		M_2 [kNm]		M_3 [kNm]	
		Bottom	Top	Bottom	Top	Bottom	Top
1	extrN	**442.93**	**442.93**	45.69	−23.75	228.98	−145.29
	extrM_2	214.22	−212.80	**94.47**	**49.43**	3.92	4.58
	extrM_3	379.41	−379.70	1.39	1.34	**267.32**	**169.48**
	−extrN	**−442.93**	**−442.93**	−45.69	23.75	−228.98	145.29
	−extrM_2	−214.22	212.80	**−94.47**	**−49.43**	−3.92	−4.58
	−extrM_3	−379.41	379.70	−1.39	−1.34	**−267.32**	**−169.48**
2	extrN	**420.46**	**420.46**	44.41	−24.46	229.53	−145.43
	extrM_2	122.39	−116.80	**152.55**	**88.07**	−64.63	−39.24
	extrM_3	348.14	−349.15	−35.57	−19.73	**277.21**	**175.14**
	−extrN	**−420.46**	**−420.46**	−44.41	24.46	−229.53	145.43
	−extrM_2	−122.39	116.80	**−152.55**	**−88.07**	64.63	39.24
	−extrM_3	−348.14	349.15	35.57	19.73	**−277.21**	**−175.14**
3	extrN	**434.56**	**434.56**	73.97	−41.25	208.45	−132.69
	extrM_2	212.51	−209.64	**151.26**	**85.51**	−18.32	−8.38
	extrM_3	347.47	−349.40	−10.63	−4.34	**260.69**	**165.03**
	−extrN	**−434.56**	**−434.56**	−73.97	41.25	−208.45	132.69
	−extrM_2	−212.51	209.64	**−151.26**	**−85.51**	18.32	8.38
	−extrM_3	−347.47	349.40	10.63	4.34	**−260.69**	**−165.03**
4	extrN	**429.16**	**429.16**	44.58	−23.69	226.53	−143.65
	extrM_2	126.48	−118.90	**151.26**	**85.51**	−79.01	−49.29
	extrM_3	345.00	−346.48	−42.41	−23.69	**281.79**	**177.94**
	−extrN	**−429.16**	**−429.16**	−44.58	23.69	−226.53	143.65
	−extrM_2	−126.48	118.90	**−151.26**	**−85.51**	79.01	49.29
	−extrM_3	−345.00	346.48	42.41	23.69	**−281.79**	**−177.94**

Tables 4.79 and 4.80 present the 6 design triads at the bottom and the top of the structural members C1 and W1, respectively, at the ground storey of building B for all four different FE models used in the analysis obtained by application of the aforementioned simplified approach. The single extreme value attained by a certain stress resultant in each design triad is noted by bold faced fonts.

(3) Finally, the seismic design triads derived in the previous step are superposed to the corresponding stress resultants of the considered structural members due to the gravitational permanent and quasi-permanent variable actions summarized in Table 4.73 (Sect. 4.2.8) to obtain the design triads for the EC8 design seismic loading combination G "+" Ψ_2Q "±" E. The thus obtained triads are reported in Tables 4.81 and 4.82.

Beams BX1 and BY2 (uni-axial bending)

The previously described three steps are applied to obtain the extreme values of the moment M_3 and of the shearing force V_2 at critical cross-sections (left end, right

Table 4.80 Design triads (expected -most probable- concurrent values of N, M_2, and M_3 stress resultants for simultaneous seismic action along axes X and Y) for wall W1 (ground storey) of building B [Extreme values in bold]

		N [kN]		M_2 [kNm]		M_3 [kNm]	
Mass position		Bottom	Top	Bottom	Top	Bottom	Top
1	extrN	**39.88**	**39.88**	0.01	0.00	−94.44	23.92
	extrM_2	0.00	0.00	**1832.56**	**94.83**	0.00	0.00
	extrM_3	−38.29	35.97	−0.02	0.00	**98.36**	**26.53**
	−extrN	**−39.88**	**−39.88**	−0.01	0.00	94.44	−23.92
	−extrM_2	0.00	0.00	**−1832.56**	**−94.83**	0.00	0.00
	−extrM_3	38.29	−35.97	0.02	0.00	**−98.36**	**−26.53**
2	extrN	**39.88**	**39.88**	0.01	0.00	−94.44	23.92
	extrM_2	0.00	0.00	**1502.90**	**102.31**	0.00	0.00
	extrM_3	−38.29	35.97	0.00	0.00	**98.36**	**26.53**
	−extrN	**−39.88**	**−39.88**	−0.01	0.00	94.44	−23.92
	−extrM_2	0.00	0.00	**−1502.90**	**−102.31**	0.00	0.00
	−extrM_3	38.29	−35.97	0.00	0.00	**−98.36**	**−26.53**
3	extrN	**39.56**	**39.56**	−56.69	10.74	−93.65	23.72
	extrM_2	−1.26	4.15	**1781.34**	**102.39**	3.12	3.24
	extrM_3	−37.97	35.66	57.02	12.59	**97.57**	**26.32**
	−extrN	**−39.56**	**−39.56**	56.69	−10.74	93.65	−23.72
	−extrM_2	1.26	−4.15	**−1781.34**	**−102.39**	−3.12	−3.24
	−extrM_3	37.97	−35.66	−57.02	−12.59	**−97.57**	**−26.32**
4	extrN	**39.56**	**39.56**	56.70	−10.75	−93.65	23.72
	extrM_2	1.26	−4.15	**1781.34**	**102.39**	−3.12	−3.24
	extrM_3	−37.97	35.66	−57.03	−12.60	**97.57**	**26.32**
	−extrN	**−39.56**	**−39.56**	−56.70	10.75	93.65	−23.72
	−extrM_2	−1.26	4.15	**−1781.34**	**−102.39**	3.12	3.24
	−extrM_3	37.97	−35.66	57.03	12.60	**−97.57**	**−26.32**

end, and at midspan) of the beams BX1 and BY2 of the ground (1st) storey of building B (Fig.4.5), which need to be designed for uni-axial bending. However, in this case, the procedure of obtaining the design seismic effects for simultaneous seismic action along the two principal axes X and Y is significantly simplified by the fact that only a single seismic effect (i.e., stress resultant M_3 and corresponding shearing force V_2) is required in the detailing of beam sections, as opposed to the vector of the three concurrently acting seismic effects (triads) N, M_2, M_3 considered for the case of vertical structural members.

Specifically, Tables 4.83 and 4.84 report the extreme values of M_3 and V_2 for the beams BX1 and BY2, respectively, for all four FE models considered in the analysis. These are obtained by first computing the maximum values of M_3 and V_2 by modal combining of the peak (seismic) modal values of these stress resultants along the directions X and Y of the seismic action, separately, using the CQC modal combination rule and, then, by application of the SRSS rule for spatial combination to the previously computed maximum values. Further, Tables 4.85 and 4.86 report

Table 4.81 Design triads for column C1 (ground storey) for the seismic design load combination G "+" ψ_2Q "±" E of building B [Extreme values in bold]

Mass position		N [kN]		M_2 [kNm]		M_3 [kNm]	
		Bottom	Top	Bottom	Top	Bottom	Top
1	extrN	**−33.26**	**−10.14**	34.06	−2.07	222.81	−135.88
	extrM_2	−261.97	−665.87	**82.84**	**71.11**	−2.25	13.99
	extrM_3	−96.78	−832.77	−10.24	23.02	**261.15**	**178.89**
	−extrN	**−919.12**	**−896.00**	−57.32	45.43	−235.15	154.70
	−extrM_2	−690.41	−240.27	**−106.10**	**−27.75**	−10.09	4.83
	−extrM_3	−855.60	−73.37	−13.02	20.34	**−273.49**	**−160.07**
2	extrN	**−55.73**	**−32.61**	32.78	−2.78	223.36	−136.02
	extrM_2	−353.80	−569.87	**140.92**	**109.75**	−70.80	−29.83
	extrM_3	−128.05	−802.22	−47.20	1.95	**271.04**	**184.55**
	−extrN	**−896.65**	**−873.53**	−56.04	46.14	−235.70	154.84
	−extrM_2	−598.58	−336.27	**−164.18**	**−66.39**	58.46	48.65
	−extrM_3	−824.33	−103.92	23.94	41.41	**−283.38**	**−165.73**
3	extrN	**−41.63**	**−18.51**	62.34	−19.57	202.28	−123.28
	extrM_2	−263.68	−662.71	**139.63**	**107.19**	−24.49	1.03
	extrM_3	−128.72	−802.47	−22.26	17.34	**254.52**	**174.44**
	−extrN	**−910.75**	**−887.63**	−85.60	62.93	−214.62	142.10
	−extrM_2	−688.70	−243.43	**−162.89**	**−63.83**	12.15	17.79
	−extrM_3	−823.66	−103.67	−1.00	26.02	**−266.86**	**−155.62**
4	extrN	**−47.03**	**−23.91**	32.95	−2.01	220.36	−134.24
	extrM_2	−349.71	−571.97	**139.63**	**107.19**	−85.18	−39.88
	extrM_3	−131.19	−799.55	−54.04	−2.01	**275.62**	**187.35**
	−extrN	**−905.35**	**−882.23**	−56.21	45.37	−232.70	153.06
	−extrM_2	−602.67	−334.17	**−162.89**	**−63.83**	72.84	58.70
	−extrM_3	−821.19	−106.59	30.78	45.37	**−287.96**	**−168.53**

the values of M_3 and V_2 for the beams BX2 and BY5, respectively, for the seismic design loading combination G "+" Ψ_2Q "±" E for which the sections of BX2 and BY5 need to be detailed. The latter values have been obtained by superposing the extreme values of the seismic effects of Tables 4.83 and 4.84 to the corresponding stress resultants of the considered structural members due to the gravitational permanent and quasi-permanent variable actions summarized in Table 4.73 (Sect. 4.2.8).

4.2.9.3 Verification Check of the Influence of Second Order Effects

The rationale for the verification check for second-order effects and its implications in the seismic design process have been discussed in detail in Sects. 2.4.3.3 and 3.2.1. This deformation-based verification check involves determination of the interstorey drift sensitivity coefficients θ_X and θ_Y along the principal directions X

Table 4.82 Design triads for wall W1 (ground storey) for the seismic design load combination G "+" ψ_2Q "±" E of building B [Extreme values in bold]

Mass position		N [kN]		M_2 [kNm]		M_3 [kNm]	
		Bottom	Top	Bottom	Top	Bottom	Top
1	extrN	**−630.96**	**−580.96**	0.01	0.00	−94.98	23.91
	extrM_2	−670.84	−620.84	**1832.56**	**94.83**	−0.54	−0.01
	extrM_3	−709.13	−584.87	−0.02	0.00	**97.82**	**26.51**
	−extrN	**−710.72**	**−660.72**	−0.01	0.00	93.90	−23.94
	−extrM_2	−670.84	−620.84	**−1832.56**	**−94.83**	−0.54	−0.01
	−extrM_3	−632.55	−656.81	0.02	0.00	**−98.90**	**−26.54**
2	extrN	**−630.96**	**−580.96**	0.01	0.00	−94.98	23.91
	extrM_2	−670.84	−620.84	**1502.90**	**102.31**	−0.54	−0.01
	extrM_3	−709.13	−584.87	0.00	0.00	**97.82**	**26.51**
	−extrN	**−710.72**	**−660.72**	−0.01	0.00	93.90	−23.94
	−extrM_2	−670.84	−620.84	**−1502.90**	**−102.31**	−0.54	−0.01
	−extrM_3	−632.55	−656.81	0.00	0.00	**−98.90**	**−26.54**
3	extrN	**−631.28**	**−581.28**	−56.69	10.74	−94.19	23.71
	extrM_2	−672.10	−616.69	**1781.34**	**102.39**	2.58	3.22
	extrM_3	−708.81	−585.18	57.02	12.59	**97.03**	**26.30**
	−extrN	**−710.40**	**−660.40**	56.69	−10.74	93.11	−23.74
	−extrM_2	−669.58	−624.99	**−1781.34**	**−102.39**	−3.66	−3.25
	−extrM_3	−632.87	−656.50	−57.02	−12.59	**−98.11**	**−26.33**
4	extrN	**−631.28**	**−581.28**	56.70	−10.75	−94.19	23.71
	extrM_2	−669.58	−624.99	**1781.34**	**102.39**	−3.66	−3.25
	extrM_3	−708.81	−585.18	−57.03	−12.60	**97.03**	**26.30**
	−extrN	**−710.40**	**−660.40**	−56.70	10.75	93.11	−23.74
	−extrM_2	−672.10	−616.69	**−1781.34**	**−102.39**	2.58	3.22
	−extrM_3	−632.87	−656.50	57.03	12.60	**−98.11**	**−26.33**

Table 4.83 Extreme values of stress resultants in beam BX1 of building B (ground storey)

	V_2 [kN]			M_3 [kNm]		
Mass position	Left end	Midspan	Right end	Left end	Midspan	Right end
1	±139.58	±139.58	±139.58	±289.63	±41.88	±205.90
2	±145.28	±145.28	±145.28	±301.62	±43.76	±214.13
3	±135.98	±135.98	±135.98	±282.24	±40.89	±200.50
4	±147.69	±147.69	±147.69	±306.65	±44.51	±217.66

and Y, respectively, for all storeys defined in Eq. (3.17). The "rigorous approach" presented in Sect. 3.2.1.2 and delineated in FC-3.10b is herein followed to estimate the coefficients θ_X and θ_Y via Eq. (3.20) using displacements obtained by means of the MRSM of analysis for building B with centers of mass positioned at location 3, as defined in Table 4.74. This choice of mass location serves the purpose of presenting numerical results for a non-symmetric structure in which the floor

Table 4.84 Extreme values of stress resultants in beam BY2 of building B (ground storey)

	V_2 [kN]			M_3 [kNm]		
Mass position	Left end	Midspan	Right end	Left end	Midspan	Right end
1	±145.49	±145.49	±145.49	±113.82	±17.12	±148.06
2	±116.34	±116.34	±116.34	±91.05	±13.66	±118.36
3	±138.65	±138.65	±138.65	±108.51	±16.28	±141.07
4	±138.50	±138.50	±138.50	±108.38	±16.26	±140.91

Table 4.85 Design effects of beam BX1 (ground storey) for the seismic design load combination G "+" ψ_2Q "±" E of building B

		V_2 [kN]			M_3 [kNm]		
Mass position	Loading combination	Left end	Midspan	Right end	Left end	Midspan	Right end
1	$G+\psi_2Q+E$	110.23	139.55	165.24	275.52	55.47	195.26
	$G+\psi_2Q-E$	−168.94	−139.62	−113.93	−303.73	−28.29	−216.54
2	$G+\psi_2Q+E$	115.92	145.25	170.93	287.51	57.35	203.49
	$G+\psi_2Q-E$	−174.63	−145.31	−119.62	−315.72	−30.17	−224.76
3	$G+\psi_2Q+E$	106.63	135.95	161.64	268.13	54.48	189.87
	$G+\psi_2Q-E$	−165.34	−136.02	−110.33	−296.35	−27.29	−211.14
4	$G+\psi_2Q+E$	118.34	147.66	173.35	292.55	58.11	207.02
	$G+\psi_2Q-E$	−177.05	−147.72	−122.04	−320.76	−30.92	−228.29

Table 4.86 Design effects of beam BY2 (ground storey) for the seismic design load combination G "+" ψ_2Q "±" E of building B

		V_2 [kN]			M_3 [kNm]		
Mass position	Loading combination	Left end	Midspan	Right end	Left end	Midspan	Right end
1	$G+\psi_2Q+E$	136.74	148.67	168.08	113.34	20.07	139.59
	$G+\psi_2Q-E$	−154.23	−142.3	−122.89	−114.29	−14.17	−156.53
2	$G+\psi_2Q+E$	107.59	119.52	138.93	90.57	16.6	109.89
	$G+\psi_2Q-E$	−125.08	−113.15	−93.74	−91.52	−10.71	−126.83
3	$G+\psi_2Q+E$	129.91	141.84	161.25	108.03	19.23	132.6
	$G+\psi_2Q-E$	−147.4	−135.47	−116.06	−108.98	−13.34	−149.54
4	$G+\psi_2Q+E$	129.75	141.68	161.09	107.91	19.21	132.44
	$G+\psi_2Q-E$	−147.24	−135.31	−115.9	−108.86	−13.32	−149.38

diaphragms are displaced along all three possible degrees of freedom: translation along the X and Y axes, and rotation about the Z axis. Note that selecting the mass location 4 would have a similar effect. The lateral absolute and relative floor translations due to the gravitational loads of the design seismic loading combination (G "+" ψ_2Q) are ignored, since they are negligible compared with the corresponding translations due to the design seismic action.

The first step of the adopted approach for determining the interstorey drift sensitivity coefficients θ_X and θ_Y involves calculating their values along directions X and Y for each mode i considered in the analysis and at each building storey k, for design seismic action along direction X, that is, $\theta^{(k)}_{X,EXi}$ and $\theta^{(k)}_{Y,EXi}$, and for design seismic action along direction Y, that is, $\theta^{(k)}_{X,EYi}$ and $\theta^{(k)}_{Y,EYi}$. Tables 4.87, 4.88, 4.89, and 4.90 present, in tabular form, the required calculations for determining the four peak (seismic) "modal" design interstorey drifts for the second mode shape (i = 2) and for the ground storey (k = 1). The values of these drifts are required for the determination of the interstorey drift sensitivity coefficients. Note that this particular mode shape has been selected since it is activated for seismic action along both principal directions, X and Y, in a non-trivial manner, as can be deduced from the modal participation mass ratios of Table 4.76. The following results for the peak (seismic) "modal" interstorey drift sensitivity coefficient are reached:

- Table 4.87: $\theta^{(1)}_{X,EX2} = \left|\frac{P^{(1)}_{tot}\cdot d^{(1)}_{rX,EX2}}{V^{(1)}_{X(tot),EX2}\cdot h^{(1)}}\right| = \left|\frac{-4879.05\cdot 0.00022}{7.38\cdot 4.00}\right| = 0.03675;$
- Table 4.88: $\theta^{(1)}_{Y,EX2} = \left|\frac{P^{(1)}_{tot}\cdot d^{(1)}_{rY,EX2}}{V^{(1)}_{Y(tot),EX2}\cdot h^{(1)}}\right| = \left|\frac{-4879.92\cdot 0.000366}{38.12\cdot 4.00}\right| = 0.0117;$
- Table 4.89: $\theta^{(1)}_{X,EY2} = \left|\frac{P^{(1)}_{tot}\cdot d^{(1)}_{rX,EY2}}{V^{(1)}_{X(tot),EY2}\cdot h^{(1)}}\right| = \left|\frac{-4879.92\cdot 0.00115}{38.12\cdot 4.00}\right| = 0.03675$; and
- Table 4.90: $\theta^{(1)}_{Y,EY2} = \left|\frac{P^{(1)}_{tot}\cdot d^{(1)}_{rY,EY2}}{V^{(1)}_{Y(tot),EY2}\cdot h^{(1)}}\right| = \left|\frac{-4879.92\cdot 0.00189}{196.97\cdot 4.00}\right| = 0.0117.$

Tables 4.91 and 4.92 report the modal interstorey drift sensitivity coefficients at all storeys for the first 7 mode shapes of building B for seismic excitation along the X and Y directions, respectively. These results are obtained by performing similar calculations as those presented in Tables 4.87 to 4.91 to determine the ground storey interstorey drift sensitivity coefficients for the 2nd mode shape. Further, Table 4.93 collects the values of the products $v_{iX}\cdot S_{ai}/\omega_i^2$ and $v_{iY}\cdot S_{ai}/\omega_i^2$ used for the calculation of the modal interstorey drift sensitivity coefficients of Tables 4.87 to 4.91 for all the 7 modes considered (i = 1,2,...,7). Notably, the coefficients reported in Tables 4.91 and 4.92 are identical. In other words, the modal interstorey drift sensitivity coefficients are independent of the considered direction of the seismic action. This is because the product $v_i\cdot S_{ai}/\omega_i^2$, by which the influence of the seismic action is expressed enters in both the numerator and the denominator of the θ_X and θ_Y ratios and, therefore, its effect cancels out.

Having determined the modal interstorey drift sensitivity coefficients, the maximum value of the interstorey drift sensitivity coefficients is found at each storey

Table 4.87 Determination of the peak (seismic) modal interstorey drift, $d^{(1)}_{rX,EX2}$ along direction X (ground storey of building B, 2nd mode shape, Seismic excitation along X)

Vertical member j	Axial load $(G+\psi_2 Q)$ [kN]	Shear force $V^{(1)}_{X2j}$ [kN/m]	$V^{(1)}_{Xj,EX2} = \left(v_{2X}\frac{S_{a2}}{\omega_2^2}\right)\cdot V^{(1)}_{X2j}$ [kN]	u^t_{X2j} [−	u^b_{X2j} [−]	$d^{(1)}_{rX2,j} = q(u^t_{X2j} - u^b_{X2j})$ [−]	$\left(v_{2X}\frac{S_{a2}}{\omega_2^2}\right)\cdot d^{(1)}_{rX,j}$ [cm]
C1	−476.19	140.84	5.26	1.48	0.00	2.97	0.11
C2	−295.64	90.83	3.39	1.48	0.00	2.97	0.11
C3	−372.40	86.15	3.22	1.48	0.00	2.97	0.11
C4	−578.40	128.55	4.80	1.48	0.00	2.97	0.11
C5	−476.19	−82.95	−3.10	−0.89	0.00	−1.78	−0.07
C6	−295.64	−54.68	−2.04	−0.89	0.00	−1.78	−0.07
C7	−372.40	−51.52	−1.92	−0.89	0.00	−1.78	−0.07
C8	−578.40	−74.97	−2.80	−0.89	0.00	−1.78	−0.07
W1	−670.84	7.61	0.28	0.30	0.00	0.59	0.02
W2	−763.80	7.59	0.28	0.30	0.00	0.59	0.02
	$P^{(1)}_{tot}$	$V^{(1)}_{X2(tot)}$	$V^{(1)}_{X(tot),EX2}$				$d^{(1)}_{rX,EX2}$
	−4879.92	197.45	7.38				0.0222

The third column reports the modal values of shear forces corresponding to the second mode shape if the normalized mode shape components are considered as real displacements

v_{2X} is the modal participation factor for the second mode shape and excitation along axis X

The values of columns 5–7 are multiplied by 100. The mode shapes are normalized so that the generalized mass has unit value

Table 4.88 Determination of the peak (seismic) modal interstorey drift $d^{(1)}_{rY,EX2}$ along direction Y (ground storey of building B, 2nd mode shape, Seismic excitation along X)

Vertical member j	Axial load $(G+\psi_2 Q)$ [kN]	Shear force $V^{(1)}_{Y2j}$ [kN/m]	$V^{(1)}_{Yj,EX2} = \left(v_{2X}\frac{S_{a2}}{\omega_2^2}\right)\cdot V^{(1)}_{Y2j}$ [kN]	u^t_{Y2j} [–]	u^b_{Y2j} [–]	$d^{(1)}_{rY2,j} = q(u^t_{Y2j} - u^b_{Y2j})$ [–]	$\left(v_{2X}\frac{S_{a2}}{\omega_2^2}\right)\cdot d^{(1)}_{rY,j}$ [cm]
C1	−476.19	−159.86	−5.97	−1.71	0.00	−3.41	−0.13
C2	−295.64	−7.59	−0.28	−0.12	0.00	−0.25	−0.01
C3	−372.40	50.44	1.88	0.87	0.00	1.73	0.06
C4	−578.40	276.33	10.32	3.04	0.00	6.08	0.23
C5	−476.19	−159.86	−5.97	−1.71	0.00	−3.41	−0.13
C6	−295.64	−7.78	−0.29	−0.12	0.00	−0.25	−0.01
C7	−372.40	50.46	1.89	0.87	0.00	1.73	0.06
C8	−578.40	276.33	10.32	3.04	0.00	6.08	0.23
W1	−670.84	−208.79	−7.80	−0.12	0.00	−0.25	−0.01
W2	−763.80	910.63	34.02	0.87	0.00	1.73	0.06
	$P^{(1)}_{tot}$	$V^{(1)}_{Y2(tot)}$	$V^{(1)}_{Y(tot),EX2}$				$d^{(1)}_{rY,EX2}$
	−4879.92	1020.3	38.12				0.0366

The third column reports the modal values of shear forces corresponding to the second mode shape if the normalized mode shape components are considered as real displacements

v_{2X} is the modal participation factor for the second mode shape and excitation along axis X

The values of columns 5–7 are multiplied by 100. The mode shapes are normalized so that the generalized mass has unit value

Table 4.89 Determination of the peak (seismic) modal interstorey drift $d^{(1)}_{rX,EY2}$ along direction X (ground storey of building B, 2nd mode shape, Seismic excitation along Y)

Vertical member j	Axial load (G + ψ$_2$Q) [kN]	Shear force $V^{(1)}_{X2j}$ [kN/m]	$V^{(1)}_{Xj,EY2} = \left(v_{2Y}\frac{S_{a2}}{\omega_2^2}\right) \cdot V^{(1)}_{X2j}$ [kN]	u^t_{X2j} [−]	u^b_{X2j} [−]	$d^{(1)}_{rX2,j} = q(u^t_{X2j} - u^b_{X2j}$) [−]	$\left(v_{2Y}\frac{S_{a2}}{\omega_2^2}\right) \cdot d^{(1)}_{rX,j}$ [cm]
C1	−476.19	140.84	27.19	1.48	0.00	2.97	0.57
C2	−295.64	90.83	17.54	1.48	0.00	2.97	0.57
C3	−372.40	86.15	16.63	1.48	0.00	2.97	0.57
C4	−578.40	128.55	24.82	1.48	0.00	2.97	0.57
C5	−476.19	−82.95	−16.01	−0.89	0.00	−1.78	−0.34
C6	−295.64	−54.68	−10.56	−0.89	0.00	−1.78	−0.34
C7	−372.40	−51.52	−9.95	−0.89	0.00	−1.78	−0.34
C8	−578.40	−74.97	−14.47	−0.89	0.00	−1.78	−0.34
W1	−670.84	7.61	1.47	0.30	0.00	0.59	0.11
W2	−763.80	7.59	1.47	0.30	0.00	0.59	0.11
	$P^{(1)}_{tot}$	$V^{(1)}_{X2(tot)}$ [a]	$V^{(1)}_{X(tot),EY2}$			$d^{(1)}_{rX2}$ [b]	$d^{(1)}_{rX,EY2}$
	−4879.92	197.45	38.12				0.115

The values of columns 5–7 are multiplied by 100. The mode shapes are normalized so that the generalized mass has unit value

[a]The term $v_{2Y} \cdot S_{a2}/\omega_2^2$ (2nd mode shape; excitation along axis Y) is equal to 0.19305 (see also Table 4.93)

[b]The average (mean) value of the modal design (inelastic) interstorey drifts from all j elements for the 2nd mode shape and for excitation along axis Y, $d^{(1)}_{rX2}$, is equal to zero

Table 4.90 Determination of the peak (seismic) modal interstorey drift $d_{rY,EY2}^{(1)}$ along direction Y (ground storey of building B, 2nd mode shape, Seismic excitation along Y)

Vertical member j	Axial load (G + ψ_2Q) [kN]	Shear force $V_{Y2j}^{(1)}$ [kN/m]	$V_{Yj,EY2}^{(1)} = \left(v_{2Y}\frac{S_{a2}}{\omega_2^2}\right) \cdot V_{Y2j}^{(1)}$ [kN]	u_{Y2j}^{t} [−]	u_{Y2j}^{b} [−]	$d_{rY2,j}^{(1)} = $ q $(u_{Y2j}^{t} - u_{Y2j}^{b})$ [−]	$\left(v_{2Y}\frac{S_{a2}}{\omega_2^2}\right) \cdot d_{rY,j}^{(1)}$ [cm]
C1	−476.19	−159.86	−30.86	−1.71	0.00	−3.41	−0.66
C2	−295.64	−7.59	−1.47	−0.12	0.00	−0.25	−0.05
C3	−372.40	50.44	9.74	0.87	0.00	1.73	0.33
C4	−578.40	276.33	53.35	3.04	0.00	6.08	1.17
C5	−476.19	−159.86	−30.86	−1.71	0.00	−3.41	−0.66
C6	−295.64	−7.78	−1.50	−0.12	0.00	−0.25	−0.05
C7	−372.40	50.46	9.74	0.87	0.00	1.73	0.33
C8	−578.40	276.33	53.35	3.04	0.00	6.08	1.17
W1	−670.84	−208.79	−40.31	−0.12	0.00	−0.25	−0.05
W2	−763.80	910.63	175.80	0.87	0.00	1.73	0.33
	$P_{tot}^{(1)}$		$V_{Y(tot),EY2}^{(1)}$				$d_{rY,EY2}^{(1)}$
	−4879.92	1020.3	196.97				0.189

The product $v_{2Y} \cdot S_{a2}/\omega_2^2$ (2nd mode shape; excitation along axis Y) is equal to 0.19305 (see also Table 4.93)

Table 4.91 Peak modal (seismic) interstorey drift sensitivity coefficients for seismic excitation along the X axis for building B (mass position 3 in Table 4.74)

	1st storey		2nd storey		3rd storey		4th storey		5th storey	
Mode	$\theta_{X,EX}$	$\theta_{Y,EX}$	$\theta_{X,EX}$	$\theta_{Y,EX}$	$\theta_{X,EX}$	$\theta_{Y,EX}$	$\theta_{X,EX}$	$\theta_{Y,EX}$	$\theta_{X,EX}$	$\theta_{Y,EX}$
1	0.037	0.017	0.034	0.019	0.025	0.015	0.016	0.010	0.009	0.006
2	0.037	0.012	0.034	0.014	0.025	0.012	0.016	0.009	0.009	0.006
3	0.037	0.010	0.034	0.013	0.025	0.011	0.017	0.008	0.009	0.006
4	0.032	0.013	0.034	0.012	0.022	0.013	0.014	0.008	0.007	0.004
5	0.031	0.008	0.039	0.008	0.022	0.010	0.014	0.006	0.007	0.003
6	0.025	0.008	0.022	0.021	0.017	0.008	0.012	0.004	0.006	0.003
7	0.032	0.006	0.031	0.006	0.023	0.008	0.014	0.005	0.007	0.003

Table 4.92 Peak modal (seismic) interstorey drift sensitivity coefficients for seismic excitation along the Y axis for building B (mass position 3 in Table 4.74)

	1st storey		2nd storey		3rd storey		4th storey		5th storey	
Mode	$\theta_{X,EY}$	$\theta_{Y,EY}$	$\theta_{X,EY}$	$\theta_{Y,EY}$	$\theta_{X,EY}$	$\theta_{Y,EY}$	$\theta_{X,EY}$	$\theta_{Y,EY}$	$\theta_{X,EY}$	$\theta_{Y,EY}$
1	0.037	0.017	0.034	0.019	0.025	0.015	0.016	0.010	0.009	0.006
2	0.037	0.012	0.034	0.014	0.025	0.012	0.016	0.009	0.009	0.006
3	0.037	0.010	0.034	0.013	0.025	0.011	0.017	0.008	0.009	0.006
4	0.032	0.013	0.034	0.012	0.022	0.013	0.014	0.008	0.007	0.004
5	0.031	0.008	0.039	0.008	0.022	0.010	0.014	0.006	0.007	0.003
6	0.025	0.008	0.022	0.021	0.017	0.008	0.012	0.004	0.006	0.003
7	0.032	0.006	0.031	0.006	0.023	0.008	0.014	0.005	0.007	0.003

Table 4.93 Values of the product $v_{ik} \cdot S_{ai}/\omega_i^2$ for each mode shape i and for the two considered directions of the seismic action k = X or Y, assuming mode shape normalization with respect to the generalized mass of each mode (building B; mass position 3 in Table 4.74)

Mode shape i	Seismic excitation along the X axis $v_{iX} \cdot S_{ai}/\omega_i^2$	Seismic excitation along the Y axis $v_{iY} \cdot S_{ai}/\omega_i^2$
1	0.61704	−0.012
2	0.037	0.19305
3	−0.006	0.29471
4	0.03141	−0.001
5	0.001	0.00718
6	−0.00495	0.000187
7	0.000089	−0.01002

Table 4.94 Expected extreme values of the interstorey drift sensitivity coefficients θ_X and θ_Y for building B (mass position 3 in Table 4.74)

Storey	extrθ_X	extrθ_Y
1	**0.145**	0.046
2	**0.144**	0.061
3	0.100	0.048
4	0.065	0.032
5	0.034	0.021

and for each direction of seismic excitation by modal combination using the CQC rule. Next, spatial combination using the SRSS rule is considered to evaluate the expected extreme values of the interstorey drift sensitivity coefficients θ_X and θ_Y for all storeys of building B due to simultaneous design seismic action along the principal directions X and Y. These extreme values are given in Table 4.94.

As expected, the extreme values of the interstorey drift sensitivity coefficients θ_Y are smaller than the θ_X for all storeys due to the walls W1 and W2, which render the structure stiffer along the Y principal direction. Further, in all storeys, the criterion of clause §4.4.2.2 (2) of EC8 is satisfied along direction Y, that is, extr$\theta_Y \leq 0.1$, and, therefore, second-order effects need not be accounted for. However, this is not the case for direction X, for which it is found that $0.1 \leq$ extr$\theta_X \leq 0.2$ for the first two storeys. Therefore, according to clauses §4.4.2.2(2)-(4) of EC8, all seismic effects derived from the analysis step for these particular storeys (i.e., stress resultants and deformations) must be increased by a factor of 1/(1-extrθ_X).

4.2.9.4 Verification Check for Maximum Interstorey Drift Demands

The aims of and rationale for the verification check for the maximum allowed interstorey drifts (or damage limitation verification check) and its implications in the seismic design process have been discussed in detail in Sect. 3.2.2. This deformation-based verification check relies on Eq. (3.23) and involves determination of the design interstorey drifts d_{rX} and d_{rY} along the principal axes X and Y, respectively, for all building storeys and for simultaneous design seismic action along the X and Y directions. The computational steps that need to be taken to estimate the expected extreme values of d_{rX} and d_{rY} from displacements (seismic effects) derived from the MRSM of analysis are provided in FC-3.11b (see also Sect. 4.1.9.4).

Herein, the computational steps for obtaining the interstorey drifts d_{rX} and d_{rY} are presented and the verification check of Eq. (3.23) is undertaken for the FE structural model of building B with centers of mass positioned at location 3, as defined in Table 4.74. As in all previous deformation checks, the horizontal translations due to the gravitational loads of the design seismic loading combination (G "+" ψ_2Q) are ignored as being negligible.

Table 4.95 reports the modal ordinates (normalized mode shape displacements), u_X and u_Y, along principal directions X and Y, respectively, of the upper end of all vertical structural members at all storeys for the 2nd mode shape obtained from standard modal analysis. The average (mean) values of the above quantities, u_{Xm} and u_{Ym}, corresponding to each floor slab are computed in the last row of Table 4.95. Next, Table 4.96 provides the average elastic relative floor slab displacements (interstorey drifts), $d_{eX}^{(k)}$ and $d_{eY}^{(k)}$, and the corresponding inelastic (design) interstorey drifts $d_{rX}^{(k)}$ and $d_{rY}^{(k)}$ for the 2nd mode shape and for all k storeys. By performing similar operations, the design interstorey drifts $d_{rX}^{(k)}$ and $d_{rY}^{(k)}$ for the 7 first mode shapes considered in the analysis of building B are obtained and reported in Table 4.97.

The seismic modal values of the interstorey drifts for each direction of the seismic excitation X and Y are obtained by multiplying the modal interstorey drift values of Table 4.97 with the products $v_{iX} \cdot S_{ai}/\omega_i^2$ and $v_{iY} \cdot S_{ai}/\omega_i^2$, respectively, given in Table 4.93. The thus obtained seismic modal values are reported in Tables 4.98 and 4.99 for seismic excitation along directions X and Y, respectively.

Finally, the expected extreme values of interstorey drifts for simultaneous seismic action along both principal directions X and Y are computed (Table 4.100) by first applying the CQC rule for modal combination to the seismic modal interstorey drift values of Tables 4.98 and 4.99 and, subsequently, by applying the SRSS rule for spatial combination.

Having determined the expected extreme values of the interstorey drifts, the damage limitation verification check of clause §4.4.3.2 of EC8 can be performed by comparing the ratios of Eq. (3.23) reported in Table 4.101 with pertinent limiting values. Assuming that building B has brittle non-structural infill walls, the maximum allowed interstorey drift according to clause 4.4.3.2(1) of EC8 is equal to 0.5 % of the storey height, i.e., (v· extr$d_{rX}^{(k)}$)/h <0.005 and (v· extr$d_{rY}^{(k)}$)/h <0.005 for all k storeys. The above condition is met at all storeys of building B.

Table 4.95 Modal ordinates [values*100] at the upper end (top) of vertical members for the 2nd mode shape of building B (mass position 3 in Table 4.74)

	1st storey		2nd storey		3rd storey		4th storey		5th storey	
Vertical member	u^t_{X2j}	u^t_{Y2j}	u^t_{X2j}	u^t_{Y2j}	u^t_{X2j}	u^t_{Y2j}	u^t_{X2j}	u^t_{Y2j}	u^t_{X2j}	u^t_{Y2j}
C1	1.48	−1.71	2.83	−3.21	3.93	−4.39	4.70	−5.18	5.12	−5.53
C2	1.48	−0.12	2.83	−0.15	3.93	−0.12	4.70	−0.05	5.12	0.06
C3	1.48	0.87	2.83	1.75	3.93	2.55	4.70	3.16	5.12	3.55
C4	1.48	3.04	2.83	5.95	3.93	8.42	4.70	10.21	5.12	11.24
C5	−0.89	−1.71	−1.75	−3.21	−2.48	−4.39	−2.99	−5.18	−3.27	−5.53
C6	−0.89	−0.12	−1.75	−0.15	−2.48	−0.12	−2.99	−0.05	−3.27	0.06
C7	−0.89	0.87	−1.75	1.75	−2.48	2.55	−2.99	3.16	−3.27	3.55
C8	−0.89	3.04	−1.75	5.95	−2.48	8.42	−2.99	10.21	−3.27	11.24
W1	0.30	−0.12	0.54	−0.15	0.73	−0.12	0.86	−0.05	0.92	0.06
W2	0.30	0.87	0.54	1.75	0.73	2.55	0.86	3.16	0.92	3.55
Average values	0.30	0.49	0.54	1.03	0.73	1.53	0.86	1.94	0.92	2.22

Table 4.96 Modal mean interstorey drifts [values*100] for the 2nd mode shape of building B (mass position 3 in Table 4.74)

Storey k	u^t_{Xm2}	u^b_{Xm2}	$d^{(k)}_{eX2} = u^t_{Xm2} - u^b_{Xm2}$	$d^{(k)}_{rX2} = q \cdot d^{(k)}_{eX2}$	u^t_{Ym2}	u^b_{Ym2}	$d^{(k)}_{eY2} = u^t_{Ym2} - u^b_{Ym2}$	$d^{(k)}_{rY2} = q \cdot d^{(k)}_{eY2}$
1	0.30	0.00	0.30	0.59	0.49	0.00	0.49	0.98
2	0.54	0.30	0.24	0.48	1.03	0.49	0.54	1.08
3	0.73	0.54	0.19	0.38	1.53	1.03	0.50	1.01
4	0.86	0.73	0.13	0.26	1.94	1.53	0.40	0.81
5	0.92	0.86	0.07	0.14	2.22	1.94	0.29	0.57

Table 4.97 Modal mean interstorey drifts [values*100] of building B (mass position 3 in Table 4.74)

	1st storey		2nd storey		3rd storey		4th storey		5th storey	
Mode i	$d_{rXi}^{(1)}$	$d_{rYi}^{(1)}$	$d_{rXi}^{(2)}$	$d_{rYi}^{(2)}$	$d_{rXi}^{(3)}$	$d_{rYi}^{(3)}$	$d_{rXi}^{(4)}$	$d_{rYi}^{(4)}$	$d_{rXi}^{(5)}$	$d_{rYi}^{(5)}$
1	3.85	−0.03	3.12	−0.03	2.48	−0.03	1.73	−0.02	0.90	−0.01
2	0.59	0.98	0.48	1.08	0.38	1.01	0.26	0.81	0.14	0.57
3	−0.21	2.56	−0.17	3.01	−0.13	2.95	−0.09	2.47	−0.04	1.87
4	9.58	−0.08	1.91	−0.03	−6.40	0.05	−9.95	0.09	−7.27	0.07
5	1.54	2.16	0.24	0.83	−1.05	−1.16	−1.48	−2.45	−1.02	−2.34
6	−11.33	0.14	9.28	−0.04	13.35	−0.13	−6.12	−0.03	−16.49	0.05
7	0.35	−7.93	0.10	−3.47	−0.21	3.98	−0.40	9.67	−0.32	10.37

*The values in Tables 4.95, 4.96, and 4.97 are obtained by mode shape normalization so that the generalized mass has unit value

Table 4.98 Seismic modal interstorey drifts [in cm] for seismic excitation along the X direction of building B (mass position 3 in Table 4.74)

	1st storey		2nd storey		3rd storey		4th storey		5th storey	
Mode i	$d_{rXi,EX}^{(1)}$	$d_{rYi,EX}^{(1)}$	$d_{rXi,EX}^{(2)}$	$d_{rYi,EX}^{(2)}$	$d_{rXi,EX}^{(3)}$	$d_{rYi,EX}^{(3)}$	$d_{rXi,EX}^{(4)}$	$d_{rYi,EX}^{(4)}$	$d_{rXi,EX}^{(5)}$	$d_{rYi,EX}^{(5)}$
1	2.37	−0.02	1.93	−0.02	1.53	−0.02	1.07	−0.01	0.56	−0.01
2	0.02	0.04	0.02	0.04	0.01	0.04	0.01	0.03	0.01	0.02
3	0.00	−0.02	0.00	−0.02	0.00	−0.02	0.00	−0.02	0.00	−0.01
4	0.30	0.00	0.06	0.00	−0.20	0.00	−0.31	0.00	−0.23	0.00
5	0.00	0.00	0.00	0.00	0.00	0.00	0.00	0.00	0.00	0.00
6	0.06	0.00	−0.05	0.00	−0.07	0.00	0.03	0.00	0.08	0.00
7	0.00	0.00	0.00	0.00	0.00	0.00	0.00	0.00	0.00	0.00

Table 4.99 Seismic modal interstorey drifts [in cm] for seismic excitation along the Y direction of building B (mass position 3 in Table 4.74)

	1st storey		2nd storey		3rd storey		4th storey		5th storey	
Mode i	$d_{rXi,EY}^{(1)}$	$d_{rYi,EY}^{(1)}$	$d_{rXi,EY}^{(2)}$	$d_{rYi,EY}^{(2)}$	$d_{rXi,EY}^{(3)}$	$d_{rYi,EY}^{(3)}$	$d_{rXi,EY}^{(4)}$	$d_{rYi,EY}^{(4)}$	$d_{rXi,EY}^{(5)}$	$d_{rYi,EY}^{(5)}$
1	−0.05	0.00	−0.04	0.00	−0.03	0.00	−0.02	0.00	−0.01	0.00
2	0.11	0.19	0.09	0.21	0.07	0.19	0.05	0.16	0.03	0.11
3	−0.06	0.75	−0.05	0.89	−0.04	0.87	−0.03	0.73	−0.01	0.55
4	−0.01	0.00	0.00	0.00	0.00	0.00	0.01	0.00	0.00	0.00
5	0.01	0.02	0.00	0.01	−0.01	−0.01	−0.01	−0.02	−0.01	−0.02
6	0.00	0.00	0.00	0.00	0.00	0.00	0.00	0.00	0.00	0.00
7	0.00	0.08	0.00	0.03	0.00	−0.04	0.00	−0.10	0.00	−0.10

Table 4.100 Expected (most probable) extreme values of the relative floor displacements [in cm] of building B (mass position 3 in Table 4.74)

Storey	$\text{extrd}_{rX}^{(k)}$	$\text{extrd}_{r}^{(k)}$
1	2.40	0.81
2	1.93	0.95
3	1.54	0.92
4	1.11	0.78
5	0.60	0.59

Table 4.101 Expected extreme values of ratios ($v \cdot d_{rX}$)/h and ($v \cdot d_{rY}$)/h of Eq. (3.23) for building B (mass position 3 in Table 4.74)

Storey k	$\text{extrd}_{rX}^{(k)}$ [cm]	$\text{extrd}_{r}^{(k)}$ [cm]	v	h_k [cm]	$\frac{v \cdot \text{extrd}_{rX}^{(k)}}{h_k}$	$\frac{v \cdot \text{extrd}_{r}^{(k)}}{h_k}$
1	2.40	0.81	0.5	400	0.0030	0.0010
2	1.93	0.95	0.5	300	0.0032	0.0016
3	1.54	0.92	0.5	300	0.0026	0.0015
4	1.11	0.78	0.5	300	0.0019	0.0013
5	0.60	0.59	0.5	300	0.0010	0.0010

4.3 Example C: Four-Storey Building with Central R/C Core and a Basement on Compliant Supporting Ground

4.3.1 Geometric, Material, and Seismic Action Data

Units

Length: m; Force: kN; Mass t (1 t = 10^3 kg), Time: s.

Material properties for reinforced concrete

Modulus of Elasticity: $E = 2.9 \cdot 10^7$ kN/m^2; Poisson ratio: $v = 0.2$; weight per unit volume: $\gamma = 25$ kN/m^3.

Description and geometry of building C superstructure and foundation

The example building C has four storeys above the ground and one underground storey (basement). All storeys have the same plan shown in Fig. 4.9. The building has at least one planar r/c wall at every side of its perimeter and a single-cell U-shaped r/c core at its interior enclosing a staircase and an elevator. The walls have locally increased out-of-plane dimensions at the joints with the transverse beams to allow for sufficient anchorage length for the longitudinal reinforcement of beams. All vertical structural members have a constant sectional geometry and are continuous from the foundation to the full height of the building, with the exception

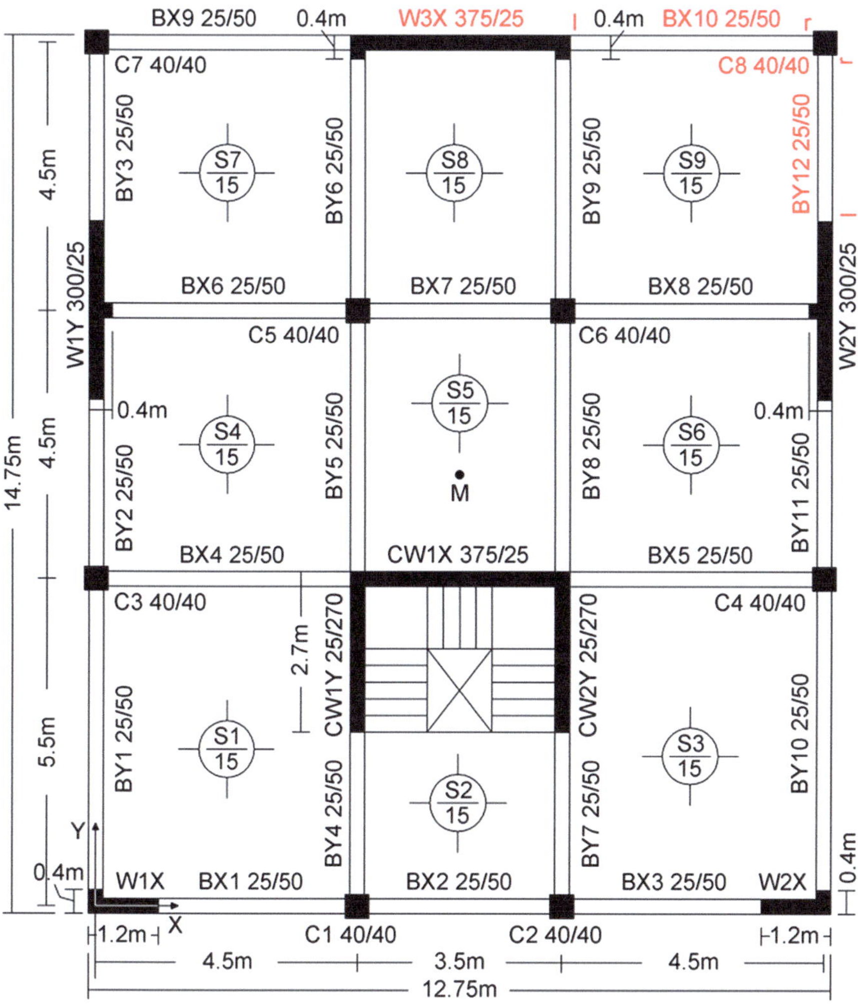

Fig. 4.9 Typical floor plan of building C (structural members for which analysis results are reported in detail are marked in red)

of the core that extends 3 m higher from the building roof to allow for terrace access through the staircase and to house the mechanical equipment of the elevator. The basement has a 0.25 m thick concrete wall around its perimeter resting on strip footings forming a box foundation.

The building foundation comprises (Fig. 4.10)

- three rectangular pad footings, F1, F2, and F3, which support the r/c core (Fig. 4.11c) and the internal columns C5 and C6 (Fig. 4.11b);
- strip footings of 1.40 m width and 0.85 m height (see also Fig. 4.11a) along the perimeter of the building supporting the basement walls; and

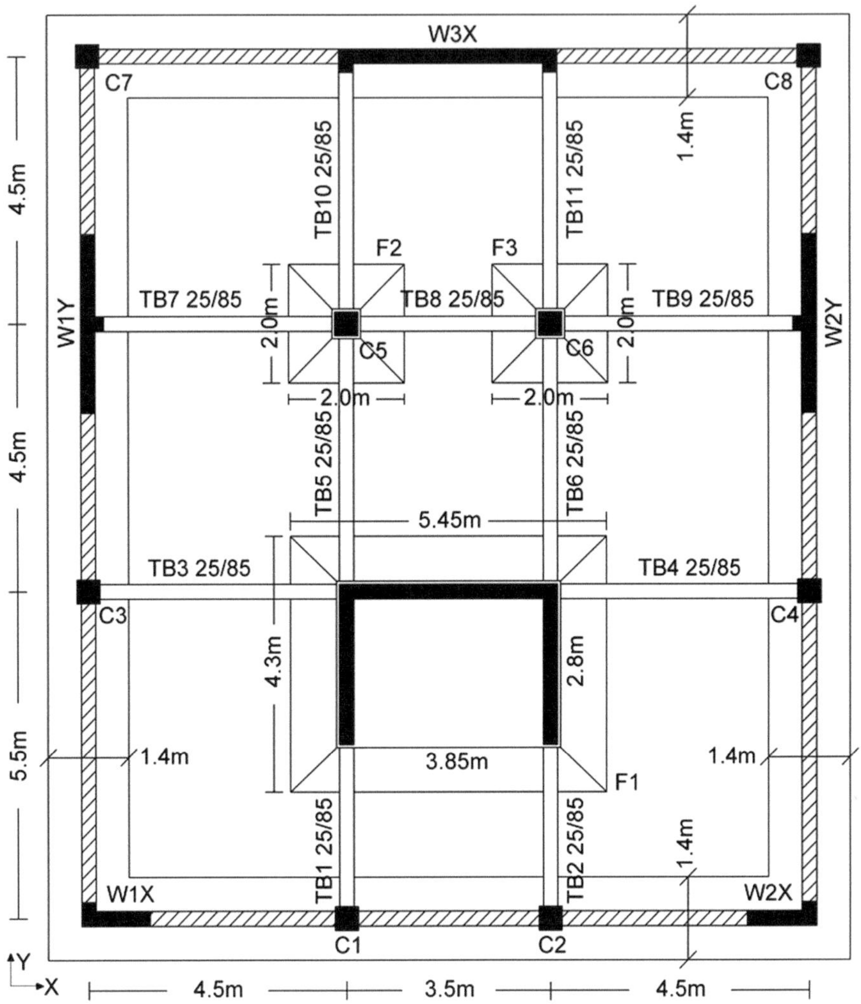

Fig. 4.10 Plan view of building C at the foundation level

- two-way foundation tie beams 0.25 m × 0.85 m, connecting the interior pad footings with each other and with the perimetric strip footings forming a grid.

The foundation slab (basement floor) is 0.15 m thick and lies 0.85 m above the foundation level (Fig. 4.11a).

Storey heights

Basement: 3 m; Ground storey: 4.5 m; 1st to 3rd storeys: 3 m

Gravity loads imposed on beams and slabs

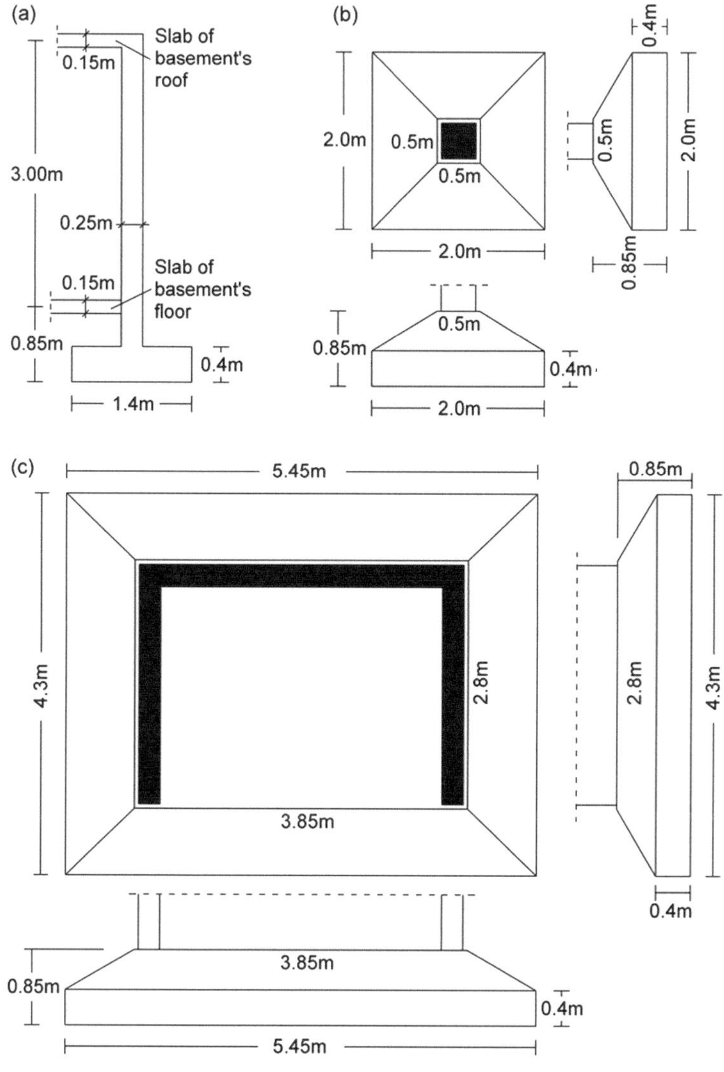

Fig. 4.11 Sectional geometry of foundation members: (**a**) strip footing of basement wall; (**b**) pad footings F2 and F3 of internal columns C5 and C6, respectively: (**c**) pad footing F1 of r/c core

– Double-layered masonry walls occupy the full storey height at all storeys along the perimeter of the building. These infill walls impose 3.6 kN/m^2 of "permanent" weight on all exterior beams except the beams of the top storey. The exterior beams of the top storey accommodate a "permanent" uniform distributed load of 3.6 kN/m, corresponding to a double-layered masonry roof parapet 1 m in height.
– Single-layered masonry walls occupy the full height of all overground storeys along the full length of all interior beams. These infill walls impose 2.1 kN/m^2 of

"permanent" weight on all interior beams except the beams of the top storey and the basement.

- Permanent floor finishings of 1.3 kN/m^2 weight evenly distributed in plan is assumed for all floor slabs and the staircase.
- The assumed "live" gravity loads (variable action) is taken as evenly distributed in plan with values Q = 2 kN/m^2 for the floor slabs and Q = 3.5 kN/m^2 for the staircase.

Directions of the horizontal seismic action (clause §4.3.3.1(11)P of EC8)

Axes X and Y shown in Fig. 4.9 can be unambiguously identified as the "principal" orthogonal axes along which the input seismic action is assumed to act for design purposes.

Design spectrum data (clause §3.2.2.5 of EC8)

- Peak ground acceleration:$a_{gR} = 0.20$ g
- Ground type: B ($S = 1.2$, $T_B = 0.15$ s, $T_C = 0.5$ s)
- Importance category: II (residential building) → $\gamma_I = 1$
- Damping coefficient: $\zeta = 5$ %
- Behaviour factor q: to be determined in Sect. 4.3.6 below

Supporting ground conditions

- Medium density sand
- Coefficient of subgrade reaction: $K_s = 90{,}000$ kN/m^3 (Terzhaghi 1955)
- Spring coefficients modeling soil compliance: to be determined in Sect. 4.3.2.2 below

4.3.2 Modeling Assumptions

4.3.2.1 Structural Modeling Assumptions

A spatial (three-dimensional) finite element (FE) model is considered which accounts for flexural, shear, axial, and torsional deformations of r/c structural members. The infill walls are not included in the FE model, assuming their influence on the lateral stiffness and strength of the building structure to be negligible (see clause §4.3.1(8) of EC8). The beam-column joints are modeled as perfectly rigid (see clause §4.3.1(2) of EC8) using rigid offsets (arms) at the end of FE members, as shown in Fig. 2.36a (see Sect. 2.3.3.2).

Modeling of floor slabs

Floor slabs are assumed to act as perfectly rigid diaphragms in their plane (see Sect. 2.3.3.1). The actual height level of these diaphragms is defined in the considered FE model as shown in Fig. 2.37a.

Table 4.102 Assumed effective width of the beams of the FE model for building C

Beam	Length (m)	b_w [m]	Bay	Shape	b_{eff} [m]
BX1, BX3	3.225	0.250	Internal	Γ	0.476
BX2	3.1	0.250	Internal	Γ	0.467
BX4, BX5	4.175	0.250	External	T	0.960
BX6, BX8	4.025	0.250	External	T	0.934
BX7	3.1	0.250	Internal	T	0.684
BX9, BX10	4.175	0.250	External	Γ	0.605
BY1, BY10	5.025	0.250	External	Γ	0.677
BY2, BY11	2.8	0.250	Internal	Γ	0.446
BY3, BY12	2.8	0.250	External	Γ	0.488
BY4, BY7	2.725	0.250	External	T	0.713
BY5,BY8	4.175	0.250	Internal	T	0.835
BY6, BY9	4.025	0.250	External	T	0.934

Effective flange width of beams

The effective width b_{eff} of the upper flange of the beams included in the FE model is given in Table 4.102. It is computed according to clause §5.3.2.1 of EC2 as

- $b_{eff} = b_w + 0.2 \cdot l_o$, for T-shaped beams, and
- $b_{eff} = b_w + 0.1 \cdot l_o$, for L-shaped beams,

where b_w is the width of the beam web, $l_0 = 0.85 \cdot L$ for beams occupying external frame bays, and $l_0 = 0.70 \cdot L$ for beams occupying internal frame bays in which L is the length of the beam excluding its rigid offsets. The left end of beam BX1 and the right end of beam BX3 are assumed to be clamped to the r/c walls W1X and W2X, respectively. Therefore, these beams are taken as internal and their effective width is computed by assuming $l_0 = 0.70 \cdot L$.

Effective rigidities of structural members (see §2.3.2.1)

The flexural rigidity (EI) and the shear rigidity (GA_s) are assumed equal to 50 % of the values corresponding to uncracked gross section properties for all r/c members (clause §4.3.1(7) of EC8). The torsional rigidity (GJ) is taken to be equal to 10 % of the value corresponding to uncracked gross section properties for all r/c members, as discussed in Sect. 2.3.2.1 (see also Fardis 2009). However, the axial rigidity (EA) of structural members is not reduced compared to the value corresponding to uncracked gross section properties given that the vertical structural members are under compression due to the gravity loads and all beam members are considered to be part of the perfectly rigid diaphragms within the plane of the floor slabs. It is noted that clause §5.4(2) of EC2 allows for considering the uncracked gross section properties to compute the rigidity of r/c structural members under gravity loads. Nevertheless, given that the reduction is generally considered towards safety, it also preserved for the vertical loads, exactly as in the analyses for the seismic loads. Therefore, in all analyses, static and seismic, the same model of the structure is employed.

Table 4.103 Section properties of the equivalent columns for the modeling of the planar r/c walls of building C

Wall member	A	I_{XX}	I_{YY}	I_T	A_{SX}	A_{SY}
W1X, W2X	0.3	0.036	0.001563	0.00543	0.25	0.25
W3X	0.9375	1.0986	0.004883	0.0187	0.7813	0.7813
W1Y, W2Y	0.75	0.5625	0.003906	0.0148	0.625	0.625

Fig. 4.12 Modeling of the central r/c core of building C using an equivalent frame model

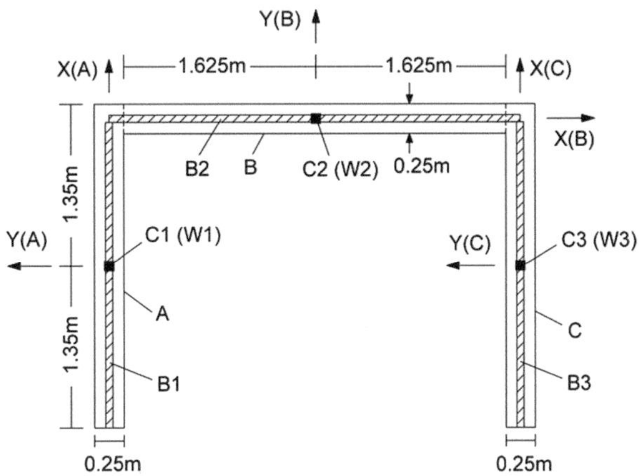

Modeling of planar r/c walls

The r/c walls of building C are modeled by means of an equivalent frame model, as discussed in Sect. 2.3.3.3 and shown in Fig. 2.38d. The model comprises an **equivalent column positioned at the center of gravity of the actual shear wall,** which is connected to the beams at each floor level by means of "virtual" perfectly rigid arms of length equal to half the sectional length of each wall. The uncracked gross section properties of the equivalent columns for all walls are listed in Table 4.103. Note that, for simplification purposes, the local out-of-plane extensions of walls to ensure adequate anchorage of the longitudinal rebar of transverse beams are ignored.

The effective rigidities of the equivalent column computed from the above properties are reduced to account for concrete cracking according to the previously mentioned assumptions.

Modeling of r/c core of building C

The U-shaped r/c core of building C is modeled via an equivalent frame model, as detailed in Sect. 2.3.3.4 (Fig. 2.44). Specifically, the model comprises three equivalent columns positioned at the center of gravity of the web and the flanges of the U-shaped core and horizontal rigid arms (beam FE members) connecting these columns with each other and with the actual beams of the building supported to the core at each floor slab level, as shown in Fig. 4.12. The section properties of the three equivalent column members used to model the core are derived as indicated in

Table 4.104 Section properties of the equivalent columns for the modeling of the r/c core of building C

Equivalent column	A	I_{XX}	I_{YY}	I_T	A_{SX}	A_{SY}
C1(W1)	0.675	0.003516	0.4101	0.0132	0.675	0.0
C2(W2)	0.8125	0.004232	0.7152	0.0161	0.8125	0.0
C3(W3)	0.675	0.003516	0.4101	0.0132	0.675	0.0

Table 4.105 Torsional rigidity of rigid arms used in the equivalent frame model of the r/c core of building C

Storey	$I_{T(B1)}$	$I_{T(B2)}$	$I_{T(B3)}$
extension above the roof	0.007	0.007	0.007
2nd, 3rd and 4th	0.0148	0.0148	0.0148
Ground	0.0187	0.0187	0.0187
Basement	0.0265	0.0265	0.0265

Fig. 2.44 and are given in Table 4.104. Further, the rigid arms are taken as perfectly rigid (i.e., non-deformable) in bending and shearing but with finite torsional rigidity to capture properly the torsional deformation of the cross-section of the core, as discussed in detail in Sect. 2.3.3.4. The finite values of the torsional rigidities of the beams depend on the storey height, as indicated in Fig. 2.44, and are given for each floor in Table 4.105.

Finally, it is mentioned that all the geometrical properties of the items with which the core is simulated are reduced with the same ratios that the geometric properties of the remaining structural elements are reduced.

Modeling of pad foundations and foundation tie beams resting on compliant soil

The FE modeling of the rectangular pad foundations F2 and F3 involves the consideration of a node positioned at the center of gravity of the pad at the foundation level (i.e., 3.85 m below the ground level). This node is connected via a perfectly rigid arm (i.e., one-dimensional FE beam member) to the base of the column (bottom FE member end located 3 m below the ground level) supported by the pad footing, as shown in Fig. 4.13. In this manner, the footing itself is modeled as being perfectly rigid. This is a reasonable assumption given the significantly larger size of the footing compared to the sectional dimensions of the structural members converging to it, namely the columns C5 and C6 and the foundation tie beams. For the special case of the pad footing F1 of the core, the node at the center of gravity of the footing is connected to all three bases of the equivalent columns (C1, C2, and C3 in Fig. 4.12) via three different rigid arms.

Further, the compliance of the ground supporting the pad foundations is modeled via a vertical translational linear spring, k_3, and two rotational linear springs about the horizontal principal axes X and Y, k_{r1} and k_{r2}, respectively, placed at the nodes located at the center of gravity of each pad footing at the foundation level, as shown in Fig. 4.13. The adopted method for determining the stiffness of these springs (spring coefficients) is detailed in the following Sect. 4.3.2.2.

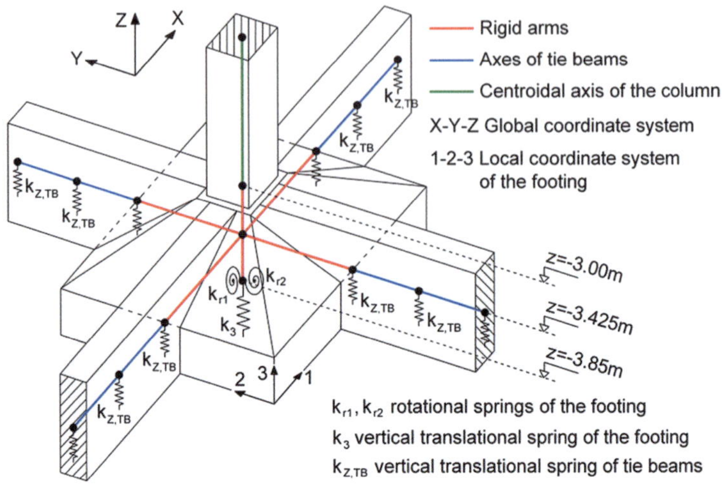

Fig. 4.13 Three-dimensional drawing of pad foundation and tie beams resting on compliant soil

Lastly, the 0.25 × 0.85 (m) foundation tie beams are modeled via standard linear one-dimensional beam finite elements placed in a location 3.425 m below the ground level, that is, at the place where the center of gravity of the tie beam section lies: ((-3.85) + (-3))/2 = -3.425 m (Fig. 4.13). Each tie beam is discretized by several beam finite elements along the longitudinal direction. Vertical translational linear springs $k_{z,TB}$ are considered at the nodes of the above FE grid to model the compliance of the supporting ground, as shown in Fig. 4.13. The adopted method for determining the stiffness of these springs is detailed in the following Sect. 4.3.2.2. The connection of the tie beams to the pad footings is accomplished by perfectly rigid arms, as shown in Fig. 4.13.

Modeling of the basement wall supported by strip foundations on compliant soil

The wall on the perimeter of the underground storey (basement) depicted in Fig. 4.14 is modeled via two-dimensional shell finite elements, as shown in Fig. 4.15. The shell elements have the same thickness as the wall (0.25 m) and in-plane dimensions of approximately 1 m × 1 m extending from the basement floor slab (3 m below the ground level) to the basement roof at the ground level.

The strip foundation of the basement perimetric wall is modeled via one-dimensional beam finite elements with inverted T-shaped cross-section of height 0.85 m located at 3.58 m below the ground level, that is, at the center of gravity of the cross-section of the strip foundation. Each segment of the strip foundation is discretized by several beam elements along the longitudinal direction of the strip (see also Fig. 4.19 in the following Sect. 4.3.2.2). The length of each beam finite element is equal to the side dimension of the shell elements modeling the supported wall. Perfectly rigid arms are used to connect these beam finite elements with the shell elements, as shown in Fig. 4.15. Further, the compliance of the supporting ground underneath the strip foundations are modelled via vertical

Fig. 4.14 Three-dimensional drawing of the perimetric basement wall

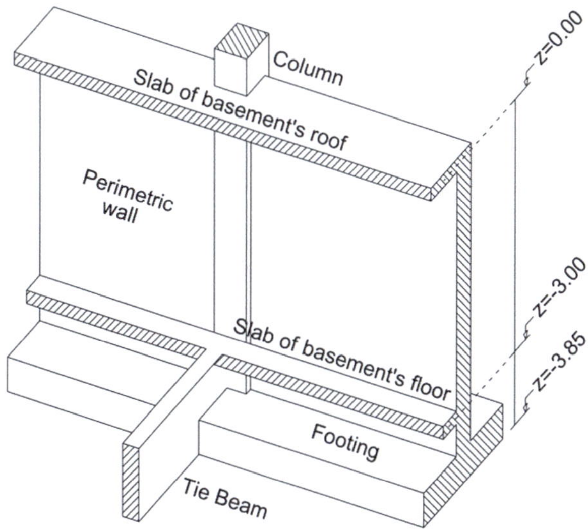

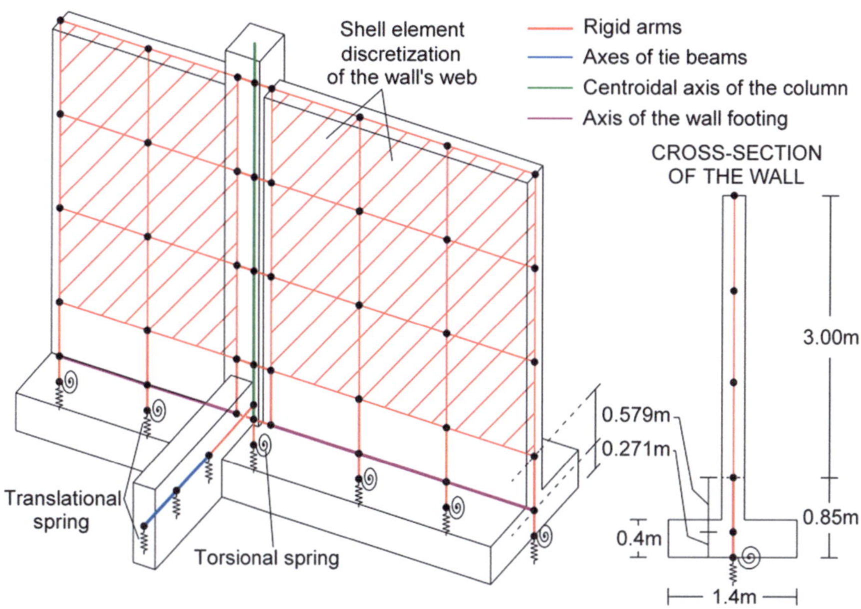

Fig. 4.15 FE Modeling of the perimetric basement wall resting on compliant soil

translational linear springs and rotational linear springs. These springs are placed at the foundation level (3.85 m below the ground) and are linked to the beam finite elements of the strip foundation, which lie at 3.58 m below the ground, via vertically oriented rigid arms of length 3.85-358 = 0.27 m, as shown in Fig. 4.15. A suitable method for determining the stiffness of the aforementioned springs is

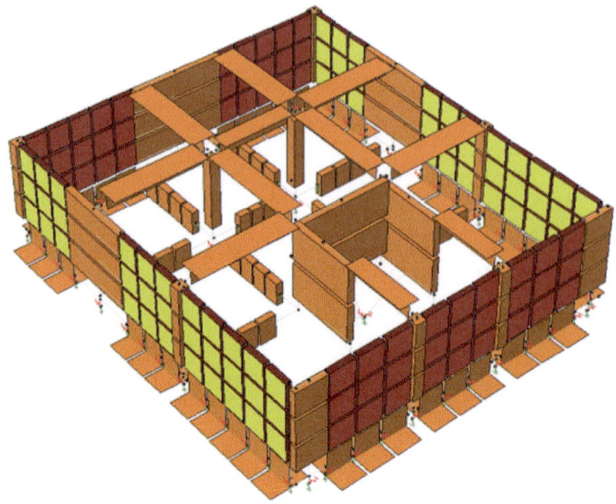

Fig. 4.16 Extruded three-dimensional view of the FE model of the underground storey (basement)

detailed in the following Sect. 4.3.2.2. The connection of the tie beams to the pad footings is accomplished by perfectly rigid arms, as shown in Fig. 4.13.

Lastly, it is noted that the basement floor and rood slabs are assumed to act as a perfectly rigid diaphragm in their plane similarly as the floor slabs of the superstructure.

For visualization purposes, Figs. 4.16 and 4.17 display three-dimensional extruded views of the FE model adopted in the seismic analysis of building C for the foundation and basement (underground storey) and the full structure, respectively.

As a final remark, it is noted that the consideration of shell finite elements to model parts of structures for seismic design requires using a structural analysis/design software package that offers modeling capabilities beyond one-dimensional beam finite elements and a user/design engineer experienced in the finite element method. Since the above requirements may not always be satisfied in practice, equivalent frame models, such as the ones used to model the planar walls and the core of building C, can be considered to bypass the need for using shell elements in the superstructure of buildings. Such models offer adequate accuracy for the seismic design of ordinary r/c building structures. However, in this particular example, shell elements were preferred to model the basement walls, as equivalent frame model adequate to capture the contribution of basement walls realistically (box-type foundation) is significantly more complex compared to the modeling of planar walls and simple cores.

4.3.2.2 Supporting Ground Modeling Using Linear Springs (Compliant Soil)

In this section, the values of the linear translational and rotational spring (stiffness) coefficients used to account for soil compliance of the supporting ground for pad

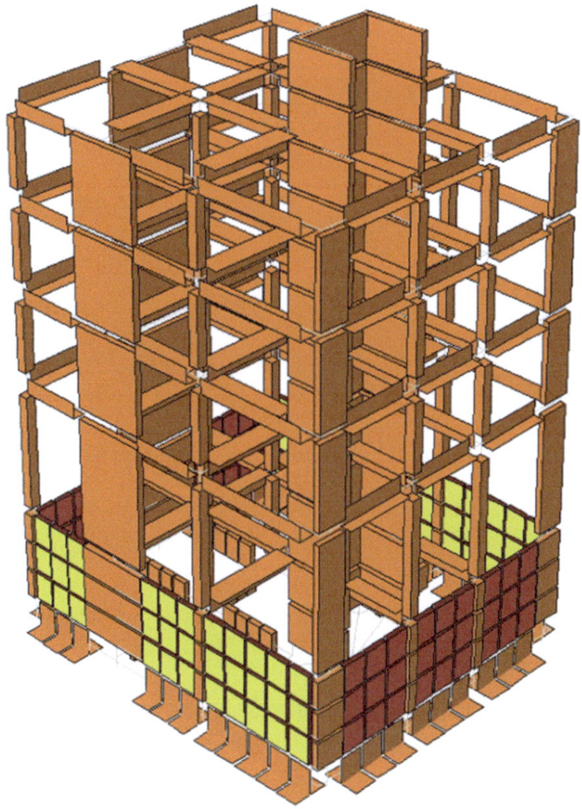

Fig. 4.17 Extruded three-dimensional view of the FE model of building C

footings, foundation tie beams, and strip footings (see Figs. 4.13 and 4.15) are determined using a standard procedure of geotechnical engineering. The first step of the procedure is to determine the value of the so-called coefficient or modulus of subgrade reaction K_s (in kN/m^3). Next, the spring stiffness coefficients are derived from standard formulae accounting for the geometry of the foundation elements.

The coefficient K_s is a phenomenological empirical soil property. It can be derived by first adopting a K_{s1} value of the coefficient derived from field tests on different types of soil loaded by a rectangular plate of 1.0 ft × 1.0 ft size. According to previous studies (Terzhaghi 1955), K_{s1} ranges from 19,200 kN/m^3 to 96,200 kN/m^3 for medium density sands (ground type B of EC8). Herein, the value of $K_{s1} = 90{,}000$ kN/m^3 is adopted. The above value needs to be corrected twice following certain empirical formulae to account for (i) the different characteristic size of each considered foundation from the assumed 1 ft adopted in deriving K_{s1}, (size correction) and (ii) the deviation from the rectangular shape of each considered foundation (shape correction). Specifically, the characteristic size, b_c, of a rectangular footing with dimensions L_X and L_Y is defined to be equal to the smallest dimension, that is, $b_c = \min\{L_X, L_Y\}$. Then, according to Terzhaghi (1955), the size correction to the K_{s1} value is accomplished for sandy soils using the expression

Table 4.106 Coefficients of subgrade reaction for pad footings of building C

Footing	L_X [m]	L_Y [m]	b_c [m]	K_{s2} [kN/m^3]	m [-]	K_s [kN/m^3]
F1	5.45	4.30	4.30	25,749.05	1.267	23,938.0
F2, F3	2.00	2.00	2.00	29,756.25	1.00	29,756.2

Table 4.107 Spring coefficients of pad footings of building C

Footing	k_3 [kN/m]	k_{r1} [kNm/rad]	k_{r2} [kNm/rad]
F1	560,986	864,386	1,388,557
F2, F3	119,025	39,675	39,675

$$K_{s2} = K_{s1}\left(\frac{b_c + 0.3}{2b_c}\right)^2. \tag{4.1}$$

Next, the above value is further corrected using the formula (shape correction)

$$K_s = K_{s2}\left(\frac{m + 0.3}{1.5m}\right), \tag{4.2}$$

which holds for any soil type. In the last equation, m is the aspect ratio of the foundation given as max{L_X, L_Y}/min{L_X, L_Y}.

In the remainder of this section, the above corrections and the calculation of the corresponding spring coefficients is undertaken separately for each different type of footing of building C, namely pad, strip, and tie beams.

(a) **Coefficient of subgrade reaction and spring coefficients for pad footings**

Table 4.106 details all required calculations for deriving the coefficient of subgrade reaction K_s for the pad footings of the core (F1) and of the two internal columns (F2 and F3) of building C by application of Eqs. (4.1) and (4.2).

Consider a rectangular pad footing with the local coordinate system shown in Fig. 4.18. The stiffness of the translational spring, k_3, oriented along local axis "3" and located at the center of gravity of the footing, is given by the expression

$$k_3 = K_s L_1 L_2. \tag{4.3}$$

Further, the stiffness of the rotational springs, k_{r1} and k_{r2}, about local axis "1" and "2", respectively, located at the center of gravity of the footing, is given by the expressions

$$k_{r1} = \frac{K_s L_1 L_2^3}{12} \tag{4.4}$$

and

$$k_{r2} = \frac{K_s L_1^3 L_2}{12}. \tag{4.5}$$

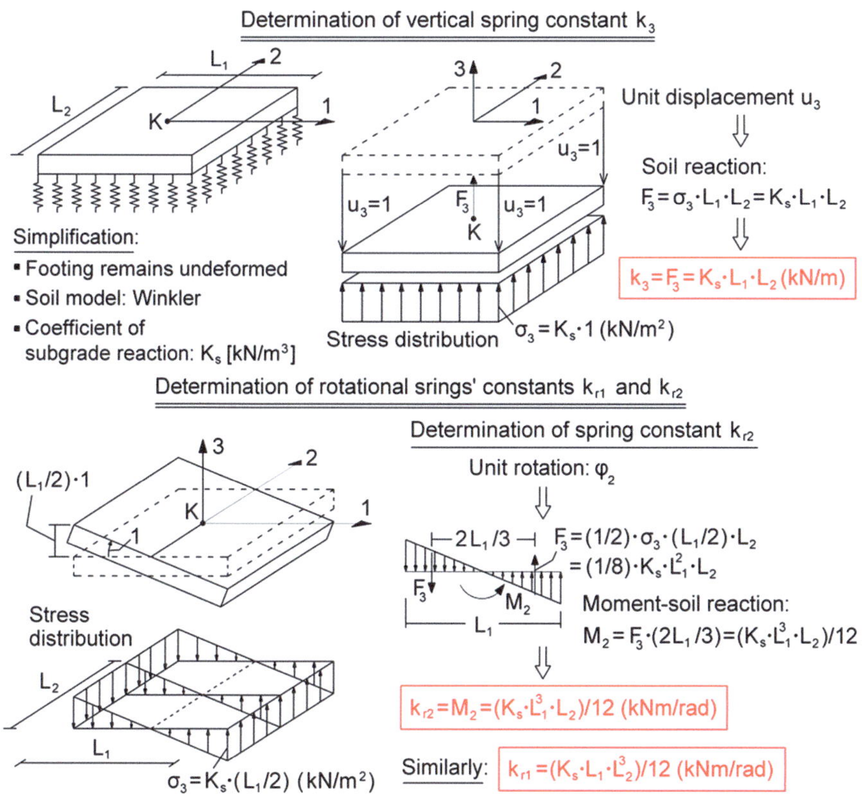

Fig. 4.18 Derivation of spring coefficients for pad footings resting on compliant soil

The rationale of Eqs. (4.3), (4.4), and (4.5) is explained in Fig. 4.18. Table 4.107 reports the values of the spring coefficients for all pad footings of building C obtained by application of the last three equations.

(b) **Coefficient of subgrade reaction and spring coefficients for strip footings**

The basement walls in the perimeter of building C are supported by strip foundations of width b = 1.4 m, which are modelled via beam elements resting on linear translational and rotational springs (Fig. 4.15). The stiffness of these spring coefficients are determined following a similar procedure as in the case of the pad footing springs. In particular, the coefficient of subgrade reaction is computed first by assuming two different strip footings along each principal direction X and Y. Along X, the dimensions of the two perimetric strip footings are taken as 13.9 m × 1.4 m, and along Y, the dimensions of the two perimetric strip footings are taken as 13.1 m × 1.4 m. The length of the strip footings along each direction is computed such that there is no overlapping of tributary areas at the corners of the building. That is, the longitudinal dimension along the X direction is computed as 1.4/2+ 4.5+ 3.5+ 4.5+ 1.4/2 = 13.9 m, and the longitudinal dimension along the Y direction is computed as (5.5 − 1.4/2) + 4.5+ (4.5 − 1.4/2) = 13.91 m (see Fig. 4.10).

Table 4.108 Coefficients of subgrade reaction for strip footings of building C

Longitudinal direction of strips	L [m]	b [m]	K_{s2} [kN/m^3]	m = L/b [−]	K_s [kN/m^3]
X	13.9	1.4	33,176.0	9.93	23,231.2
Y	13.1	1.4	33,176.0	9.36	23,300.0

The coefficients of subgrade reaction are computed for the above footings, as detailed in Table 4.108, using Eqs. (4.1) and (4.2). It is seen that the resulting corrected K_s values for the two strips are very similar. Thus, for simplification purposes, a single K_s value is assumed for all strip footings equal to 23,265.6 kN/m^3, that is, the average of the two K_s values found in Table 4.108.

Upon determination of K_s, the values of the vertical translational and rotational spring coefficients modeling the soil compliance underneath the strip footings are computed by means of Eqs. (4.3) and (4.4), respectively, as in the case of the pad footings. That is, $k_3 = K_s \cdot L_{eff} \cdot b = 23{,}265.5 \cdot 1.4 \cdot L_{eff} = 32{,}571.25 \cdot L_{eff}$ and $k_r = K_s \cdot b^3 \cdot L_{eff}/12 = 23{,}265.2 \cdot 1.4^3 \cdot L_{eff}/12 = 5320 \cdot L_{eff}$, where L_{eff} is the total tributary length contributed by each of the beam finite elements modeling the strip footings linked at each joint of the FE strip footing grid where the k_3 and k_r springs are assigned. This effective (tributary) length is taken as being equal to the sum of half of the lengths of the beams linked at each joint. Therefore, it depends on the number of the beam finite elements used in modeling the strip footings. Figure 4.19 shows the in-plan view of the FE grid used to model the strip footing and includes the length of all beam finite elements used. Based on these lengths, the L_{eff} corresponding to each joint of the grid and the coefficients of the springs assigned to each joint are collected in Table 4.109.

(c) **Coefficient of subgrade reaction and spring coefficients for tie beams**

Each 0.25 × 0.85 (m) foundation tie beam, tying the pad and strip footings of building C in a two-way grillage, as shown in Fig. 4.10, is modelled by two or more linear one-dimensional beam elements of equal length resting on vertical translational springs to account for soil compliance (see also Fig. 4.13). Therefore, these springs, placed at the joints of the FE grid, are equally spaced with distances dependent on the total length of each tie beam, L, and on the number of beam finite elements, n, used in the FE discretization. Table 4.110 reports the length L, aspect ratio m (used in Eq. (4.2)), and number of elements n corresponding to each tie beam. Based on this data, translational spring coefficients at the outer and the inner joints of tie beams are computed in Table 4.110 using Eqs. (4.1) (4.2) (4.3). Note that the outer joints are the common nodes of tie bars with rigid arm elements modeling pad footings (see Fig. 4.13) and strip footings (see Fig. 4.15).

4.3.2.3 Vertical Load Modeling Assumptions

- The permanent (self weight and finishings) and variable evenly distributed area loads carried by the slabs are transferred to the beams using triangular and trapezoidal tributary areas (rule of 45° or 60°). In this manner, they are

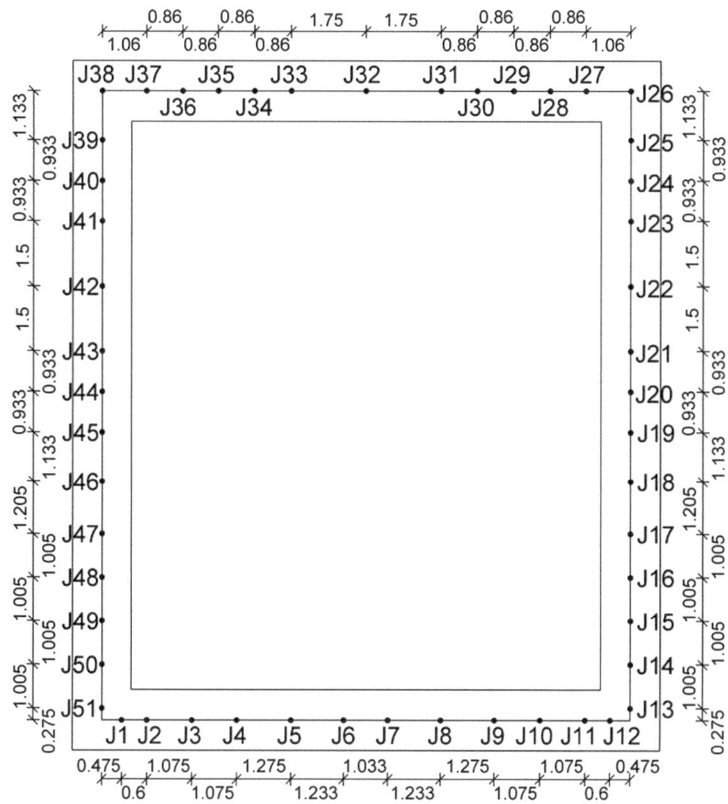

Fig. 4.19 FE discretization grid of the strip foundation of the basement walls

distributed along the length of the beams following triangular or trapezoidal distributions.

- The masonry infill self weight (permanent action) is transferred directly onto the beams and is computed without accounting for architectural openings to the infill walls.
- The self weight of the r/c beams and the infill walls carried by the beams is considered to be uniformly distributed along the length of the beams.
- The self weight of the r/c columns is modeled as a uniformly distributed axial load along the height of the columns.
- The permanent (self weight and finishings) and variable evenly distributed area loads carried by the staircase is transferred to the load-resisting structural system as gravitational point forces acting at mid-storey height to the three equivalent columns, C1, C2, C3, used in the equivalent frame model of the r/c core, as shown in Fig. 4.12.
- The self weight of the basement wall and its foundation and of the foundation tie beams is evenly distributed (lumped) at the nodes of the FE discretization grid

Table 4.109 Spring coefficients at the joints of the basement wall strip foundation of Fig. 4.19

Joint	L_{eff} [m]	k_3 [kN/m]	k_r [kNm/rad]
J1/J12	0.675	21,985.6	3591.0
J2/J11	0.838	27,278.4	4455.5
J3/J10	1.075	35,014.1	5719.0
J4/J9	1.175	38,271.2	6251.0
J5/J8	1.254	40,850.9	6672.3
J6/J7	1.133	36,914.1	6029.3
J13/J51	0.878	28,581.3	4668.3
J14/J15/J16/J17	1.005	32,734.1	5346.6
J18/J46	1.169	38,080.4	6219.8
J19/J25/J39/J45	1.033	33,657.0	5497.3
J20/J24/J40/J44	0.933	30,399.8	4965.3
J21/J23/ J41/J43	1.217	39,628.4	6472.6
J22/J42	1.500	48,856.9	7980.0
J26/J38	1.097	35,719.8	5834.2
J27/J37	0.960	31,268.4	5107.2
J28/J29/J30/J34/J35/J36	0.860	28,011.3	4575.2
J31/J33	1.305	42,505.5	6942.6
J32	1.750	56,999.7	9309.9
J47/J48/J49/J50	1.105	35,991.2	5878.6

used, as seen in Fig. 4.15, based on the weight per unit volume of r/c and on actual cross-sectional dimensions of structural members.

– The self weight of the pad footings is accounted for by applying gravitational point loads at the nodes positioned at the center of gravity of each pad at the foundation level, as seen in Fig. 4.13. These point loads are computed as the product of the volume of the pad footings, whose geometry is detailed in Fig. 4.11, times the weight per unit volume of r/c.

4.3.2.4 Mass/Inertial Modeling Assumptions

– The total mass of each storey is lumped at the center of gravity M (geometrical center) of the corresponding rigid floor diaphragm.
– The total mass of each storey comprises:

  • The self mass of the storey slab and beams, including all finishings;
  • The self mass of the masonry infill walls resting on the storey beams (ignoring any openings);
  • The self mass of the vertical r/c members (columns and walls) extending above and below the considered storey slab up to the middle of their total storey height; and

Table 4.110 Translational spring coefficients at the joints of the tie beams

Tie beam	L [m]	b [m]	m = L/b	n[a]	K_{s2} [kN/m^3]	K_s [kN/m^3]	$L_{eff,m}$[b] [m]	$L_{eff,o}$[c] [m]	$k_{3,m}$[d] [kN/m]	$K_{3,o}$[e] [kN/m]
TB1/TB2	1.425	0.25	5.7	2	108,900	78,968	0.713	0.356	14,066	7033
TB3/TB4	2.825	0.25	11.3	4	108,900	75,812	0.706	0.353	13,386	6693
TB5/TB6	2.575	0.25	10.3	4	108,900	76,124	0.644	0.322	12,251	6126
TB7/TB9	2.8	0.25	11.2	4	108,900	75,841	0.700	0.350	13,272	6636
TB8	1.5	0.25	6	3	108,900	78,650	0.500	0.250	9831	4916
TB10/TB11	2.8	0.25	11.2	4	108,900	75,841	0.700	0.350	13,272	6636

[a] n: number of one-dimensional beam elements used to model each tie beam
[b] $L_{eff,m}$: tributary effective length of inner joints
[c] $L_{eff,o}$: tributary effective length of outer joints
[d] $k_{3,m}$: spring coefficient at inner joints
[e] $k_{3,o}$: spring coefficient at outer joints

Table 4.111 Storey mass and gravity loads of the seismic design load combination for building C

Storey	Dead load G_k [kN]	Variable action Q_k [kN]	ψ_2 [a]	φ [b]	$\psi_E = \varphi \cdot \psi_2$	Combination of actions $G_k + \psi_E \cdot Q_k$ [c] [kN]	mass m [t]
1st	2298.70	376.125	0.3	0.8	0.24	2389.00	243.52
2nd	2173.00	376.125	0.3	0.8	0.24	2263.25	230.71
3rd	2173.00	376.125	0.3	0.8	0.24	2263.25	230.71
4th	1738.55	376.125	0.3	1.0	0.30	1851.39	188.72[d]
Total sum	8383.25	1504.50				8766.89	893.66

[a]The combination coefficient ψ_2 of the quasi-permanent value of the variable action Q_k ("live" gravity loads) is given in Eurocode 0 (CEN 2002) Annex A1 and is taken to be equal to 0.3 assuming that building A is an ordinary occupancy residential or office building

[b]The reduction factor φ is given in Table 4.2 of EC8 (clause §4.2.4 of EC8). It is herein assumed to be equal to 1.0 for the top storey and equal to 0.8 for the rest of the storeys, which are assumed to have correlated occupancies. This assumption is particularly valid for residential buildings at night time and for office buildings during day hours

[c]The storey masses are computed from this gravity load combination, as specified in clause §3.2.4 (2)P of EC8

[d]The mass of the core and the staircase extending above the roof of the building by 3 m is added to the mass of the 4th floor

- The mass that corresponds to the variable gravity load of the seismic design load combination, as defined in clauses §3.2.4(2)P, 4.2.4(2)P and 4.3.1(10)P of EC8.

The mass of each storey of building C is computed from the gravity loads of the seismic design load combination, as detailed in Table 4.111.

4.3.3 Verification Checks for Regularity for Building C

The rationale for the regularity verification checks has been discussed in detail in Sects. 3.1.1 and 4.1.3. Herein, the EC8 regularity checks in plan and elevation are carried out for building C following the FC-3.2 and FC-3.3, respectively.

4.3.3.1 Verification Check for Regularity in Elevation

Following the required steps of FC-3.3 to check for regularity in elevation, it is readily verified that building C is regular in elevation. In particular, the path of the relevant logical and computational steps closely following FC-3.3 is indicated in Fig. 4.20 in red. The above classification allows for

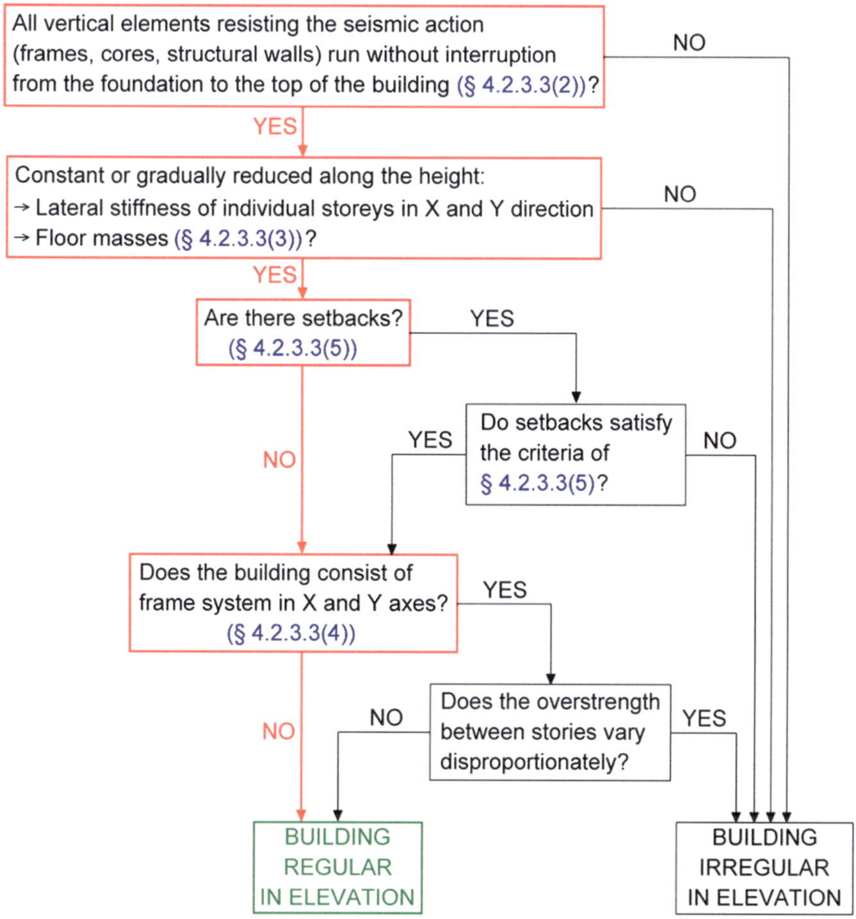

Fig. 4.20 Regularity in elevation verification check for building C following FC-3.3

– using the LFM of analysis provided that condition (a) of clause §4.3.3.2.1(2) of EC8 is satisfied as well. Still, only the MRSM of analysis will be employed for the seismic analysis of building C in this example.
– a reduction of 20 % to the basic (reference) value of the behaviour factor, q_0, (see clause §5.2.2.2(3) of the EC8).

4.3.3.2 Verification Check for Regularity in Plan

The verification check procedure for regularity in plan follows FC-3.2 and FC-3.2a. The outcomes from the pertinent logical and computational steps for building C are summarized in Fig. 4.21, which closely follows FC-3.2 (the path taken is marked in red). Focusing on checking the validity of the inequalities of Eqs. (3.1) and (3.2)

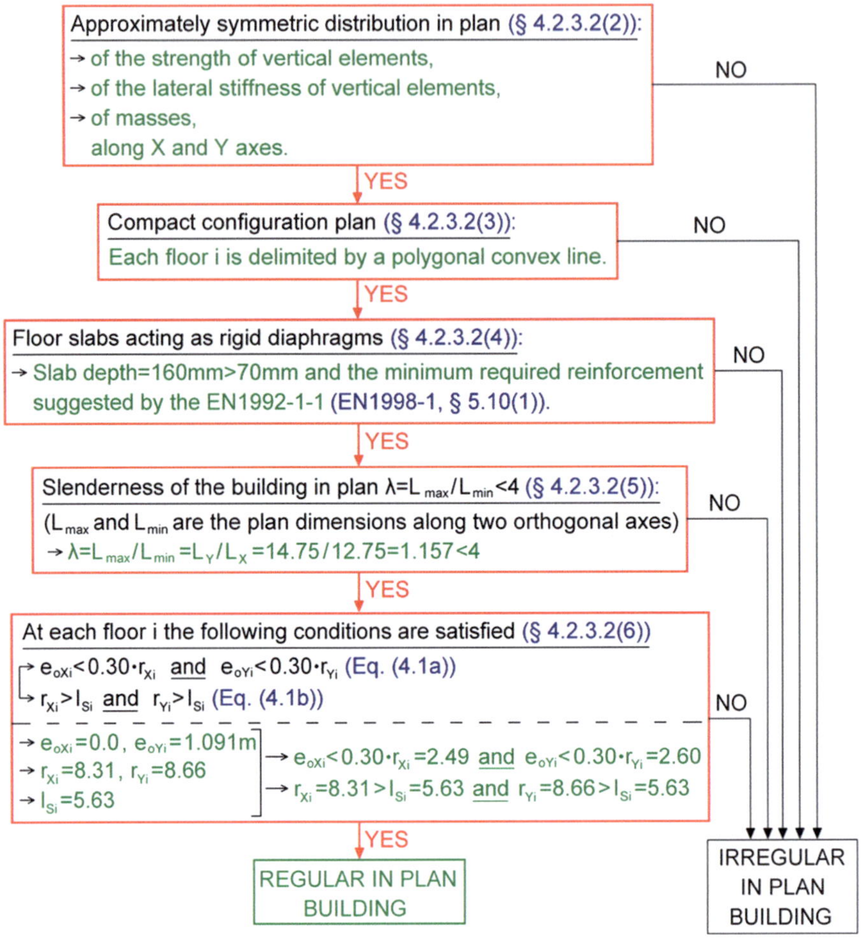

Fig. 4.21 Regularity in plan verification check for building C following FC-3.2

(clause §4.2.3.2(6) of EC8), the same procedure is herein followed as in Sect. 4.1.3.1 for building A, to compute the structural eccentricities, e_{oX} and e_{oY}, and the torsional radii, r_X and r_Y, along the principal axes X and Y, respectively, for each floor diaphragm of building C with the aid of the fictitious elastic vertical axis.

Specifically, the determination of the structural eccentricities requires that a linear static analysis be performed first for torsional moments, M_{Zk}, about the gravitational axis Z at each floor diaphragm k (loading case "M"). These torsional moments can follow a "triangular" distribution along the height of the building by application of equation (4.11) of EC8. Table 4.112 reports the values of the torsional moments along with the pertinent steps for computing these values for building C, assuming an arbitrarily taken total base shear equal to 5000 kN (see Sect. 4.2.3.2).

Table 4.112 Triangular distribution of torsional moments and lateral forces along the height of building C for an arbitrary base shear equal to 5000 kN (Equation (4.11) of EC8)

Storey k	m [t]	J_m [t · m²]	z_k [m]	$m_k \cdot z_k$	F_{Xk} [kN] / F_{Yk} [kN] / M_{Zk} [kNm]
1st	243.52	7714.12	4.50	1095.84	702.79
2nd	230.71	7308.17	7.50	1730.33	1109.70
3rd	230.71	7308.17	10.50	2422.46	1553.58
4th	188.72	5978.24	13.50	2547.72	1633.92
Total sum	893.66			7796.34	5000.00

Next, the coordinates X_{Po} and Y_{Po} of the "center of twist" P_o at the storey k lying closer to the 80 % level of the total height of the building (3rd storey for building C), through which the fictitious elastic vertical axis passes, are determined by the relationships $X_{Po} = X_{Mk} - (u_{Y(Mk)}/\theta_{Z(Mk)})$ and $Y_{Po} = Y_{Mk} - (u_{X(Mk)}/\theta_{Z(Mk)})$ included in FC-3.2. In the above relationships, X_{Mk} and Y_{Mk} are the coordinates of the center of mass of storey k ($X_{Mk} = 6.25$ m and $Y_{Mk} = 7.25$ m for building C), and $u_{X(Mk)}$, $u_{Y(Mk)}$, and $\theta_{Z(Mk)}$ are the translations along axes X, Y, and the rotation about axis Z, respectively, of point M (Fig. 4.9) due to the loading case "M". For building C and the torsional moments of Table 4.112, the aforementioned nodal displacements are: $u_{X(Mk)} = 0.000597$ m, $u_{Y(Mk)} = 0.0$ m, and $\theta_{Z(Mk)} = 0.000547$ m. Therefore, the coordinates of point P_o for building C (with regard to the global coordinate system shown in Fig. 4.9), through which the fictitious elastic vertical axis passes, are computed as:

- $X_{Po} = X_{Mk} - (u_{Y(Mk)}/\theta_{Z(Mk)}) = 6.25 - 0/0.000547 = 6.25$ m; and
- $Y_{Po} = Y_{Mk} - (u_{X(Mk)}/\theta_{Z(Mk)}) = 7.25 - 0.000597/0.000547 = 6.159$ m

and the structural eccentricities common to all stories of building C are:

- $e_{oX} = X_{Mk} - X_{Po} = 6.25 - 6.25 = 0$ m; and
- $e_{oY} = Y_{Mk} - Y_{Po} = 7.25 - 6.159 = 1.091$ m.

Furthermore, the torsional stiffness radii r_X and r_Y corresponding to the fictitious elastic vertical axis are determined by the relationships $r_X = (u_{Y(FY)}/\theta_{Z(Mk)})^{1/2}$ and $r_Y = (u_{X(FX)}/\theta_{Z(Mk)})^{1/2}$ included in FC-3.2. In the above relationships, $u_{Y(FY)}$ and $u_{X(FX)}$ are the translations of point P_o along axis Y due to the loading case "F_Y"and along axis X due to the loading case "F_X", respectively. Loading cases "F_X" and "F_Y" involve lateral forces applied at the traces of the previously defined fictitious elastic axis at each floor diaphragm of the building along the assumed directions of the seismic action, that is, axes X and Y, respectively. The considered lateral forces may follow a "triangular" distribution along the height of the building by application of equation (4.11) of EC8, as calculated in Table 4.112. For building C and the lateral forces of Table 4.112, the aforementioned nodal displacements are: u_Y

Table 4.113 Verification check of inequality in Eq. (3.1), Sect. 3.1.1.1, for building C

Storey	e_{oX} [m]	e_{oY} [m]	r_X [m]	r_Y [m]	$e_{oX} < 0.3 \cdot r_X$	$e_{oY} < 0.3 \cdot r_Y$
1st	0.00	1.091	8.31	8.66	YES	YES
2nd	0.00	1.091	8.31	8.66	YES	YES
3rd	0.00	1.091	8.31	8.66	YES	YES
4th	0.00	1.091	8.31	8.66	YES	YES

Table 4.114 Verification check of inequality in Eq. (3.2), Sect. 3.1.1.1, for building C

Storey	l_s[a] [m]	r_X [m]	r_Y [m]	$r_X \geq l_s$	$r_Y \geq l_s$
1st	5.63	8.31	8.66	YES	YES
2nd	5.63	8.31	8.66	YES	YES
3rd	5.63	8.31	8.66	YES	YES
4th	5.63	8.31	8.66	YES	YES

[a]The radius of gyration l_s for a rectangular floor plan such as that of building C is obtained using the following relationship: $l_s = [(L_X^2 + L_Y^2)/12]^{1/2} = [(14.75^2 + 12.75^2)/12]^{1/2} = 5.63$ m

$_{(FY)} = 0.0378$ m and $u_{X(FX)} = 0.0411$ m. Therefore, the torsional stiffness radii common to all stories of building C are:

- $r_X = (u_{Y(FY)}/\theta_{Z(Mk)})^{1/2} = (0.0378/0.000547)^{1/2} = 8.31$; and
- $r_Y = (u_{X(FX)}/\theta_{Z(Mk)})^{1/2} = (0.0411/0.000547)^{1/2} = 8.66$.

Having computed the structural eccentricities and torsional radii for the fictitious elastic axis, the inequalities of Eqs. (3.1) and (3.2) are checked in Tables 4.113 and 4.114, respectively, from which it is seen that building C is regular in plan.

4.3.4 *Classification of the Lateral Load-resisting Structural System of Building C*

The rationale for classifying building structures according to the properties of their lateral load-resisting structural system and the implications of this classification in the seismic design process have been discussed in detail in Sect. 3.1.2. The classification procedure follows the steps delineated in FC-3.4. The outcomes for building C are summarized in Fig. 4.22 (the path taken is marked in red). The first step of this procedure is the torsional sensitivity verification check, which follows FC-3.2a. This check has already been undertaken in the previous section during the regularity in elevation verification check, and it was found that building C is not torsionally sensitive (see Table 4.114).

Subsequently, an additional check is required to quantify the percentage of the total base shear resisted by the r/c walls and the core of building C along the principal directions X and Y. To this end, two independent static analyses along X

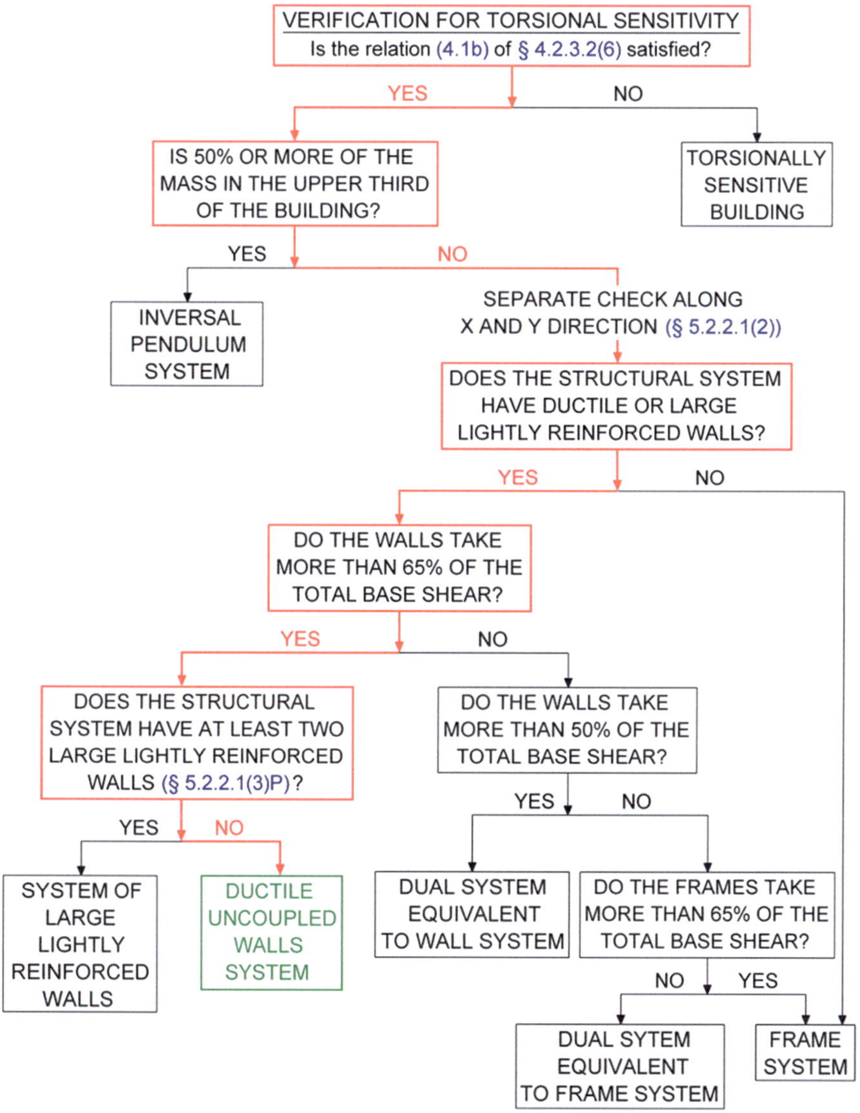

Fig. 4.22 Classification of the lateral load-resisting system of building C following FC-3.4

and Y are undertaken for the horizontal system of forces F_{Xk} and F_{Yk}, respectively, of Table 4.112, acting at the center of gravity of the floors. The percentage of the total base shear, V_X and V_Y, resisted at the base of walls and columns is computed in Table 4.115 independently for each of the above analyses. It is seen that building C is classified as a wall structural system along both principal axes, since the walls and the core resist more than 65 % of the total base shear applied to the structure.

Table 4.115 Verification check of the percentage of the total base shear resisted by r/c walls for the loading cases F_{Xk} and F_{Yk} defined in Table 4.112 for building C

	Analysis along direction X (F_{Xk})		Analysis along direction Y (F_{Yk})	
Member	V_X,walls [kN]	V_X,columns [kN]	V_Y,walls [kN]	V_Y,columns [kN]
W1X	245.74	–	–	16.84
C1	–	45.02	–	31.78
C2	–	45.02	–	31.78
W2X	245.74	–	–	16.84
CW1Y	–	0.75	990.51	–
CW2Y	–	0.75	990.51	–
C3	–	33.83	–	37.4
CW1X	2604.68	–	–	0.518
C4	–	33.83	–	37.4
W1Y	–	33.75	1333.61	–
C5	–	29.3	–	36.5
C6	–	29.3	–	36.5
W2Y	–	33.75	1333.61	–
C7	–	22.58	–	31.05
W3X	1573.39	–	–	44.11
C8	–	22.58	-	31.05
Total Sum	4669.55	330.46	4648.24	351.77
Percentage	**93.39 %**	**6.61 %**	**92.96 %**	**7.04 %**

4.3.5 *Selection of Ductility (Capacity) Class of Building C*

The rationale for deciding upon the desirable ductility capacity class according to EC8 has been discussed in Sect. 3.1.3, and its relevant points have been highlighted in Sect. 4.1.5. For building C, it is decided to adopt the medium ductility class (DCM).

4.3.6 *Determination of the Maximum Allowed Behaviour Factor for Building C*

A detailed presentation of the procedure for determining the maximum allowed value, $\max q_{allow}$, of the behaviour factor q has been provided in Sect. 3.1.4, and its relevant points have been highlighted in Sect. 4.1.6.

The procedure of determining the $\max q_{allow}$ follows FC-3.5. The path taken (marked in red) and the outcomes of the pertinent logic and computational steps (in green fonts) for building C are given in Fig. 4.23, which closely follows the above flowchart. It is seen that the $\max q_{allow}$ value of the behaviour factor is equal to 3.0. However, herein, the lower value $q = 2.0$ is chosen.

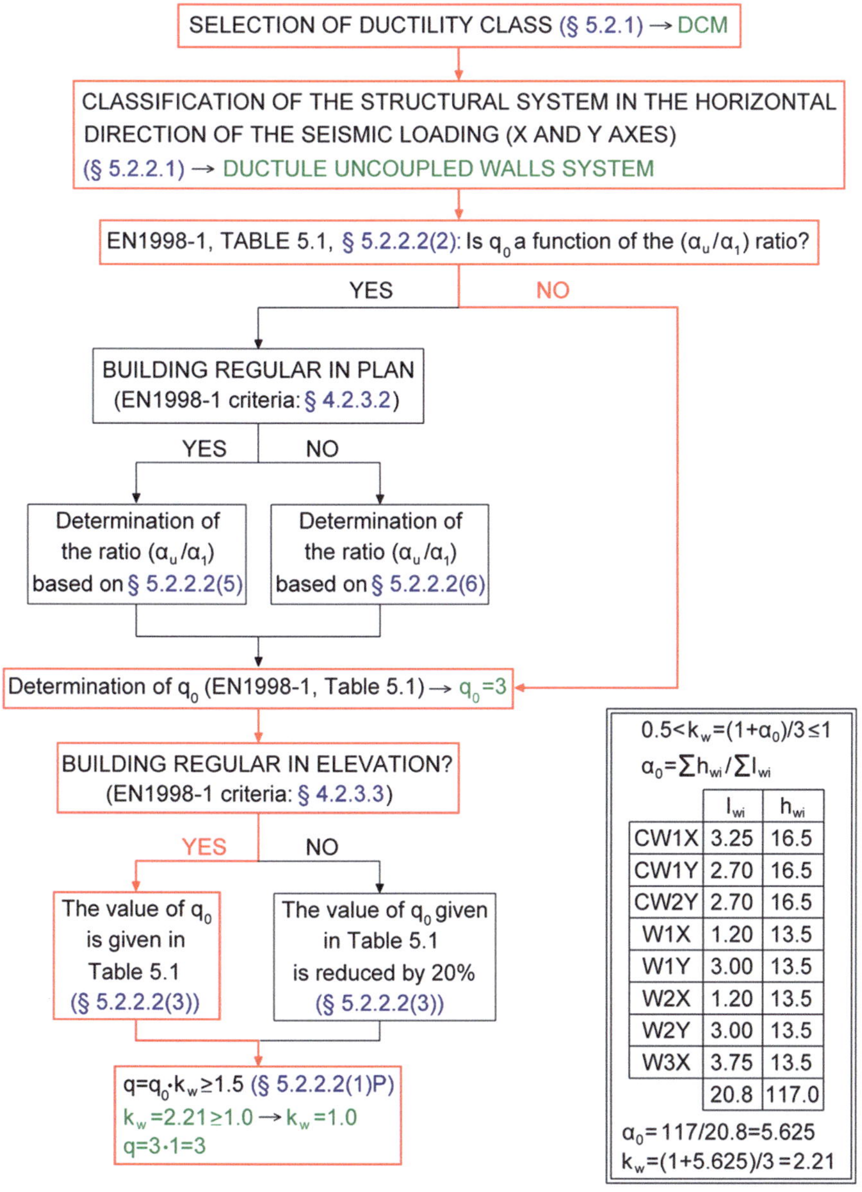

Fig. 4.23 Determination of the maximum allowable value of behaviour factor q for building C

4.3.7 *Selection of an Equivalent Linear Method of Seismic Analysis for Building C*

The modal response spectrum method (MRSM) is of general use for the seismic analysis of any building structure regardless of regularity conditions in plan and elevation. In fact, being more accurate and more straightforward to apply, the MRSM is recommended to be used even in the seismic analysis of buildings for which the simpler lateral force method (LFM) can be used (see also Sect. 2.4.3.1). Therefore, the MRSM is used for the seismic analysis of building C to obtain the seismic effects reported in Sect. 4.3.9.

4.3.8 *Static Analysis for Gravity Loads of the Design Seismic Loading Combination (G "+" ψ_2Q) for Building C*

The design seismic loading combination involves gravitational (statically applied) permanent and quasi-permanent variable actions (see Eq. (2.12)). Effects due to these actions can be derived separately by means of standard static analysis and superposed to the effects due to the seismic (accidental) action. Table 4.116 reports stress resultants (effects) due to the gravity loads of the design seismic loading combination at critical cross-sections of selected r/c structural members, namely, the wall W3X, the column C8 and the beams BX10 and BY12 at the ground (1st) storey.

Table 4.116 Sectional stress resultants of wall W3X, column C8 and beams BX10 and BY12 at the ground storey for the permanent and quasi-permanent variable actions of the design seismic loading combination (G "+" ψ_2Q) [Sign convention follows Fig. 2.32]

Member	Position	N [kN]	V_2 [kN]	V_3 [kN]	T [kNm]	M_2 [kNm]	M_3 [kNm]
W3X	Bottom	−1072.6	≈0.0	15.72	0.0	−34.46	≈0.0
	Top	−958.65	≈0.0	15.72	0.0	36.29	≈0.0
C8	Bottom	−243.06	−1.49	0.455	0.0	−0.073	1.35
	Top	−226.06	−1.49	0.455	0.0	1.86	−5.00
BX10	Left end	0.0	−36.16	0.0	−0.204	0.0	−29.78
	Midspan	0.0	−4.85	0.0	−0.204	0.0	15.50
	Right end	0.0	31.35	0.0	−0.204	0.0	−14.92
BY12	Left end	0.0	−29.43	0.0	0.115	0.0	−16.33
	Midspan	0.0	−3.73	0.0	0.115	0.0	7.62
	Right end	0.0	19.05	0.0	0.115	0.0	−4.40

Table 4.117 Accidental eccentricities and polar moment of inertia for seismic excitation along X and Y axes for building C

		Accidental eccentricities [m]		Polar moment of inertia [tm^2]		
Storey	Mass [t]	e_{aX}[a]	e_{aY}[b]	J_{mX}[c]	J_{mY}[d]	
1st	243.52	0.6375	0.7375	7812.97	7846.46	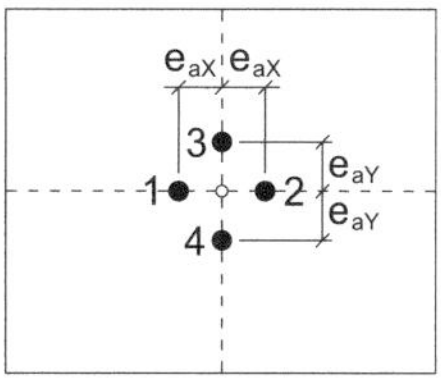
2nd–3rd	230.71	0.6375	0.7375	7401.98	7433.71	
4th	188.72	0.6375	0.7375	6054.80	6080.75	

[a] $e_{aX} = 0.05L_X$
[b] $e_{aY} = 0.05L_Y$
[c] $J_{mX} = J_m + me_{aX}^2$
[d] $J_{mY} = J_m + me_{aY}^2$

4.3.9 *Seismic Analysis of Building B Using the Modal Response Spectrum Method and Deformation-Based Verification Checks*

The implementation of the modal response spectrum method (MRSM) of seismic analysis follows FC-3.7. Four different spatial FE models are considered in the analysis corresponding to the positioning of the center of mass of each floor diaphragm at four different sets of locations (positions 1 to 4, as shown in Table 4.117) to account for accidental mass eccentricity (see also Sect. 3.1.5.1 and Fig. 3.4). The accidental eccentricities, $\pm e_{aX}$ and $\pm e_{aY}$, along the principal axes X and Y, respectively, define the four displaced locations of the mass center measured from the geometric center of each slab. These eccentricities are computed in Table 4.117 taken to be equal to 5 % of the length of building A, $L_X = 12.75$ m and $L_Y = 14.75$ m (Fig. 4.9), along axes X and Y, respectively, assuming that the masonry infill walls are evenly distributed in plan. Further, the polar moment of inertia about the gravitational axis of each floor diaphragm is computed with respect to the displaced position as $J_{mi} = J_m + m \cdot e_{ai}^2$, $i = X,Y$, where $J_m = m \cdot (L_X^2 + L_Y^2)/12$ is the polar moment of inertia with respect to the geometric center of each slab, as reported in Table 4.117.

4.3.9.1 Modal Analysis Results

Table 4.118 lists the natural periods corresponding to the first 6 mode shapes of vibration for the four considered FE structural models (position of center of mass 1 to 4, as shown in Table 4.117) of building C derived from standard modal analysis.

Table 4.118 Natural periods of the four considered FE structural models (center of mass displaced by $\pm e_{aX}$ and $\pm e_{aY}$ as shown in Table 4.117) for building C

	Natural period [s]			
Mode shape	Mass position 1	Mass position 2	Mass position 3	Mass position 4
1	0.498	0.498	0.491	0.508
2	0.472	0.472	0.470	0.470
3	0.312	0.312	0.318	0.308
4	0.102	0.102	0.097	0.106
5	0.095	0.095	0.0952	0.095
6	0.077	0.077	0.081	0.074

Table 4.119 Modal participation mass ratios and cumulative participation mass ratios as percentages of the total mass of building C

	Mass position 1				Mass position 2			
	Individual mode		Cumulative sum		Individual mode		Cumulative sum	
Mode shape	X	Y	X	Y	X	Y	X	Y
1	87.17	1.89	**87.17**	**1.89**	87.17	1.89	**87.17**	**1.89**
2	2.20	86.81	**89.37**	**88.70**	2.20	86.81	**89.37**	**88.70**
3	1.42	0.73	**90.79**	**89.43**	1.42	0.73	**90.79**	**89.43**
4	5.54	1.39	96.33	**90.82**	5.54	1.39	96.33	**90.82**
5	1.57	7.96	97.90	98.78	1.57	7.96	97.90	98.78
6	1.48	0.36	99.38	99.14	1.48	0.36	99.38	99.14
	Mass position 3				Mass position 4			
	Individual mode		Cumulative sum		Individual mode		Cumulative sum	
Mode shape	X	Y	X	Y	X	Y	X	Y
1	90.58	0.00	**90.58**	**0.00**	87.13	0.00	**87.13**	**0.00**
2	0.00	89.42	90.58	**89.42**	0.00	89.42	**87.13**	**89.42**
3	0.13	0.00	90.71	**89.42**	3.76	0.00	**90.89**	**89.42**
4	7.97	0.00	98.68	**89.42**	6.30	0.00	97.19	**89.42**
5	0.00	9.71	98.68	**99.13**	0.00	9.71	97.19	**99.13**
6	0.68	0.00	99.36	99.13	2.24	0.00	99.43	99.13

Further, Table 4.119 reports the modal participation mass ratios for each mode (i.e., ratios of effective modal mass over the total mass of building C) along axes X and Y, as well as the corresponding cumulative modal participation mass ratios for the four considered FE models. The latter results suggest that at least the first 4 mode shapes need to be considered to satisfy the criterion of clause 4.3.3.3.1(3) of EC8 for the FE models with mass positions 1 and 2, that is, a sufficient number of modes are considered in the MRSM such that 90 % or more of the total oscillatory mass along both principal axes is activated. Further, for the FE models with mass positions 3 and 4, the first 5 mode shapes need to be considered in the analysis to satisfy the above criterion. Nevertheless, in all ensuing numerical results, all 6 mode shapes reported are utilized in implementing the MRSM of analysis.

4.3.9.2 Selected Design Seismic Effects (Sectional Stress Resultants)

In this section, the design seismic effects at critical sections (sectional stress resultants) for the column C8, the wall W3X and the beams BX10 and BY12 of the ground (1st) storey obtained by means of the MRSM are presented in tabular form.

Vertical structural members C8 and W3X (bi-axial bending with axial force)

In the case of the vertical structural members C8 and W3X, which need to be designed for bi-axial bending with axial force, the design values of the three concurrent pertinent stress resultants ("design triads"), namely moments M_2, M_3 and axial force N, as defined in Fig. 2.32, are reported. Herein, the design triads are compiled by considering the extreme value of a single stress resultant together with the expected (most probable) values of the other two stress resultants attained concurrently. Specifically, the 6 design triads of Eq. (3.13) are considered derived by means of a simplified approach detailed in the Greek Seismic Code EAK2000 (Gupta 1992; Anastassiadis 1993; Earthquake Planning and Protection Organization (EPPO) 2000) and assuming simultaneous seismic action along the two principal axes X and Y. The following three computational steps are taken to determine the design triads for the considered vertical members (see also Sect. 3.1.5.1):

(a) The peak (seismic) modal values of the considered stress resultants are obtained separately by application of the MRSM for each of the directions of the seismic excitation X and Y. Next, these modal values are superposed by means of the CQC rule for modal combination (clause §4.3.3.3.2(3)P of EC8), to derive the (non-concurrent) maximum values of stress resultants for seismic excitation along axes X and Y, independently.
(b) The SRSS rule for spatial combination (clause §4.3.3.5.1(2)b of EC8) is employed to obtain the extreme values of the considered stress resultants from the maximum values derived in the previous step for simultaneous seismic action along the X and Y horizontal directions. Tables 4.120 and 4.121 report the thus obtained extreme values of the M_2, M_3, and N stress resultants developing at the base (bottom) and the top of the structural members C8 and W3X, respectively, at the ground (1st) storey of building C for all four different FE models used in the analysis. Further, Tables 4.122 and 4.123

Table 4.120 Extreme values of stress resultants in column C8 (ground storey) of building C

	N [kN]		M_2 [kNm]		M_3 [kNm]	
Mass position	bottom	top	bottom	top	bottom	top
1	±298.93	±298.93	±28.17	±26.30	±21.91	±18.23
2	±329.08	±329.08	±37.98	±35.23	±23.52	±19.29
3	±306.20	±306.20	±32.17	±29.84	±27.19	±22.34
4	±322.87	±322.87	±34.91	±32.51	±17.28	±14.39

Table 4.121 Extreme values of stress resultants in column W3X (ground storey) of building C

	N [kN]		M_2 [kNm]		M_3 [kNm]	
Mass position	Bottom	Top	Bottom	Top	Bottom	Top
1	±194.78	±194.78	±52.47	±39.85	±3371.41	±365.00
2	±194.78	±194.78	±52.47	±39.85	±3371.41	±365.00
3	±196.14	±196.14	±52.84	±40.14	±4042.73	±464.04
4	±196.14	±196.14	±52.84	±40.14	±2542.74	±236.82

Table 4.122 Design triads (expected -most probable- concurrent values of N, M_2, and M_3 stress resultants for simultaneous seismic action along axes X and Y) for column C8 (ground storey) of building C [Extreme values in bold]

		N [kN]		M_2 [kNm]		M_3 [kNm]	
Mass position		Bottom	Top	Bottom	Top	Bottom	Top
1	extrN	**298.93**	**298.93**	27.01	−25.34	−11.92	10.19
	extrM_2	286.65	−288.01	**28.17**	**26.30**	−6.68	−6.24
	extrM_3	−162.61	167.17	−8.59	−9.00	**21.91**	**18.23**
	−extrN	**−298.93**	**−298.93**	−27.01	25.34	11.92	−10.19
	−extrM_2	−286.65	288.01	**−28.17**	**−26.30**	6.68	6.24
	−extrM_3	162.61	−167.17	8.59	9.00	**−21.91**	**−18.23**
2	extrN	**329.08**	**329.08**	36.33	−33.83	−2.02	2.32
	extrM_2	314.80	−316.06	**37.98**	**35.23**	4.53	2.75
	extrM_3	−28.29	39.56	7.32	5.03	**23.52**	**19.29**
	−extrN	**−329.08**	**−329.08**	−36.33	33.83	2.02	−2.32
	−extrM_2	−314.80	316.06	**−37.98**	**−35.23**	−4.53	−2.75
	−extrM_3	28.29	−39.56	−7.32	−5.03	**−23.52**	**−19.29**
3	extrN	**306.20**	**306.20**	29.72	−27.78	−8.26	7.25
	extrM_2	282.85	−285.01	**32.17**	**29.84**	1.93	0.63
	extrM_3	−93.08	99.33	2.28	0.84	**27.19**	**22.34**
	−extrN	**−306.20**	**−306.20**	−29.72	27.78	8.26	−7.25
	−extrM_2	−282.85	285.01	**−32.17**	**−29.84**	−1.93	−0.63
	−extrM_3	93.08	−99.33	−2.28	−0.84	**−27.19**	**−22.34**
4	extrN	**322.87**	**322.87**	34.11	−31.83	−4.80	4.54
	extrM_2	315.47	−316.12	**34.91**	**32.51**	−1.62	−2.14
	extrM_3	−89.61	101.91	−3.28	−4.83	**17.28**	**14.39**
	−extrN	**−322.87**	**−322.87**	−34.11	31.83	4.80	−4.54
	−extrM_2	−315.47	316.12	**−34.91**	**−32.51**	1.62	2.14
	−extrM_3	89.61	−101.91	3.28	4.83	**−17.28**	**−14.39**

present the 6 design triads at the bottom and the top of the above structural members using the simplified approach of EAK2000 (see also Fardis, 2009; Sect. 4.3.2.3). The single extreme value attained by a certain stress resultant in each design triad is noted by bold faced fonts.

Table 4.123 Design triads (expected -most probable- concurrent values of N, M_2, and M_3 stress resultants for simultaneous seismic action along axes X and Y) for wall W3X (ground storey) of building C [Extreme values in bold]

Mass position		N [kN]		M_2 [kNm]		M_3 [kNm]	
		Bottom	Top	Bottom	Top	Bottom	Top
1	extrN	**194.78**	**194.78**	52.27	−39.65	−754.02	−188.95
	extrM_2	194.07	−193.77	**52.47**	**39.85**	−757.60	183.56
	extrM_3	−43.56	−100.84	−11.79	20.04	**3371.41**	**365.00**
	−extrN	**−194.78**	**−194.78**	−52.27	39.65	754.02	188.95
	−extrM_2	−194.07	193.77	**−52.47**	**−39.85**	757.60	−183.56
	−extrM_3	43.56	100.84	11.79	−20.04	**−3371.41**	**−365.00**
2	extrN	**194.78**	**194.78**	52.27	−39.65	754.02	188.95
	extrM_2	194.07	−193.77	**52.47**	**39.85**	757.59	−183.56
	extrM_3	43.56	100.84	11.79	−20.04	**3371.41**	**365.00**
	−extrN	**−194.78**	**−194.78**	−52.27	39.65	−754.02	−188.95
	−extrM_2	−194.07	193.77	**−52.47**	**−39.85**	−757.59	183.56
	−extrM_3	−43.56	−100.84	−11.79	20.04	**−3371.41**	**−365.00**
3	extrN	**196.14**	**196.14**	52.64	−39.92	0.00	0.00
	extrM_2	195.38	−195.07	**52.84**	**40.14**	0.00	0.00
	extrM_3	0.00	0.00	0.00	0.00	**4042.73**	**464.03**
	−extrN	**−196.14**	**−196.14**	−52.64	39.92	0.00	0.00
	−extrM_2	−195.38	195.07	**−52.84**	**−40.14**	0.00	0.00
	−extrM_3	0.00	0.00	0.00	0.00	**−4042.73**	**−464.03**
4	extrN	**196.14**	**196.14**	52.64	−39.92	0.00	0.00
	extrM_2	195.38	−195.07	**52.84**	**40.14**	0.00	0.00
	extrM_3	0.00	0.00	0.00	0.00	**2542.74**	**236.81**
	−extrN	**−196.14**	**−196.14**	−52.64	39.92	0.00	0.00
	−extrM_2	−195.38	195.07	**−52.84**	**−40.14**	0.00	0.00
	−extrM_3	0.00	0.00	0.00	0.00	**−2542.74**	**−236.81**

(c) Finally, the seismic design triads derived in the previous step are superposed to the corresponding stress resultants of the considered structural members due to the gravitational permanent and quasi-permanent variable actions summarized in Table 4.116 (Sect. 4.3.8) to obtain the design triads for the EC8 design seismic loading combination G "+" Ψ_2Q "±" E. The thus obtained triads are reported in Tables 4.124 and 4.125.

Beams BX10 and BY12 (uni-axial bending)

The previously described three steps are applied to obtain the extreme values of the moment M_3 and of the shearing force V_2 at critical cross-sections (left end, right end, and at midspan) of the beams BX10 and BY12 of the ground (1st) storey of building C (Fig. 4.9), which need to be designed for uni-axial bending. However, in this case, the procedure of obtaining the design seismic effects for simultaneous seismic action along the two principal axes X and Y is significantly simplified by the fact that only a single seismic effect (i.e., stress resultant M_3 and corresponding

Table 4.124 Design triads for column C8 (ground storey) for the seismic design load combination G "+" ψ_2Q "±" E of building C [Extreme values in bold]

Mass position		N [kN]		M_2 [kNm]		M_3 [kNm]	
		Bottom	Top	Bottom	Top	Bottom	Top
1	extrN	**55.87**	**72.87**	26.94	−23.48	−10.57	5.19
	extrM_2	43.59	−514.07	**28.10**	**28.16**	−5.33	−11.24
	extrM_3	−405.67	−58.89	−8.66	−7.14	**23.26**	**13.23**
	−extrN	**−541.99**	**−524.99**	−27.09	27.20	13.27	−15.19
	−extrM_2	−529.71	61.95	**−28.24**	**−24.44**	8.03	1.24
	−extrM_3	−80.45	−393.23	8.52	10.86	**−20.56**	**−23.23**
2	extrN	**86.02**	**103.02**	36.26	−31.97	−0.67	−2.68
	extrM_2	71.74	−542.12	**37.90**	**37.09**	5.88	−2.25
	extrM_3	−271.35	−186.50	7.25	6.89	**24.87**	**14.29**
	−extrN	**−572.14**	**−555.14**	−36.40	35.69	3.37	−7.32
	−extrM_2	−557.86	90.00	**−38.05**	**−33.37**	−3.18	−7.75
	−extrM_3	−214.77	−265.62	−7.39	−3.17	**−22.17**	**−24.29**
3	extrN	**63.14**	**80.14**	29.65	−25.92	−6.91	2.25
	extrM_2	39.79	−511.07	**32.10**	**31.70**	3.28	−4.37
	extrM_3	−336.14	−126.73	2.21	2.70	**28.54**	**17.34**
	−extrN	**−549.26**	**−532.26**	−29.79	29.64	9.61	−12.25
	−extrM_2	−525.91	58.95	**−32.25**	**−27.98**	−0.58	−5.63
	−extrM_3	−149.98	−325.39	−2.35	1.02	**−25.84**	**−27.34**
4	extrN	**79.81**	**96.81**	34.04	−29.97	−3.45	−0.46
	extrM_2	72.41	−542.18	**34.84**	**34.37**	−0.27	−7.14
	extrM_3	−332.67	−124.15	−3.35	−2.97	**18.63**	**9.39**
	−extrN	**−565.93**	**−548.93**	−34.19	33.69	6.15	−9.54
	−extrM_2	−558.53	90.06	**−34.99**	**−30.65**	2.97	−2.86
	−extrM_3	−153.45	−327.97	3.21	6.69	**−15.93**	**−19.39**

shearing force V_2) is required in the detailing of beam sections, as opposed to the vector of the three concurrently acting seismic effects (triads) N, M_2, M_3 considered for the case of vertical structural members.

Specifically, Tables 4.126 and 4.127 report the extreme values of M_3 and V_2 for the beams BX10 and BY12, respectively, for all four FE models considered in the analysis. These are obtained by first computing the maximum values of M_3 and V_2 by modal combining the peak (seismic) modal values of these stress resultants along the directions X and Y of the seismic action, separately, using the CQC modal combination rule and, then, by application of the SRSS rule for spatial combination to the previously computed maximum values. Further, Tables 4.128 and 4.129 report the values of M_3 and V_2 for the beams BX10 and BY12, respectively, for the seismic design loading combination G "+" Ψ_2Q "±" E for which the sections of BX10 and BY12 need to be detailed. The latter values have been obtained by superposing the extreme values of the seismic effects of Tables 4.126 and 4.127 to the corresponding stress resultants of the considered structural members due to the

Table 4.125 Design triads for wall W3X (ground storey) for the seismic design load combination G "+" ψ_2Q "±" E of building C [Extreme values in bold]

Mass position		N [kN]		M_2 [kNm]		M_3 [kNm]	
		Bottom	Top	Bottom	Top	Bottom	Top
1	extrN	**−877.78**	**−763.87**	17.81	−3.36	−754.02	−188.95
	extrM_2	−878.49	−1152.42	**18.01**	**76.14**	−757.60	183.56
	extrM_3	−1116.12	−1059.49	−46.25	56.33	**3371.41**	**365.00**
	−extrN	**−1267.34**	**−1153.43**	−86.73	75.94	754.02	188.95
	−extrM_2	−1266.63	−764.88	**−86.93**	**−3.56**	757.60	−183.56
	−extrM_3	−1029.00	−857.81	−22.67	16.25	**−3371.41**	**−365.00**
2	extrN	**−877.78**	**−763.87**	17.81	−3.36	754.02	188.95
	extrM_2	−878.49	−1152.42	**18.01**	**76.14**	757.59	−183.56
	extrM_3	−1029.00	−857.81	−22.67	16.25	**3371.41**	**365.00**
	−extrN	**−1267.34**	**−1153.43**	−86.73	75.94	−754.02	−188.95
	−extrM_2	−1266.63	−764.88	**−86.93**	**−3.56**	−757.59	183.56
	−extrM_3	−1116.12	−1059.49	−46.25	56.33	**−3371.41**	**−365.00**
3	extrN	**−876.42**	**−762.51**	18.18	−3.63	0.00	0.00
	extrM_2	−877.18	−1153.72	**18.38**	**76.43**	0.00	0.00
	extrM_3	−1072.56	−958.65	−34.46	36.29	**4042.73**	**464.03**
	−extrN	**−1268.70**	**−1154.79**	−87.10	76.21	0.00	0.00
	−extrM_2	−1267.94	−763.58	**−87.30**	**−3.85**	0.00	0.00
	−extrM_3	−1072.56	−958.65	−34.46	36.29	**−4042.73**	**−464.03**
4	extrN	**−876.42**	**−762.51**	18.18	−3.63	0.00	0.00
	extrM_2	−877.18	−1153.72	**18.38**	**76.43**	0.00	0.00
	extrM_3	−1072.56	−958.65	−34.46	36.29	**2542.74**	**236.81**
	−extrN	**−1268.70**	**−1154.79**	−87.10	76.21	0.00	0.00
	−extrM_2	−1267.94	−763.58	**−87.30**	**−3.85**	0.00	0.00
	−extrM_3	−1072.56	−958.65	−34.46	36.29	**−2542.74**	**−236.81**

Table 4.126 Extreme values of stress resultants in beam BX10 of building C (ground storey)

	V_2 [kN]			M_3 [kNm]		
Mass position	Left end	Midspan	Right end	Left end	Midspan	Right end
1	±28.84	±28.84	±28.84	±65.07	±4.87	±55.32
2	±30.34	±30.34	±30.34	±68.41	±5.08	±58.26
3	±36.08	±36.08	±36.08	±81.36	±6.05	±69.26
4	±21.54	±21.54	±21.54	±48.61	±3.64	±41.33

gravitational permanent and quasi-permanent variable actions summarized in Table 4.116 (Sect. 4.3.8).

Table 4.127 Extreme values of stress resultants in beam BY12 of building C (ground storey)

	V_2 [kN]			M_3 [kNm]		
Mass position	Left end	Midspan	Right end	Left end	Midspan	Right end
1	±65.39	±65.39	±65.39	±102.89	±11.35	±80.19
2	±89.56	±89.56	±89.56	±140.83	±15.45	±109.93
3	±75.53	±75.53	±75.53	±118.78	±13.04	±92.70
4	±81.82	±81.82	±81.82	±128.70	±14.16	±100.39

Table 4.128 Design effects beam BX10 (ground storey) for the seismic design load combination G "+" ψ_2Q "±" E of building C

		V2 [kN]			M3 [kNm]		
Mass position	Loading combination	Left end	Midspan	Right end	Left end	Midspan	Right end
1	G + ψ2Q + E	−7.33	23.99	60.18	35.29	20.37	40.4
	G + ψ2Q − E	−65	−33.68	2.51	−94.84	10.63	−70.25
2	G + ψ2Q + E	−5.82	25.49	61.69	38.64	20.58	43.33
	G + ψ2Q − E	−66.5	−35.19	1.01	−98.19	10.42	−73.18
3	G + ψ2Q + E	−0.086	31.23	67.42	51.58	21.55	54.33
	G + ψ2Q − E	−72.24	−40.92	−4.73	−111.13	9.45	−84.18
4	G + ψ2Q + E	−14.62	16.7	52.89	18.84	19.14	26.4
	G + ψ2Q − E	−57.7	−26.39	9.8	−78.39	11.86	−56.25

Table 4.129 Design effects beam BY12 (ground storey) for the seismic design load combination G "+" ψ_2Q "±" E of building C

		V_2 [kN]			M_3 [kNm]		
Mass position	Loading combination	Left end	Midspan	Right end	Left end	Midspan	Right end
1	$G+\psi_2Q+E$	35.96	61.66	84.44	86.56	18.97	75.79
	$G+\psi_2Q-E$	−94.82	−69.12	−46.34	−119.23	−3.73	−84.59
2	$G+\psi_2Q+E$	60.13	85.83	108.61	124.5	23.07	105.53
	$G+\psi_2Q-E$	−118.98	−93.28	−70.5	−157.16	−7.83	−114.33
3	$G+\psi_2Q+E$	46.1	71.8	94.58	102.45	20.66	88.3
	$G+\psi_2Q-E$	−104.96	−79.26	−56.47	−135.11	−5.42	−97.1
4	$G+\psi_2Q+E$	52.39	78.09	100.87	112.37	21.77	95.99
	$G+\psi_2Q-E$	−111.25	−85.55	−62.76	−145.03	−6.54	−104.79

4.3.9.3 Verification Check of the Influence of Second Order Effects

The rationale for the verification check for second-order effects and its implications in the seismic design process have been discussed in detail in Sects. 2.4.3.3 and 3.2.1. This deformation-based verification check involves determination of the interstorey drift sensitivity coefficients θ_X and θ_Y along the principal directions X and Y, respectively, for all storeys defined in Eq. (3.17). The "rigorous approach"

Table 4.130 Expected extreme values of the interstorey drift sensitivity coefficients θ_X and θ_Y for building C

Storey	extrθ_X	extrθ_Y
1	0.0326	0.0345
2	0.0370	0.0510
3	0.0261	0.0302
4	0.0227	0.0260

Table 4.131 Expected extreme values of ratios $(v \cdot d_{rX})/h$ and $(v \cdot d_{rY})/h$ of Eq. (3.23) for building C

Storey k	extr$d_{rX}^{(k)}$ [cm]	extr$d_{rY}^{(k)}$ [cm]	v	h_k [cm]	$\frac{v \cdot \text{extr}d_{rX}^{(k)}}{h_k}$	$\frac{v \cdot \text{extr}d_{rY}^{(k)}}{h_k}$
1	1.44	1.23	0.5	450	0.0016	0.0014
2	1.02	0.95	0.5	300	0.0017	0.0016
3	0.97	0.93	0.5	300	0.0016	0.0015
4	0.87	0.85	0.5	300	0.0015	0.0014

presented in Sect. 3.2.1.2 and delineated in FC-3.10b is herein followed to estimate the coefficients θ_X and θ_Y via Eq. (3.20) using displacements obtained by means of the MRSM of analysis for the building C. The step-by-step numerical implementation of this approach has been exemplified for buildings A and B in Sects. 4.1.9.3 and 4.2.9.3, respectively. Herein, the expected extreme values of the interstorey drift sensitivity coefficients θ_X and θ_Y for all storeys of building C due to simultaneous design seismic action along the principal directions X and Y are directly reported in Table 4.130 for mass position 1 obtained by the aforementioned approach. It is seen that the criterion of clause §4.4.2.2 (2) of EC8 is satisfied at all floors along both considered directions Y, that is, extr$\theta_X \leq 0.1$ and extr$\theta_Y \leq 0.1$. Therefore, no special provisions need to be taken to account for second-order effects for building C.

4.3.9.4 Verification Check for Maximum Interstorey Drift Demands

The aims of and rationale for the verification check for the maximum allowed interstorey drifts (or damage limitation verification check) and its implications in the seismic design process have been discussed in detail in Sect. 3.2.2. This deformation-based verification check relies on Eq. (3.25) and involves determination of the design interstorey drifts d_{rX} and d_{rY} along the principal axes X and Y, respectively, for all building storeys and for simultaneous design seismic action along the X and Y directions. The computational steps that need to be taken to estimate the expected extreme values of d_{rX} and d_{rY} from displacements (seismic effects) derived from the MRSM of analysis are provided in FC-3.11b (see also Sect. 4.1.9.4). Detailed illustrations of the underlying numerical procedure and involved calculations have been provided for the cases of buildings A and B in Sects. 4.1.9.4 and 4.2.9.4, respectively. Herein, the expected extreme values of the interstorey drifts for all storeys of building C due to simultaneous design seismic action along the principal directions X and Y are directly reported in Table 4.131. It is noted that the values obtained for mass position 1 are listed. Assuming that building C has brittle non-structural infill walls, the maximum allowed interstorey drift according to clause 4.4.3.2(1) of EC8 is equal to 0.5 % of the storey height, i.e., $\left(\text{v extr}d_{rX}^{(k)}\right)/\text{h} < 0.005$ and $\left(\text{v extr}d_{rY}^{(k)}\right)/\text{h} < 0.005$ for all k storeys. The above condition is met at all storeys of building C along both principal directions.

4.3.10 Determination of Normal Stresses Transferred from Pad Footings to Supporting Ground

In this section, two alternative methods for the calculation of vertical (normal) stresses transferred from pad footings to the supporting ground are considered to determine the stress undertaken by the ground supporting the r/c core footing of building C due to the seismic design load combination G "+" ψ_2Q "±" E. The first method uses the Meyerhof stress distribution (Arnold and Fenton 2013). This method assumes that a reduced ("effective") part of the total footing area transfers normal stresses to the supporting ground. The method yields an average value for the normal stresses assuming uniform distribution over the effective footing area. The second method assumes that normal stresses are linearly distributed over the total footing area of the pad and yields a normal stress/ pressure diagram as a function of the X and Y coordinates.

Irrespective of the method used to determine the peak (design) value of normal stresses transmitted to the ground, these stresses need to be compared with the bearing capacity of the supporting soil to verify the adequacy of the footing size. The soil bearing capacity depends on the soil characteristics and its macro-mechanical properties. It can be determined either by empirical formulae, such as those specified in Eurocode 7, hereafter EC7 (CEN 2003), or by pertinent tables proposed in the literature. The soil properties required to determine the soil bearing capacity are usually estimated from standard geotechnical field tests for soil characterization. Nevertheless, the verification check for soil bearing capacity is not undertaken herein, as the focus is on determining the peak normal stresses transmitted to the ground, i.e., the design effects due to the seismic design load combination computed independently by the two considered methods.

4.3.10.1 Determination of Pad Footing Normal Stresses Assuming Uniform Distribution Over a Reduced Footing Area

Consider a rectangular pad footing (e.g., footing F2 in Figs. 4.10 and 4.11(b)) carrying at its center of gravity an axial (gravitational) force F_3 and bending moments M_1 and M_2 about the two horizontal principal axes X and Y. A method commonly used in practice to quantify the normal stresses induced by such a footing to the supporting ground assumes uniformly distributed stress/pressure acting over a reduced ("effective") footing area (Meyerhof stress distribution, see

Table 4.132 Spring support reactions due to various design actions at the center of gravity of pad footings required for the determination of normal stresses transmitted to the supporting ground

Action	N [kN]	M_1 [kNm]	M_2 [kNm]
Permanent (G)	$F_{3,G}$	$M_{1,G}$	$M_{2,G}$
Variable (Q)	$F_{3,Q}$	$M_{1,Q}$	$M_{2,Q}$
Seismic (E)	$\pm$extrF_3	$\pm M_{1,F3}$	$\pm M_{2,F3}$
	$\pm N_{3,M1}$	$\pm$extrM_1	$\pm M_{2,M1}$
	$\pm N_{3,M2}$	$\pm M_{1,M2}$	$\pm$extrM_2

Arnold and Fenton (2013). The dimensions of the effective area depend on the load eccentricity along the X and Y directions defined as the ratio of the bending moment about the Y and X directions, respectively, over the axial force F_3. Apart from its computational simplicity, this method is quite attractive, since it yields a single (average) value of the design normal stresses for which the soil bearing capacity can be checked for (see e.g., Fardis 2009). Specifically, the considered method comprises the following three steps.

Step 1: Determination of support reactions F_3, M_1, M_2 due to permanent (G), variable (Q) and design seismic actions ($\pm$E).

The calculation of the normal stresses transferred by a pad footing to the supporting ground necessitates knowledge of the support reactions at the center of gravity of the footing. These reactions coincide with the force F_3 and the moments M_1 and M_2 carried by the vertical translational spring, k_3, and the two rotational springs, k_{r1} and k_{r2}, used in a typical FE model to capture the soil compliance for pad footings, as depicted in Fig. 4.13. Therefore, similar to the case of determining design seismic effects for vertical structural members subject to bi-axial bending under axial force (see Sect. 3.1.5.1), 6 different design triads (F_3, M_1, M_2) for each of the four FE models considered in the context of the MRSM (four different assumed mass locations in Table 4.117) need to be determined. Each seismic design triad comprises the expected extreme value of a particular reaction with alternating signs due to simultaneous seismic action along the X and Y directions and the corresponding concurrent values of the other two reactions, as delineated in Table 4.132.

Step 2: Determination of the design triads F_3, M_1, M_2 (support reactions) due to the seismic design combination of actions G "+" ψ_2Q "$\pm$" E.

The design triads of support reactions F_3, M_1, M_2 due to the design seismic combination of actions G "+" ψ_2Q "$\pm$" E are computed by utilizing the action effects of Table 4.132 in conjunction with equation (4.30) of EC8 (clause §4.4.2.6 (4) EC8) applied to compute the design values of M_1 and M_2 reactive moments. Specifically, the aforementioned equation reads as

$$E_{Fd} = E_{F,G+\psi 2Q} + \gamma_{Rd}\Omega E_{F,E}, \tag{4.6}$$

where

- E_{Fd} is the design value of the support reaction E_F at the footing;
- $E_{F,G+\psi 2Q}$ is the value of the support reaction E_F due to the combination of the permanent and variable actions, G"+"ψ_2Q, of the design seismic action combination;
- $E_{F,E}$ is the value of the support reaction E_F due to the design seismic action;
- γ_{Rd} is the overstrength factor determined by

$$\gamma_{Rd} = \begin{cases} 1.0 & \textit{if } q \leq 3 \\ 1.2, & \text{otherwise} \end{cases}; \text{ and}$$

- Ω is the minimum value of the moment capacity over the moment demand ratio (M_{Rd}/M_{Ed}) $\leq$ q in the two orthogonal principal directions X and Y at the lowest cross-section where a plastic hinge can form in the vertical structural member for the seismic design combination. The procedure of determining the values Ω_1 and Ω_2 of the above ratio given by the expressions

$$\Omega_1' = \min\left\{\Omega_1\left(N_{Ed(+)}^{(v.e)}\right), \Omega_1\left(N_{Ed(-)}^{(v.e)}\right)\right\} \leq q \tag{4.7}$$

and

$$\Omega_2' = \min\left\{\Omega_2\left(N_{Ed(+)}^{(v.e.)}\right), \Omega_2\left(N_{Ed(-)}^{(v.e.)}\right)\right\} \leq q \tag{4.8}$$

along the principal directions "1" and "2" in the local coordinate system of the footing (see Fig. 4.13) is delineated in Figs. 4.24 and 4.25, respectively. It is noted that the moment capacity M_{RD} of vertical r/c structural members is significantly influenced by the value of the axial force $N_{Ed}^{(v.e.)}$ they carry. In this regard, two different values of the moment capacity M_{RD} are considered in determining the Ω_1 and Ω_2 ratios in the context of the MRSM: the first value, $M_{RD(+)}$, is computed by assuming that the design value of the axial load N carried by the structural member supported by the considered pad footing attains the positive sign, $N_{Ed(+)}^{(v.e.)}$, and the second value, $M_{RD(-)}$, is computed by assuming that the design value of the axial load attains the negative sign, $N_{Ed(-)}^{(v.e.)}$.

Table 4.133 collects the 6 design triads (for each mass position) of support reactions F_3, M_1, M_2 due to the design seismic combination of actions G "+" ψ_2Q "±" E. Note that these design triads are compiled by the values of the spring support reactions for different types of actions of Table 4.132, while Eq. (4.6) is applied to determine only the expected extreme values of the reactive moments M_1 and M_2 for simultaneous seismic action along both principal axes X and Y (noted in bold-faced typeset in Table 4.133).

It is noted in passing that the two design triads shown in Table 4.134 (for each mass position) solely comprising the expected extreme values due to the design

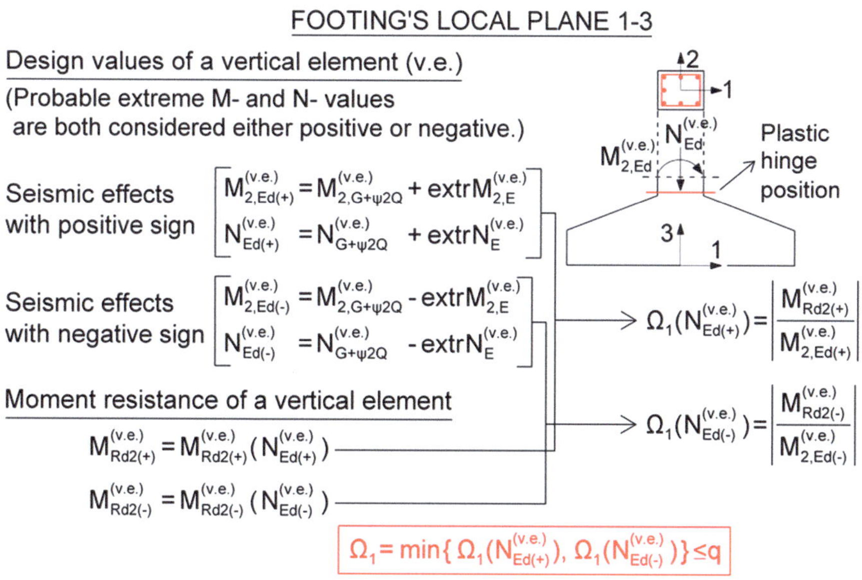

Fig. 4.24 Calculation of the factor Ω for the vertical plane 1–3 of the footing in local coordinates

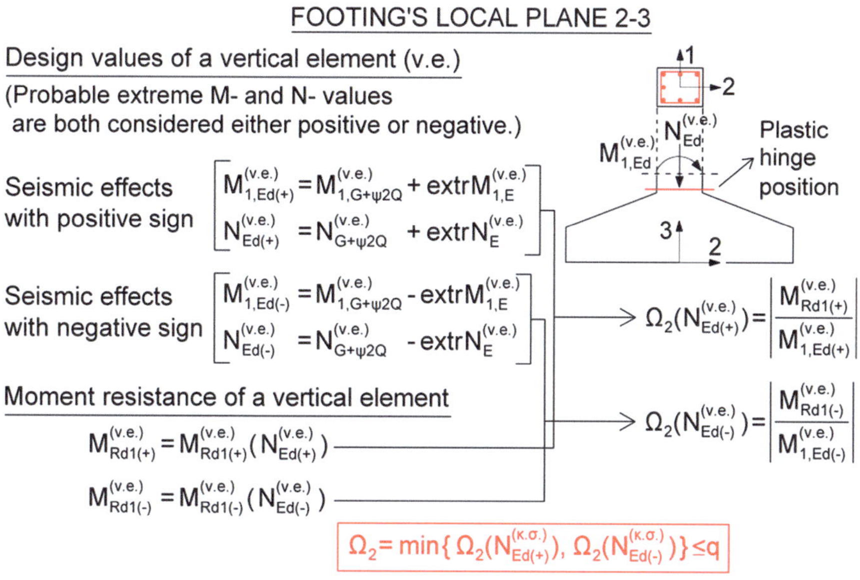

Fig. 4.25 Calculation of the factor Ω for the vertical plane 2–3 of the footing in local coordinates

Table 4.133 Design triads of support reactions F_3, M_1, M_2 due to the seismic design combination of actions G "+" ψ_2Q "±" E

Design triad	F_3 [kN]	M_1 [kNm]	M_2 [kNm]
(1)	$F_{3,G+\psi 2Q} + extrF_3$	$M_{1,G+\psi 2Q} + M_{1,F3}$	$M_{2,G+\psi 2Q} + M_{2,F3}$
(2)	$F_{3,G+\psi 2Q} + F_{3,M1}$	$M_{1,G+\psi 2Q} + \gamma_{Rd} \cdot \Omega_1 \cdot extrM_1$	$M_{2,G+\psi 2Q} + M_{2,M1}$
(3)	$F_{3,G+\psi 2Q} + F_{3,M2}$	$M_{1,G+\psi 2Q} + M_{1,M2}$	$M_{2,G+\psi 2Q} + \gamma_{Rd} \cdot \Omega_2 \cdot extrM_2$
(4)	$F_{3,G+\psi 2Q} - extrF_3$	$M_{1,G+\psi 2Q} - M_{1,F3}$	$M_{2,G+\psi 2Q} - M_{2,F3}$
(5)	$F_{3,G+\psi 2Q} - F_{3,M1}$	$M_{1,G+\psi 2Q} - \gamma_{Rd} \cdot \Omega_1 \cdot extrM_1$	$M_{2,G+\psi 2Q} - M_{2,M1}$
(6)	$F_{3,G+\psi 2Q} - F_{3,M2}$	$M_{1,G+\psi 2Q} - M_{1,M2}$	$M_{2,G+\psi 2Q} - \gamma_{Rd} \cdot \Omega_2 \cdot extrM_2$

Table 4.134 Design triads comprising only the expected extreme values of support reactions F_3, M_1, M_2 due to the seismic design combination of actions G "+" ψ_2Q "±" E

Design triad	F_3 [kN]	M_1 [kNm]	M_2 [kNm]
(1)	$F_{3,G+\psi 2Q} + extrF_3$	$M_{1,G+\psi 2Q} + \gamma_{Rd} \cdot \Omega_1 \cdot extrM_1$	$M_{2,G+\psi 2Q} + \gamma_{Rd} \cdot \Omega_2 \cdot extrM_2$
(2)	$F_{3,G+\psi 2Q} - extrF_3$	$M_{1,G+\psi 2Q} - \gamma_{Rd} \cdot \Omega_1 \cdot extrM_1$	$M_{2,G+\psi 2Q} - \gamma_{Rd} \cdot \Omega_2 \cdot extrM_2$

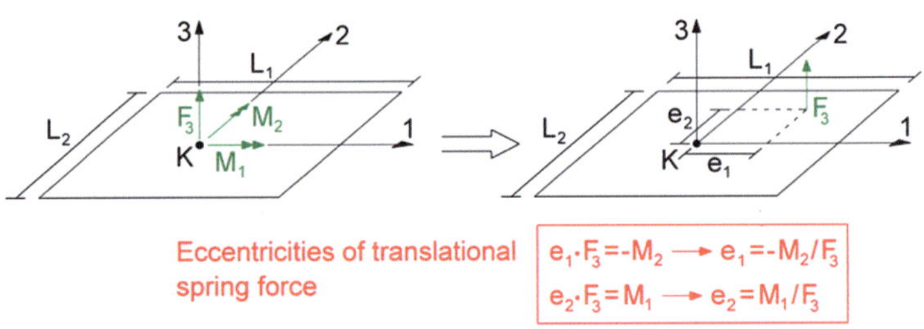

Fig. 4.26 Calculation of load eccentricities e_1 and e_2 for a typical rectangular pad footing

seismic combination of actions G "+" ψ_2Q "±" E may be alternatively used in place of the design triads of Table 4.133. However, this consideration yields, in general, overly conservative results and it is not recommended.

Step 3: Calculation of the design (effective) value of normal stresses

The average constant value of normal stresses σ_M assumed to act over an effective area A' of a rectangular pad footing is computed for each of the F_3, M_1, M_2 support (spring) reaction design triads reported in Table 4.133 (or in Table 4.134) by means of the following procedure.

First, the load eccentricities e_1 and e_2 along the directions of the local axes "1" and "2" of the footing are calculated by the expressions

$$e_1 = -M_2/F_3 \quad \text{and} \quad e_2 = M_1/F_3 \tag{4.9}$$

as shown in Fig. 4.26.

Table 4.135 Spring support reactions due to different design actions at the center of gravity of footing F1 (mass position 1) supporting the r/c core of building C

Action	F_3 [kN]	M_1 [kNm]	M_2 [kNm]
Permanent (G)	2686.16	30.42	0.0007
Variable (Q)	451.51	7.92	0.0003
Seismic (+E)	399.30	−451.54	14.07
	−399.28	451.56	−14.07
	5.66	−6.40	992.87
Seismic (−E)	−399.30	451.54	−14.07
	399.28	−451.56	14.07
	−5.66	6.40	−992.87

Table 4.136 Design triads of support reactions F_3, M_1, M_2 for footing F1 of building C due to the seismic design combination of actions G "+" ψ_2Q "±" E for mass position 1 [Extreme values in bold]

Design triad	F_3 [kN]	M_1 [kNm]	M_2 [kNm]
(1)	3220.91	−418.74	14.07
(2)	2422.33	935.91	−14.07
(3)	2827.28	26.40	1985.74
(4)	2422.32	484.34	−14.07
(5)	3220.90	−870.32	14.07
(6)	2815.96	39.20	−1985.74

Next, the reduced (effective) area A′ of the footing, over which the normal stresses transferred to the ground are non-zero, is computed by

$$A^{'} = L_1^{'} \cdot L_2^{'}, \tag{4.10}$$

where $L_1^{'}$ and $L_2^{'}$ are the effective dimensions along the local axes "1" and "2", respectively, defined by the expressions

$$L_1^{'} = L_1 - 2|e_1| \quad \text{and} \quad L_2^{'} = L_2 - 2|e_2|. \tag{4.11}$$

Finally, the value of the average normal stress σ_M assumed to be constant over the effective footing area A′ is given by

$$\sigma_M = F_3/A^{'}. \tag{4.12}$$

Calculation of the average normal stress σ_M for the r/c core footing F1 of building C

For numerical illustration purposes, the above described three-step procedure is applied to compute the average normal stress σ_M for the footing F1 supporting the r/c core of building C (see Figs. 4.9 and 4.10(c)) due to the seismic design load combination G "+" ψ_2Q "±" E for the FE model with masses positioned at location 1 in Table 4.117.

Table 4.135 collects the forces F_3 and the moments M_1 and M_2 carried by the vertical translational spring and the two rotational springs, respectively, located at the center of gravity of footing F1 for different action effects.

Next, Table 4.136 provides the 6 design triads of support reactions F_3, M_1, M_2 due to the seismic design combination of actions G "+" ψ_2Q "±" E for the footing

Table 4.137 Calculation of the average normal stress σ_M via Eqs. (4.9), (4.10), (4.11), and (4.12) for footing F1 of building C due to the seismic design combination of actions G"+"ψ_2Q"±"E for mass position 1

Design triad	$e_1 = -M_2/F_3$ [m]	$e_2 = -M_1/F_3$ [m]	$L_1' = L_1 - 2 \cdot \lvert e_1 \rvert$ [m]	$L_2' = L_2 - 2 \cdot \lvert e_2 \rvert$ [m]	$A' = L_1' \cdot L_2'$ [m^2]	σ_M [kN/m^2]
(1)	−0.0044	−0.1300	5.441	4.040	21.983	146.52
(2)	0.0058	0.386	5.438	3.527	19.183	126.28
(3)	−0.7024	0.009	4.045	4.281	17.319	163.24
(4)	0.0058	0.200	5.438	3.900	21.210	114.21
(5)	−0.0044	−0.270	5.441	3.760	20.457	157.45
(6)	0.7052	0.014	4.040	4.272	17.258	163.17

Table 4.138 Triads of the design values of the reactions of the springs of the footing consisting only of the extreme values (mass position 1)

Design triad	F_3 [kN]	M_1 [kNm]	M_2 [kNm]
(1)	3220.91	935.91	1985.74
(2)	2422.32	−870.32	−1985.74

Table 4.139 Developed active stresses of the soil for footing P1 with the use of the possible extreme values of the reaction of the springs (mass position 1)

Design triad	$e_1 = -M_2/F_3$ [m]	$e_2 = -M_1/F_3$ [m]	$L_1' = L_1 - 2 \cdot \lvert e_1 \rvert$ [m]	$L_2' = L_2 - 2 \cdot \lvert e_2 \rvert$ [m]	$A' = L_1' \cdot L_2'$ [m^2]	σ_M [kN/m^2]
(1)	−0.6165	0.2906	4.217	3.719	15.682	205.39
(2)	0.8198	−0.359	3.810	3.581	13.647	177.50

F1 (Fig. 4.10), as defined in Table 4.133. The extreme values of the support reactions are noted with bold-faced typeset. In applying Eq. (4.6), the maximum value for the ratio Ω is adopted, that is, $\Omega_1 = \Omega_2 = q = 2.0$, to simplify calculations (see Eqs. (4.7) and (4.8)). Moreover, the overstrength factor γ_{RD} is equal to 1.0, since $q = 2.0 < 3.0$.

Finally, Table 4.137 reports all required intermediate calculations for determining the average normal stress σ_M of the considered footing. It is seen (Table 4.137) that the maximum value of load eccentricity e_1 is equal to 0.7052 m and is smaller than the 1/3 of the footing dimension L_1 which is equal to $5.45/3 = 1.817$ m. Similarly, the maximum value of load eccentricity e_2 is equal to 0.386 m and is smaller than the 1/3 of the footing dimension L_2 which is equal to $4.3/3 = 1.433$ m. Therefore, footing F1 does not observe large eccentricities along any of the local principal axes, as defined in clause §6.5.4(1)P of EC7 (CEN 2003).

For the sake of comparison, Table 4.139 reports all required intermediate calculations for determining the average normal stress σ_M of the considered footing in the case of the two design triads comprising only the extreme values reported in Table 4.138 (see also Table 4.134). It is seen that the maximum value of the normal

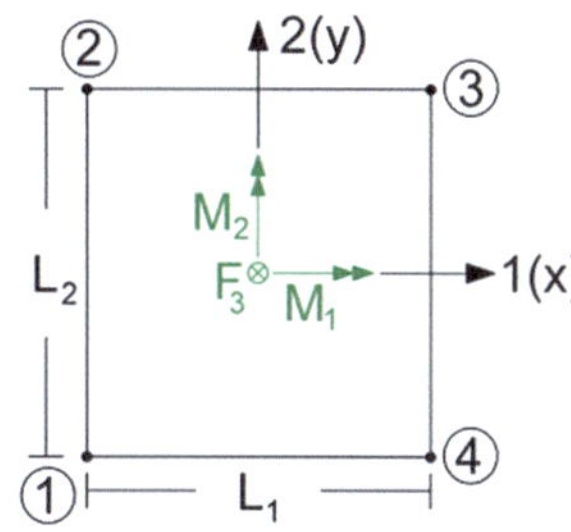

Fig. 4.27 Treatment of rectangular pad footing as a rectangular cross-section of linear structural members subject to bi-axial bending under axial force in the local coordinate system

stress σ_M obtained by the latter design triads is larger by (205.39-163.24)/163.24 = 25.82 % compared with the case of the 6 design triads defined in Table 4.133. Clearly, the consideration of only the extreme values of support reactions for the soil bearing capacity verification check yields quite conservative results in exchange for computational simplicity.

4.3.10.2 Determination of Pad Footing Normal Stresses Assuming Linear Distribution Over the Total Footing Area

This method of quantifying normal stresses applied by a rectangular footing to the supporting ground considers the standard normal stress distribution formula applicable to cross-sections of any prismatic linear structural member subject to bi-axial bending under axial force. That is,

$$\sigma(x, y) = \frac{F_3}{A} + \frac{M_1}{I_{11}}y - \frac{M_2}{I_{22}}x, \tag{4.13}$$

where $I_{11} = L_1 \cdot L_2^3/12, I_{22} = L_1^3 \cdot L_2/12$, and $A = L_1 \cdot L_2$ (Fig. 4.26). In the above expression, the coordinates $\times$ and y are measured along the local principal axes "1" and "2" of the footing, as defined in Fig. 4.27, whose origin is located at the center of gravity. It is important to note that Eq. (4.13) applies only for the case of compressive normal stresses (i.e., no uplifting/tensile stresses develop in any region of the footing). The values of the reactions F_3, M_1 and M_2 to be considered in Eq. (4.13) correspond to the design triads of Table 4.133 or, alternatively, of Table 4.134. Further, the stress distribution over the total area of the footing can be retrieved by application of Eq. (4.13) at the four corner points indicated in Fig. 4.27 and, then, by linear interpolation along axes "1" and "2". Therefore, for each mass position in Table 4.117 and for each design triad F_3, M_1 and M_2 considered, Eq. (4.13) is applied to find the normal stresses at the four corner points of each footing.

Calculation of linearly distributed normal stress σ(x,y) for the r/c core footing F1 of building C

For numerical illustration purposes, Eq. (4.13) is applied in conjunction with the 6 design triads of Table 4.136 to obtain the linearly distributed normal stresses of footing F1 due to the seismic design combination of actions G "+" ψ_2Q "±" E for building C with mass position 1 defined in Table 4.117. The obtained normal stress distributions are shown in Fig. 4.28. It is seen that normal stresses are compressive along the total area of the footing for 6 design triads considered. This is because the associated load eccentricities do not exceed the 1/6 of the footing dimensions along axes "1" and "2" for any of the design triads (see Table 4.137). Therefore, Eq. (4.13) yields valid numerical results (normal stress distribution) for the herein considered footing. Further, for the sake of comparison, the average (constant) normal stress values σ_M computed in Table 4.137 are also given for each of the examined design triads in Fig. 4.28. It is observed that, in all cases, σ_M attain smaller values than the maximum compressive stresses developed in certain corners of the footing, as computed by Eq. (4.13).

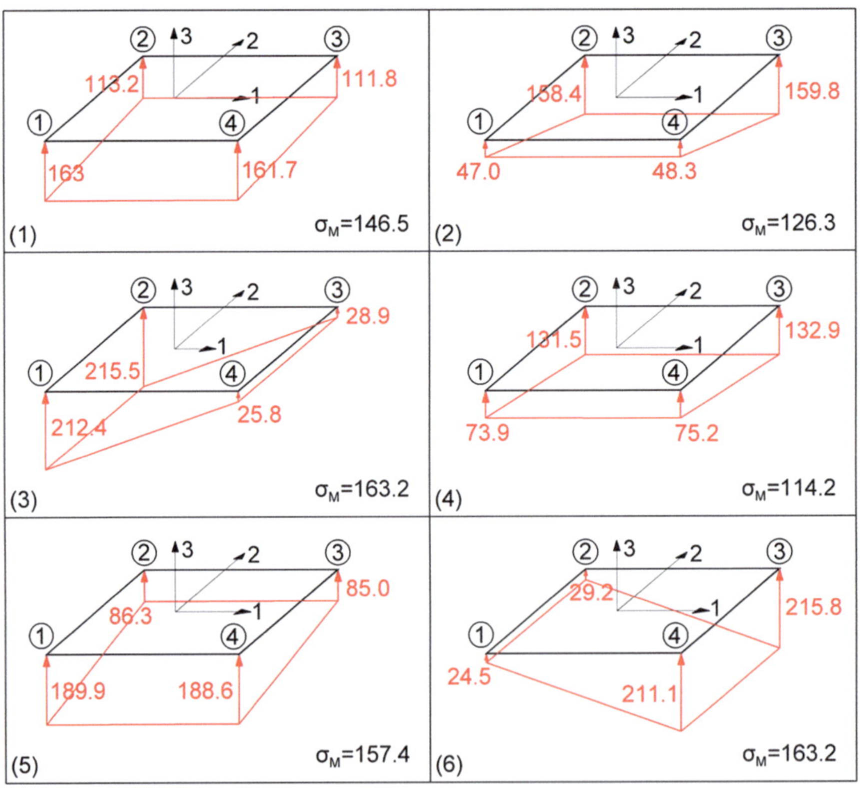

Fig. 4.28 Solid stresses of the footing P1 for the six combinations of Table 4.136 (numbers in parentheses denote the corresponding design triad of Table 4.136)

4.3.11 Detailing and Design Verifications of Typical Structural Members of Building C

In this section, the required detailing and design verification checks for ultimate limit state under bending and shear, including the capacity design provisions, are provided for certain structural members of building C in self-contained and self-explanatory tabular and graphical formats. The considered structural members are the beams BX10 and BY12, the column C8, and the planar ductile wall TY2 (see Fig. 4.9) at the ground storey. Note that beam BY12 connects the wall TY2 with the corner column C8, while both the considered beams converge to column C8. The critical cross-sections of the above structural members are detailed and verified for the effects (stress resultants) of the seismic design combination of actions G"+"ψ_2Q"$\pm$"E assuming that the centers of mass at each floor slab are located at position 1 of Table 4.117.

In particular, beams BX10 and BY12 are first detailed for bending to determine the longitudinal reinforcement. Next, column C8 is designed for bending accounting for special capacity design provisions (i.e., "weak beam-strong column" rule, as discussed in Sect. 2.2.4.1). Subsequently, the considered beams and the column are designed for shear using the capacity design shear forces. The column is also verified for adequate concrete confinement. Finally, wall TY2 is designed for bending and shear, including concrete confinement verification check for the critical regions. It is noted that the considered design action effects, obtained for mass position 1, are not necessarily the most critical for all structural members examined (see, e.g., Tables 4.124 and 4.125). The complete seismic design of the examined structural members should involve their detailing and verification, following the aforementioned sequence, for seismic design combination of actions G"+"ψ_2Q"$\pm$"E for all four different mass positions of Table 4.117.

Furthermore, it is also noted that the capacity design verification checks are only undertaken for illustration purposes and can be omitted for building C. This is because more than 50 % of the seismic base shear is undertaken by the r/c walls and, therefore, the capacity design verification check of columns is not critical, as their contribution in resisting the design seismic action is relatively small.

Finally, it is noted that pertinent clarification comments and footnotes are included to explain the underlying assumptions and computations involved in the derivation of all the ensuing tabulated numerical data associated with the various detailing and design verification checks. Moreover, numerical data typesetting is color-coded as follows: black is used to indicate data and numerical values chosen or selected by the design engineer, while blue is reserved for numerical data and quantities obtained by means of automated computations from previous data. In this manner, a distinction is made between the quantities and data values free to be chosen by the design engineer based on their judgement and experience and those quantities and data that are obtained by application of EC8 specific computation steps.

DESIGN OF BEAM BY12 TO BENDING - Input data

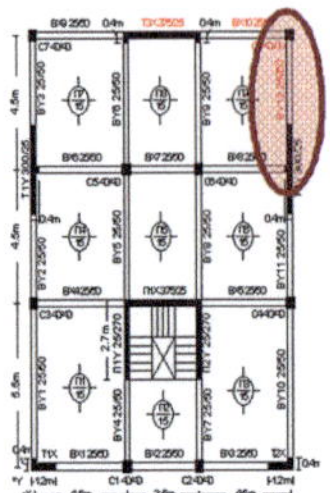

Building data

	XC2	environmental exposure class (EN 1992-1-1, Table 4.1)
	S4	structural class
	DCM	ductility class
T (sec) =	0.47	uncoupled transl. period at the direction of the beam
	B	soil class
T_c (sec) =	0.50	corner period
q_0 =	2.00	basic value of the behaviour factor (EN1998-1, Table 5.1)

Beam section BY12

b_w (m) =	0.25	beam section width
h_w (m) =	0.50	beam section height
h_f (m) =	0.15	slab thickness
L (m) =	2.80	beam length (clear distance between faces of the supports)
	L	section type: single-sided (edge) beam / flange beam

Material properties

f_{ck} (MPa) =	20.00	characteristic compressive concrete strength
f_{yk} (MPa) =	500.00	characteristic yield strength of reinforcement
a_{cc} =	0.85	coefficient accounting for long term effects on the compressive strength
γ_c =	1.50	partial factor for concrete
γ_s =	1.15	partial factor for steel
E_s (MPa) =	2E+05	Modulus of Elasticity for Steel

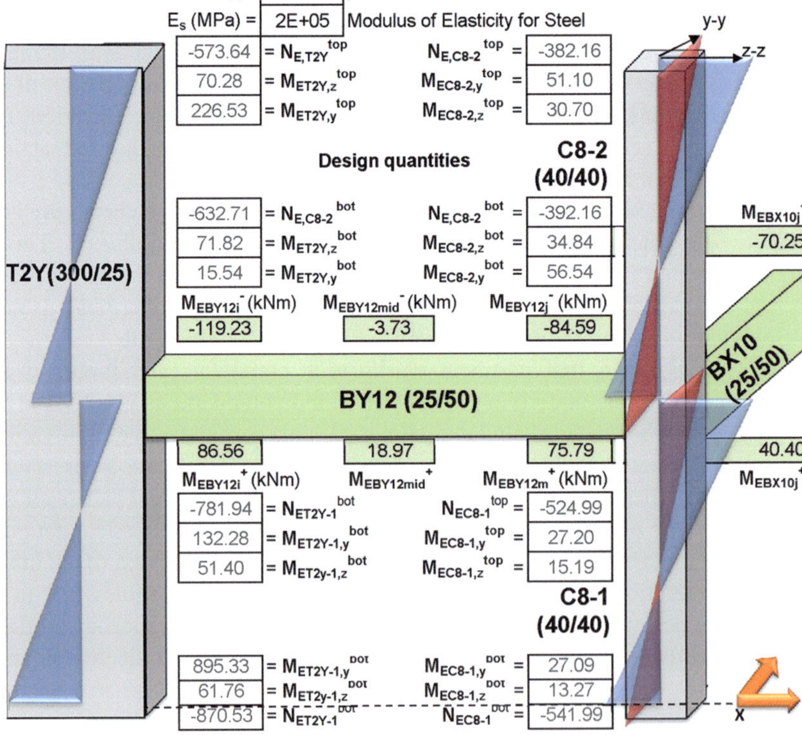

Column sections C8-1 & C8-2

b_c (m) =	0.40	column section width along the local axis z-z
h_c (m) =	0.40	column section width along the local axis x-x
H_c (m) =	3.00	column member height

Shear wall section T2Y

L_w (m) =	3.00	wall length
b_{wall} (m) =	0.25	wall thickness

Design values for concrete and steel

f_{cm} (MPa) = f_{ck} + 8MPa =	**28.00**	mean value of concrete cylinder compressive strength
f_{ctm} (MPa) =	**2.21**	mean value of axial tensile strength of concrete
a_{cc} =	**0.85**	coefficient accounting long term effects on the compressive strength
f_{cd} (MPa) = a_{cc} f_{ck} / γ_c =	**11.33**	design value of concrete compressive strength
f_{yd} (MPa) = f_{yk} / γ_s =	**434.78**	design yield strength of reinforcement
f_{ywd} (MPa) = f_{yk} / γ_s =	**434.78**	design yield of shear reinforcement

Required concrete cover - Durability

Initial bar diameter

d_{bL} (mm) =	14	Φ14	Desirable diameter of longit. reinforcement
d_w (mm) =	8	Φ8	Desirable diameter of shear reinforcement

Concrete cover and effective section depth

$c_{min,b}$ (mm) =	**14.00**	minimum cover
$\Delta c_{dur,\gamma}$ (mm) =	0.00	additive safety element
$\Delta c_{dur,st}$ (mm) =	0.00	reduction of min cover for use of stainless steel
$\Delta c_{dur,add}$ (mm) =	0.00	reduction of min cover for use of additional protection
$c_{min,dur}$ (mm) =	25.00	minimum cover due to environmental conditions

EN 1992-1-1, §4.4.1.2 Table 4.4

Environmental Requirement for $c_{min,dur}$ (mm)							
Structural Class	Exposure Class according to Table 4.1						
	X0	XC1	XC2 / XC3	XC4	XD1 / XS1	XD2 / XS2	XD3 / XS3
S1	10	10	10	15	20	25	30
S2	10	10	15	20	25	30	35
S3	10	10	20	25	30	35	40
S4	10	15	25	30	35	40	45
S5	15	20	30	35	40	45	50
S6	20	25	35	40	45	50	55

c_{min} (mm) =	**25.00**	minimum concrete cover
Δc_{dev} (mm) =	0.00	allowance in design for deviation
c_{nom} (mm) = c_{min} + Δc_{dev} =	**25.00**	nominal concrete cover
d (mm) = h - c_{nom} - d_w - d_{bL} /2=	**460.00**	effective section depth

Geometrical requirements

	DCM	check not required
b_w (m) =	0.25	check b_w>0.2m (for DCH)
b_w (m) =	0.25	check b_w<min(b_c+h_w, 2b_c)
b_w (m) =	0.25	check h_w/b_w<3.5 (for DCH)

Effective width of beam

l_1 (m) =	4.50	transverse spacing between beams at the left
b_1 (m) = (l_1 - b_w)/2 =	2.13	spacing between support face and transv. mid-span
l_2 (m) =	0.00	transverse spacing between beams at the right
b_2 (m) = (l_2 - b_w)/2 =	0.00	spacing between support face and transv. mid-span
l_0 (m) =	3.83	spacing between the points of zero moment

EN 1992-1-1, §5.3.2.1, Figure 5.2

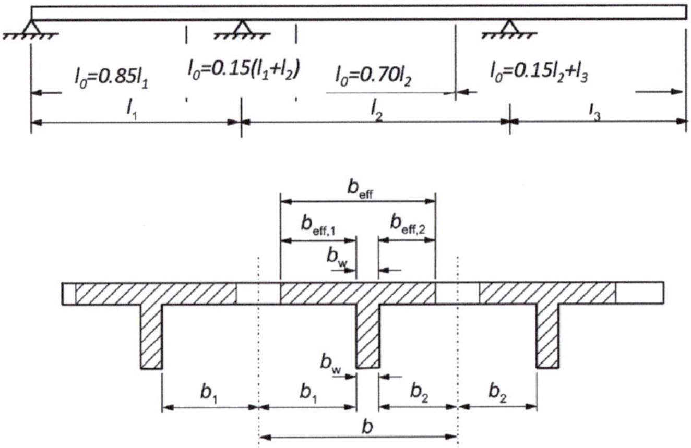

$b_{eff,1}$ (m) = 0.2b_1 + 0.1l_0 =	0.81		
$b_{eff,1}$ (m) =	0.77	check $b_{eff,1}$ < 0.2l_0 =	0.77
$b_{eff,2}$ (m) = 0.2b_2 + 0.1l_0 =	0.00		
$b_{eff,2}$ (m) =	0.00	check $b_{eff,2}$ < 0.2l_0 =	0.77
b_{eff} (m) = Σ$b_{eff,i}$ + b_w =	1.02	effective beam width	
b(m) = b_1 + b_2 + b_w =	2.38		
b_{eff} (m) =	1.02	check b_{eff} < b	

Minimum longitudinal reinforcement

	DCM	ductility class
ρ_{min} = 0.5f_{ctm}/f_{yk} =	0.22%	uniform minimum reinforcement ratio (DCM/DCH)
$A_{s,min}$ (cm^2) = ρ_{min} $b_w d$ =	2.54	minimum longitudinal reinforcement
cm^2	0.00	additional requirement 2Ø14 top and bottom (for DCH)
$A_{s,min}$ (cm^2) = ρ_{min} $b_w d$ =	2.54	final minimum longitudinal reinforcement
ρ_{min} =	0.22%	minimum ratio of longitudinal reinforcement

Maximum diameter of longitudinal reinforcement

	DCM	ductility class	EN 1998-1, §5.6.2.2
γ_{Rd} =	1.00	modeling uncertainty factor	

Left beam-wall joint (BY12 - T2Y)

joint:	interior	(adequate anchorage within the shear wall)
$v_d = N_{E,T2Y}/(b_{wall}L_w f_{cd})$ =	**-0.07**	axial load of the vertical member
k_D =	**0.67**	coefficient dependent on ductility class
$k_2 = (1+0.75k_D\rho'/\rho_{max})$ =	**1.25**	joint-dependent coefficient (interior/exterior)
$d_{bL} \leq$		
$(7.5/\gamma_{Rd})h_c{*}(1+0.8v_d)f_{ctm}/(f_{yd}k_2)$ =	**96.45**	max bar diameter to be achored within the left joint[1]

Right beam-column joint (BY12 - C8)

joint:	exterior	
$v_d = N_{E,C8}/(b_c h_c f_{cd})$ =	**-0.22**	axial load of the vertical member
$k_2 = (1+0.75k_D\rho'/\rho_{max})$ =	**1.00**	joint-dependent coefficient (interior/exterior)
k_D =	**0.00**	coefficient dependent on ductility class
$d_{bL} \leq$		
$(7.5/\gamma_{Rd})h_c{*}(1+0.8v_d)f_{ctm}/(f_{yd}k_2)$ =	**17.89**	max bar diameter to be achored within the right joint
d_{bL} (mm) ≤	**17.89**	most critical max bar diameter between the two joints[1]

Longitudinal reinforcement at the beam left support

Left support - top (rectangular section)

$M_{E,B12Yi}^{-}$ (kNm) =	**-119.23**	design bending moment
$K = M_{E,B12Yi}^{-}/(b_w d^2 f_{cd})$ =	**0.20**	joint-dependent coefficient (interior/exterior)
K_{lim} =	0.29	Diagram 5.6 (Designer's guide to EN1992-1-1)
K =	**0.20**	check for comp. reinforcement requirement ($K<K_{lim}$)
$A_s f_{yd}/b_w d f_{cd}$ =	**0.18**	Diagram 5.7 (Designer's guide to EN1992-1-1)
$A_{s,req}$ (cm^2) =	**5.39**	area of required reinforcement
$A_{s,req}$ (cm^2) =	**5.39**	check vs min required reinforcement ($A_{s,req}>A_{s,min}$)
	4Φ14	provided reinforcement
$A_{s1,i}$ (cm^2) =	**6.16**	area of provided reinforcement $A_{s1,i}>A_{s,req}$

Left support - bottom (flange beam)

$M_{E,B12Yi}^{+}$ (kNm) =	**86.56**	design bending moment
$K = M_{E,B12Yi}^{-}/(b_{eff} d^2 f_{cd})$ =	**0.04**	joint-dependent coefficient (interior/exterior)
K_{lim} =	0.29	Diagram 5.6 (Designer's guide to EN1992-1-1)
K =	**0.04**	check for comp. reinforcement requirement (K<Klim)
$A_s f_{yd}/b_{eff} d f_{cd}$ =	**0.04**	Diagram 5.7 (Designer's guide to EN1992-1-1)
$A_{s,req}$ (cm^2) =	**4.35**	area of required reinforcement
$A_{s,req}$ (cm^2) =	**4.35**	check vs min required reinforcement ($A_{s,req}>A_{s,min}$)
	3Φ14	provided reinforcement
$A_{s2,i}$ (cm^2) =	**4.62**	area of provided reinforcement $A_{s2,i}>A_{s,req}$

Check maximum ratio of longitudinal reinforcement

$\rho_{max} = \rho' + 0.0018f_{cd}/(\mu_{\varphi}\varepsilon_{sy,d}f_{yd})$ =	**0.92%**	maximum ratio of longitudinal reinforcement
where:		
T_c/T =	**1.06**	
$\mu_{\varphi} = 2q_0\text{-}1$ =	**3.00**	curvature ductility (for $T \geq T_c$)
$\mu_{\varphi} = 1+(2q_0\text{-}1)T_c/T$ =	**4.19**	curvature ductility (for $T<T_c$)
μ_{φ} =	**4.19**	final value of curvature ductility
$\varepsilon_{sy,d} = f_{yd}/E_s$ =	**0.2%**	design value of tension steel strain at yield
$\rho' = A_{s2,i}/(b_w d)$ =	**0.40%**	ratio of compression reinforcement
$A_{s,max}$ (cm^2) =	**10.54**	area of provided reinforcement $A_{s,prov} > A_{s,req}$
	Valid	area of provided reinf. top: $A_{s,max} > A_{s1,i} > A_{s,req}$
	Valid	area of provided reinf. bottom: $A_{s,max} > A_{s2,i} > A_{s,req}$

Longitudinal reinforcement at the beam right support

Right support - top (rectangular section)

$M_{E,B12Yj}^{-}$ (kNm) =	**-84.59**	design bending moment
$K = M_{E,B12Yi}^{-}/(b_w d^2 f_{cd})$ =	**0.14**	joint-dependent coefficient (interior/exterior)
K_{lim} =	0.29	Diagram 5.6 (Designer's guide to EN1992-1-1)
K =	**0.14**	check for comp. reinforcement requirement ($K<K_{lim}$)
$A_s f_{yd}/b_w d f_{cd}$ =	**0.13**	Diagram 5.7 (Designer's guide to EN1992-1-1)
$A_{s,req}$ (cm^2) =	**3.96**	area of required reinforcement
$A_{s,req}$ (cm^2) =	**3.96**	check vs min required reinforcement ($A_{s,req}>A_{s,min}$)
	3Φ14	provided reinforcement
$A_{s1,j}$ (cm^2) =	**4.62**	area of provided reinforcement $A_{s1,j} > A_{s,req}$

Right support - bottom (flange beam)

$M_{E,B12Yj}^{+}$ (kNm) =	**75.79**	design bending moment
$K = M_{E,B12Yi}^{-}/(b_{eff} d^2 f_{cd})$ =	**0.03**	joint-dependent coefficient (interior/exterior)
K_{lim} =	0.29	Diagram 5.6 (Designer's guide to EN1992-1-1)
K =	**0.03**	check for comp. reinforcement requirement (K<Klim)
$A_s f_{yd}/b_{eff} d f_{cd}$ =	**0.03**	Diagram 5.7 (Designer's guide to EN1992-1-1)
$A_{s,req}$ (cm^2) =	**3.82**	area of required reinforcement
$A_{s,req}$ (cm^2) =	**3.82**	check vs min required reinforcement ($A_{s,req}>A_{s,min}$)
	3Φ14	provided reinforcement
$A_{s2,j}$ (cm^2) =	**4.62**	area of provided reinforcement $A_{s2,j} > A_{s,req}$

Check maximum ratio of longitudinal reinforcement

$\rho_{max} = \rho' + 0.0018f_{cd}/(\mu_{\varphi}\varepsilon_{sy,d}f_{yd})$ =	**0.92%**	maximum ratio of longitudinal reinforcement
όπου:		
T_c/T =	**1.06**	
$\mu_{\varphi} = 2q_0\text{-}1$ =	**3.00**	curvature ductility (for $T \geq T_c$)
$\mu_{\varphi} = 1+(2q_0\text{-}1)T_c/T$ =	**4.19**	curvature ductility (for $T<T_c$)
μ_{φ} =	**4.19**	final value of curvature ductility (for $T<T_c$)
$\varepsilon_{sy,d} = f_{yd}/E_s$ =	**0.2%**	design value of tension steel strain at yield

$\rho' = A_{s2,j}/(b_w d) =$	**0.40%**	ratio of compression reinforcement
$A_{s,max}$ (cm^2) =	**10.54**	area of provided reinforcement $A_{s,prov} > A_{s,req}$
	Valid	area of provided reinf. top: $A_{s,max} > A_{s1,j} > A_{s,req}$
	Valid	area of provided reinf. bottom: $A_{s,max} > A_{s2,j} > A_{s,req}$

Longitudinal reinforcement at the beam mid-span

Mid-span - bottom (flange beam)

$M_{E,B12Ym}^{+}$ (kNm) =	**18.97**	design bending moment
$K = M_{E,B12Yi}^{-}/(b_{eff}\, d^2 f_{cd}) =$	**0.01**	joint-dependent coefficient (interior/exterior)
$K_{lim} =$	0.29	Diagram 5.6 (Designer's guide to EN1992-1-1)
$K =$	**0.01**	check for comp. reinforcement requirement (K<Klim)
$A_s f_{yd}/b_{eff}\, d\, f_{cd} =$	**0.01**	Diagram 5.7 (Designer's guide to EN1992-1-1)
$A_{s,req}$ (cm^2) =	**1.01**	area of required reinforcement
$A_{s,req}$ (cm^2) =	**2.54**	check vs min required reinforcement ($A_{s,req} > A_{s,min}$)
	2Φ14	provided reinforcement
$A_{s1,m}$ (cm^2) =	**3.08**	area of provided reinforcement $A_{s1,m} > A_{s,req}$

Mid-span - top (rectangular section)

$M_{E,B12Ym}^{-}$ (kNm) =	**-3.73**	design bending moment
$K = M_{E,B12Yi}^{-}/(b_w\, d^2 f_{cd}) =$	**0.01**	joint-dependent coefficient (interior/exterior)
$K_{lim} =$	0.29	Diagram 5.6 (Designer's guide to EN1992-1-1)
$K =$	**0.01**	check for comp. reinforcement requirement (K<Klim)
$A_s f_{yd}/b_w\, d\, f_{cd} =$	**0.01**	Diagram 5.7 (Designer's guide to EN1992-1-1)
$A_{s,req}$ (cm^2) =	**0.20**	area of required reinforcement
$A_{s,req}$ (cm^2) =	**2.54**	check vs min required reinforcement ($A_{s,req} > A_{s,min}$)
	2Φ14	provided reinforcement
$A_{s2,m}$ (cm^2) =	**3.08**	area of provided reinforcement $A_{s2,m} > A_{s,req}$

Check maximum ratio of longitudinal reinforcement

$\rho_{max} = \rho' + 0.0018 f_{cd}/(\mu_\varphi \varepsilon_{sy,d} f_{yd}) =$	**0.78%**	maximum ratio of longitudinal reinforcement
where:		
$T_c/T =$	**1.06**	
$\mu_\varphi = 2q_0 - 1 =$	**3.00**	curvature ductility (for T≥Tc)
$\mu_\varphi = 1 + (2q_0 - 1)T_c/T =$	**4.19**	curvature ductility (for T<Tc)
$\mu_\varphi =$	**4.19**	final value of curvature ductility (for T<Tc)
$\varepsilon_{sy,d} = f_{yd}/E_s =$	**0.2%**	design value of tension steel strain at yield

$\rho' = A_{s2,m}/(b_w d) =$	**0.27%**	ratio of compression reinforcement
$A_{s,max}$ (cm^2) =	**9.00**	area of provided reinforcement $A_{s,prov} > A_{s,req}$
	Valid	area of provided reinf. bottom: $A_{s,max} > A_{s1,m} > A_{s,req}$
	Valid	area of provided reinf. top: $A_{s,max} > A_{s2,m} > A_{s,req}$

General checks for compression reinforcement

Valid	comp. reinf. > 50% tens. at support i ($A_{s2,i} > A_{s1,i}/2$)
Valid	comp. reinf. > 50% tens. at support j ($A_{s2,j} > A_{s1,j}/2$)
Valid	comp. reinf. at support i > 25% tens. at mid-span ($A_{s2,i} > A_{s1,m}/4$)

Valid	comp. reinf. at support j > 25% tens. at mid-span ($A_{s2,j} > A_{s1,m}/4$)
Valid	comp. reinf. at mid-span ($A_{s2,m} > \max\{A_{s1,i}, A_{s1,j}\}/4$)

NOTES

[1] In cases where the maximum diameter is found very small (e.g. Ø<14) it is either possible to apply detailing provisions for non-confronting diameter limits, or, to increase the dimensions of the column.

DESIGN OF BEAM BX10 TO BENDING - Input data

Building data

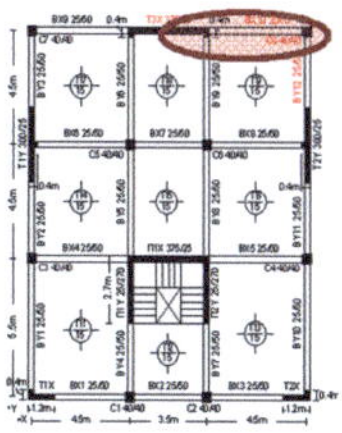

	XC2	environmental exposure class
	S4	structural class
	DCM	ductility class
T (sec) =	0.47	uncoupled transl. period at the direction of the beam
	B	soil class
T_c (sec) =	0.50	corner period
q_0 =	2.00	basic value of the behaviour factor

Beam section BX10

b_w (m) =	0.25	beam section width
h_w (m) =	0.50	beam section height
h_f (m) =	0.15	slab thickness
L (m) =	4.50	beam length
	L	section type: single-sided (edge) beam/flange beam

Material properties

f_{ck} (MPa) =	20.00	characteristic compressive concrete strength
f_{yk} (MPa) =	500.00	characteristic yield strength of reinforcement
a_{cc} =	0.85	coef. to account long term effects on comp. strength
γ_c =	1.50	partial factor for concrete
γ_s =	1.15	partial factor for steel
E_s (MPa) =	200000	Modulus of Elasticity for Steel

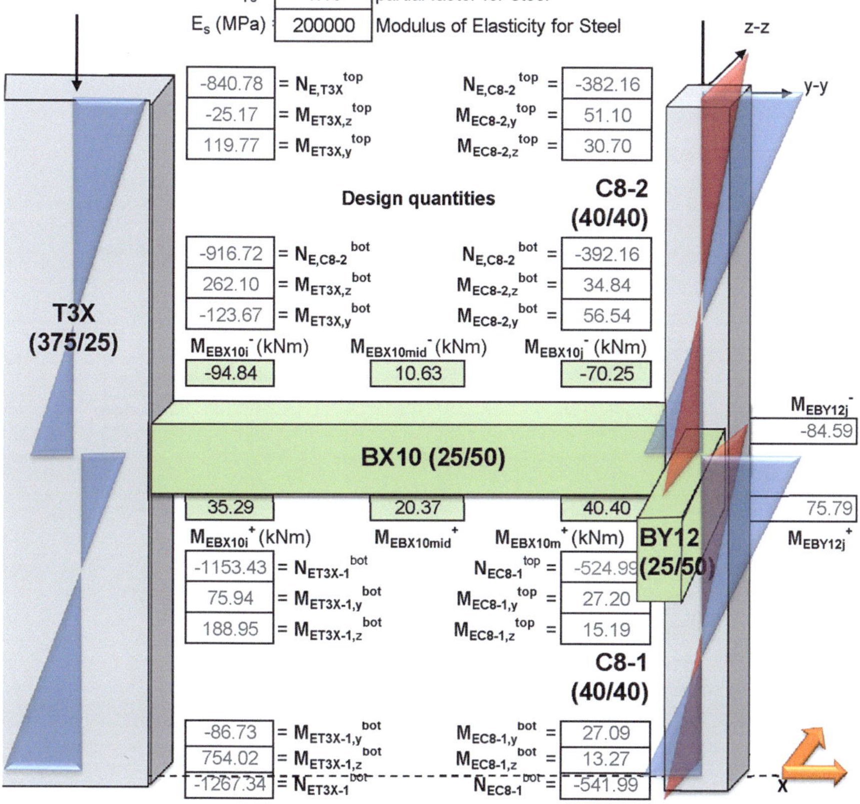

Column sections C8-1 & C8-2

b_c (m) =	0.40	column section width along the local axis z-z
h_c (m) =	0.40	column section width along the local axis x-x
H_c (m) =	3.00	column section height

Shear wall section T3X

L_w (m) =	3.75	wall length
b_{wall} (m) =	0.25	wall thickness

Design values for concrete and steel

f_{cm} (MPa) = f_{ck} + 8MPa =	**28.00**	mean value of concrete cylinder comp. strength
f_{ctm} (MPa) =	**2.21**	mean value of axial tensile strength of concrete
a_{cc} =	**0.85**	coefficient accounting long term effects
f_{cd} (MPa) = $a_{cc}\, f_{ck} / \gamma_c$ =	**11.33**	design value of concrete compressive strength
f_{yd} (MPa) = f_{yk} / γ_s =	**434.78**	design yield strength of reinforcement
f_{ywd} (MPa) = f_{yk} / γ_s =	**434.78**	design yield of shear reinforcement

Required concrete cover - Durability

Initial bar diameter

d_{bL} (mm) =	14	Φ14	Desirable diameter of long. reinf.
d_w (mm) =	8	Φ8	Desirable diameter of shear reinf.

Concrete cover and effective section depth

$c_{min,b}$ (mm) =	**14.00**	minimum cover
$\Delta c_{dur,\gamma}$ (mm) =	0.00	additive safety element
$\Delta c_{dur,st}$ (mm) =	0.00	reduction of min cover for use of stainless steel
$\Delta c_{dur,add}$ (mm) =	0.00	reduction of min cover for use of added protection
$c_{min,dur}$ (mm) =	25.00	minimum cover due to environmental conditions

EN 1992-1-1, §4.4.1.2 Table 4.4

Environmental Requirement for $c_{min,dur}$ (mm)							
Structural Class	Exposure Class according to Table 4.1						
	X0	XC1	XC2 / XC3	XC4	XD1 / XS1	XD2 / XS2	XD3 / XS3
S1	10	10	10	15	20	25	30
S2	10	10	15	20	25	30	35
S3	10	10	20	25	30	35	40
S4	10	15	25	30	35	40	45
S5	15	20	30	35	40	45	50
S6	20	25	35	40	45	50	55

c_{min} (mm) =	**25.00**	minimum concrete cover
Δc_{dev} (mm) =	0.00	allowance in design for deviation
c_{nom} (mm) = c_{min} + Δc_{dev} =	**25.00**	nominal concrete cover
d(mm)=h $-c_{nom}-d_w-d_{bL}/2$=	**460.00**	effective section depth

Geometrical requirements

	DCM	Check not required
b_w (m) =	0.25	check b_w>0.2m (for DCH)
b_w (m) =	0.25	check b_w<min(b_c+h_w, 2b_c)
b_w (m) =	0.25	check h_w/b_w<3.5 (for DCH)

Effective width of beam

l_1 (m) =	4.50	transverse spacing between beams at the left
b_1 (m) = (l_1 - b_w)/2 =	2.13	spacing between support face and transv. mid-span
l_2 (m) =	0.00	transverse spacing between beams at the right
b_2 (m) = (l_2 - b_w)/2 =	0.00	spacing between support face and transv. mid-span
l_0 (m) =	3.83	spacing between the points of zero moment

EN 1992-1-1, §5.3.2.1, Figure 5.2

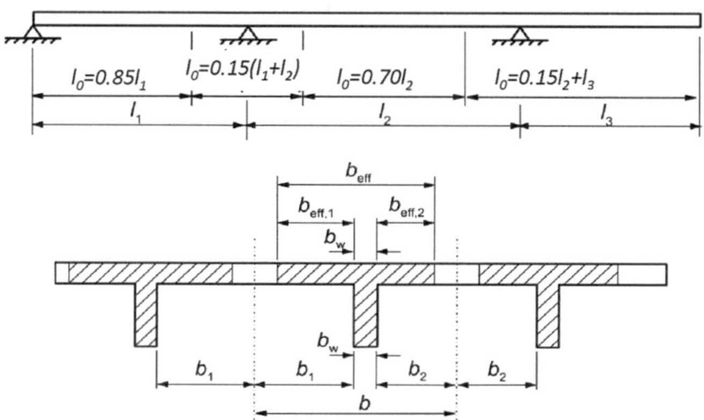

$b_{eff,1}$ (m) = 0.2b_1 + 0.1l_0 =	0.81	
$b_{eff,1}$ (m) =	0.77	check $b_{eff,1}$ < 0.2l_0
$b_{eff,2}$ (m) = 0.2b_2 + 0.1l_0 =	0.00	
$b_{eff,2}$ (m) =	0.00	check $b_{eff,1}$ < 0.2l_0
b_{eff} (m) = Σ$b_{eff,i}$ + b_w =	1.02	effective beam width
b(m) = b_1 + b_2 + b_w =	2.38	
b_{eff} (m) =	1.02	check b_{eff} < b

Minimum longitudinal reinforcement

	DCM	ductility class
ρ_{min} = 0.5f_{ctm}/f_{yk} =	0.22%	uniform minimum reinforcement ratio (DCM/DCH)
$A_{s,min}$ (cm^2) = ρ_{min} b_wd =	2.54	Minimum longitudinal reinforcement
cm^2	0.00	additional requirement 2Ø14 top & bottom (for DCH)
$A_{s,min}$ (cm^2) = ρ_{min} b_wd =	2.54	final Minimum longitudinal reinforcement
ρ_{min} =	0.22%	minimum ratio of longitudinal reinforcement

Maximum diameter of longitudinal reinforcement

	DCM	ductility class	EN 1998-1, §5.6.2.2

Left beam-column joint (BX10 - T3X)

joint:	interior	(adequate anchorage within the shear wall)
$v_d = N_{E,T2Y}/(b_{wall}L_w f_{cd}) =$	-0.08	axial load of the vertical member
$k_1 =$	7.50	coefficient dependent on ductility class
$k_D =$	0.67	coefficient dependent on ductility class
$k_2 = (1+0.75k_D\rho'/\rho_{max}) =$	1.25	joint-dependent coefficient (interior/exterior)
(mm) $\leq k_1 h_c{}^*(1+0.8v_d)f_{ctm}/(f_{yd}k_2) =$	107.15	max bar diameter to be achored within the left joint[1]

Right beam-column joint (BX10 - C8)

joint:	interior	
$v_d = N_{E,C8}/(b_c h_c f_{cd}) =$	-0.21	axial load of the vertical member
$k_1 =$	7.50	coefficient dependent on ductility class
$k_D =$	0.67	coefficient dependent on ductility class
$k_2 = (1+0.75k_D\rho'/\rho_{max}) =$	1.25	joint-dependent coefficient (interior/exterior)
(mm) $\leq k_1 h_c{}^*(1+0.8v_d)f_{ctm}/(f_{yd}k_2) =$	10.14	max bar diameter to be achored within the right joint
d_{bL} (mm) $\leq$	10.14	most critical max bar diameter between the joints[1]

Longitudinal reinforcement at the beam left support

Left support - top (rectangular section)

$M_{E,B12Yi}^-$ (kNm) =	-94.84	design bending moment
$K = M_{E,B12Yi}^-/(b_w d^2 f_{cd}) =$	0.16	joint-dependent coefficient (interior/exterior)
$K_{lim} =$	0.29	Diagram 5.6 (Designer's guide to EN1992-1-1)
$K =$	0.16	check for comp. reinforcement requirement (K<Klim)
$A_s f_{yd}/b_w d f_{cd} =$	0.15	Diagram 5.7 (Designer's guide to EN1992-1-1)
$A_{s,req}$ (cm^2) =	4.40	area of required reinforcement
$A_{s,req}$ (cm^2) =	4.40	check versus min required reinf. ($A_{s,req}>A_{s,min}$)
	3Φ14	provided reinforcement
$A_{s1,i}$ (cm^2) =	4.62	area of provided reinforcement $A_{s1,i}>A_{s,req}$

Left support - bottom (flange beam)

$M_{E,B12Yi}^+$ (kNm) =	35.29	design bending moment
$K = M_{E,B12Yi}^-/(b_{eff} d^2 f_{cd}) =$	0.01	joint-dependent coefficient (interior/exterior)
$K_{lim} =$	0.29	Diagram 5.6 (Designer's guide to EN1992-1-1)
$K =$	0.01	check for comp. reinforcement requirement (K<Klim)
$A_s f_{yd}/b_{eff} d f_{cd} =$	0.01	Diagram 5.7 (Designer's guide to EN1992-1-1)
$A_{s,req}$ (cm^2) =	1.82	area of required reinforcement
$A_{s,req}$ (cm^2) =	2.54	check versus min. required reinf. ($A_{s,req}>A_{s,min}$)
	2Φ14	provided reinforcement
$A_{s2,i}$ (cm^2) =	3.08	area of provided reinforcement $A_{s2,i}>A_{s,req}$

Check maximum ratio of longitudinal reinforcement

$\rho_{max} = \rho' + 0.0018f_{cd}/(\mu_\varphi\varepsilon_{sy,d}f_{yd}) =$	0.78%	maximum ratio of longitudinal reinforcement
$T_c/T =$	1.06	
$\mu_\varphi = 2q_0-1 =$	3.00	curvature ductility (for $T\geq T_c$)
$\mu_\varphi = 1+(2q_0-1)T_c/T =$	4.19	curvature ductility (for $T\geq T_c$)
$\mu_\varphi =$	4.19	final value of curvature ductility (for $T<T_c$)
$\varepsilon_{sy,d} = f_{yd}/E_s =$	0.2%	design value of tension steel strain at yield

$\rho' = A_{s2,i}/(b_w d)$ =	**0.27%**	ratio of compression reinforcement
$A_{s,max}$ (cm^2) =	**9.00**	area of provided reinforcement $A_{s,prov} > A_{s,req}$
	Valid	area of provided reinf. top: $A_{s,max} > A_{s1,i} > A_{s,req}$
	valid	area of provided reinf. bottom: $A_{s,max} > A_{s2,i} > A_{s,req}$

Longitudinal reinforcement at the beam right support

Right support - top (rectangular section)

$M_{E,B12Yj}^-$ (kNm) =	**-70.25**	design bending moment
$K = M_{E,B12Yi}^- / (b_w d^2 f_{cd})$ =	**0.12**	joint-dependent coefficient (interior/exterior)
K_{lim} =	0.29	Diagram 5.6 (Designer's guide to EN1992-1-1)
K =	**0.12**	check for comp. reinforcement requirement ($K<K_{lim}$)
$A_s f_{yd} / b_w d f_{cd}$ =	**0.11**	Diagram 5.7 (Designer's guide to EN1992-1-1)
$A_{s,req}$ (cm^2) =	**3.34**	area of required reinforcement
$A_{s,req}$ (cm^2) =	**3.34**	check vs min required reinforcement ($A_{s,req}>A_{s,min}$)
	3Φ14	provided reinforcement
$A_{s1,j}$ (cm^2) =	**4.62**	area of provided reinforcement $A_{s1,j} > A_{s,req}$

Right support - bottom (flange beam)

$M_{E,B12Yj}^+$ (kNm) =	**40.40**	design bending moment
$K = M_{E,B12Yi}^- / (b_{eff} d^2 f_{cd})$ =	**0.02**	joint-dependent coefficient (interior/exterior)
K_{lim} =	0.29	Diagram 5.6 (Designer's guide to EN1992-1-1)
K =	**0.02**	check for comp. reinforcement requirement ($K<K_{lim}$)
$A_s f_{yd} / b_{eff} d f_{cd}$ =	**0.02**	Diagram 5.7 (Designer's guide to EN1992-1-1)
$A_{s,req}$ (cm^2) =	**2.08**	area of required reinforcement
$A_{s,req}$ (cm^2) =	**2.54**	check vs min required reinforcement ($A_{s,req}>A_{s,min}$)
	2Φ14	provided reinforcement
$A_{s2,j}$ (cm^2) =	**3.08**	area of provided reinforcement $A_{s2,j} > A_{s,req}$

Check maximum ratio of longitudinal reinforcement

$\rho_{max} = \rho' + 0.0018 f_{cd}/(\mu_\varphi \varepsilon_{sy,d} f_{yd})$ =	**0.78%**	maximum ratio of longitudinal reinforcement
όπου:		
T_c/T =	**1.06**	
$\mu_\varphi = 2q_0 - 1$ =	**3.00**	curvature ductility (for $T \geq T_c$)
$\mu_\varphi = 1 + (2q_0 - 1)T_c/T$ =	**4.19**	curvature ductility (for $T \geq T_c$)
μ_φ =	**4.19**	final value of curvature ductility (for $T<T_c$)
$\varepsilon_{sy,d} = f_{yd}/E_s$ =	**0.2%**	design value of tension steel strain at yield

$\rho' = A_{s2,j}/(b_w d)$ =	**0.27%**	ratio of compression reinforcement
$A_{s,max}$ (cm^2) =	**9.00**	area of provided reinforcement $A_{s,prov} > A_{s,req}$
	Valid	area of provided reinf. top: $A_{s,max} > A_{s1,j} > A_{s,req}$
	Valid	area of provided reinf. bottom: $A_{s,max} > A_{s2,j} > A_{s,req}$

Longitudinal reinforcement at the beam mid-span

Mid-span - bottom (flange beam)

$M_{E,B12Ym}^+$ (kNm) =	**20.37**	design bending moment
$K = M_{E,B12Yi}^- / (b_{eff} d^2 f_{cd})$ =	**0.01**	joint-dependent coefficient (interior/exterior)
K_{lim} =	0.29	Diagram 5.6 (Designer's guide to EN1992-1-1)
K =	**0.01**	check for comp. reinforcement requirement ($K<K_{lim}$)
$A_s f_{yd} / b_{eff} d f_{cd}$ =	**0.01**	Diagram 5.7 (Designer's guide to EN1992-1-1)
$A_{s,req}$ (cm^2) =	**1.08**	area of required reinforcement

$A_{s,req}$ (cm^2) =	**2.54**	check vs min required reinforcement ($A_{s,req}$>$A_{s,min}$)
	2Φ14	provided reinforcement
$A_{s1,m}$ (cm^2) =	**3.08**	area of provided reinforcement $A_{s1,m}$ >$A_{s,req}$

Mid-span - top (rectangular section)

$M_{E,B12Ym}^{-}$ (kNm) =	**10.63**	design bending moment
$K = M_{E,B12Yi}^{-} / (b_w d^2 f_{cd})$ =	**0.02**	joint-dependent coefficient (interior/exterior)
K_{lim} =	0.29	Diagram 5.6 (Designer's guide to EN1992-1-1)
K =	**0.02**	check for comp. reinforcement requirement (K<K_{lim})
$A_s f_{yd} / b_w d f_{cd}$ =	**0.02**	Diagram 5.7 (Designer's guide to EN1992-1-1)
$A_{s,req}$ (cm^2) =	**0.55**	area of required reinforcement
$A_{s,req}$ (cm^2) =	**2.54**	check vs min required reinforcement ($A_{s,req}$>$A_{s,min}$)
	2Φ14	provided reinforcement
$A_{s2,m}$ (cm^2) =	**3.08**	area of provided reinforcement $A_{s2,m}$ >$A_{s,req}$

Check maximum ratio of longitudinal reinforcement

$\rho_{max} = \rho' + 0.0018 f_{cd}/(\mu_\varphi \varepsilon_{sy,d} f_{yd})$ =	**0.78%**	maximum ratio of longitudinal reinforcement, where:
T_c/T =	**1.06**	
$\mu_\varphi = 2q_0-1$ =	**3.00**	curvature ductility (for T≥T_c)
$\mu_\varphi = 1+(2q_0-1)T_c/T$ =	**4.19**	curvature ductility (for T<T_c)
μ_φ =	**4.19**	final value of curvature ductility (for T<T_c)
$\varepsilon_{sy,d} = f_{yd}/E_s$ =	**0.2%**	design value of tension steel strain at yield
$\rho' = A_{s2,m}/(b_w d)$ =	**0.27%**	ratio of compression reinforcement
$A_{s,max}$ (cm^2) =	**9.00**	area of provided reinforcement $A_{s,prov}$ >$A_{s,req}$
	Valid	area of provided reinf. bottom: $A_{s,max}$ > $A_{s1,m}$ >$A_{s,req}$
	Valid	area of provided reinf. top: $A_{s,max}$ > $A_{s2,m}$ >$A_{s,req}$

General checks for compression reinforcement

Valid	comp. reinf. > 50% tens. at support i ($A_{s2,i}$ > $A_{s1,i}/2$)
Valid	comp. reinf. > 50% tens. at support j ($A_{s2,j}$> $A_{s1,j}/2$)
Valid	comp. reinf. at support i > 25% tens. at mid-span ($A_{s2,i}$ > $A_{s1,m}/4$)
Valid	comp. reinf. at support j > 25% tens. at mid-span ($A_{s2,j}$ > $A_{s1,m}/4$)
Valid	comp. reinf. at mid-span $A_{s2,m}$ > max{$A_{s1,i}$, $A_{s1,j}$}/4

NOTES

[1] In cases where the maximum diameter is found very small (e.g. Ø<14 it is either possible to apply detailing provision for non confronting diameter limits, or, to increase the dimensions of the column.

DESIGN OF COLUMN C8 TO BENDING - Input data[1]

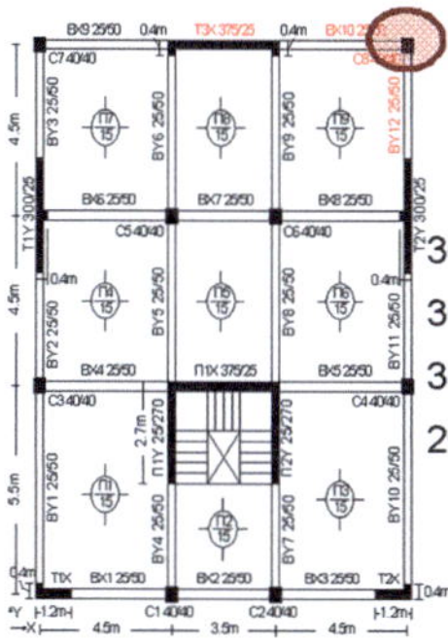

3Φ14	149.91	Moment Resistance BY12 (top): $M_{RdBY12j}^{-}$ (kNm)
3Φ14	142.02	Moment Resistance BY12 (bottom): $M_{RdBY12j}^{+}$ (kNm)
3Φ14	95.87	Moment Resistance BX10 (top): $M_{RdBX10i}^{-}$ (kNm)
2Φ14	97.70	Moment Resistance BX10 (bottom): $M_{RdBX10i}^{+}$ (kNm)
	-392.16	Axial load Column C8-2: $N_{E,C8-2}^{bot}$ (kN)*[2]
	-524.99	Axial load Column C8-1: $N_{E,C8-1}^{top}$ (kN)*[2]

Column section C8-1 (storey 1) & C8-2 (storey 2)

b_c (m) =	0.40	column section width along the local z-z axis
h_c (m) =	0.40	column section width along the local x-x axis
H_c (m) =	3.00	column section height

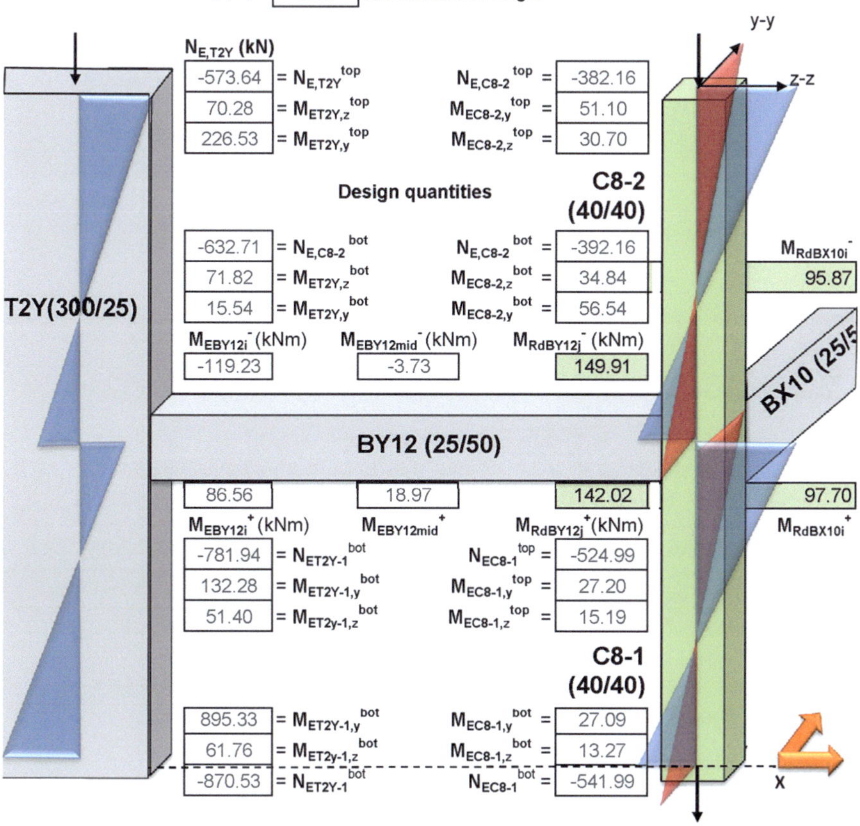

Required concrete cover - Durability

Initial bar diameter

d_{cL} (mm) =	16	Φ16	max desirable diameter of long.reinf.
d_w (mm) =	8	Φ8	desirable diameter of shear reinforcement

Concrete cover and effective section depth

$c_{min,b}$ (mm) =	**16.00**	minimum cover
$\Delta c_{dur,\gamma}$ (mm) =	0.00	additive safety element
$\Delta c_{dur,st}$ (mm) =	0.00	reduction of minimum cover for use of stainless steel
$\Delta c_{dur,add}$ (mm) =	0.00	reduction of min cover for use of additional protection
$c_{min,dur}$ (mm) =	25.00	minimum cover due to environmental conditions

c_{min} (mm) =	**25.00**	minimum concrete cover
Δc_{dev} (mm) =	10.00	allowance in design for deviation
c_{nom} (mm) = $c_{min} + \Delta c_{dev}$ =	**35.00**	nominal concrete cover
d (mm) = $h_c - c_{nom} - d_w - d_{cL}/2$ =	**349.00**	effective section depth

Capacity design requirement

93.39%	% base shear resisted by shear walls along the x-x dir.
yes	check not required[3]
92.96%	% base shear resisted by shear walls along the y-y dir.
yes	check not required[3]

Check for minimum column dimensions

	DCM	check not required
b_c (m) =	**0.40**	check b_c>0.25m (for DCH)
h_c (m) =	**0.40**	check h_c>0.25m (for DCH)

Check for minimum column normalised axial load

	DCM	check not required
$v = N_E/(bhf_{cd})$ =	**-0.29**	check v<0.55 (for DCH) at head of column C8-1
$v = N_E/(bhf_{cd})$ =	**-0.22**	check v<0.55 (for DCH) at head of column C8-2

Capacity Design Values

Capacity design x-x (+)

$M_{RC8\text{-}1,y}{}^{top} + M_{RC8\text{-}2,y}{}^{bot} > 1.3\,(M_{RdBY12j\text{-}}+0)$ => $M_{RC8\text{-}1,y}{}^{t}+M_{RC8\text{-}2,y}{}^{b}>$ 194.89

& $M_{RC8\text{-}1,y}{}^{top} = (N_{EC8\text{-}1}{}^{top}/N_{EC8\text{-}2}{}^{bot})\,M_{RC8\text{-}2,y}{}^{bot}$ => $M_{RC8\text{-}1,y}{}^{t}/M_{RC8\text{-}2,y}{}^{b}=$ 1.34

$M_{RC8\text{-}1,y}{}^{top}$ (kNm) = **111.56** **$M_{RC8\text{-}2,y}{}^{bot}$ (kNm) =** **83.33**

Capacity design x-x (-)

$M_{RC8\text{-}1,y}{}^{top} + M_{RC8\text{-}2,y}{}^{bot} > 1.3\,(M_{RdBY12j+}+0)$ => $M_{RC8\text{-}1,y}{}^{t}+M_{RC8\text{-}2,y}{}^{b}>$ 184.62

& $M_{RC8\text{-}1,y}{}^{top} = (N_{EC8\text{-}1top}/N^{EC8\text{-}2bot})\,M_{RC8\text{-}2,y}{}^{bot}$ => $M_{RC8\text{-}1,y}{}^{t}/M_{RC8\text{-}2,y}{}^{bot}=$ 1.34

$M_{RC8\text{-}1,y}{}^{top}$ (kNm) = **105.68** **$M_{RC8\text{-}2,y}{}^{bot}$ (kNm) =** **78.94**

Capacity design y-y (+)

$M_{RC8\text{-}1,z}^{top} + M_{RC8\text{-}2,z}^{bot} > 1.3\,(M_{RdBX10i\text{-}}+0)$ => $M_{RC8\text{-}1,z}^{t}+M_{RC8\text{-}2,z}^{b} >$ 124.63

& $M_{RC8\text{-}1,z}^{top} = (N_{EC8\text{-}1}^{top}/N_{EC8\text{-}2}^{bot})\,M_{RC8\text{-}2,y}^{bot}$ => $M_{RC8\text{-}1,z}^{t} / M_{RC8\text{-}2,z}^{b} =$ 1.34

$M_{RC8\text{-}1,z}^{top}$ **(kNm) =** **71.34** $M_{RC8\text{-}2,z}^{bot}$ **(kNm) =** **53.29**

Capacity design y-y (-)

$M_{RC8\text{-}1,z}^{top} + M_{RC8\text{-}2,z}^{bot} > 1.3\,(M_{RdBX10i+}+0)$ => $M_{RC8\text{-}1,z}^{t}+M_{RC8\text{-}2,z}^{b} >$ 127.01

& $M_{RC8\text{-}1,ztop} = (N_{EC8\text{-}1}^{top}/N_{EC8\text{-}2bot})\,M_{RC8\text{-}2,z}^{bot}$ => $M_{RC8\text{-}1,z}^{t} / M_{RC8\text{-}2,z}^{b} =$ 1.34

$M_{RC8\text{-}1,y}^{top}$ **(kNm) =** **72.70** $M_{RC8\text{-}2,y}^{bot}$ **(kNm) =** **54.31**

Required longitudinal reinforcement (Capacity Design) - column top C8-1

Column C8-1 (top)

	$M_{RC8\text{-}1,y}^{top}$	$M_{RC8\text{-}1,z}^{top}$	$N_{EC8\text{-}1}^{top}$	ω	
Capacity design x-x (+)	**111.56**	**0.00**	**-524.99**	0.12	from Interaction Diagram *4
Capacity design x-x (-)	**105.68**	**0.00**	**-524.99**	0.12	
Capacity design y-y (+)	**0.00**	**71.34**	**-524.99**	0.00	
Capacity design y-y (-)	**0.00**	**72.70**	**-524.99**	0.00	

Column C8-2 (bottom)

	$M_{RC8\text{-}1,y}^{top}$	$M_{RC8\text{-}1,z}^{top}$	$N_{EC8\text{-}1}^{top}$	ω	uniform ω $C_{8\text{-}1}$ & $C_{8\text{-}2}$ *5
Capacity design x-x (+)	**83.33**	**0.00**	**-392.16**	0.11	**0.12**
Capacity design x-x (-)	**78.94**	**0.00**	**-392.16**	0.10	**0.11**
Capacity design y-y (+)	**0.00**	**53.29**	**-392.16**	0.02	**0.01**
Capacity design y-y (-)	**0.00**	**54.31**	**-392.16**	0.02	**0.01**
					most critical ω
					0.12

Required longitudinal reinforcement (Action Effects) - column top C8-1

Column C8-1 (top)

	$M_{EC8\text{-}1,y}^{top}$	$M_{EC8\text{-}1,z}^{top}$	$N_{EC8\text{-}1}^{top}$	ω	
Load Combination	**27.20**	**15.19**	**-524.99**	0.00	from Interaction Diagram

Column C8-2 (bottom)

	$M_{EC8\text{-}1,y}^{bot}$	$M_{EC8\text{-}1,z}^{bot}$	$N_{EC8\text{-}1}^{bot}$	ω	uniform ω $C_{8\text{-}1}$ & $C_{8\text{-}2}$
Load Combination	**56.54**	**34.84**	**-392.16**	0.05	**0.03**
					critical $\omega_{maxC8,1\text{-}2}$
					0.12

Required, minimum & provided longitudinal reinforcement

$_{C8,1\text{-}2}\ (cm^2)=\omega_{maxC81,2}\, bh\, f_{cd}/f_{yd}$ =	**4.80**	area of required longitudinal reinforcement
ρ_{min} =	**1.00%**	uniform minimum reinforcement ratio (DCM/DCH)
$A_{s,min}\ (cm^2) = \rho_{min}\, h_c b_c$ =	**16.00**	minimum longitudinal reinforcement
$A_{s,req}\ (cm^2)$ =	**16.00**	check vs min required reinforcement ($A_{s,req} > A_{s,min}$)
	8Φ16	provided reinforcement pre-confinement)
$A_{s,prov}$ =	**16.08**	area of provided longitudinal reinforcement [*6]
	valid	check area of provided reinforcement: $A_{s,prov} > A_{s,req}$

Maximum transverse reinforcement

ρ_{max} =	**4.00%**	uniform max volumetric reinforcement ratio (DCM/DCH)
$A_{s,max}\ (cm^2) = \rho_{max}\, h_c b_c$ =	**64.00**	maximum longitudinal reinforcement
	valid	area of provided reinforcement: $A_{s,max} > A_{s,prov} > A_{s,min}$

Check for three rebars per side

	8Φ16	provided reinforcement (prior to confinement checks)
cm^2	**16.08**	final area of provided longitudinal reinforcement
	3	number of provided rebars per side
	valid	check of at least 3 rebars per side

Check rebar spacing

s (mm) =	**133.33**	spacing between rebars along y-y
s (mm) =	**133.33**	spacing between rebars along z-z
	DCM	ductility class
	150.00	maximum permissible spacing between rebars
	valid	check spacingς netween rebars along y-y (local axes)
	valid	check spacingς netween rebars along z-z (local axes)

Moment Resistance update

Column C8-1 (top)

		$A_s\ (cm^2)$	v	ω	M_{Rd} (kNm)
$M_{RdC8\text{-}1,y}^{top}$	**8Φ16**	**16.08**	**-0.29**	**0.39**	**108.80**
$M_{RdC8\text{-}1,z}^{top}$	**8Φ16**	**16.08**	**-0.29**	**0.39**	**108.80**

Column C8-2 (bottom)

		$A_s\ (cm^2)$	v	ω	M_{Rd} (kNm)
$M_{RdC8\text{-}2,y}^{bot}$	**8Φ16**	**16.08**	**-0.22**	**0.39**	**97.92**
$M_{RdC8\text{-}2,z}^{bot}$	**8Φ16**	**16.08**	**-0.22**	**0.39**	**97.92**

Verification of Capacity Design

Capacity design x-x (+)

Verify that $M_{RC8\text{-}1,y}^{top} + M_{RC8\text{-}2,y}^{bot} > 1.3\ (M_{RdBY12j-}+0)$ => **206.72 > 149.91**

valid

Capacity design x-x (-)

Verify that $M_{RC8\text{-}1,y}^{top} + M_{RC8\text{-}2,y}^{bot} > 1.3\ (M_{RdBY12j+}+0)$ => **206.72 > 142.02** valid

Capacity design y-y (+)

Verify that $M_{RC8-1,z}^{top} + M_{RC8-2,z}^{bot} > 1.3\ (M_{RdBX10i-}+0)$ => **206.72 > 95.87**

valid

Capacity design y-y (-)

Verify that $M_{RC8-1,z}^{top} + M_{RC8-2,z}^{bot} > 1.3\ (M_{RdBY10i+}+0)$ => **206.72 > 97.70**

valid

NOTES

[1] Data complementary to those defined for beam design to flexure

[2] It is noted that capacity design shall be performed for the minimum absolute compressive or tensile force at the joint (column top and bottom) that may potentially develop plastic hinge at the beam ends within the respective plane of bending. Given that it is only a single load combination that is presented herein, inevitably, calculations are perfomed with the same set of design quantities independently of whether they are simultaneously critical for all ultimate limit states and all members. It is also noted that since the contribution of shear walls is significant along both directions and capacity design is not mandatory, the column design quantities are rather minor. As a result the procedure is presented primarily for demonstration purposes.

[3] This verification, though not required, is presented for demonstration pursposes. The existing reinforcement within the beam effective width are neglected.

[4] The influence of column bending in the plane which is transverse to the plane of verification is neglected.

[5] 50% of the resulting reinforcement ratio is assumed at the top and bottom, hence further capacity verification is required.

[6] The provided reinforcement refers to the particular (not necessarily most critical) load combination pressented. The final reinforcement to be provided will result from all the load combinations particularly those involving the minimum axial load for all possible mass locations.

DESIGN OF BEAM BY12 FOR SHEAR - Input Data

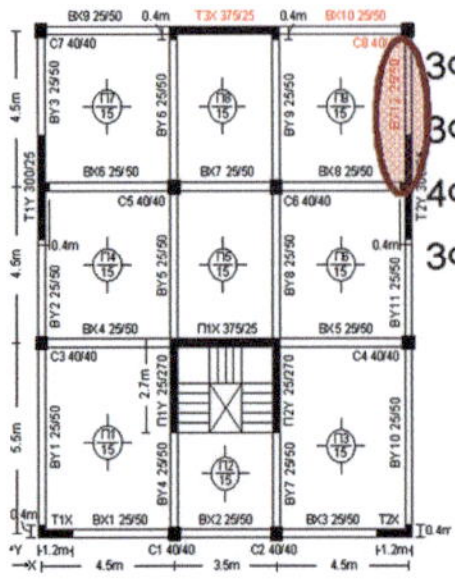

Moment Resistance

3Φ14	149.91	Moment Resistance BY12 (right, top): $M_{RdBY12j}^{-}$ (kNm)
3Φ14	142.02	Moment Resistance BY12 (right, bottom): $M_{RdBY12j}^{+}$
4Φ14	209.17	Moment Resistance BY12 (left, top): $M_{RdBY12i}^{-}$ (kNm)
3Φ14	142.02	Moment Resistance BY12 (left, bottom): $M_{RdBY12i}^{+}$
	351.19	Moment Resistance Sum BY12 (left): $\Sigma M_{Rd,bi}$ (kNm)
	291.93	Moment Resistance Sum BY12 (right): $\Sigma M_{Rd,bj}$ (kNm)
	206.72	Moment Resistance Sum C8 (right): $\Sigma M_{Rd,c8}$ (kNm)

Shear Diagram V(G+ψ$_2$Q)

-29.43	Shear force BY12 (right, kN): $V_{0,G+\psi 2Q}$ (L-d)
19.05	Shear force BY12 (left, kN): $V_{0,G+\psi 2Q}$ (d)

Shear Diagram V(1.35G+1.50Q)

-38.23	Shear force BY12 (right, kN): $V_{1.36G+1.50Q}$ (L-d)
24.75	Shear force BY12 (left, kN): $V_{1.35G+1.50Q}$ (d)

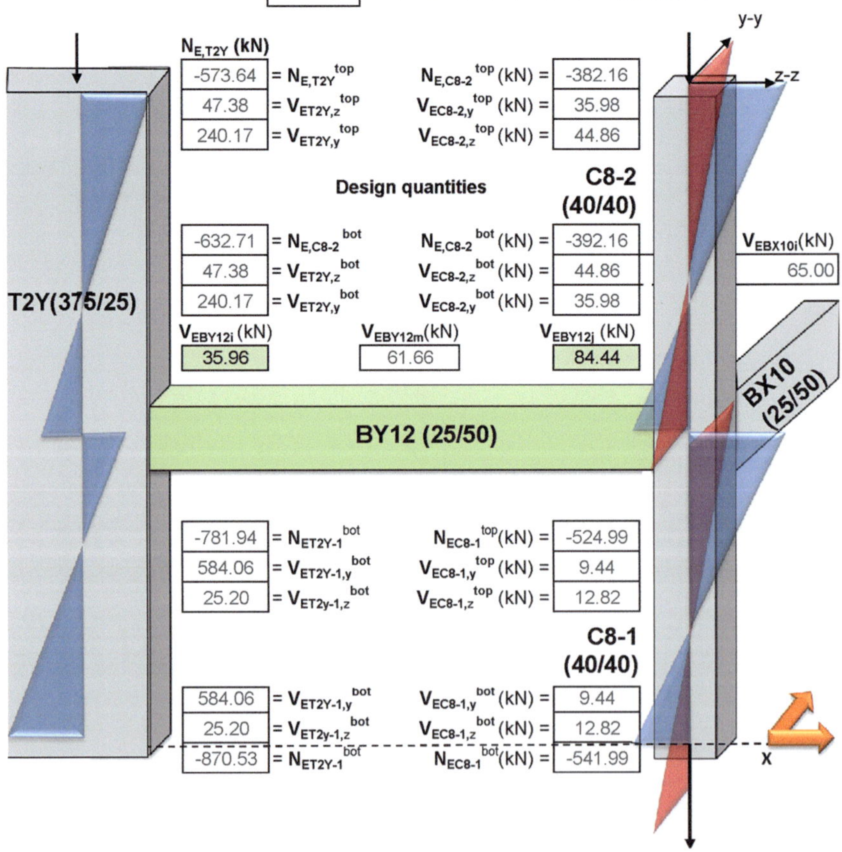

Beam BY12		
b_w (m) =	0.25	beam section width
h_w (m) =	0.50	beam section height
h_f (m) =	0.15	slab thickness
L (m) =	2.80	beam length
L_{cl} (m) =	2.60	net beam length
	L	section type: single-sided (edge) beam / flange beam
b_{eff} (m) =	1.02	
d (mm) = h - c_{nom} - d_w - d_{bL} /2=	460.00	effective section depth

Modeling uncertainty factor (overstrength)

	DCM	ductility class
γ_{Rd} =	1.2	

Derivation of capacity values of shear forces

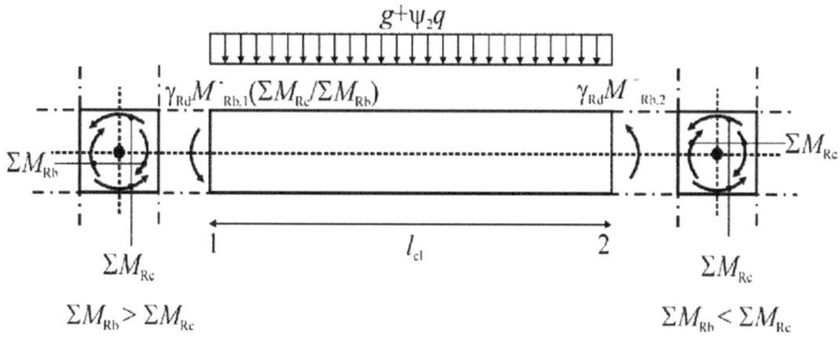

Left support

min{1.0, $\Sigma M_{Rd,c}/\Sigma M_{Rd,b}\}_i$ =	1.00	
min{1.0, $\Sigma M_{Rd,c}/\Sigma M_{Rd,b}\}_j$ =	0.71	
$M_{RdBY12i}^-$ (kNm) =	209.17	Beam Moment Resistance BY12 (left, top)
$M_{RdBY12j}^+$ (kNm) =	142.02	Beam Moment Resistance BY12 (right, bottom)
L_{cl} (m) =	2.60	net beam length
$V_{0,G+\psi 2Q}$ (d) =	19.05	beam shear BY12 (left, kN)
γ_{Rd} =	1.20	
$V_{CD,i}$ (d) (kN)=	162.01	capacity design values of shear forces BY12 (left)
$V_{1.35G+1.50Q}$ (d) =	24.75	beam shear BY12 (left, kN)
V_E(d) (kN)=	162.01	shear resistance beam BY12 (left)

Right support

min{1.0, $\Sigma M_{Rd,c}/\Sigma M_{Rd,b}\}_i$ =	1.00	
min{1.0, $\Sigma M_{Rd,c}/\Sigma M_{Rd,b}\}_j$ =	0.71	
$M_{RdBY12i}^+$ (kNm) =	142.02	Beam Moment Resistance BY12 (left, top)
$M_{RdBY12j}^-$ (kNm) =	149.91	Beam Moment Resistance BY12 (right, bottom)
L_{cl} (m) =	2.60	net beam length
$V_{0,G+\psi 2Q}$ (L-d) =	29.43	beam shear BY12 (right, kN)
γ_{Rd} =	1.20	
$V_{CD,j}$(L-d) (kN)=	143.97	capacity design values of shear forces BY12 (right)
$V_{1.35G+1.50Q}$ (L-d) =	38.23	beam shear BY12 (right, kN)
V_E(L-d) (kN)=	143.97	shear resistance beam BY12 (right)

Critical length

	DCM	ductility class
l_{cr} (m) =	0.75	critical region ($1.5h_w$ for DCH, $1.0h_w$ for DCM)

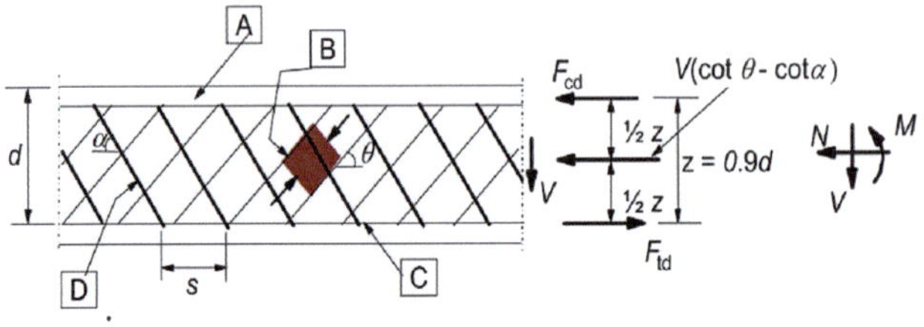

A - compression chord, B - struts, C - tensile chord, D - shear reinforcement

WITHIN CRITICAL REGION
check shear resistance against compression of the diagonal struts (at the support face)

	DCM	ductility class
θ (°) =	45.00	angle between the concrete compression strut and the beam axis perpendicular to the shear force θ=45° (cotθ=1) for DCH, recomm. θ=21.8° for DCM
α (°) =	90.00	angle of the inclined shear reinf. (α=90°: vertical)
α_{cw} =	1.00	for non-prestressed members
z (m) = 0.9d =	0.41	lever arm of internal forces
$v_1 = 0.6(1-f_{ck}/250)$ =	0.55	reduction coefficient due to shear cracking
$V_{Rd,max}$ (kN)=		
$=a_{cw}b_w z v_1 f_{cd}(\cot\theta+\cot\alpha)$		becomes $V_{Rd,max}$ (kN) $=a_{cw}b_w z v_1 f_{cd}/(\cot\theta+\tan\theta)$
$/(1+\cot^2\theta)$=	323.75	for α=90°
$V_E(d)$ (kN) =	162.01	shear resistance beam BY12 (left)
	valid	check $V_{Rd,max} > V_{CD}(d)$
$V_E(L-d)$ (kN) =	143.97	shear resistance beam BY12 (right)
	valid	check $V_{Rd,max} > V_{CD}(L-d)$

Check for shear reversal

	DCM	ductility class
		check required

Required shear reinforcement

Check shear resistance due to reinforcement (at spacing d from the support face)

	Left support	
$\rho_{w,req,i} = V_E(d)/(0.9b_w d f_{ywd})$ =	0.36%	required shear reinforcement ratio

	2	number hoop legs
	Φ8/110	provided reinforcement
A_{sw} (cm^2) =	**1.01**	area of shear reinforcement Φ8
$\rho_{w,prov,i} = A_{sw}/(b_w s_w \sin\alpha)$ =	**0.37%**	ratio of provided reinforcement
	valid	check $\rho_{w,prov,i} > \rho_{w,req}$

	DCM	ductility class
$\rho_{w,min}=0.08\sqrt{f_{ck}(MPa)}/f_{yk}$ =	**0.07%**	minimum reinforcement ratio for DCM & DCH
	valid	check $\rho_{w,prov,i} > \rho_{w,min}$
$s_w<\{8d_{bL}, h_w/4, 24d_{bw}, 225\}$=	**112**	required spacing between ties for DCM (mm)
$s_w<\{dd_{bL}, h_w/4, 24d_{bw}, 175\}$=	**84**	required spacing between ties for DCH (mm)
	valid	check $s < s_w$

Right support

$\rho_{w,req,i}=V_E(L-d)/(0.9b_w d f_{ywd})$ =	**0.32%**	required shear reinforcement ratio
	2	number hoop legs
	Φ8/110	provided reinforcement
A_{sw} (cm^2)	**1.01**	area of shear reinforcement Φ8
$\rho_{w,prov,i} = A_{sw}/(b_w s_w \sin\alpha)$ =	**0.37%**	ratio of provided reinforcement
	valid	check $\rho_{w,prov,i} > \rho_{w,req}$

	DCM	ductility class
$\rho_{w,mi} = 0.08\sqrt{f_{ck}(MPa)}/f_{yk}$ =	**0.07%**	minimum reinforcement ratio for DCM & DCH
	valid	check $\rho_{w,prov,i} > \rho_{w,min}$
$s_w<\{8d_{bL}, h_w/4, 24d_{bw}, 225\}$=	**112**	required spacing between ties for DCM (mm)
$s_w<\{dd_{bL}, h_w/4, 24d_{bw}, 175\}$=	**84**	required spacing between ties for DCH
	valid	check $s < s_w$

OUTSIDE OF CRITICAL REGION

check shear resistance due to reinforcement (at spacing d from the support face)

Left support

$V_E(l_{cr})$ (kN) =	**17.36**	
$\rho_{w,req,i} = V_E(l_{cr})/(0.9b_w d f_{ywd})$ =	**0.04%**	required shear reinforcement ratio
	2	number hoop legs
	Φ8/320	provided reinforcement
A_{sw} (cm^2) =	**1.01**	area of shear reinforcement Φ8
$\rho_{w,prov,i} = A_{sw}/(b_w s_w \sin\alpha)$ =	**0.13%**	ratio of provided reinforcement
	valid	check $\rho_{w,prov,i} > \rho_{w,req}$

	DCM	ductility class
$\rho_{w,min} = 0.08\sqrt{f_{ck}(MPa)}/f_{yk}$ =	**0.07%**	minimum reinforcement ratio for DCM & DCH
	valid	check $\rho_{w,prov,i} > \rho_{w,min}$
s_w (mm) < 0,75d =	**345**	required spacing between ties for DCM
	valid	required spacing between ties for DCH
		check $s < s_w$

Left support

$V_E(L-l_{cr})$ (kN) =	**45.63**	
$\rho_{w,req,i}=V_E(L-l_{cr})/(0.9b_w d f_{ywd})$ =	**0.10%**	required shear reinforcement ratio

	2	number hoop legs
	Φ8/320	provided reinforcement
A_{sw} (cm^2) =	**1.01**	area of shear reinforcement Φ8
$\rho_{w,prov,i} = A_{sw}/(b_w s_w \sin\alpha)$ =	**0.13%**	ratio of provided reinforcement
	valid	check $\rho_{w,prov,i} > \rho_{w,req}$

	DCM	ductility class
$\rho_{w,min} = 0.08\sqrt{f_{ck}}(\text{MPa})/f_{yk}$ =	**0.07%**	minimum reinforcement ratio for DCM & DCH
	valid	check $\rho_{w,prov,i} > \rho_{w,min}$
s_w (mm) < 0,75d =	**345**	required spacing between ties for DCM
	valid	required spacing between ties for DCH
		check $s < s_w$

NOTES

[1] Data complementary to those defined for beam design to flexure

DESIGN OF COLUMN C8 FOR SHEAR - Input Data[1]

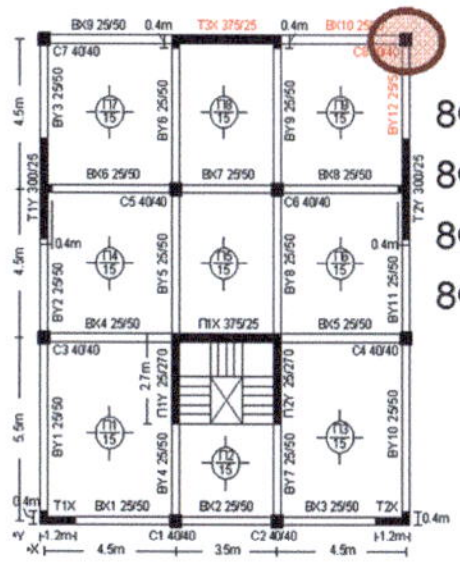

		Column edge Moment Resistance
8Φ16	108.80	Moment Resistance C8-1 $M_{RC8-1,y}^{top}$ (kNm)
8Φ16	108.80	Moment Resistance C8-1 $M_{RC8-1,z}^{top}$ (kNm)
8Φ16	108.80	Moment Resistance C8-1 $M_{RC8-1,y}^{bot}$ (kNm)
8Φ16	108.80	Moment Resistance C8-1 $M_{RC8-1,z}^{bot}$ (kNm)

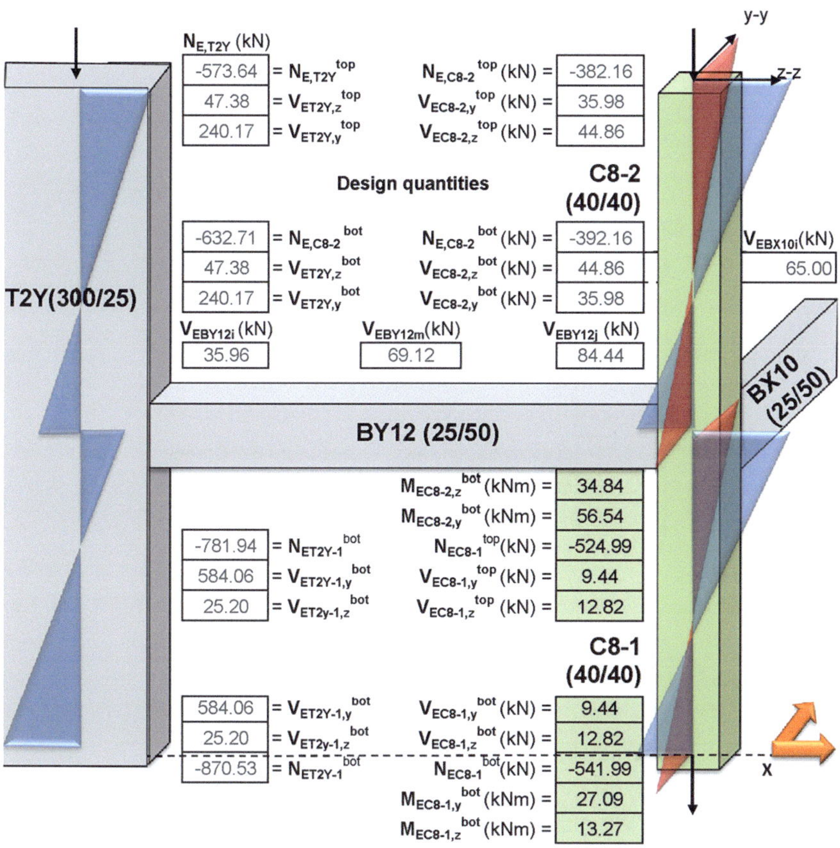

Column section C8-1

b_c (m) =	0.40	column section width along the local axis z-z
h_c (m) =	0.40	column section width along the local axis x-x
H_c (m) =	3.00	column section height
H_{cl} (m) =	2.50	net column height

Modeling uncertainty factor (overstrength)

	DCM	ductility class
γ_{Rd} =	1.3	

Derivation of capacity values of shear forces

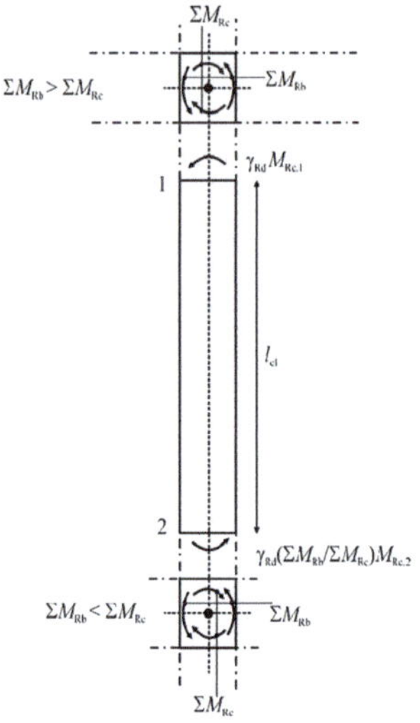

direction y-y 2

	0.71	$= \min\{1,\Sigma M_{Rd,c}/\Sigma M_{Rd,b}\}^{top}$
	1.00	$= \min\{1,\Sigma M_{Rd,c}/\Sigma M_{Rd,b}\}^{bot}$
$M_{RC8\text{-}1,y}{}^{top}$ (kNm) =	108.80	
$M_{RC8\text{-}1,y}{}^{bot}$ (kNm) =	108.80	
H_{cl} (m) =	2.50	
γ_{Rd} =	1.30	
$V_{CD,y}$(d) (kN)=	96.64	capacity design shear force
$V_{EC8\text{-}1,y}{}^{top}$ (kN) =	9.44	beam shear BY12 (left, kN)
$V_{E,C8,y,max}$ (kN)=	96.64	shear resistance of beam BY12

direction z-z 2

	0.71	$= \min\{1,\Sigma M_{Rd,c}/\Sigma M_{Rd,b}\}^{top}$
	1.00	$= \min\{1,\Sigma M_{Rd,c}/\Sigma M_{Rd,b}\}^{bot}$
$M_{RC8\text{-}1,z}{}^{top}$ (kNm) =	108.80	
$M_{RC8\text{-}1,z}{}^{bot}$ (kNm) =	108.80	
H_{cl} (m) =	2.50	
γ_{Rd} =	1.30	
$V_{CD,z}$(d) (kN)=	96.64	capacity design shear force of beam BY12 (right)
$V_{EC8\text{-}1,z}{}^{top}$ (kN) =	12.82	beam shear BY12 (right, kN)
$V_{E,C8,z,max}$ (kN)=	96.64	shear resistance of beam BY12 (right)

Critical length

	DCM	ductility class
l_{cr}(m)=max{h_c,b_c,0.45,H_{cl}/5}=	0.50	critical region (DCM)
l_{cr}= =max{1.5h_c,1.5b_c,0.6,H_{cl}/5}=	0.60	critical region (DCH)

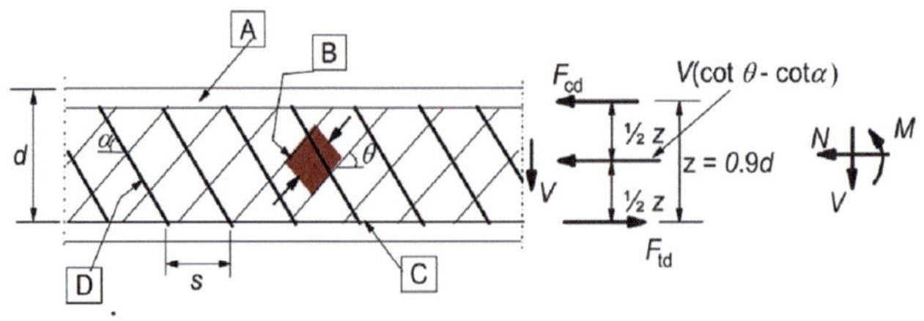

A - compression chord, B - struts, C - tensile chord, D - shear reinforcement

CONFINEMENT

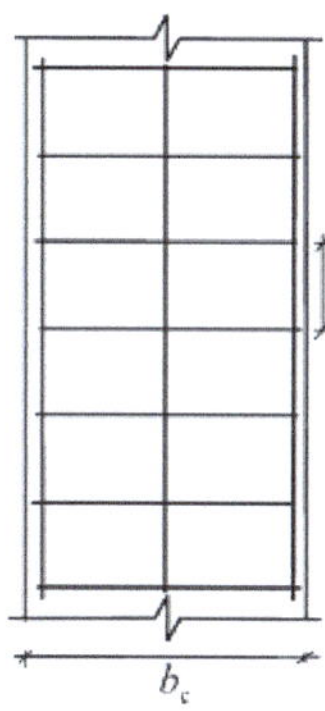

	DCM	ductility class
	yes	column at the base of the building
$\omega_{wd,min}$ >	0.120	minimum mechanical volumetric ration of confining reinforcement within critical length
v_d =	-0.30	$= N_E/(h_c b_c f_{cd})$
$v_{d,max}$ =	-0.55	maximum value of normilized axial load (0.55 for DCH, 0.65 για ΚΠΜ)
	valid	check $v_d < v_{d,max}$
T (sec) =	0.47	uncoupled transl. period at the direction of the beam
	B	soil class
$\bar{}_c$ (sec) =	0.50	maximum spectral acceleration
q_0 =	2.00	basic value of the behaviour factor
T_c/T =	1.06	
$\mu_\varphi = 2q_0-1$ =	3.00	curvature ductility (for $T \geq T_c$)
$\mu_\varphi = 1+(2q_0-1)T_c/T$ =	4.19	curvature ductility (for $T < T_c$)
μ_φ =	4.19	final value of curvature ductility
$\varepsilon_{sy,d} = f_{yd}/E_s$ =	0.22%	design value of tension steel strain at yield

WITHIN CRITICAL REGION

required mechanical volumetric confining reinforcement ratio

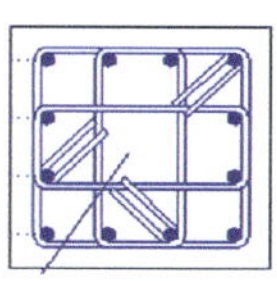

d_{bL} (mm)	16	diameter of column longitudinal reinforcement
d_{bw}(mm)=max{, $d_{bL}/4$} =	6.0	min diameter of transverse reinforcement for DCM
$_w$ = max{6mm, $0.4\sqrt{(f_{yd}/f_{wd})}d_{bL}$} =	6.4	min diameter shear reinforcement for DCH
	6.4	min diameter of shear reinforcement within cr. length
	4	number of hoop legs[3]
	Φ8/100	provided reinforcement
b_0 (m) =	0.322	width of the confined core
h_0 (m) =	0.322	width of the confined core
$\alpha_s = (1-s_w/2b_0)(1-s_w/2h_0)$ =	0.71	
n_b =	4.00	number of hoop legs parallel to b_0
n_h =	4.00	number of hoop legs parallel to h_0
	0.78	$= \alpha_{n=} 1-\{b_0/[(n_h-1)h_0]+h_0/[(n_b-1)b_0]\}/3$
$\alpha = \alpha_n \alpha_s$ =	0.55	confinement effectiveness factor
$\alpha\omega_{wd}$ >	0.07	$=30\mu_\varphi v_d \varepsilon_{sy,d} b_c/b_0 - 0.035$
$\omega_{wd,req}$ =	0.120	
$\omega_{wd,req}$ =	0.120	check $\omega_{wd,req} > \omega_{wd,min}$

provided mechanical volumetric confining reinforcement ratio

L_{wd} (m) = $(n_b b_0 + n_h h_0)$ =	2.58	length of shear reinforcement deployment
A_{sw} (cm^2) =	0.50	area of shear reinforcement
$V_0 = L_{wd} A_{sw}/s_w$ (mm^3) =	1294.84	volume of shear reinforcement
$V_c = b_0 h_0$ (mm^3) =	103684	volume of concrete core
$\omega_{wd,prov} = V_0 f_{yd}/(V_c f_{cd})$ =	0.48	
	valid	check $\omega_{wd,prov} > \omega_{wd,req}$

check of shear reinforcement spacing

	DCM	ductility class
$s_{w,max}$(mm) = $=\min\{8d_{bL},b_0/2,175\}$=	128.0	maximum spacing of shear reinforcement for DCM
$s_{w,max}$(mm) = $\min\{6d_{bL},b_0/3,125\}$=	96.0	maximum spacing of shear reinforcement for DCH
$s_{w,max}$(mm) <	128.0	maximum spacing of shear reinforcement
	valid	check $s_w<s_{w,max}$

OUTSIDE OF CRITICAL REGION

	2	number of hoop legs
	Φ8/320	provided reinforcement
d_{bL} (mm)	16	diameter of column longitudinal reinforcement
d_{bw} (mm) = max{6mm, $d_{bl}/4$} =	6.0	minimum diameter of shear reinforcement for DCM & DCH

check spacing of shear reinforcement

$_{,max}$(mm) $=\min\{20d_{bL},h_c,b_c,400\}$=	320.0	maximum spacing shear reinforcement for DCM
	valid	check $s_w<s_{w,max}$

OUTSIDE OF CRITICAL REGION
check shear resistance against compression of the diagonal struts

	DCM	ductility class
θ (°) =	45.00	angle between the concrete compression strut and the beam axis perpendicular to the shear force θ=45° (cotθ=1) for DCH, recomm. θ=21.8° for DCM
α (°) =	90.00	angle of the inclined shear reinf. (α=90°: vertical)
α_{cw} =	1.00	for non-prestressed members
z (m) = 0.9d =	0.31	lever arm of internal forces
$v_1 = 0.6(1-f_{ck}/250)$ =	0.55	reduction coefficient due to shear cracking
$V_{Rd,max}$ (kN)= $_{cw}b_wzv_1f_{cd}(\cot\theta+\cot\alpha)/(1+\cot^2\theta)$=	393.00	becomes $V_{Rd,max}$ (kN) $=a_{cw}\,b_w\,z\,v_1\,f_{cd}/(\cot\theta+\tan\theta)$ for α=90°
$V_{E,C8,y,max}$ (kN)=	96.64	shear resistance of column
	valid	check $V_{Rd,max} > V_E$
$V_{E,C8,z,max}$ (kN)=	96.64	shear resistance of column
	valid	check $V_{Rd,max} > V_E$

Required shear reinforcement

check shear resistance due to reinforcement

$\rho_{w,req,i} = V_E/(0.9b_chfy_{wd})$ =	0.18%	required shear reinforcement ratio
	4	number hoop legs
	Φ8/100	provided reinforcement
A_{sw} (cm²)	2.01	area of shear reinforcement Φ8
$\rho_{w,prov,i} = A_{sw}/(b_cs_w\sin\alpha)$ =	0.50%	ratio of provided reinforcement

	DCM	ductility class
$\rho_{w,min} = 0.08\sqrt{f_{ck}}(MPa)/f_{yk} =$	0.07%	minimum reinforcement ratio for DCM & DCH
	valid	check $\rho_{w,prov,i} > \rho_{w,min}$
$s_{w,max}(mm) =$ $=\min\{8d_{bL}, b_0/2, 175\}=$	128.0	required spacing between ties for DCM
$s_{w,max}(mm) =$ $=\min\{6d_{bL}, b_0/3, 125\}=$	96.0	required spacing between ties for DCH
$s_{w,max}(mm) <$	96.0	maximum spacing of shear reinforcement
	not valid	check $s < s_w$

NOTES

[1] Data complementary to those defined for beam design to flexure

[2] With respect to the local column axes

[3] The shape and number of the hoop legs were derived by the confinement requirements and have as such increased the demand for longitudinal reinforcement.

DESIGN OF DUCTILE WALL FOR BENDING - Input data

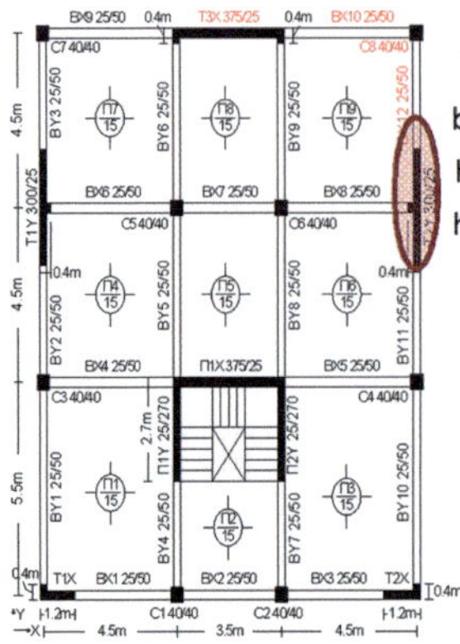

Ductile Wall T2Y

l_w (m) =	3.00	length of shear wall
b_{wl} (m) =	0.25	width of shear wall
h_w (m) =	12.00	total height of shear wall
h_{st} (m) =	3.00	storey height
n =	4	number of building storeys
$b_{wl,min}$		minimum width of shear wall
(m)=	**0.15**	= max{150mm, h_w/20}
	valid	check of minimum shear wall width
	DCM	ductility class

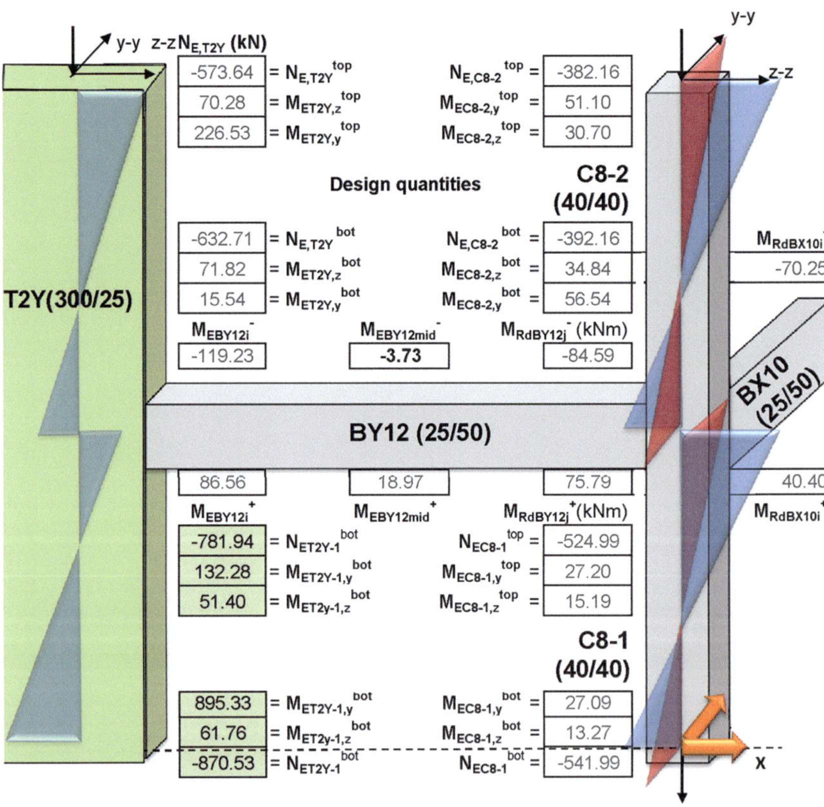

Dimensions of confined boundary elements

Critical (confined) length in boundary elements

l_c (m) = max{$0.15l_w$, $1.5b_w$, l(ε_c>0.2%)} = **0.45** width of boundary elements

Width of confined boundary elements

$\max\{0.20l_w, 2b_{wl}\}$ =	**0.60**	
	valid	check l_c (m) < $\max\{0.20l_w, 2b_{wl}\}$
	no	requirement for wider dimensions
$b_{w,min}$ (m) = max{200mm, $h_{st}/15$} =	**0.20**	minimum width if $l_c \leq \max\{0.20l_w, 2b_{wl}\}$
$b_{w,min}$ (m) = max{200mm, $h_{st}/10$} =	**0.20**	minimum width if $l_c > \max\{0.20l_w, 2b_{wl}\}$
	valid	check $b_w > b_{w,min}$

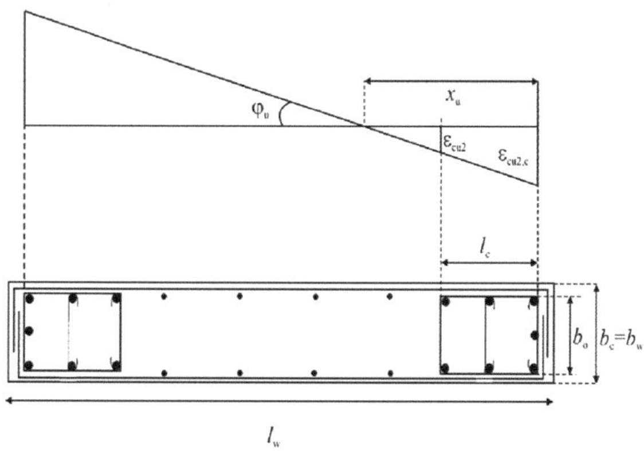

Web reinforcement

#	2	number of #
	Φ8/200	provided reinforcement
A_v (mm^2/m) =	**502.65**	web reinforcement area

Check minimum & maximum ratio of longitudinal reinforcement of the web

$\rho_{v,prov} = A_v/(1 x b_w)$ =	**0.20%**	
$\rho_{v,min}$ =	**0.20%**	minimum reinforcement ratio
$\rho_{v,max}$ =	**4.00%**	maximum reinforcement ratio
	valid	check $\rho_{v,min} < \rho_{v,prov} < \rho_{v,max}$

Maximum spacing of longitudinal reinforcement

s_v(mm) = min{$3b_{wo}$,400mm}=	**400.0**	required spacing for DCM
d_{bv} =	**8**	diameter of longitudinal reinforcement
s_v(mm) = min{$25d_{bv}$,250mm}=	**200.0**	required spacing for DCH
$s_{w,max}$(mm) <	**400.0**	maximum spacing of shear reinforcement
	valid	check $s < s_w$

Minimum and maximum diameter of longitudinal reinforcement

	DCM	ductility class
d_{bv} (mm) =	**8**	diameter of longitudinal reinforcement
$d_{bv,min}$ (mm) =	**8**	min diameter of longitudinal reinf. (DCH)
$d_{bv,max}$ (mm) = $b_{wl}/8$ =	**31**	max diameter of longitudinal reinf. (DCH)
	valid	check $d_{bv,min} < d_{bv} < d_{bv,max}$

Check minimum & maximum ratio of shear reinforcement

$\rho_{h,prov}$ =	0.20%	
$\rho_{h,min} = \max\{0.1\%, 0.25\rho_v\}$ =	0.10%	minimum reinforcement ratio for DCM
$\rho_{h,min}$ =	0.20%	minimum reinforcement ratio for DCH
$\rho_{h,max}$ =	4.00%	maximum reinforcement ratio
	valid	check $\rho_{h,min} < \rho_{h,prov} < \rho_{h,max}$

Maximum spacing of shear reinforcement

s_h(mm) =	400.0	required spacing for DCM
d_{bv} =	8	diameter of horizontal rebars
s_h(mm) $= \min\{25d_{bv}, 250\text{mm}\}$ =	200.0	required spacing for DCH
$s_{w,max}$(mm) <	400.0	maximum spacing of shear reinforcement
	valid	check $s < s_w$

Minimum and maximum diameter of shear reinforcement

	DCM	ductility class
d_{bh} (mm) =	8	diameter of longitudinal reinforcement
$d_{bh,min}$ (mm) =	8	min diameter of longitudinal reinf. (DCH)
$d_{bh,max}$ (mm) $= b_{wl}/8$ =	31	max diameter of longitudinal reinf. (DCH)
	valid	check $d_{bh,min} < d_{bh} < d_{bh,max}$

Lever arm of internal forces

z (m) $= 0.8l_w$ =	2.40	
d (m) $= 0.9l_w$ =	2.70	effective section depth
d_1 (m) $= 0.1l_w$ =	0.30	distance of comp. reinf. from the edge
d_1 / d =	0.11	

Longitudinal reinforcement of confined boundary elements

Check of normalised axial load

	DCM	ductility class
$v_d = N_E/(b_{wl} l_w f_{cd})$ =	-0.10	normalised axial load
$v_{d,max}$ =	-0.30	v_d=0.35 for DCH, v_d=0.40 for DCM
	valid	check $\lvert v_d \rvert < \lvert v_{d,max} \rvert$

Wall aspect ratio

h_w/l_w =	4.00	
	YES	requirement to verify aspect ratio (h_w/l_w>2)[1]

	$M_{ET2Y-1,y}^{bot}$	$M_{ET2y-1,z}^{bot}$	N_{ET2Y-1}^{bot}	ω
	kNm	kNm	kN	
Current Load Combination	895.33	61.76	-870.53	0.02

	$M_{ET2Y-1,y}^{bo}$	$M_{ET2y-1,z}^{bot}$	N_{ET2Y-1}^{bot}	ω	
	kNm	kNm	kN		Most critical ω_{tot}
Load Combination (minN)[2]	885.54	19.59	-618.35	0.08	0.08

Required, minimum & provided longitudinal reinforcement of the confined boundary elements

$A_{s1}\,(cm^2)=\omega_1 b_{wl} d\, f_{cd}/f_{yd}$ =	**7.04**	area of required longitudinal reinforcement
$\rho_{req} = A_{s1}/l_c b_{wl}$ =	**0.63%**	
ρ_{min} =	**0.50%**	minimum reinforcement ratio (DCM/DCH)
$A_{s,min}\,(cm^2) = \rho_{min}\, l_c b_{wl}$ =	**5.63**	minimum longitudinal reinforcement
$A_{s,req}\,(cm^2)$ =	**7.04**	check min reinforcement ($A_{s,req} > A_{s,min}$)
cm^2	6Φ14	provided reinf. per boundary element
$A_{s,prov}\,(cm^2)$ =	**9.24**	area of provided longitudinal reinf.t[3]
	valid	area of provided reinf.t: $A_{s,prov} > A_{s,req}$

Maximum longitudinal reinforcement

ρ_{max} =	**4.00%**	minimum reinforcement ratio (DCM/DCH)
$A_{s,max}\,(cm^2) = \rho_{max}\, l_c b_{wl}$ =	**45.00**	minimum longitudinal reinforcement
	valid	area of provided reinf.: $A_{s,max} > A_{s,prov}$

Moment resistance at the base of the ductile wall

$A_{s,prov}\,(cm^2)$ =	**9.24**	area of provided longitudinal reinforcement
$M_{RdT2Y\text{-}1,y}^{bot}$ (kNm) =	**1239.30**	

Overview[4]

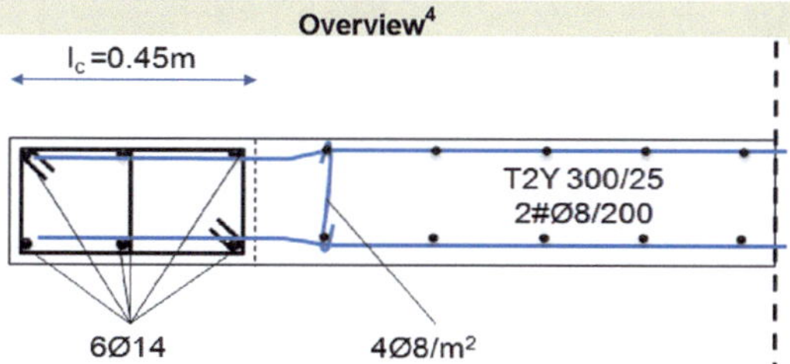

n_L =	3	number of rebars per side
$s\,(m) = (l_c-2c_{nom})/n_L$	**0.13**	spacing between restrained rebars[5]
	valid	verify at least 2 rebars per side

NOTES

[1] For aspect ratios that exceed h_w/l_w>2, verification of the ductile wall for bending is required. The design bending moment diagram along the height of the wall should be given by an envelope of the bending moment diagram from the analysis, vertically displaced (tension shift). The envelope may be assumed linear, if the structure does not exhibit significant discontinuities of mass, stiffness or resistance over its height (EN 1998-1, §5.4.2.4 (5)).

[2] GIven that the current load conbination used for the verification of all structural members (neams, columns, ductile walls) to ULS is not simultaneously critical against bending and shear, an additional load combination is herein provided for comparison.

[3] The longitudinal reinforcement derived for the confined boundary elements is approximate. Recalculation of the ductile wall moment resistance is required explicitly considering the contribution of the web longitudinal reinforcement.

[4] The above reinforcement arrangement is not final since the load combinations examined are not necessarily the most critical.

[5] There is no limitation as per the maximum spacing of the laterally restrained rebars. This calculation is complimentary.

DESIGN OF DUCTILE WALL FOR SHEAR - Input data

Shear wall section T2Y

l_w (m) =	3.00	length of shear wall
b_{wl} (m) =	0.25	width of shear wall
h_{st} (m) =	3.00	storey height
$_{wl,min}$ (m) = max{150mm, h_w/20} =	**0.15**	minimum width of shear wall
	valid	check minimum width of shear wall
	DCM	ductility class
cm^2	6Φ14	provided reinforcement per boundary element
	1239.30	Moment resistance at the wall base: $M_{RT2Y-1,y}^{bot}$ (kNm)
T (sec) =	0.47	uncoupled transl. period at the direction of the beam
	B	soil class
T_c (sec) =	0.50	corner period
q_0 =	2.00	basic value of the behaviour factor
PGA (g) =	0.24	peak ground acceleration
$S_e(T_c)$ (m/sec^2) =	0.73	spectral acceleration at period T_c (elastic spectrum)
$S_e(T_1)$ (m/sec^2) =	0.73	spectral acceleration at the fundamental period T_1
n =	4	number of building storeys

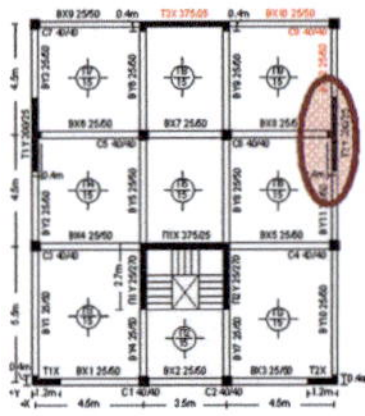

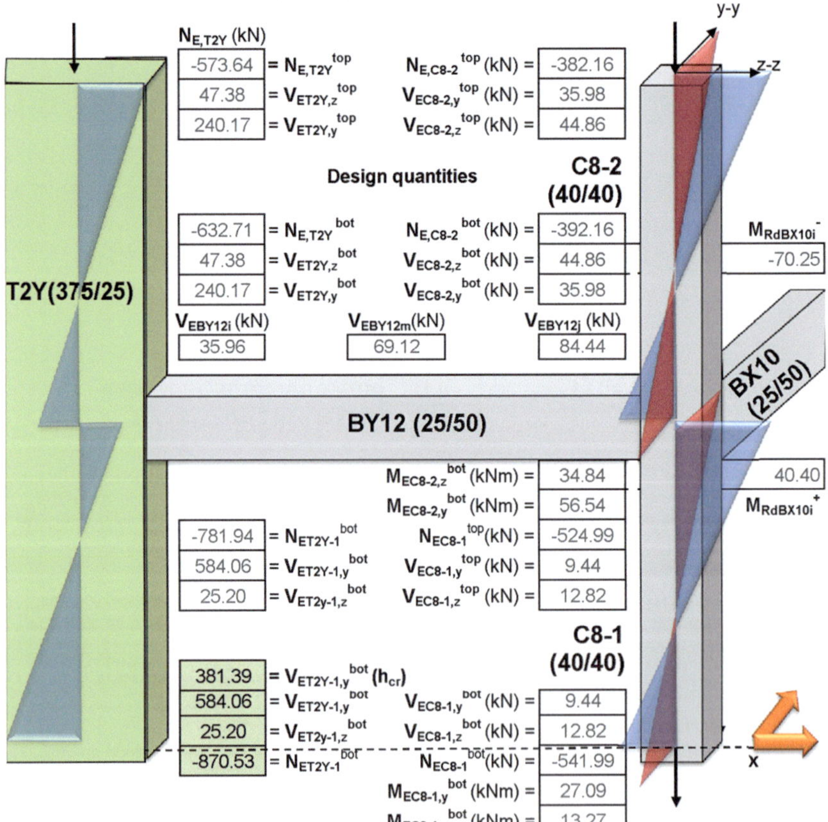

Design shear

	DCM	ductility class

Shear resistance for DCM

ε =	1.50	multiplication factor
$\varepsilon V_{ET2Y-1,y}^{bot}$ (kN) =	876.09	shear resistance for DCM

Shear resistance for DCH

check aspect ratio

h_w/l_w =	4.00	requirement for aspect ratio
	YES	verification ($h_w/l_w > 2$)
	1.50	multiplication factor ε
$\varepsilon = 1.2M_{Rd}/M_{Ed} < q$	1.66	multiplication factor ε for $H_w/l_w<2$
$\varepsilon = \sqrt{[(1.2M_{Rd}/M_{ed})^2+0.1(qS_e(T_c)/S_e(T_1))^2]}<q$	1.78	multiplication factor ε for $H_w/l_w>2$
$\varepsilon V_{ET2Y-1,y}^{bot}$ (kN) =	1038.08	shear resistance for DCH

Critical height of wall

$h_{cr} > \max\{l_w, H_w/6\}$ =	3.00	critical height of wall
$h_{cr} < \min\{2l_w, h_{st}\}$ =	3.00	critical height if storeys $n \le 6$
$h_{cr} < \min\{2l_w, 2h_{st}\}$ =	6.00	critical height if storeys $n > 6$

WITHIN CRITICAL REGION
check shear resistance against compression of the diagonal struts

	DCM	ductility class
θ (°) =	45.00	angle between the comp. strut and the beam axis perpendicular to the shear force $\theta=45°$ ($\cot\theta=1$) for DCH, recomm. $\theta=21.8°$ for DCM
$V_{Rd,max}= 0.24(1-f_{ck}/250)b_{wl}l_w f_{cd}\sin(2\theta)$=	1876.80	resistance against concrete crushing (DCM)
$V_{Rd,max}$= $= 0.4[0.24(1-f_{ck}/250)b_{wl}l_w f_{cd}\sin(2\theta)]$=	750.72	resistance against concrete crushing (DCH)
$\varepsilon V_{ET2Y-1,y}^{bot}$ (kN) =	1038.08	shear resistance of the wall
	valid	check $V_{Rd,max} > V_E$ (DCM)

Required shear reinforcement

$\alpha_s = M_{Ed}/(V_{Ed}l_w)$ =	0.51	
	not valid	check $a_s \ge 2$
	DCM	ductility class

Check shear resistance due to reinforcement

$\rho_{h,req} = \varepsilon V_{ET2Y-1,y}^{bot}/(0.8b_{wl}l_w f_{ywd})$ =	0.40%	required % of shear reinforcement
#	2	
	Φ12/200	web reinf.t (increase compared to Φ8/200)
A_{sw} (mm²/m) =	1130.97	area of web reinforcement 2#
A_{sw} (cm²)	1.13	area of shear reinforcement Φ12
$\rho_{h,prov} = A_{sw}/(b_{wl}s_w)$ =	0.45%	% of provided shear reinforcement
	valid	check $\rho_{h,prov} > \rho_{h,req}$

WITHIN CRITICAL REGION
Check shear resistance against compression of the diagonal struts

	DCM	ductility class
$\theta\ (^{o})$ =	45.00	angle between the comp. strut and the beam axis perpendicular to the shear force $\theta=45^{o}$ ($\cot\theta=1$) for DCH, recomm. $\theta=21.8^{o}$ for DCM
$V_{Rd,max}= 0.24(1-f_{ck}/250)b_{wl}l_{w}f_{cd}\sin(2\theta)$=	1876.80	resistance against concrete crushing
$\varepsilon V_{ET2Y-1,y}{}^{bot}(h_{cr})$ (kN) =	677.87	shear resistance of the wall
	valid	check $V_{Rd,max} > \varepsilon V_{ET2Y-1,y}{}^{bot}(h_{cr})$

Reverification of longitudinal reinforcement ratio

	DCM	ductility class
ρ_L (στο $A_c=l_c b_{wl}$) =	0.82%	% of longitudinal reinf. within the cr.l length
	not valid	check if ρ_L >2% within $A_c=l_c b_{wl}$

CONFINEMENT

	DCM	ductility class
$\omega_{wd,min}$ >	0.12	minimum mechanical volumetric ratio of confining reinforcement within critical length
v_d =	-0.68	$= N_{Ed}/(l_c b_{wl} f_{cd})$
$\mu_\varphi = 2q_0-1$ =	3.00	curvature ductility (for $T \geq T_c$)
$\mu_\varphi = 1+(2q_0-1)T_c/T$ =	4.19	curvature ductility (for $T<T_c$)
μ_φ =	4.19	final value of curvature ductility
$\varepsilon_{sy,d} = f_{yd}/E_s$ =	0.22%	design value of tension steel strain at yield

Required mechanical volumetric ratio of reinf. in the confined boundary elements

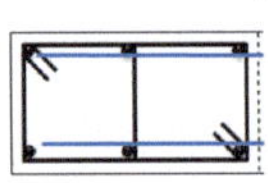

6Ø14

d_{bL} (mm)	14	diameter of boundary element long.l reinf.
d_{bw} (mm) = max{6mm, $d_{bl}/4$} =	6.0	min diameter of shear reinf. for DCM
d_{bw} = max{6mm, $0.4\sqrt{(f_{yd}/f_{wd})}d_{bL}$} =	6.0	min. diameter of shear reinf. for DCH
	6.0	min. diameter of shear reinf.within cr. length
	Φ10/70	provided shear reinf. (boundary element)
l_{c0} (m) =	0.400	width of the confined core
b_0 (m) =	0.200	width of the confined core
$\alpha_s = (1-s_w/2b_0)(1-s_w/2l_c)$ =	0.75	
n_b =	2	number of hoop legs || to b_{wl}
n_h =	3	number of hoop legs || to l_w
$\alpha_{n}= 1-\{b_0/[(n_h-1)l_{c0}]+l_{c0}/[(n_b-1)b_0]\}/3$ =	0.50	
$\alpha = \alpha_n \alpha_s$ =	0.38	confinement effectiveness factor
$\omega_v = \rho_v f_{yd}/f_{cd}$ =	0.08	mechanical ratio of the web long. reinf.
$\alpha\omega_{wd}$ >	0.22	$=30\mu_\varphi(v_d+\omega_v)\varepsilon_{sy,d}b_{wl}/b_0-0.035$

$\omega_{wd,req}$ =	**0.60**	
	valid	check $\omega_{wd,req} > \omega_{wd,min}$

Provided mechanical volumetric ratio of reinf. in the confined boundary elements

L_{wd} (m) = $(n_b b_0 + n_h h_0)$ =	**1.40**	length of shear reinforcement deployment
A_{sw} (cm^2) =	**0.79**	area of shear reinforcement
$V_0 = L_{wd} A_{sw} / s_w$ (mm^3) =	**1570.80**	volume of the confinement reinforcement
$V_c = b_0 l_c$ (mm^3) =	**80000**	volume of the concrete core
$\omega_{wd,prov} = V_0 f_{yd} / (V_c f_{cd})$ =	**0.75**	
	valid	check $\omega_{wd,prov} > \omega_{wd,req}$

Overview

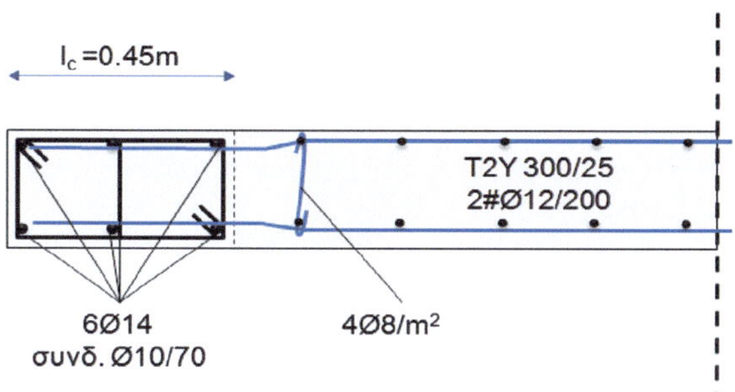

DESIGN OF DUCTILE WALL FOR BENDING - Input data

Ductile Wall T3X

l_w (m) =	3.75	length of shear wall
b_{wl} (m) =	0.25	width of shear wall
h_w (m) =	12.00	total height of shear wall
h_{st} (m) =	3.00	storey height
n =	4	number of building storeys
$b_{wl,min}$ (m) = max{150mm, h_w/20} =	**0.15**	minimum width of shear wall
	valid	check of minimum shear wall width
	DCM	ductility class

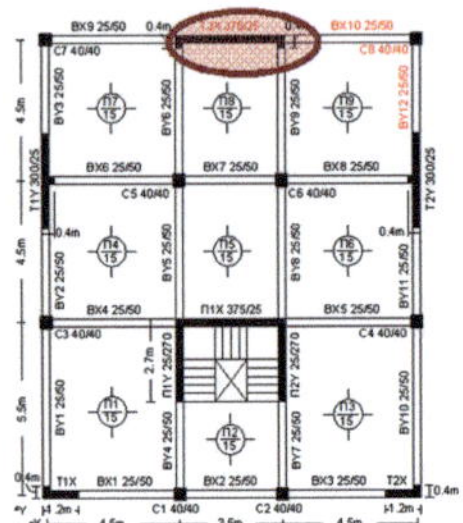

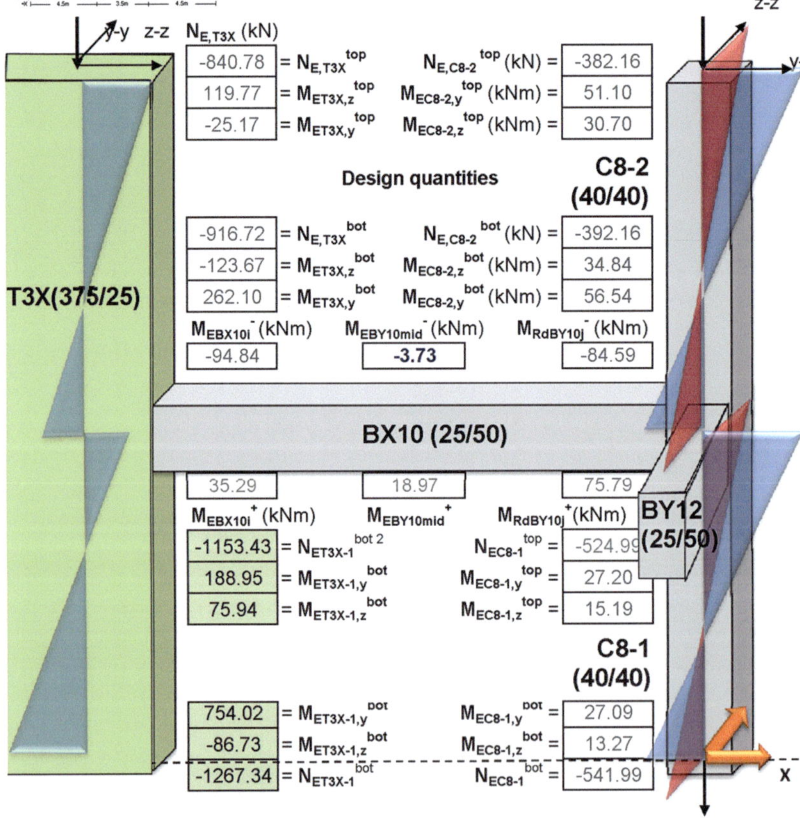

Dimensions of confined boundary elements

Critical (confined) length in boundary elements

l_c (m) = max{0.15l_w, 1.5b_w, $l(\varepsilon_c$>0.2%)} =	0.56	width of boundary elements

Width of confined boundary elements

max{0.20l_w, 2b_{wl}) =	0.75	
	valid	check l_c (m) < max{0.20l_w, 2b_{wl})
	no	requirement for wider dimensions
$b_{w,min}$ (m) = max{200mm, h_{st}/15) =	0.20	minimum width if $l_c \leq$ max{0.20l_w, 2b_{wl})
$b_{w,min}$ (m) = max{200mm, h_{st}/10) =	0.20	minimum width if l_c > max{0.20l_w, 2b_{wl})
	valid	check $b_w > b_{w,min}$

Web reinforcement

#	2	number of #
	Φ8/200	provided reinforcement
A_v (mm^2/m) =	502.65	web reinforcement area

Check minimum & maximum ratio of longitudinal reinforcement of the web

$\rho_{v,prov} = A_v/(1xb_w)$ =	0.20%	
$\rho_{v,min}$ =	0.20%	minimum reinforcement ratio
$\rho_{v,max}$ =	4.00%	maximum reinforcement ratio
	valid	check $\rho_{v,min} < \rho_{v,prov} < \rho_{v,max}$

Maximum spacing of longitudinal reinforcement

s_v(mm) =min{3b_{wo},400mm}=	400.0	required spacing for DCM
d_{bv} =	8	diameter of longitudinal reinforcement
s_v(mm) =min{25d_{bv},250mm}=	200.0	required spacing for DCH
$s_{w,max}$(mm) <	400.0	maximum spacing of shear reinforcement
	valid	check s < s_w

Minimum and maximum diameter of longitudinal reinforcement

	DCM	ductility class
d_{bv} (mm) =	8	diameter of longitudinal reinforcement
$d_{bv,min}$ (mm) =	8	min diameter of longitudinal reinf. (DCH)
$d_{bv,max}$ (mm) = b_{wl}/8 =	31	max diameter of longitudinal reinf. (DCH)
	valid	check $d_{bv,min} < d_{bv} < d_{bv,max}$

Check minimum & maximum ratio of shear reinforcement

$\rho_{h,prov}$ =	0.20%	
$\rho_{h,min}$ = max{0.1%, 0.25ρ_v} =	0.10%	minimum reinforcement ratio for DCM
$\rho_{h,min}$ =	0.20%	minimum reinforcement ratio for DCH
$\rho_{h,max}$ =	4.00%	maximum reinforcement ratio
	valid	check $\rho_{h,min} < \rho_{h,prov} < \rho_{h,max}$

Maximum spacing of shear reinforcement

s_h(mm) =	**400.0**	required spacing for DCM
d_{bv} =	**8**	diameter of horizontal rebars
s_h(mm) = min{$25d_{bv}$, 250mm} =	**200.0**	required spacing for DCH
$s_{w,max}$(mm) <	**400.0**	maximum spacing of shear reinforcement
	valid	check $s < s_w$

Minimum and maximum diameter of shear reinforcement

	DCM	ductility class
d_{bh} (mm) =	**8**	diameter of longitudinal reinforcement
$d_{bh,min}$ (mm) =	**8**	min diameter of longitudinal reinf. (DCH)
$d_{bh,max}$ (mm) = $b_{wl}/8$ =	**31**	max diameter of longitudinal reinf. (DCH)
	valid	check $d_{bh,min} < d_{bh} < d_{bh,max}$

Lever arm of internal forces

z (m) = $0.8l_w$ =	**3.00**	
d (m) = $0.9l_w$ =	**3.38**	effective section depth
d_1 (m) = $0.1l_w$ =	**0.38**	distance of compression reinf. from the edge
d_1 / d =	**0.11**	

Longitudinal reinforcement of confined boundary elements

Check of normalised axial load

	DCM	ductility class
$v_d = N_E/(b_{wl}l_w f_{cd})$ =	**-0.12**	normalised axial load
$v_{d,max}$ =	**-0.30**	v_d=0.35 for DCH, v_d=0.40 for DCM
	valid	check $\|v_d\|<\|v_{d,max}\|$

Wall aspect ratio

h_w/l_w =	**3.20**	
	YES	requirement to verify aspect ratio ($h_w/l_w > 2$)[1]

	$M_{ET2Y-1,y}^{bot}$	$M_{ET2y-1,z}^{bot}$	N_{ET2Y-1}^{bot}	ω
	kNm	**kNm**	**kN**	
Current Load Combination	**754.02**	**-86.73**	**-1267.34**	0.020

	$M_{ET2Y-1,y}^{bot}$	$M_{ET2y-1,z}^{bot}$	N_{ET2Y-1}^{bot}	ω	
	kNm	**kNm**	**kN**		Most critical ω_{tot}
Load Combination (minN)[2]	**-188.95**	**3.30**	**-703.87**	0.01	0.02

Required, minimum & provided longitudinal reinforcement of the confined boundary elements

A_{s1} (cm^2) = $\omega_1 b_{wl} d\, f_{cd}/f_{yd}$ =	**2.20**	area of required longitudinal reinforcement
$\rho_{req} = A_{s1}/l_c b_{wl}$ =	**0.16%**	
ρ_{min} =	**0.50%**	minimum reinforcement ratio (DCM/DCH)
$A_{s,min}$ (cm^2) = $\rho_{min}\, l_c b_{wl}$ =	**7.03**	minimum longitudinal reinforcement
$A_{s,req}$ (cm^2) =	**7.03**	check minimum reinforcement ($A_{s,req}>A_{s,min}$)
cm^2	6Φ14	provided reinf. per boundary element
$A_{s,prov}$ (cm^2) =	**9.24**	area of provided longitudinal reinforcement[3]
	valid	area of provided reinforcement: $A_{s,prov}>A_{s,req}$

Maximum longitudinal reinforcement

ρ_{max} =	**4.00%**	minimum reinforcement ratio (DCM/DCH)
$A_{s,max}$ (cm^2) = $\rho_{max}\, l_c b_{wl}$ =	**56.25**	minimum longitudinal reinforcement
	valid	area of provided reinf.: $A_{s,max}$>$A_{s,prov}$

Moment resistance at the base of the ductile wall

$A_{s,prov}$ (cm^2) =	**9.24**	area of provided longitudinal reinforcement
$M_{RdT2Y\text{-}1,y}^{bot}$ (kNm) =	**1290.94**	

Overview[4]

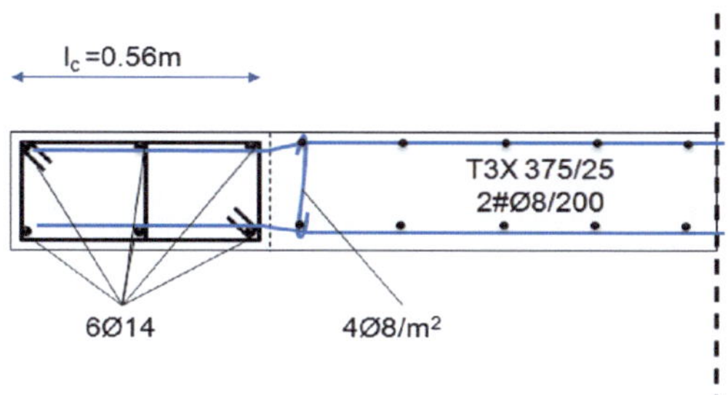

n_L =	3	number of rebars per side
s (m) = (l_c-2c_{nom})/n_L	**0.17**	spacing between restrained rebars[5]
	valid	verify at least 2 rebars per side

NOTES

[1] For aspect ratios that exceed h_w/l_w>2, verification of the ductile wall for bending is required. The design bending moment diagram along the height of the wall should be given by an envelope of the bending moment diagram from the analysis, vertically displaced (tension shift). The envelope may be assumed linear, if the structure does not exhibit significant discontinuities of mass, stiffness or resistance over its height (EN 1998-1, §5.4.2.4 (5)).

[2] Given that the current load conbination used for the verification of all structural members (neams, columns, ductile walls) to ULS is not simultaneously critical against bending and shear, an additional load combination is herein provided for comparison.

[3] The longitudinal reinforcement derived for the confined boundary elements is approximate. Recalculation of the ductile wall moment resistance is required explicitly considering the contribution of the web longitudinal reinforcement.

[4] the above reinforcement arrangement is not final since the load combinations examined are not necessarily the most critical.

[5] There is no limitation as per the maximum spacing of the laterally restrained rebars. This calculation is complimentary.

DESIGN OF DUCTILE WALL FOR SHEAR - Input data

Shear wall section T3X

l_w (m) =	3.75	length of shear wall
b_{wl} (m) =	0.25	width of shear wall
h_{st} (m) =	3.00	storey height
$b_{wl,min}$(m)=max{150, h_w/20}=	**0.15**	minimum width of shear wall
	valid	check minimum width of shear wall
	DCM	ductility class
cm^2	6Φ14	provided reinforcement per boundary element
	1290.94	Moment resistance at the wall base: $M_{RT3X-1,y}^{bot}$ (kNm)
T (sec) =	0.47	uncoupled transl.l period at the direction of the beam
	B	soil class
T_c (sec) =	0.50	maximum spectral acceleration
q_0 =	2.00	basic value of the behaviour factor
PGA (g) =	0.24	peak ground acceleration
$S_e(T_c)$ (sec) =	0.73	spectral acceleration at period T_c (elastic spectrum)
$S_e(T_1)$ (sec) =	0.73	spectral acceleration at the fundamental period T_1
n =	4	number of building storeys

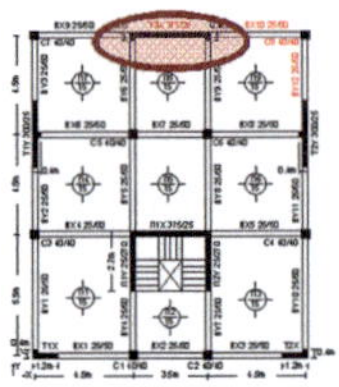

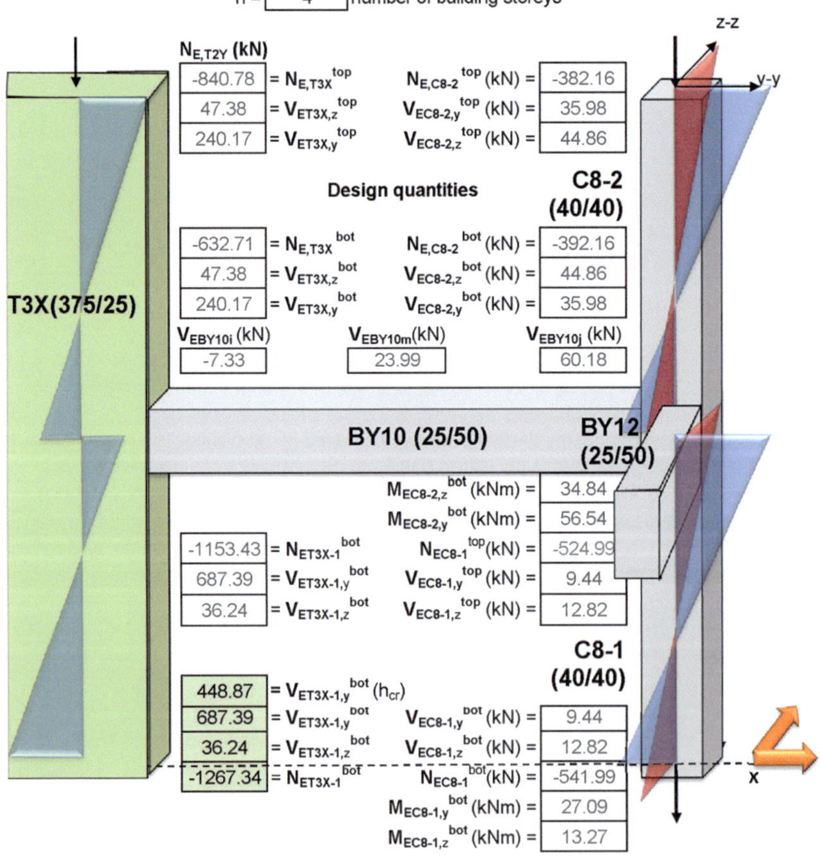

Design shear

	DCM	ductility class

Shear resistance for DCM

ε =	1.50	multiplication factor
$\varepsilon V_{ET3X-1,y}^{bot}$ (kN) =	1031.09	shear resistance for DCM

Shear resistance for DCH

	check aspect ratio	
h_w/l_w =	4.00	
requirement for aspect ratio verification	YES	(hw/lw > 2)
	1.50	multiplication factor ε
$\varepsilon = 1.2M_{Rd}/M_{Ed} < q$	1.73	multiplication factor ε for H_w/l_w<2
$\varepsilon = \sqrt{[(1.2M_{Rd}/M_{ed})^2 + 0.1(qS_e(T_c)/S_e(T_1))^2]} < q$	1.84	multiplication factor ε for H_w/l_w>2
$\varepsilon V_{ET3X-1,y}^{bot}$ (kN) =	1266.31	shear resistance for DCH

Critical height of wall

$h_{cr} > \max\{l_w, H_w/6\}$ =	3.75	critical height of wall
$h_{cr} < \min\{2l_w, h_{st}\}$ =	3.00	critical height if storeys n ≤ 6
$h_{cr} < \min\{2l_w, 2h_{st}\}$ =	6.00	critical height if storeys n > 6

WITHIN CRITICAL REGION
Check shear resistance against compression of the diagonal struts

	DCM	ductility class
θ (°) =	45.00	angle between the comp. strut and the beam axis perpendicular to the shear force θ=45° (cotθ=1) for DCH, recomm. θ=21.8° for DCM
$V_{Rd,max} = 0.24(1-f_{ck}/250)b_{wl}l_wf_{cd}\sin(2\theta)$=	1903.35	resistance against concrete crushing
$V_{Rd,max} = 0.4[0.24(1-f_{ck}/250)b_{wl}l_wf_{cd}\sin(2\theta)]$ =	761.34	
$\varepsilon V_{ET3X-1,y}^{bot}$ (kN) =	1266.31	shear resistance of the wall
	valid	check $V_{Rd,max} > \varepsilon V_{ET2}$ (DCM)

Reverification of longitudinal reinforcement ratio

$\alpha_s = M_{Ed}/(V_{Ed}l_w)$ =	0.35	
	not valid	% of long. reinf. within the critical length
	DCM	check if ρ_L >2% within $A_c=l_cb_{wl}$

Check shear resistance due to reinforcement

$\rho_{h,req} = \varepsilon V_{ET3X-1,y}^{bot}/(0.8b_{wl}l_wf_{ywd})$ =	0.39%	required % of shear reinforcement
#	2	
	Φ12/200	web reinf. (increase compared to Φ8/200)
A_{sw} (mm²/m) =	1130.97	area of web reinforcement 2#
A_{sw} (cm²)	1.13	area of shear reinforcement Φ12
$\rho_{h,prov} = A_{sw}/(b_{wl}s_w)$ =	0.45%	% of provided shear reinforcement
	valid	check $\rho_{h,prov} > \rho_{h,req}$

OUTSIDE CRITICAL REGION
Check shear resistance against compression of the diagonal struts

	DCM	ductility class
θ (°) =	45.00	angle between the comp. strut and the beam axis perpendicular to the shear force θ=45° (cotθ=1) for DCH, recomm. θ=21.8° for DCM
$V_{Rd,max}$= 0.24(1-f_{ck}/250)$b_{wl}$$l_w$$f_{cd}$sin(2θ)=	**1903.35**	resistance against concrete crushing
$\varepsilon V_{ET3X-1,y}^{bot}(h_{cr})$ (kN) =	**826.90**	shear resistance of the shear wall
	valid	check $V_{Rd,max} > \varepsilon V_{ET2Y-1,y}^{bot}(h_{cr})$

Reverification of longitudinal reinforcement ratio

	DCM	ductility class
ρ_L (στο $A_c=l_c b_{wl}$) =	**0.82%**	% of longitudinal reinf. within the cr. length
	not valid	check if ρ_L >2% within $A_c=l_c b_{wl}$

CONFINEMENT

	DCM	ductility class
$\omega_{wd,min}$ >	**0.12**	minimum mechanical volumetric ratio of confining reinforcement within critical length
ν_d =	**-0.99**	= $N_{Ed}/(l_c b_{wl} f_{cd})$
$\mu_\varphi = 2q_0-1$ =	**3.00**	curvature ductility (for T≥Tc)
$\mu_\varphi = 1+(2q_0-1)T_c/T$ =	**4.19**	curvature ductility (for T<Tc)
μ_φ =	**4.19**	final value of curvature ductility
$\varepsilon_{sy,d} = f_{yd}/E_s$ =	**0.22%**	design value of tension steel strain at yield

Required mechanical volumetric ratio of reinf. in the confined boundary elements

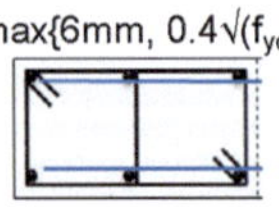

6Ø14

d_{bL} (mm)	**14**	diameter of boundary element long.l reinf.
d_{bw} (mm) = max{6mm, d_{bl}/4} =	**6.0**	min diameter of shear reinf. for DCM
d_{bw} = max{6mm, 0.4√(f_{yd}/f_{wd})d_{bL}} =	**6.0**	min. diameter of shear reinf. for DCH
	6.0	min. diameter of shear reinf.within cr. length
	Φ12/70	provided shear reinf. (boundary element)
l_{c0} (m) =	**0.513**	width of the confined core
b_0 (m) =	**0.200**	width of the confined core
$\alpha_s = (1-s_w/2b_0)(1-s_w/2l_c)$ =	**0.77**	
n_b =	2	number of hoop legs ǁ to b_{wl}
n_h =	3	number of hoop legs ǁ to l_w
$\alpha_{n=}$ 1-{b_0/[(n_h-1)l_{c0}]+l_{c0}/[(n_b-1)b_0]}/3 =	**0.44**	
$\alpha = \alpha_n \alpha_s$ =	**0.34**	confinement effectiveness factor
$\omega_v = \rho_v f_{yd}/f_{cd}$ =	**0.08**	mechanical ratio of the web long. reinf.
$\alpha\omega_{wd}$ >	**0.33**	$=30\mu_\varphi(\nu_d+\omega_v)\varepsilon_{sy,d}b_{wl}/b_0-0.035$

$\omega_{wd,req}$ =	**0.97**	
	valid	check $\omega_{wd,req} > \omega_{wd,min}$

Provided mechanical volumetric ratio of reinf. in the confined boundary elements

L_{wd} (m) = $(n_b b_0 + n_h h_0)$ =	**1.63**	length of shear reinforcement deployment
A_{sw} (cm^2) =	**1.13**	area of shear reinforcement
$V_0 = L_{wd} A_{sw} / s_w$ (mm^3) =	**2625.47**	volume of the confinement reinforcement
$V_c = b_0 l_c$ (mm^3) =	**102500**	volume of the concrete core
$\omega_{wd,prov} = V_0 f_{yd} / (V_c f_{cd})$ =	**0.98**	
	valid	check $\omega_{wd,prov} > \omega_{wd,req}$

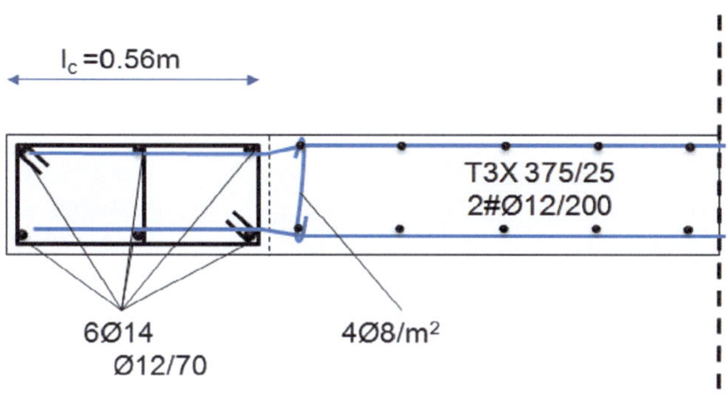

4.3.12 *Envelope Bending Moment and Shear Force Diagram for the Ductile Wall W3X of Building C*

This section serves to provide a numerical application for determining the design (envelope) in-plane bending moment and shear force diagrams of a typical ductile planar r/c wall according to EC8 following the computational steps detailed in Sect. 3.5. Specifically, the envelope bending moment and shear force diagrams for wall W3X of building C are derived for the seismic design combination of actions G"+"ψ_2Q"±"E assuming that the centers of mass at each floor slab are located at position 1 of Table 4.116.

4.3.12.1 Calculation of Bending Moment Diagram for Wall W3X

Figure 4.29 summarizes all the required calculations presented in Fig. 3.5 for determining the design envelope bending moment diagram (shown as a red line) of wall W3X for design seismic actions combination G"+"ψ_2Q"+"E. The corresponding envelope for the actions combination G"+"ψ_2Q"−"E will have the

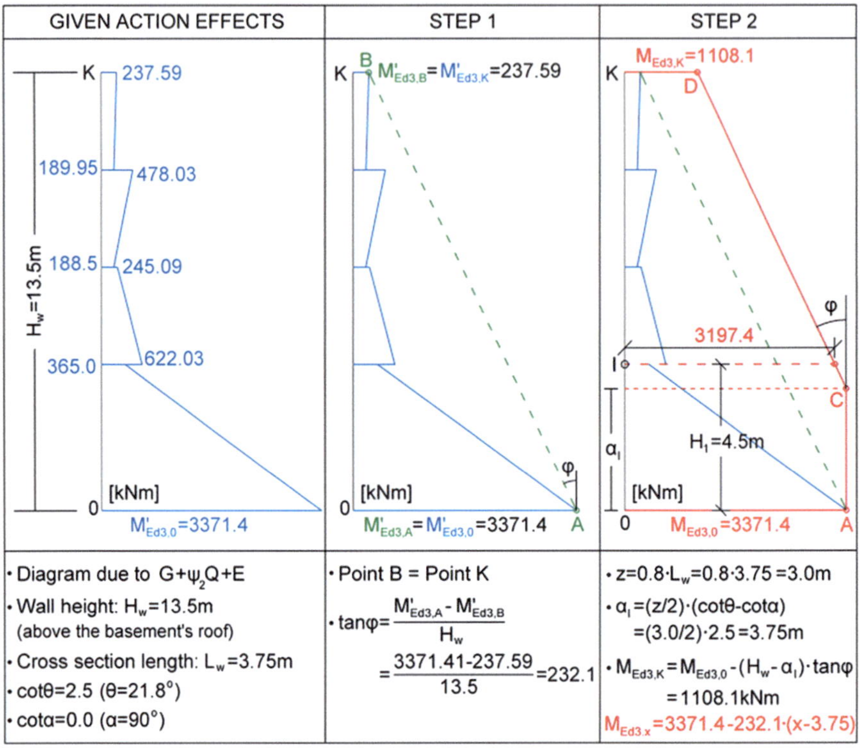

Fig. 4.29 Calculation of envelope bending moment diagram for wall W3X of building C (design seismic action combination G "+" ψ_2Q "+" E for mass position 1) following Fig. 3.5

same shape and amplitude as in the case of G"+"ψ_2Q"+"E but the opposite sign, since the in-plane bending moment M_3 for the gravitational permanent and variable actions combination G"+"ψ_2Q is negligible (see Table 4.116).

Having derived the design envelope bending moment diagram M_3, the design values of the in-plane bending moment, M_{ED3}, can be readily determined by linear interpolation for any cross-section of the W3X wall. For example, M_{ED3} is equal to 3197.4 kNm at the top end of the considered wall at the ground storey (height level $H_1 = 4.5$ m), as indicated in Fig. 4.29. Consequently, the design triad N, M_2, $\pm$extrM_3 for which this particular cross-section of the wall needs to be detailed and verified are shown in Table 4.140 (see also Table 3.7), where the values for N and M_2 are obtained from Table 4.125 (mass position 1 and design triad -extrM_3). It is observed that consideration of the envelope bending moment diagram M_3 prescribed by EC8 leads to an increase of more than 750 % in the in-plane moment for which the examined cross-section of the W3X wall needs to be designed compared to the in-plane moment derived from the structural analysis step.

However, the design triads N, M_2, $\pm$extrM_3 for the base of wall W3X at the ground floor derived from the application of the MRSM, as reported in Table 4.125, will remain the same, since the value of the envelope bending moment diagram at

Table 4.140 Design triads N, M_2, ±extrM_3 for the top of wall W3X at the ground floor upon consideration of the design envelop in-plane bending moment diagram of Fig. 4.29 (mass position 1)

$N'_{Ed,1}$ [kN]	$M'_{Ed2,1}$ [kNm]	$M'_{Ed3,1}$ [kNm]
−1059.49	56.33	365.0
−857.81	16.25	−365.0
$N'_{Ed,1}$ [kN]	$M'_{Ed2,1}$ [kNm]	$M'_{Ed3,1}$ [kNm]
−1059.49	56.33	3197.4
−857.81	16.25	−3197.4

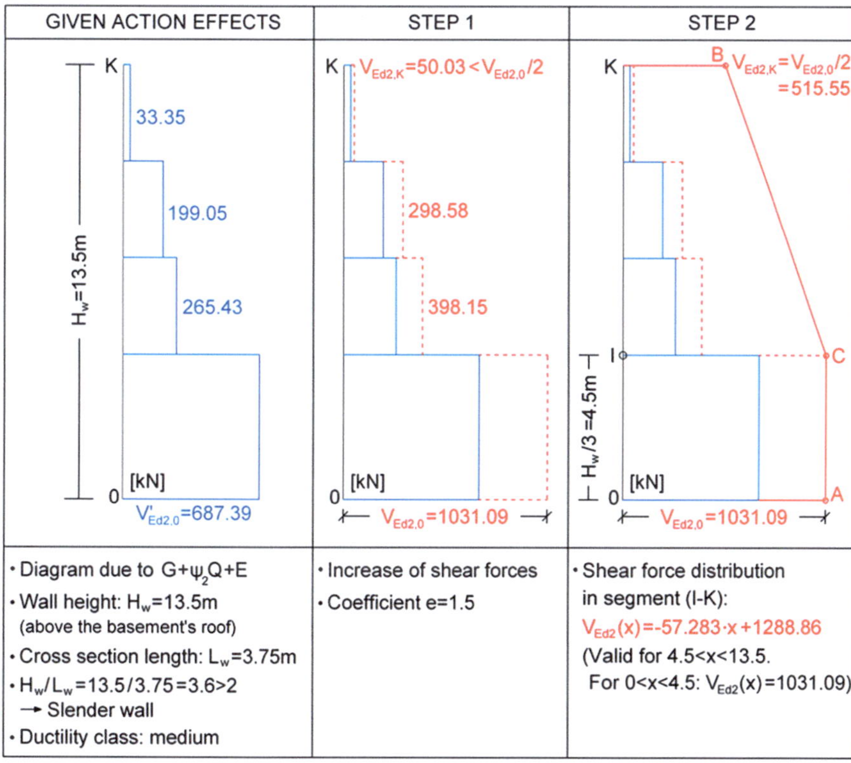

Fig. 4.30 Calculation of envelope shear force diagram for wall W3X of building C (design seismic action combination G "+" ψ_2Q "+" E for mass position 1) following Fig. 3.6

the base of walls at the ground floor is equal to the bending moment value derived from the structural analysis step for the G"+"ψ_2Q"±"E design actions combination.

4.3.12.2 Calculation of Shearing Force Diagram for Wall W3X

Figure 4.30 summarizes all the required calculations presented in Fig. 3.6 for determining the design envelope shearing force diagram (shown as a red solid line) of wall W3X for design seismic actions combination G"+"ψ_2Q"+"E. The corresponding envelop for the actions combination G“+”ψ_2Q“−”E will have the same shape and amplitude as in the case of G“+”ψ_2Q“+”E but the opposite sign, since the in-plane shearing force V_2 for the gravitational permanent and variable actions combination G“+”ψ_2Q is negligible (see Table 4.116).

Having derived the design envelope shearing force diagram V_2, the design values of the in-plane shears, V_{ED2}, can be readily determined for any cross-section of the W3X wall. For example, the following design values for V_2 are obtained at the bottom and the top of the considered wall at the ground floor:

- Actions combination $G+\psi_2Q+E$: $V_{Ed2}(0)=1031.09$ kN; $V_{Ed2}(4.5)=1031.09$ kN
- Actions combination $G+\psi_2Q\text{-}E$: $V_{Ed2}(0)=-1031.09$ kN; $V_{Ed2}(4.5)=-1031.09$ kN

In Table 4.141, all symbols and notations used in Sects. 4.1, 4.2, and 4.3 are summarized.

Table 4.141 Symbols and notations used in Sects. 4.1, 4.2, and 4.3

Glossary	
Peak modal (or seismic) value:	value corresponding to a specific mode shape using a design spectrum for excitation along a specific axis
Maximum value:	value produced by modal combination (CQC or SRSS rule) for each direction of excitation
Extreme value:	value produced by spatial combination (e.g., using the SRSS combination rule) of the corresponding maximum values obtained for each independent direction of excitation (two horizontal orthogonal and one vertical)
Fictitious elastic axis:	vertical axis in reference to which the sum of squares of floor rotations caused by torsional moments of triangular distribution along the height becomes a minimum
Notation	
1, 2, 3	: local reference axes
X, Y, Z	: global reference axes
$V_{i,walls}$	: the shear force along i axis (i = X, Y) taken by the walls
$V_{i,columns}$	: the shear force along i axis (i = X, Y) taken by the columns
S_{ai}	: design spectral acceleration for mode i
r_X	: torsional stiffness radius with respect to X axis
r_Y	: torsional stiffness radius with respect to Y axis
l_s	: radius of gyration
J_{mk} (k = X, Y)	: polar moment of inertia with regard to the center of mass displaced by the accidental eccentricity e_{ak} (k = X, Y)
h_k	: height of storey k
e_{oX}	: static (structural) eccentricity along the X axis
e_{oY}	: static (structural) eccentricity along the Y axis

(continued)

Table 4.141 (continued)

$V^{(k)}_{X(tot),EXi}$	: total shear force along the X axis at storey (k) corresponding to mode i for seismic excitation along the X axis (EX)
$V^{(k)}_{X(tot),EYi}$	: total shear force along the X axis at storey (k) corresponding to mode i for seismic excitation along the Y axis (EY)
$V^{(k)}_{Y(tot),EXi}$	: total shear force along the Y axis at storey (k) corresponding to mode i for seismic excitation along the X axis (EX)
$V^{(k)}_{Y(tot),EYi}$	: total shear force along the Y axis at storey (k) corresponding to mode i for seismic excitation along the Y axis (EY)
$d^{(k)}_{rXi,j}$	: modal design (inelastic) interstorey drift along the X axis of the vertical member j at storey k, corresponding to mode i. The values depend on the mode normalization procedure and do not correspond to any seismic excitation.
$d^{(k)}_{rXi}$	: average (mean) value of the modal (mode i) design (inelastic) interstorey drifts $d^{(k)}_{rXi,j}$ of the vertical members at storey k
$d^{(k)}_{rYi,j}$	: modal design (inelastic) interstorey drift along the Y axis of the vertical member j at storey k, corresponding to mode i. The values depend on the mode normalization procedure and do not correspond to any seismic excitation.
$d^{(k)}_{rYi}$	: average (mean) value of the modal (mode i) design (inelastic) interstorey drifts $d^{(k)}_{rYi,j}$ of the vertical members at storey k
u^{t}_{Xij}	: modal displacement along the X axis at the top, t, of the vertical member j for mode i. The values depend on the mode normalization procedure and do not correspond to any seismic excitation.
u^{b}_{Xij}	: modal displacement along the X axis at the bottom (base), b, of the vertical member j for mode i. The values depend on the mode normalization and do not correspond to any seismic excitation.
u^{t}_{Yij}	: modal displacement along the Y axis at the top, t, of the vertical member j for mode i. The values depend on the mode normalization procedure and do not correspond to any seismic excitation.
u^{b}_{Yij}	: modal displacement along the Y axis at the bottom (base), b, of the vertical member j for mode i. The values depend on the mode normalization procedure and do not correspond to any seismic excitation.
ν_{iX}	: modal participation factor for mode i and excitation along the X axis
ν_{iY}	: modal participation factor for mode i and excitation along the Y axis
$d^{(k)}_{rX,EXi}$	: average (mean) design interstorey drift along the X axis at storey k under seismic excitation along the X axis (EX) for mode i
$d^{(k)}_{rY,EXi}$	: average (mean) design interstorey drift along the Y axis at storey k under seismic excitation along the X axis (EX) for mode i
$d^{(k)}_{rX,EYi}$	: average (mean) design interstorey drift along the X axis at storey k under seismic excitation along the Y axis (EY) for mode i
$d^{(k)}_{rY,EYi}$	: average (mean) design interstorey drift along the Y axis at storey k under seismic excitation along the Y axis (EY) for mode i
$\theta^{(k)}_{X,EXi}$	: interstorey drift sensitivity coefficient along the X axis at storey k under seismic excitation along the X axis (EX) for mode i
$\theta^{(k)}_{Y,EXi}$	: interstorey drift sensitivity coefficient along the Y axis at storey k under seismic excitation along the X axis (EX) for mode i

(continued)

Table 4.141 (continued)

$\theta_{X,EYi}^{(k)}$	: interstorey drift sensitivity coefficient along the X axis at storey k under seismic excitation along the Y axis (EY) for mode i
$\theta_{Y,EYi}^{(k)}$	: interstorey drift sensitivity coefficient along the Y axis at storey k under seismic excitation along the Y axis (EY) for mode i
$V_{Xij}^{(k)}$	: modal shear along the X axis for mode i in the vertical element j of storey k. It depends on normalization of modes and does not correspond to any seismic excitation.
$V_{Yij}^{(k)}$	: modal shear along the Y axis for mode i in the vertical element j of storey k. It depends on normalization of modes and does not correspond to any seismic excitation.
$V_{X,Lci,j}^{(k)}$	: shear force along the X axis in vertical member j of storey k for loading case i (i = 1, 2, 3, 4)
$u_{X,Lci,j}^{t}$	: displacement along the X axis at the top, t, of vertical member j for the loading case i (i = 1, 2, 3, 4)
$u_{X,Lci,j}^{b}$	: displacement along the X axis at the bottom (base), b, of vertical member j for loading case i (i = 1, 2, 3, 4)
$u_{Xm,Lci}^{t(k)}$	: average mean value of the displacements $u_{X,Lci,j}^{t}$ (i = 1, 2, 3, 4; j = number of vertical member) for storey k
$u_{Xm,Lci}^{b(k)}$	: average mean value of the displacements $u_{X,Lci,j}^{b}$ (i = 1, 2, 3, 4; j = number of vertical member) for storey k
$d_{eX,Lci,j}^{(k)}$	: elastic interstorey drift along the X axis at vertical member j of storey k for the loading case Lci
$d_{eX,Lci}^{(k)}$	: elastic interstorey drift of storey k along the X axis for the loading case Lci
$d_{eY,Lci,j}^{(k)}$	: elastic interstorey drift along the Y axis at vertical member j of storey k for the loading case Lci
$d_{eY,Lci}^{(k)}$	: elastic interstorey drift of storey k along the Y axis for the loading case Lci
$d_{rX,Lci}^{(k)}$	: design interstorey drift of storey k along the X axis for the loading case Lci
$d_{rY,Lci}^{(k)}$	: design interstorey drift of storey k along the Y axis for the loading case Lci

References

Anastassiadis K (1993) Directions Seismic Defavorables et Combinaisons Defavorables des Efforts. Ann l' ITBTP 512:82–99

Arnold C, Fenton GA (2013) Modern geotechnical design codes of practice. In: Advances in soil mechanics and geotechnical engineering, vol 1. IOS Press, Amsterdam

CEN (2002) European Standard EN1990. Eurocode 0: basis of structural design. Committee for Standarization. Brussels

CEN (2003) European Standard EN 1997-1. Eurocode 7: Geotechnical design – Part 1: General, Committee for Standarization, Brussels

CSI (2012) SAP2000: integrated building design software, v. 14 – user's manual, Berkeley, California, USA

Earthquake Planning and Protection Organization (EPPO) (2000) Greek Seismic Code EAK2000 (amended in 2003), Athens, Greece (in Greek)
Fardis MN (2009) Seismic design, assessment and retrofit of concrete buildings, based on Eurocode 8. Springer, Dordrecht
Gupta AK (1992) Response spectrum method in seismic analysis and design of structures. CRC Press, New York/Weiler
Hellenic Organization for Standarization (2009) Standard 1498-1. Greek National Annex to Eurocode 8 – Part 1: general rules, seismic actions and rules for buildings (in Greek), Athens
Makarios TK, Anastassiadis K (1998) Real and fictitious elastic axes of multi-storey buildings: applications. Struct Des Tall Spec Build 7:57–71
Oden JT (1967) Mechanics of elastic structures. McGraw-Hill, New York
Terzhaghi K (1955) Evaluation of coefficient of subgrade reaction. Geotechnique 5:297–326

Appendix A – Qualitative Description of EC8 Non-Linear Static (Pushover) Analysis Method

The non-linear static (pushover) analysis method prescribed by EC8 is based on the so-called "N2" pushover method (Fajfar 2000). The method has been developed to assess the seismic vulnerability of planar frame structures accounting for their non-linear behavior in an approximate and computationally efficient manner compared to non-linear response history analysis. Aiming for simplicity and practical usability (not offered by non-linear response history analysis) rather than rigorousness, the method -despite being non-linear- consciously makes use of the superposition principle, which is normally valid exclusively within the framework of linear mechanics. It further includes various simplistic assumptions, especially when applied to asymmetric buildings, which contradict the otherwise rigorous theory of structural dynamics to a certain extent (see also Sects. 2.4.1.2 and 2.4.3.4). In general, static analysis -in any of its possible variants- is considered less informative and less reliable than non-linear response history analyses.

However, parametric studies conducted during the past decades have proved that the method provides good estimates for the floor inelastic displacements of medium rise planar frames (Krawinkler and Seneviratna 1998). Hence, this analysis method gained significant popularity for the seismic assessment of *existing* under-designed (in refer to modern seismic codes) structures. Furthermore, EC8 permits the use of this method as a standalone analysis tool for the design of *new* structures (§ 4.3.3.4.2.1(1)d), even though such a practice is not encouraged by the authors for reasons discussed in some detail in Sect. 2.4.3.4.

Overall it can be claimed that both nonlinear static (i.e., pushover) and response history analyses have their own pros and cons and, as such, they have to be applied with due consideration of their respective assumptions. A brief summary of the limitations of the two methods is presented in the Table A.1 below.

In this appendix, a *step-wise* general description of the method is given aiming to clarify its basic steps. For further application details, the reader may refer to standard specialized texts (Fardis 2009). The practical implementation of the method can be divided into three stages as follows.

I. Avramidis et al., *Eurocode-Compliant Seismic Analysis and Design of R/C Buildings*, Geotechnical, Geological and Earthquake Engineering 38,
DOI 10.1007/978-3-319-25270-4

Table A.1 Limitations of non-linear analysis methods

Criterion	Non-linear static (Pushover) analysis	Non-linear Response History Analysis (NRHA)
Consideration of higher modes (particularly important for higher buildings)	None, unless modal pushover analysis (Chopra and Goel 2002) is applied	Explicit
Applicability to irregular structures	Limited to buildings with a predominant translational mode of vibration	Yes
3-Dimensional structural response	Structure is only assessed along the two principal directions	Explicit computational-wise, though the orientation of ground motion introduces a degree of uncertainty (Athanatopoulou 2005)
Compatibility with Uniform Hazard Spectrum used in design	Implicit, through the target displacement or the performance point (based on equivalent SDOF system)	Selection, scaling and spectral matching is required, often leading to discrepancies in structural response (Sextos et al. 2011)
Hysteretic behavior of concrete members (stiffness degradation, strength deterioration and pinching)	Neglected	Explicitly accounted for, however, different assumptions affect the structural response predicted
Equivalent viscous damping	Approximate	Explicit though model-dependent
Shear capacity of r/c sections	Neglected unless implicitly accounted for, through appropriate moment resistance cut off corresponding to shear failure	Depending on the software used
Modeling of interactions (shear-moment, shear-tension, and biaxial bending)	Neglected	Depending on the software used
Computational time	Low	Moderate, depending on system inelastic demand and number of ground motions needed to reach a stable mean of structural response
Post-processing	Limited	Extensive. Requires statistical processing (e.g., maximum, mean, median) of time-variant response quantities
Interpretation	Straightforward	Requires engineering judgment re. the decisive Engineering Demand Parameters (moment, ductility demand, drift) and their proxies (mean, median, max)

Stage 1: Derivation of the pushover (capacity) curve of the building

At the first stage, a full-fledged standard pushover analysis (SPA) is undertaken involving a non-linear finite element (FE) model of the building (see also Sect. 2.4.1.2). This analysis consists of gradually increasing static lateral loads, which are applied following a predefined distribution pattern along the height of the building model that remains constant with time throughout the analysis. The assumption is fundamentally made that the entire structure mass is activated on the first mode of vibration, as if the structure essentially responds as a single-degree-of-freedom (SDOF) system along the direction of loading. At each loading increment, certain structural response quantities of interest are calculated (e.g., internal forces and displacements/deformations). The outcome of SPA is expressed in terms of a "*pushover curve*", also called "*capacity curve*", which is a plot of the horizontal displacement of a pre-specified reference point on the FE model (typically the center of gravity of the top storey) versus the increasing base shear (see e.g., Fig. 2.13).

Stage 2: Determination and analysis of an "equivalent" SDOF system

Next, the previously obtained pushover curve is firstly considered to determine the properties of an "*equivalent*" SDOF inelastic system. The peak displacement of this SDOF system subject to an elastic response or design spectrum is obtained. The derived peak displacement of the SDOF system is "converted" into a "target displacement" corresponding to the FE model of the actual structure. In the "N2 method" adopted by EC8, this conversion relies on the "equal displacement rule", which assumes that the peak inelastic response of a non-linear seismically excited SDOF is equal to the peak elastic response of a SDOF with natural period equal to the pre-yield natural period of the non-linear system (see also Fig. 1.9).

Stage 3: Seismic performance level assessment

In the last (third) stage, the previously derived target displacement is compared against a predefined maximum allowable value, which depends on the desired/targeted level of seismic performance for the considered building structure. Each of the above three stages comprises several steps. A pertinent flowchart is shown in Fig. A.1 in which each stage is noted by a different color and all the involved steps are listed. In the remainder of this appendix, a *qualitative* description of each of these steps is provided without making any particular reference to the underlying mathematical formulae, which can be found in Annex B of EC8-part 1.

It is noted that the pushover curve is often converted into a capacity diagram (Freeman 1978) and assessment is performed from a dual capacity-demand perspective. The procedure consists of finding the inelastic displacement demand by overlapping the response spectrum of the ground motion used, represented in acceleration-displacement (ADRS) terms with the previously derived capacity curve of the structure as obtained by pushover analysis. For compatibility purposes, the latter is plotted by dividing the base shear V by the weight W of the structure, that is, in units of acceleration. The point where both the demand and the capacity

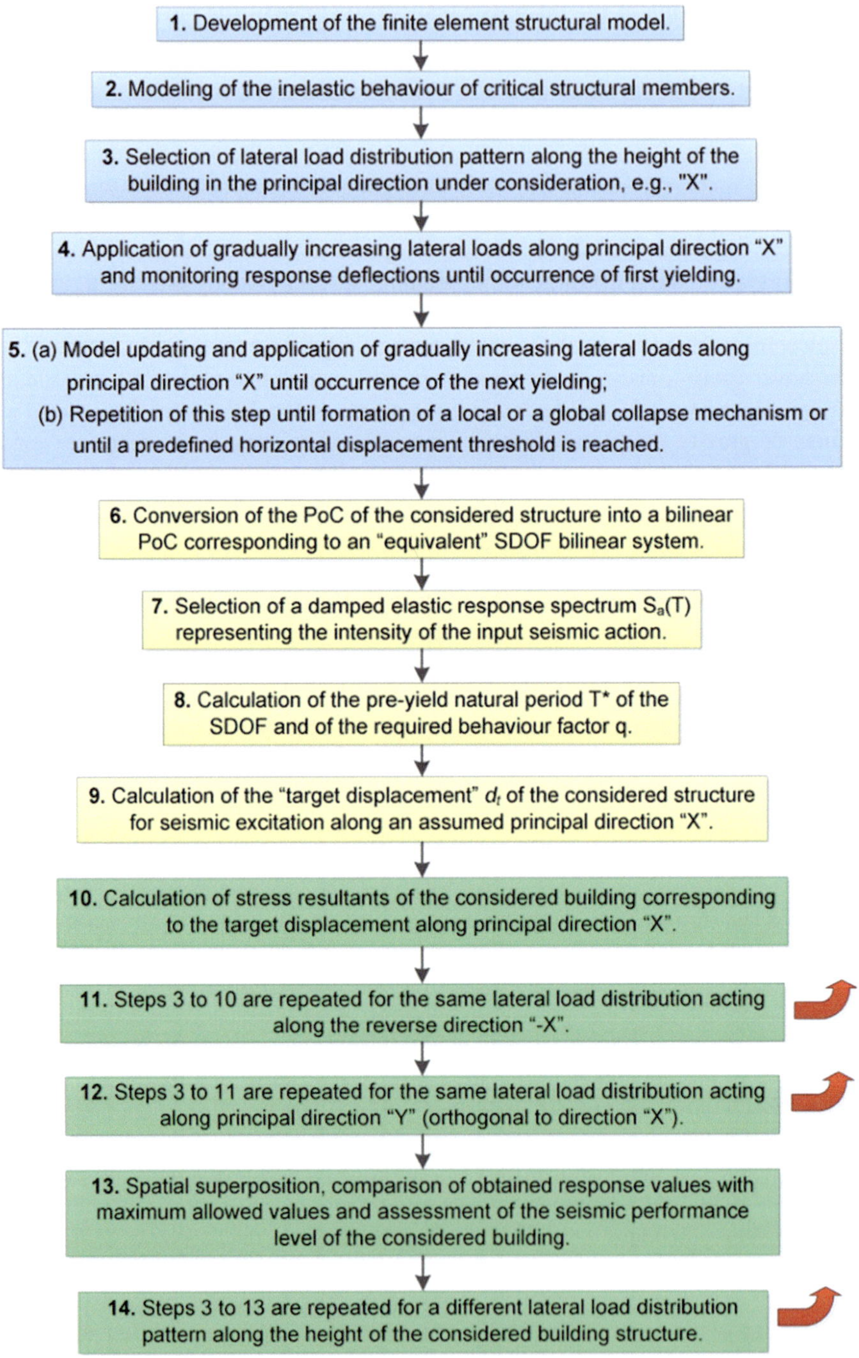

Fig. A.1 Main steps of non-linear static (Pushover) analysis method

curves intersect corresponds to the expected displacement demand during the reference ground motion used for assessment. Various improved procedures have also been proposed (Chopra and Goel 1999, 2000; Fajfar 1999), however, further insight is deemed out of the scope of this Appendix.

A.1 Derivation of the Pushover Curve of the Building

Step 1: Development of the numerical finite element (FE) structural model

The adopted FE model in undertaking SPA can be either planar (two dimensional) or spatial (three dimensional), depending on the regularity criteria of EC8 (§ 4.3.3.4.2.1(2)P and (3)). Further, it may include non-structural members (such as infill walls), and it may account for soil compliance, if deemed essential, by considering flexible supports at the foundation level. In principle, similar modeling practices applicable in developing FE models for equivalent linear types of analysis discussed in Sect. 2.3.3 can be employed to develop FE models for SPA. The only additional requirement is to account for potential material non-linear behavior expected at "critical" structural members or cross-sections, as identified through capacity design considerations (see also Sect. 2.2.4).

Step 2: Modeling of the inelastic behaviour of critical structural members

The following assumptions are commonly made in modeling the post-yielding (inelastic) behaviour at critical cross-sections of structural members in r/c buildings in a practical, efficient and effective manner, i.e., lumped or distributed plasticity modelling (Scott and Fenves 2006):

- The inelastic deformations develop at *plastic hinges* considered concentrated (lumped) at singular points within critical zones of structural elements (typically at the ends of beams and columns and at the base of walls and cores).
- The inelastic behaviour of plastic hinges is modeled by means of simplified bilinear or multi-linear force – displacement (e.g., moment-rotation) or force – deformation (e.g., moment – curvature) relationships. The properties of these simplified models depend on the materials used, on the geometry of the cross-section, on the detailing practices (i.e., assumed ductility class- see Sect. 3.1.3), and on the level of applied axial forces. The inelastic displacement or deformation capacity of plastic hinges (i.e., the maximum allowable displacements/ deformation values) can be either determined by means of specialized software or obtained from pertinent tables included in structural seismic assessment guidelines for practitioners, see FEMA-356 (FEMA 2000).
- The commonly desirable, within a capacity design approach, flexural type of inelastic behavior at plastic hinges (flexural failure mode) is usually modeled via moment versus cross-sectional rotation, or moment versus structural member chord rotation, or moment versus cross-sectional curvature relationships. The influence of the axial force applied simultaneously with bending moment to the

above relationships should be taken into account using appropriate bending moment – axial force interaction diagrams.

- Predominantly shear types of inelastic response are usually not considered in FE models. In principle, capacity design considerations at the conceptual design stage and the detailing stage of code-compliant seismic design ensure that shear modes of local failure are excluded, an assumption which, of course, does not hold for most existing, under-designed structures. In this respect, it is recommended that, at each incremental step of the SPA, shearing force demands at critical cross-sections are monitored. In case these demands reach the shear capacity of certain structural members, the moment-rotation or moment-curvature relationship shall be appropriately adapted to replace the section yield moment with the moment corresponding to shear failure.

Step 3: Selection of lateral load distribution pattern along the height of the building

Ideally, the assumed distribution of the horizontal static loads considered in SPA should closely trace the distribution of the actual seismic loads. However, the latter distribution is unknown (and varying due to the expected non-linear structural behavior), even under the simplistic assumption of a single dominant response mode. To this end, several different assumed distributions have been proposed to be used in conjunction with SPA in the scientific literature and in various seismic codes of practice and guidelines. The commonly adopted ones are the uniform (rectangular), the triangular, and the proportional to the equivalent seismic loads according to the dominant mode shape assuming linear elastic structural response. According to clause § 4.3.3.4.2.1 (2)P of EC8, in the case of applying SPA to a spatial FE model, two independent analyses (along the two "principal directions" of the structure) should be considered, with the seismic action (static lateral loads) applied only along the considered direction of analysis. Therefore, in defining a lateral load distribution according to the respective predominant mode shape, only the components of the mode shape parallel to the considered (principal) direction of the seismic action should be accounted for.

Step 4: Application of gradually increasing lateral loads along principal direction "X" and monitoring response deflections until occurrence of first yielding

The developed non-linear FE model is initially subjected to the gravitational loads of the seismic load combination, that is, $\sum_j (G_{k,j})$ "+" $\sum_i (\psi_{2,i} Q_{k,i})$ (see Sect. 2.3.1.3). Next, it is subjected to incrementally increasing lateral loads observing the distribution pattern along the height of the building defined in the previous step. These lateral loads are usually applied at the locations where the total slab masses are assumed to be "lumped" (concentrated) accounting for accidental mass eccentricities (see Fig. 3.3). At this step, lateral loads increase until the first cross-section yields (i.e., the first plastic hinge develops). Under this loading intensity, response values of interest are recorded and stored for later use.

Step 5: Model updating and application of gradually increasing lateral loads along principal direction "X" until occurrence of the next yielding; Repetition of this step until formation of a local or a global collapse mechanism or until a predefined horizontal displacement threshold is reached.

- 5(a) The FE model is updated by reducing the stiffness at the cross-section that has yielded in the previous step according to the adopted local inelastic behaviour model. For the special case of an assumed perfectly elasto-plastic model having zero post-yield stiffness, a "mechanical hinge" can be introduced at the plastic hinge location such that no further increase in the internal resisting moment is allowed as the external lateral loads are increased. The same lateral load pattern is applied to the updated model along the considered principal direction and further increased until the next cross-section yields (i.e., a new plastic hinge develops), at which point response values of interest are recorded and stored for later use.
- 5(b) The response values (stress resultants and structural deformations) recorded at the previous step are appropriately superposed to the stored response values corresponding to all previous plastic hinges.
- 5(c) Steps 5(a) and 5(b) are repeated until a sufficient number of plastic hinges develops to form a local or global collapse mechanism, or until a target horizontal displacement reaches a specific predefined value.

It is noted that:

- in between two consecutive plastic hinge formations, the FE model behaves elastically and, therefore, the calculation of response values (stress resultants and structural deformations) is straightforward, and
- any desired structural response quantity corresponding to any given lateral external load intensity can be retrieved by means of superposing recorded response quantity values corresponding to plastic hinge formations.

The above quantitative description of the SPA suggests that the method is very similar to the "step by step" inelastic static analysis used since the early 1950s for the calculation of the ultimate collapse load of, primarily, steel moment resisting frame structures (Beedle 1958). However, in the latter method, the (design) lateral load distribution pattern is independent of the properties of the structure. This is not the case for the distribution of the lateral seismic loads. In this regard, certain simplified assumptions are made regarding the lateral load distribution in the context of the SPA, as discussed in Sect. 2.4.1.2. Ultimately, the SPA results in a pushover (capacity) curve.

A.2 Determination and Analysis of an Equivalent SDOF System

Step 6: Conversion of the pushover curve of the considered structure into a bilinear pushover curve corresponding to an "equivalent" SDOF bilinear system

The previously obtained *pushover curve* of the considered building structure along a principal direction "X" is converted into a bilinear base shear-deformation curve with the aid of specific analytical formulae (Annex B of EC8-part 1). These expressions have been proposed in the context of the N2 static inelastic (pushover) method (Fajfar 1999) and rely on a series of simplified assumptions. Based on the derived bilinear curve, a perfectly elasto-plastic SDOF system is defined, having a specific yielding strength and deformation.

Step 7: Selection of a damped elastic response spectrum $S_a(T)$ representing the intensity of the input seismic action

Step 8: Calculation of the pre-yield natural period T* of the SDOF and of the required behaviour factor

The pre-yield natural period T* of the perfectly elasto-plastic SDOF defined in step 6 is computed from the pre-yielding stiffness of the SDOF system and from an "equivalent" mass. The latter is determined by an expression involving the assumed distribution of the lateral loads and the distribution of the floor masses along the height of the considered building. In case T* is shorter than the corner period T_c of the adopted EC8 response spectrum (see also Sect. 2.3.2.1), a "behaviour factor" q_u or strength reduction factor of the SDOF also needs to be obtained. This involves the use of the spectral ordinate at T* (i.e., $S_a(T^*)$) determined from the seismic spectrum adopted in step 7.

Step 9: Calculation of the "target displacement" d_t of the considered structure for seismic excitation along an assumed principal direction "X"

The "target displacement" d_t^* of the considered equivalent SDOF system is found by making use of the response spectrum of step 7 in conjunction with the T* period (step 8) and the q_u behaviour factor if $T^* < T_C$. Then, the displacement d_t^* is transformed into the displacement d_t of an MDOF system. This is taken to be the peak inelastic displacement of the "control node" of the building under consideration subjected to the seismic response spectrum of step 7 along the direction "X". It represents the seismic demand of the actual structure along the principal direction "X".

A.3 Seismic Performance Level Assessment

Step 10: Calculation of stress resultants of the considered building corresponding to the target displacement along principal direction "X"

The target displacement d_t found in the previous step is used in conjunction with the *pushover curve* obtained in step 5 to determine the intensity/amplitude of the lateral static loads that induces a displacement equal to d_t at the control node of the considered building along principal axis "X". The response values (stress resultants and deformations at all structural members) corresponding to the above lateral load intensity (or increment) are determined using the results obtained in step 5. These are treated as the seismic effects to the considered building subject to the seismic action represented by the response spectrum adopted in step 7 along the principal axis "X".

Step 11: Steps 3 to 10 are repeated for the same lateral load distribution acting along the reverse direction "-X"

Step 12: Steps 3 to 11 are repeated for the corresponding lateral load distribution acting along principal direction "Y" (orthogonal to direction "X")

In most of the practical cases, the above described non-linear inelastic (push-over) analysis needs to be performed at least twice along two horizontal and orthogonal ("principal") axes, say "X" and "Y".

Step 13: Spatial superposition, comparison of obtained response values with maximum allowed values and assessment of the seismic performance level of the considered building

The obtained response values in terms of stress resultants and deformations for seismic excitations along the "X", "-X", "Y" and "-Y" directions are spatially superposed to obtain the seismic demands ("design effects"). The latter are compared with the capacity (maximum allowable values) of various critical structural members provided by seismic codes of practice (e.g., EC8-Part 3). The underlying verification/acceptance criteria are defined in terms of stress resultants (e.g., shear forces and moments with or without axial force) for non-ductile (primarily shear types of) failure modes and in terms of displacements or deformations (e.g., rotations or curvature) for ductile (primarily flexural types of) failure modes. In the latter case, the acceptance criteria are expressed in terms of demand capacity ratios (DCRs) of the inelastic displacement or deformation accounting for the desired/targeted seismic performance level of the structure.

The assessment of the seismic performance of the considered structure is based on the severity and the extent of the observed "failures" (i.e., level and number of exceedances of the maximum allowed/capacity values of stress resultants and/or displacements/deformations). Note that this assessment is valid only for the lateral load distribution pattern assumed in step 3.

Step 14: Steps 3 to 13 are repeated for a different lateral load distribution pattern along the height of the considered building structure

In recognition of the significant influence of the assumed height-wise lateral load distribution pattern in step 3 on the results of the static inelastic analysis method, EC8 mandates that the analysis is undertaken for at least two different lateral load distributions (§ 4.3.3.2.1(1) of EC8). Structural members of the considered structure should be designed for the most unfavorable results derived by using two (or more) different lateral load distribution patterns.

References

Athanatopoulou AM (2005) Critical orientation of three correlated seismic components. Eng Struct 27:301–312

Beedle LS (1958) Plastic design of steel frames. Wiley, New York

Chopra AK, Goel RK (1999) Capacity-demand-diagram methods based on inelastic design spectrum. Earthq Spectra 15:637–656

Chopra AK, Goel RK (2000) Evaluation of NSP to estimate seismic deformation: SDF systems. J Struct Eng 126:482–490

Chopra AK, Goel RK (2002) A modal pushover analysis procedure for estimating seismic demands for buildings. Earthq Eng Struct Dyn 31:561–582

Fajfar P (1999) Capacity spectrum method based on inelastic demand spectra. Earthq Eng Struct Dyn 28:979–993

Fajfar P (2000) A nonlinear analysis method for performance-based seismic design. Earthq Spectra 16:573–592

Fardis MN (2009) Seismic design, assessment and retrofit of concrete buildings, based on Eurocode 8. Springer, Dordrecht

FEMA (2000) Prestandard and commentary for the seismic rehabilitation of buildings, FEMA-356. Federal Emergency Management Agency, Washington, DC

Freeman SA (1978) Prediction of response of concrete buildings to severe earthquake motion, Publication SP-55. American Concrete Institute, Detroit, pp 589–605

Krawinkler H, Seneviratna GDPK (1998) Pros and cons of a pushover analysis of seismic performance evaluation. Eng Struct 20:452–464

Scott M, Fenves G (2006) Plastic hinge integration methods for force-based beam-column elements. J Struct Eng ASCE 132:244–252

Sextos AG, Katsanos EI, Manolis GD (2011) EC8-based earthquake record selection procedure evaluation: validation study based on observed damage of an irregular R/C building. Soil Dyn Earthq Eng 31:583–597

Appendix B – A Note on Torsional Flexibility and Sensitivity

In this appendix, the differences between torsional flexibility and torsional sensitivity of building structures are elucidated by reviewing the relevant background theory (Sect. B.1) and by furnishing simple numerical examples (Sect. B.3). It is noted that the concept of torsional sensitivity is used to quantify torsional effects and in-plan regularity of building structures in various clauses of EC8, as discussed in Sect. B.2.

B.1 Definitions of Torsional Stiffness, Torsional Radius, and Radius of Gyration

Consider the model of a one-storey building having a rectangular in-plan layout with a single axis of symmetry X shown in Fig. B.1.

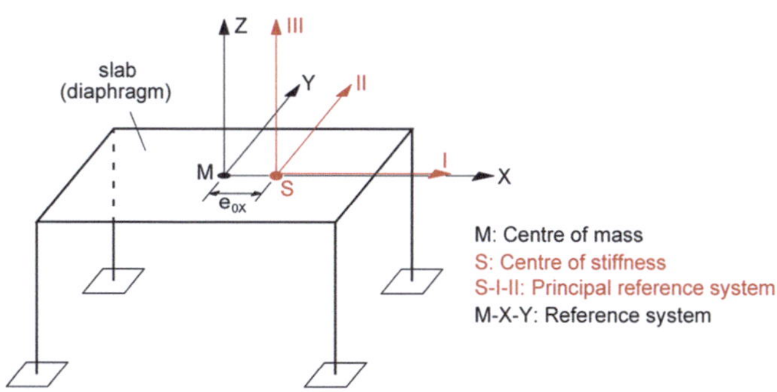

Fig. B.1 Single-storey one-way symmetrical system (simplified model)

I. Avramidis et al., *Eurocode-Compliant Seismic Analysis and Design of R/C Buildings*, Geotechnical, Geological and Earthquake Engineering 38,
DOI 10.1007/978-3-319-25270-4

The locations of the center of mass (or center of gravity) M and the center of stiffness (or shear center or center of rigidity or center of twist) S lying on the axis of symmetry are assumed to be known. Further, the following assumptions are made for simplicity:

1. All vertical structural members are axially inextensible and massless.
2. There are no horizontal structural members (beams).
3. The slab is perfectly rigid within its plane (X-Y) and behaves as a diaphragm.
4. The slab stiffness within the X-Z and Y-Z planes is negligible and, thus, the top ends of vertical structural members are free to rotate (cantilever behavior).

Since the floor slab does not deform within the X-Y plane, the motion of the considered building model can be uniquely represented by means of three generalized coordinates (degrees of freedom) at any point on the slab: a translation u_X along the X axis, a translation u_Y along the Y axis, and a rotation θ_Z about the Z axis passing through the considered point. The stiffness matrix of the model with respect to the principal coordinate system originating at the center of stiffness S (S-I-II-III) is by definition diagonal:

$$\mathbf{K}_S = \begin{bmatrix} k_I & 0 & 0 \\ 0 & k_{II} & 0 \\ 0 & 0 & k_{III} \end{bmatrix} \tag{B.1}$$

Further, the stiffness and the mass matrix of the model with respect to the coordinate system M-X-Y-Z, originating at the center of mass M of the slab, can be written as (Chopra 2007):

$$\mathbf{K}_M = \begin{bmatrix} k_X & 0 & 0 \\ 0 & \boxed{\begin{matrix} k_Y & e_{oX}k_Y \\ e_{oX}k_Y & k_Z \end{matrix}} \end{bmatrix} = \begin{bmatrix} k_I & 0 & 0 \\ 0 & \boxed{\begin{matrix} k_{II} & e_{oX}k_{II} \\ e_{oX}k_{II} & k_{III}+e_{oX}^2k_{II} \end{matrix}} \end{bmatrix} \tag{B.2}$$

and

$$\mathbf{M} = \begin{bmatrix} m & 0 & 0 \\ 0 & m & 0 \\ 0 & 0 & J_m \end{bmatrix} = \begin{bmatrix} m & 0 & 0 \\ 0 & \boxed{\begin{matrix} m & 0 \\ 0 & ml_s^2 \end{matrix}} \end{bmatrix} \tag{B.3}$$

respectively. In the above formulae:

- m is the total mass of the slab,
- l_S is the *radius of gyration* of the slab with respect to the center of mass M,
- $J_m = ml_s^2$ is the polar moment (or, simply, moment) of inertia of the slab,
- e_{oX} is the *structural eccentricity* (distance between the center of stiffness S and the center of mass M) along the X axis,

- k_I and k_X ($k_I = k_X$) are the lateral stiffnesses along the I and X axes, respectively. That is, they represent the force that needs to be applied at point *S* or at *M*, respectively, to cause a unit displacement along the X axis ($u_X = 1$), while $u_Y = \theta_Z = 0$.
- k_{II} and k_Y ($k_{II} = k_Y$) are the lateral stiffnesses along the II and Y axes, respectively. That is, they represent the force that needs to be applied at point *S* or at *M*, respectively, to cause a unit displacement along the Y axis ($u_Y = 1$), while $u_X = \theta_Z = 0$.
- k_{III} and k_Z ($k_Z = k_{III} + e_{oX}{}^2 k_{II}$) are the *torsional stiffnesses* with respect to the center of stiffness S and to the center of mass M, respectively. That is, the moment that needs to be applied at S or at M, respectively, to cause a unit rotation about the III or the Z axis, respectively, while $u_X = u_Y = 0$.

The last definition suggests that torsional stiffness is a measure of the resistance of the structure to rotational (torsional) displacements about a gravitational axis. For a given external horizontal force couple, the higher the torsional stiffness of a structure, the smaller the observed rotation of the slab about a gravitational axis of interest. *Torsional flexibility is the reciprocal of torsional stiffness* (i.e., $1/k_Z$). Clearly, for a given external horizontal force couple, the higher the torsional flexibility of a structure, the higher the observed rotation of the slab about a gravitational axis of interest.

Furthermore, note that, for a horizontal ground excitation along the X axis, the structural model of Fig. B.1 performs a purely translational motion along the X axis without rotation. This is because X is an axis of symmetry. Therefore, the torsional response of the structure is decoupled from its horizontal translation along the X axis. Hence, its torsional behavior will only depend on the 2×2 lower right sub-matrices marked in red in Eqs. (B.2), (B.3). Focusing on these sub-matrices, and, therefore, on the coupled degrees of freedom, u_Y and θ_Z, the following definitions from the theory of structural dynamics apply (Annigeri et al. 1996; Hejal and Chopra 1989).

The *uncoupled translational natural frequency* ω_Y along the Y axis is given as:

$$\omega_Y = \sqrt{\frac{k_Y}{m}} = \omega_{II} = \sqrt{\frac{k_{II}}{m}}. \tag{B.4}$$

Further, the *uncoupled rotational natural frequencies* with respect to the center of mass M and to the center of stiffness S are expressed as:

$$\omega_Z = \sqrt{\frac{k_Z}{J_m}} \quad \text{and} \quad \omega_{III} = \sqrt{\frac{k_{III}}{J_m}}, \tag{B.5}$$

respectively. It is also recalled that the natural frequency ω (in rad/s) is related to the natural period *T* (in s) by the well known expression $\omega = 2\pi/\text{T}$.

Moreover, the *torsional radius* r_{PX} is defined as the square root of the torsional stiffness about a gravitational axis passing through the point P of the X-Y plane (lying on the X axis) over the translational stiffness along the Y axis. Thus, the value of the torsional radius depends on the point of reference P. In this context, *the torsional radii with respect to the center of mass M and to the center of stiffness S* are given as:

$$r_{MX} = \sqrt{\frac{k_Z}{k_Y}} \quad \text{and} \quad r_{SX} = \sqrt{\frac{k_{III}}{k_{II}}}, \tag{B.6}$$

respectively. Notably, the following expression relates r_{MX} with r_{SX}

$$r_{MX}^2 = e_{oX}^2 + r_{SX}^2. \tag{B.7}$$

Finally, it is reminded that the *radius of gyration* for rectangular slabs of dimensions L_X and L_Y is determined by

$$l_S = \sqrt{\frac{J_m}{m}} = \sqrt{\frac{L_X^2 + L_Y^2}{12}}. \tag{B.8}$$

As shown in the next section, the ratio ω_{III}/ω_{II} is used in EC8 as an index (criterion) of structural in-plan regularity related to the concept of *torsional sensitivity.*

B.2 The Concept of Torsional Sensitivity

Generally, a building is characterized as *torsionally sensitive* if one of the first two mode shapes is dominated by rotational displacements with respect to a vertical axis of reference. A building is characterized as *torsionally non-sensitive* in the case in which its first two mode shapes are dominated by translational displacements. From a practical viewpoint, a torsionally sensitive building is highly – yet not necessarily – likely to exhibit severe rotational displacements about a vertical axis of reference under horizontal seismic excitation. Such displacements induce increased deformation demands on structural members lying close to the perimeter of the building. This non-uniform in-plan distribution of deformation demands and, consequently, of stress demands among structural members should either be avoided or be taken explicitly into account during the analysis stage. The latter can be accomplished by undertaking response spectrum based analysis using three dimensional finite element models of the structure.

In this regard, EC8 includes an explicit in-plan regularity criterion which is related to the concept of torsional sensitivity. Specifically, it requires that the torsional radius corresponding to the center of stiffness along both principal axes is larger than the radius of gyration. That is, (§ 4.2.3.2(6) and (7) of EC8),

$$r_{SX} \geq l_S \quad \text{and} \quad r_{SY} \geq l_S. \tag{B.9}$$

Focusing on the principal axis Y, it can be readily shown by considering Eqs. (B.4), (B.5), (B.6) and (B.8) that the above in-plan regularity criterion of EC8 for the one-storey model shown in Fig. B.1 can be written as

$$\frac{\omega_{III}}{\omega_{II}} \geq 1. \tag{B.10}$$

Similarly, it can be shown that $r_{SY} \geq l_s$ is equivalent to $\omega_{III}/\omega_I \geq 1$. The latter equation implies that the EC8 requirement in Eq. (B.9) ensures that the *uncoupled translational natural period along a principal axis is longer than the uncoupled rotational natural period*. Hence, it ensures that a building will be torsionally non-sensitive.

To this end, it is important to emphasize that the concept of torsional sensitivity is very different from the concept of torsional flexibility defined in the previous section. Torsional sensitivity depends on the natural vibration properties (i.e., stiffness and mass properties) of building structures, while torsional flexibility depends only on their stiffness (i.e., geometrical and material) properties. In fact, a torsionally flexible building may or may not be torsionally sensitive, while a torsionally sensitive building may have a high torsional stiffness about a vertical axis of reference. Pertinent numerical evidence is provided in the next section to illustrate further the above point involving three different structural layouts.

In view of the above remarks, it becomes evident that the torsional radius, the radius of gyration, and the structural eccentricity are critical parameters characterizing the elastic dynamic behaviour of a building exposed to strong horizontal ground motions. However, these parameters can only be unambiguously defined for one-storey buildings (such as the prototype shown in Fig. B.1) and for the following special classes of multistorey systems:

(a) buildings which are symmetric with regard to two horizontal orthogonal axes, and
(b) buildings which (i) consist of planar subsystems with proportional lateral stiffness matrices (i.e., buildings with lateral load resisting systems consisting either solely of moment frames or solely of walls), (ii) possess a vertical mass axis (i.e., the concentrated floor masses are aligned along a straight vertical line), and (iii) have the same radius of gyration at all floors (Athanatopoulou et al. 2006).

In the general case of multistorey r/c buildings comprising both walls and frames (dual systems), the above parameters can only be heuristically defined (see also Sect. 3.1.1.1 and clause §4.2.3.2(8) of EC8). A viable approach for determining these parameters for dual multistorey structures can be found elsewhere (Earthquake Planning and Protection Organization (EPPO) 2000; Makarios and Anastassiadis 1998a, b; Marino and Rossi 2004).

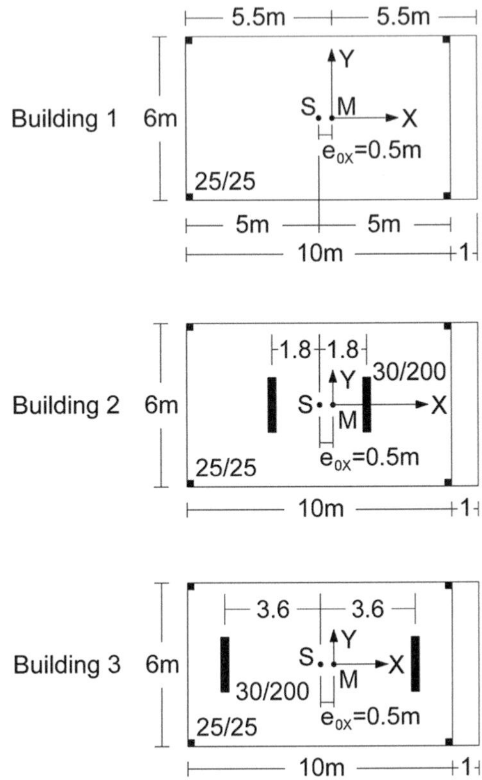

Fig. B.2 Indicative floor plans of three mono-symmetrical single-storey buildings satisfying the four assumptions listed in Sect. B.1

B.3 Illustrative Numerical Examples

Consider the three one-storey r/c building structures with one axis of symmetry (X) shown in Fig. B.2 which satisfy all four assumptions listed in Sect. B.1. All the structures have four columns with dimensions 25×25(cm), while the walls of buildings 2 and 3 have dimensions 30×200(cm). The vertical structural members of all three buildings are arranged such that the center of stiffness S is located at the same point; namely, the geometric center of the rectangle whose corners coincide with the location of the four columns of each building. The floor slab extends by 1 m beyond the rightmost columns. Therefore, assuming uniform mass distribution across the slab, the center of mass M lies $e_{oX} = 0.5$ m to the right of the center of stiffness S, as indicated in Fig. B.2. The floor height is $h = 4$ m. The mass of the slab is equal to 60 t and, thus, the moment of inertia (see Eq. (B.3)) is $J_m = 60(6^2 + 11^2)/12 = 785$ tm^2. The modulus of elasticity, E, is taken equal to 2.8×10^7 kN/m^2.

Since there are no beams in the considered structures and the slab has negligible flexural stiffness within the planes X-Z and Y-Z, structural members behave as cantilevers. Thus, the lateral stiffness along the X and Y axes are given as:

$$k_I = k_X = \sum_i k_{Xi} = \sum_i \left(3EI_{Yi}/h^3\right) = \left(3EI/h^3\right)\sum_i I_{Yi} \qquad \text{(B.11)}$$

and

$$k_{II} = k_Y = \sum_i k_{Yi} = \sum_i \left(3EI_{Xi}/h^3\right) = \left(3EI/h^3\right)\sum_i I_{Xi}, \qquad \text{(B.12)}$$

respectively, where

- $k_{Xi} = 3EI_{Yi}/h^3$ and $k_{Yi} = 3EI_{Xi}/h^3$ is the lateral stiffness of the structural member i along the X and Y axis, respectively,
- I_{Xi} and I_{Yi} is the cross-sectional second moment of inertia of the structural member i about the X and Y axis, respectively,
- E is the modulus of elasticity (assumed to be common for all members), and
- h is the storey height.

Further, by assuming the torsional stiffness of each individual member about its own center of mass to be negligible, the torsional stiffness about the center of mass of the considered building structures is determined by the expression

$$k_Z = \sum_i \left(k_{Yi}X_i^2 + k_{Xi}Y_i^2\right) = \left(3EI/h^3\right)\sum_i \left(I_{Xi}X_i^2 + I_{Yi}Y_i^2\right). \qquad \text{(B.13)}$$

In the above equation, X_i and Y_i are the coordinates of the center of mass of the i vertical structural member's cross-section on the X-Y plane with respect to the center of mass of the floor M. Making use of Eqs. (B.11, B.12 and B.13), the torsional radii r_{MX} and r_{MY} can be determined by the expressions

$$r_{MX} = \sqrt{k_Z/k_Y} = \sqrt{\sum_i \left(I_{Xi}X_i^2 + I_{Yi}Y_i^2\right)/\sum_i I_{Xi}}, \qquad \text{(B.14)}$$

and

$$r_{MY} = \sqrt{k_Z/k_X} = \sqrt{\sum_i \left(I_{Xi}X_i^2 + I_{Yi}Y_i^2\right)/\sum_i I_{Yi}}, \qquad \text{(B.15)}$$

respectively.

Focusing on the horizontal ground excitation along the Y axis (because this excitation activates the coupled DOFs, u_Y and θ_Z), Table B.1 collects the lateral stiffness k_Y, the torsional stiffness k_Z, the torsional radii r_{MX} and r_{SX}, and the radius of gyration of the three building structures in Fig. B.2 using Eqs. (B.13), (B.14), (B.15), (B.7) and (B.8). Further, the considered structures are classified as being

Table B.1 Torsional and dynamic properties of the structures considered in Fig. B.2

	Building 1	Building 2	Building 3
Lateral stiffness	$k_Y = 1709$ kN/m	$k_Y = 526{,}709$ kN/m	$k_Y = 526{,}709$ kN/m
Torsional stiffness	$k_Z = 58{,}533$ kN	$k_Z = 1{,}890{,}783$ kN	$k_Z = 6{,}993{,}783$ kN
Torsional radius with respect to center of mass	$r_{MX} = \sqrt{k_Z/k_Y} = 5.85\,\text{m}$	$r_{MX} = \sqrt{k_Z/k_Y} = 1.895\,\text{m}$	$r_{MX} = \sqrt{k_Z/k_Y} = 3.644\,\text{m}$
Torsional radius with respect to shear center	$r_{SX} = \sqrt{5.85^2 - 0.5^2} = 5.828\,\text{m}$	$r_{SX} = \sqrt{1.895^2 - 0.5^2} = 1.828\,\text{m}$	$r_{SX} = \sqrt{3.644^2 - 0.5^2} = 3.609\,\text{m}$
Radius of gyration	$l_S = 3.617$ m	$l_S = 3.617$ m	$l_S = 3.617$ m
EC8 verification	$5.828 > 3.617$	$1.828 < 3.617$	$3.609 \sim 3.617$
	Torsionally non-sensitive	Torsionally sensitive	Torsionally sensitive
Natural periods	$T_1 = 1.184$ s	$T_1 = 0.1344$ s	$T_1 = 0.0719$ s
	$T_2 = 0.726$ s	$T_2 = 0.066$ s	$T_2 = 0.0626$ s
Modal participating mass ratios	mode 1: 99 %	mode 1: 3.3 %	mode 1: 52.7 %
	mode 2: 1 %	mode 2: 96.7 %	mode 2: 47.3 %
Dominant mode shape	1st: translational along Y	2nd: translational along Y	both

torsionally sensitive or non-sensitive according to EC8 verification check of Eq. (B.9). Moreover, the natural periods and modal participating mass ratios of these structures are reported as having been obtained from standard modal analysis based on the mass and stiffness matrices of Eqs. (B.2) and (B.3).

It is observed that the torsional stiffness of building 2 is about 32 times larger than the torsional stiffness of building 1. Nevertheless, the torsional radius of building 2 is only about 1/3 of the torsional radius of building 1 and, in fact, building 2 is classified as "torsional sensitive". By examining the modal participating mass ratios of buildings 1 and 2, it becomes evident that satisfying the EC8 criterion of Eq. (B.9) ensures that the fundamental mode shape is predominantly translational (see Eq. (B.10)). However, this criterion is not related to the observed torsional stiffness/rigidity of structures: a building with a relatively small torsional radius may be torsionally stiff (e.g., building 2) and vice versa (e.g., building 1). Therefore, caution needs to be exercised in interpreting the ambiguous §5.2.2.1(4)P of EC8 stating that "the first four types of systems (i.e., frame, dual and wall systems of both types) shall possess a minimum torsional rigidity that satisfies expression (4.1b) in both horizontal directions". In general, the torsional radius of a structure should not be interpreted as an index of its torsional stiffness/rigidity, unless comparison is made between structural forms of the same lateral stiffness (e.g., building 2 vis-à-vis building 3). Only in the latter special case does increased torsional radius entail increased torsional stiffness.

Furthermore, it is noted that "torsional sensitivity" relates to the dynamic (vibration) properties of structural systems (e.g., natural frequencies and mode shapes) rather than to static ones (e.g., lateral and torsional stiffness), as can be inferred from Eqs. (B.9) and (B.10). Consequently, torsional sensitivity influences the structural response behaviour for dynamic/seismic loads and not for static loads. Hence, the application of the lateral force method for the analysis of torsionally sensitive structures shall be made with caution.

For example, the mode shapes in buildings with $r_S << l_S$ (see, e.g., building 2 in Table B.1) are practically decoupled (natural periods are well apart) and the fundamental mode shape (dominantly torsional) will have a relatively small contribution to the overall dynamic response due to horizontal base excitation. For such excitation, the second natural period will dominate the response of building 2. Thus, for horizontal base excitation, the peak observed rotation about a vertical axis of reference will be highly dependent on the torsional stiffness and structural eccentricity of the building since the first (dominantly torsional) vibration mode has small influence. However, this will not be the case for torsional base excitation for which large amplitude torsional vibration of the building is expected, since the dynamic response of a building with $r_S << l_S$ will be dominated by the fundamental, dominantly torsional mode shape.

Moreover, special attention needs to be focused on structures with $r_S \sim l_s$, such as building 3 of Fig. B.2, when analysed by means of the lateral force method. In such structures, there is a severe coupling of mode shapes (e.g., $T_1/T_2 = 1.15$ for building 3) and the values of the translational and torsional components are of the same order. Moreover, the modal participating mass ratios have similar values. Consequently, large rotations about a vertical axis of reference may potentially develop (depending on the level of torsional stiffness) due to horizontal ground excitation. For instance, the peak rotation θ_Z of building 3 computed by application of the response spectrum based method is only 11 % smaller than that of building 2, even though building 3 has 3.7 times larger torsional stiffness than building 2. In other words, the lateral force method of analysis may potentially significantly underestimate seismic deformation demands of structures with $r_S \sim l_S$.

As a closure, it is emphasized that Eqs. (B.14) and (B.15) hold only under the rather simplistic assumptions made herein to illustrate the differences between the concepts of torsional stiffness (or its reciprocal: torsional flexibility) versus torsional sensitivity. These equations do not hold for common single-storey and multistorey buildings designed for earthquake resistance which include horizontal beam elements forming moment resisting frames with columns and connecting these frames with walls in the case of dual structural systems. Specifically, the lateral stiffness of moment resisting frames depends not only on the cross-sectional geometry and the height of columns, but also on the cross-sectional geometry of the beams and on the distance between columns (length of the beams). Thus, in general, the contribution of every single vertical (or horizontal) structural member to the lateral and torsional stiffness cannot be explicitly determined by expressions similar to Eqs. (B.11), (B.12), and (B.13); these contributions are rather taken into account in an implicit manner via the stiffness matrices of the corresponding finite elements within the framework of the seismic analysis method applied.

References

Annigeri S, Mittal AK, Jain AK (1996) Uncoupled frequency ratio in asymmetric buildings. Earthq Eng Struct Dyn 25:871–881

Athanatopoulou AM, Makarios TK, Anastassiadis K (2006) Elastic earthquake analysis of isotropic asymmetric multistory buildings. Struct Des Tall Spec Build 15:417–443

Chopra AK (2007) Dynamics of structures – theory and applications to earthquake engineering, 3rd edn. Pearson, Prentice-Hall, Upper Saddle River

Earthquake Planning and Protection Organization (EPPO) (2000) Greek seismic code EAK2000 (amended in 2003), Athens, Greece (in Greek)

Hejal R, Chopra AK (1989) Earthquake analysis of a class of torsionally coupled buildings. Earthq Eng Struct Dyn 18:305–323

Makarios TK, Anastassiadis K (1998a) Real and fictitious elastic axis of multi-storey buildings: theory. Struct Des Tall Spec Build 7:33–55

Makarios TK, Anastassiadis K (1998b) Real and fictitious elastic axes of multi-storey buildings: applications. Struct Des Tall Spec Build 7:57–71

Marino EM, Rossi PP (2004) Exact evaluation of the location of the optimum torsion axis. Struct Des Tall Spec Build 13:277–290

Appendix C – Chart Form of Eurocode 2 and 8 Provisions with Respect to the Sectional Dimensions and the Reinforcement of Structural Members

This Appendix presents the Eurocode 2 (EN1992-1-1) and Eurocode 8 (EN1998-1) guidelines which relate to the limiting requirements for the sectional dimensions of structural members and the corresponding requirements for the longitudinal and shear reinforcement. The following structural members are addressed:

- Solid slabs
- Beams
- Columns
- Ductile walls
- Beam-column joints
- Foundation elements.

It is clarified that the guidelines referring to beams, columns and ductile walls refer to buildings of both Ductility Class Medium (DCM) and High (DCH), distinguished appropriately. All guidelines refer to primary members. Secondary structural members are not discussed.

Note:

References to clauses of EC2 (EN1992-1-1) are indicated by the clause number in a green-colored font followed by EN1992-1-1.

References to particular clauses of EC8 part-1 (EN1998-1) within flowcharts are indicated simply by the clause number in a blue-colored font.

I. Avramidis et al., *Eurocode-Compliant Seismic Analysis and Design of R/C Buildings*, Geotechnical, Geological and Earthquake Engineering 38,
DOI 10.1007/978-3-319-25270-4

Symbols

A_c	Area of section of concrete member
α	Confinement effectiveness factor
b_w	Thickness of confined parts of a wall section, or width of the web of a beam
b_c	Cross-sectional dimension of column
b_o	Width of confined core in a column or in the boundary element of a wall (to centreline of hoops)
d	Effective depth of a cross-section
E_s	Design value of modulus of elasticity of reinforcing steel
ε_c	Compressive strain in the concrete
$\varepsilon_{sy,d}$	Design value of steel strain at yield
f_{cd}	Design value of concrete compressive strength
f_{yd}	Design value of yield strength of steel
f_{ywd}	Design value of yield strength of transverse reinforcement
f_{yk}	Characteristic yield strength of reinforcement
f_{ctm}	Mean value of axial tensile strength of concrete
f_{ctd}	Design value of axial tensile strength of concrete
h	Cross-sectional depth
h_w	Height of wall or cross-sectional depth of beam
L_{cr}	Length of critical region
L_{cl}	Clear length of a beam or a column
μ_ϕ	Curvature ductility factor
N_{Ed}	Design axial force from the analysis for the seismic design situation
ν_d	Normalised design axial force in the column, for the seismic design situation $[\nu_d = N_{Ed}/(f_{cd} \cdot A_c)]$
ρ	Tension reinforcement ratio
ρ'	Compression steel ratio in beams
ρ_w	Shear reinforcement ratio
s	Spacing of transverse reinforcement
s_t	The transverse spacing of the legs in a series of shear links in a cross-section
s_h	Spacing of horizontal reinforcing bars in the web of ductile walls
s_v	Spacing of vertical reinforcing bars in the web of ductile walls
Φ_L	Diameter of a longitudinal reinforcing bar
ω_{wd}	Mechanical volumetric ratio of confining reinforcement

C.1 Floor Slabs

EN1992-1-1 requirements concerning the reinforcement of floor slabs are presented in the following figure.

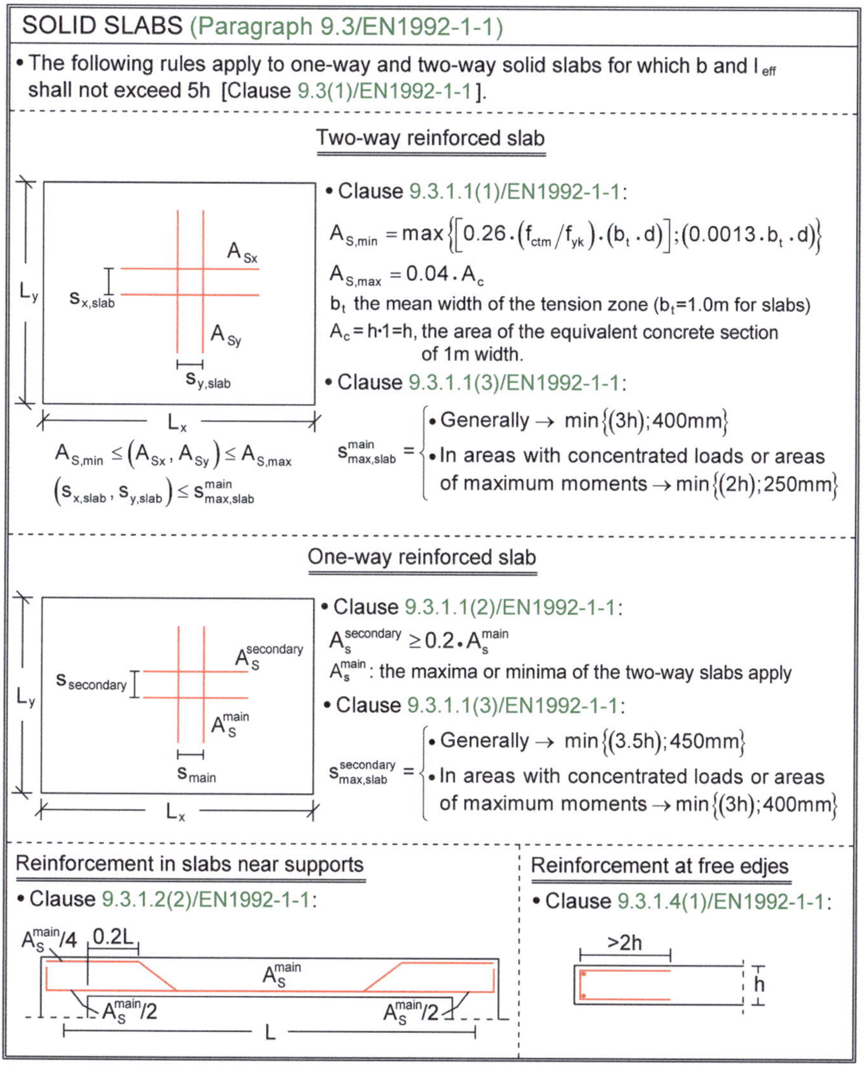

C.2 Beams

C.2.1 Sectional Dimensions

EN1998-1 requirements for the sectional dimensions of beams and for beam-column joint eccentricities are presented in the following figure.

SECTION DIMENSIONS

DCM	DCH
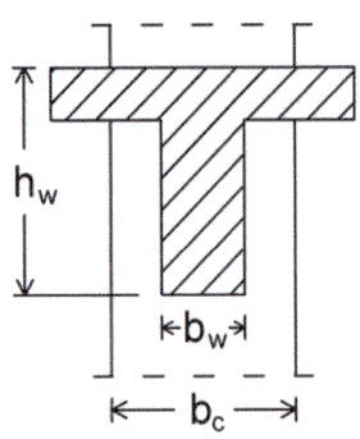	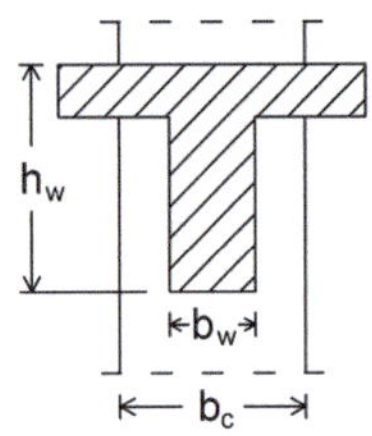
$b_w \leq \min\{(b_c + h_w); (2b_c)\}$ [Clause 5.4.1.2.1(3)P]	$b_w \leq \min\{(b_c + h_w); (2b_c)\}$ [Clause 5.5.1.2.1(5)P] $b_w \geq 0.20\text{m}$ [Clause 5.5.1.2.1(1)P] $h_w / b_w \geq 3.50^*$ [Clause 5.5.1.2.1(2)P]

JOINT ECCENTRICITY TO THE COLUMNS

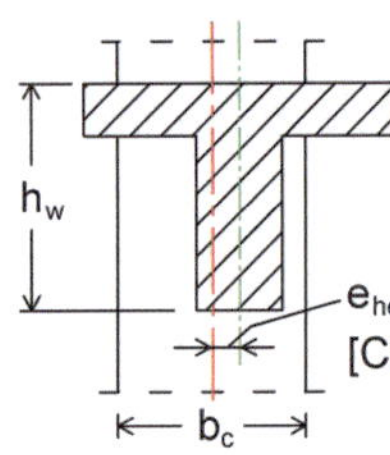

$e_{hor} < b_c/4$ [Clause 5.4.1.2.1(2)]

RULES FOR BEAMS SUPPORTING COLUMNS IN THE UPPER STOREY

- Construction of columns supported on beams is prohibited [Clause 5.5.1.2.2(2)].
- No eccentricity is allowed between the beam and column (or ductile wall) axes [Clause 5.4.1.2.5(2)P(a), Clause 5.5.1.2.4(2)P].
- The beam shall be supported on at least two indirect supports (columns or ductile walls) [Clause 5.4.1.2.5(2)P(b), Clause 5.5.1.2.4(2)P].

* This limitation is imposed in clause 5.9(3) of EN1992-1-1 (eq. 5.40b) to avoid verification against 2nd order effects related to distortion

C.2.2 Longitudinal Reinforcement

Minimum and maximum requirements for the longitudinal reinforcement of beams and their critical regions are prescribed in clauses 5.4.3.1.2(1)P-(5)P and 5.5.3.1.3 (1)P-(5)P of EN1998-1, as illustrated in the following figure.

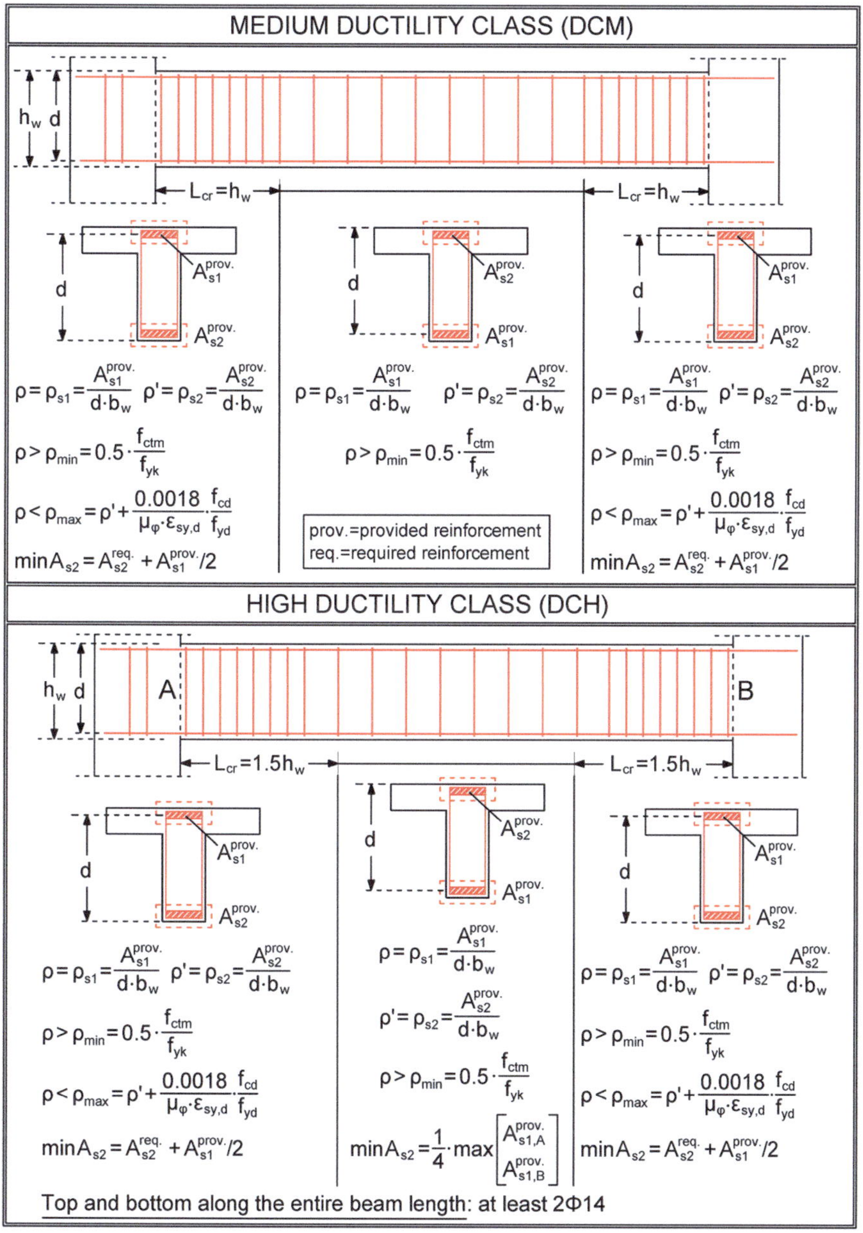

C.2.3 Shear Reinforcement

Minimum and maximum requirements for the shear reinforcement of beams and their critical regions are prescribed in clauses 5.4.3.1.2(6)P and 5.5.3.1.3(6)P of EN1998-1, as illustrated in the following figure.

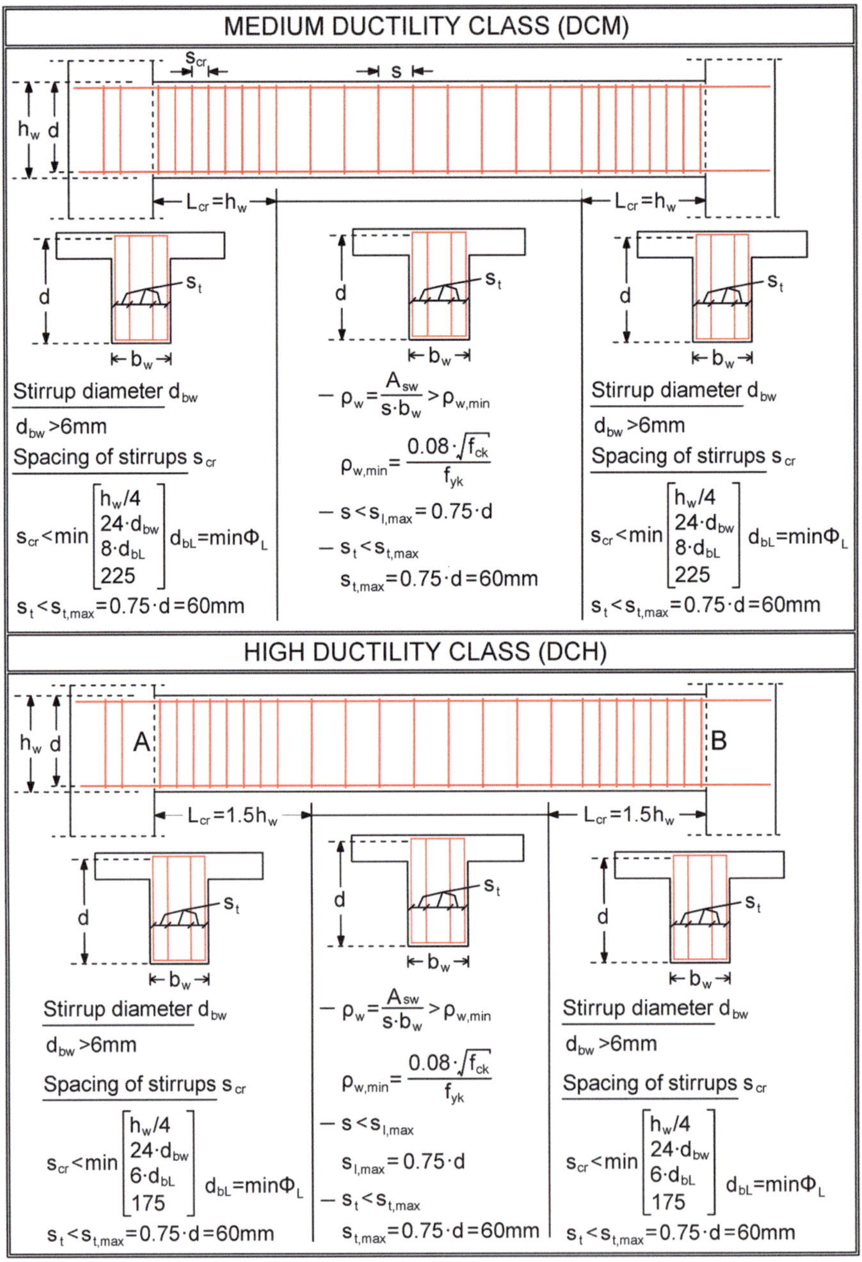

C.3 Columns

C.3.1 Sectional Dimensions

EN1998-1 requirements for the sectional dimensions of columns are illustrated in the following figure.

	DCM	DCH
SECTION DIMENSIONS	Clause 5.4.1.2.2(1) and 5.5.1.2.2(2): For principal earthquake resistant columns provided that θ>0.1 (Definition of θ: Clause 4.4.2.2(2)) NO restriction when θ<0.1. 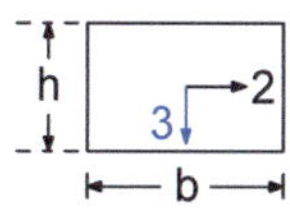 $b \geq \frac{1}{10} \cdot \max\{L_{u,2}; L_{d,2}\}$ $h \geq \frac{1}{10} \cdot \max\{L_{u,3}; L_{d,3}\}$ 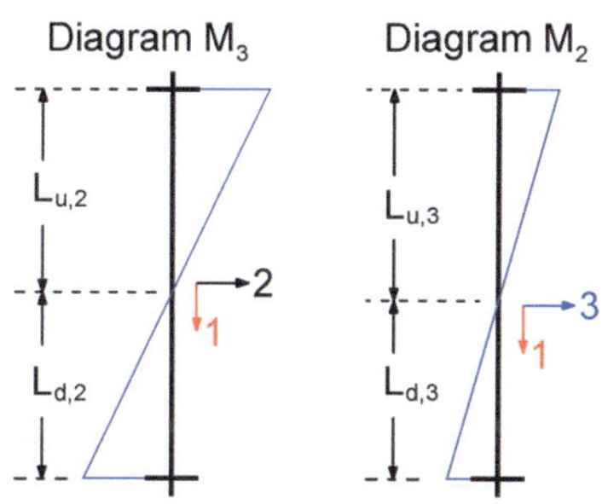	
	—	Clause 5.5.1.2.2(1)P: b≥25cm h≥25cm For principal earthquake resistant columns
MINIMUM SECTION AREA	Clause 5.4.3.2.1(3)P For principal earthquake resistant columns: $v_d = \frac{N_{Ed}}{A_c \cdot f_{cd}} \leq 0.65$ $\Rightarrow A_c \geq \frac{N_{Ed}}{0.65 \cdot f_{cd}}$	Clause 5.5.3.2.1(3)P For principal earthquake resistant columns: $v_d = \frac{N_{Ed}}{A_c \cdot f_{cd}} \leq 0.55$ $\Rightarrow A_c \geq \frac{N_{Ed}}{0.55 \cdot f_{cd}}$

C.3.2 Longitudinal Reinforcement

Minimum and maximum requirements for the longitudinal reinforcement of columns and their critical regions are prescribed in clauses 5.4.3.2.2(1)P-(2)P and 5.5.3.2.2(1)P-(2)P of EN1998-1, as illustrated in the following figure.

	DCM	DCH
(1)	Minimum/Maximum amount of total longitudinal reinforcement	
	$\rho_{min} = 0.01 \leq \rho = \left(\frac{A_{s,tot}}{b \bullet h}\right) \leq \rho_{max} = 0.04$	
	Clause 5.4.3.2.2(1)P	Clause 5.5.3.2.2(1)P
(2)	Minimum number of rebars per side =3	
	Clause 5.4.3.2.2(1)P	Clause 5.5.3.2.2(2)P
(3)	Maximum allowable distance between longitudinal rebars restrained by transverse reinforcement (ties). 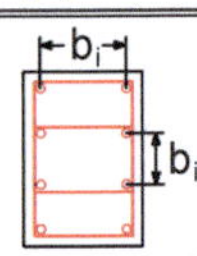	
	b_i=20cm	b_i=15cm
	Clause 5.4.3.2.2(11)	Clause 5.5.3.2.2(12)
(4)	Minimum diameter of longitudinal bars: $\min\Phi_L$=8mm	
	Clause 9.5.2(1) / EN1992-1-1	
(5)	Critical length	
	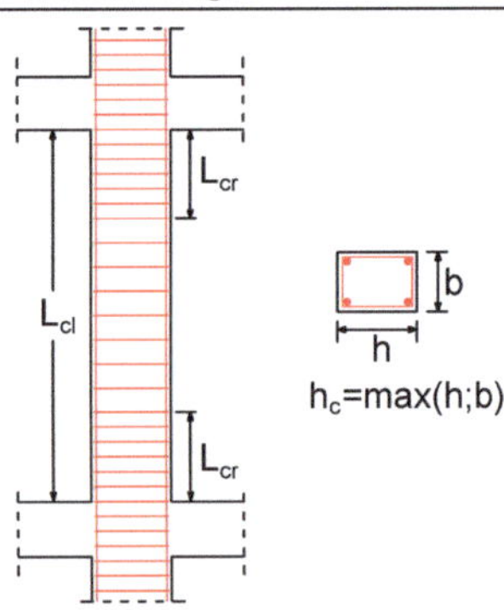	
	$L_{cr}=\max\{h_c; (L_{cl}/6); 0.45\}$ (m) Clause 5.4.3.2.2(4)	$L_{cr}=\max\{(1.5 \cdot h_c); (L_{cl}/6); 0.60\}$ (m) Clause 5.5.3.2.2(4)
	$*$ If $L_{cl}/h_c<3 \Rightarrow L_{cr}=L_{cl}$	
	Clause 5.4.3.2.2(5)P	Clause 5.5.3.2.2(5)P

Clause 5.5.3.2.2(14): The longitudinal reinforcement ratio prescribed at the base of the lower storey columns of DCH buildings (wherein the column is connected to the foundation) shall not be smaller than that prescribed at the head of the column in the same storey.

C.3.3 Shear and Confining Reinforcement

Minumum and maximum requirements for the shear reinforcement columns and their critical regions are prescribed in clauses 5.4.3.2.2(10)P-(11)P and 5.5.3.2.2 (11)P-(12)P of EN1998-1, as illustrated in the following figure.

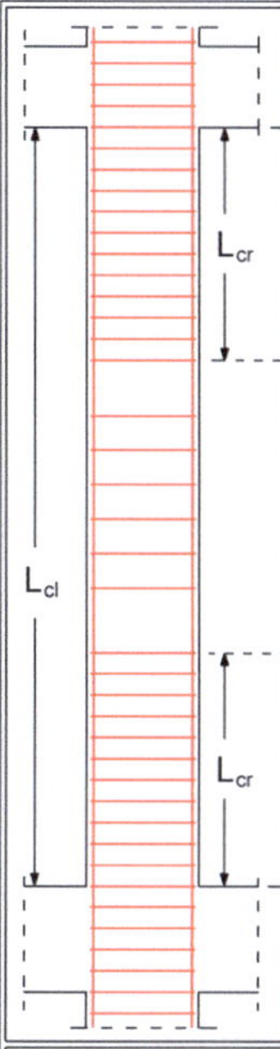

	DCM	DCH
Critical region (L_{cr}, top)	Hoop size d_{bw} $d_{bw}>\max\{d_{bL}^{max}/4; 6mm\}$ Spacing of hoops s_{cr} $s_{cr}<\{b_0/2; 8d_{bL}^{min}; 175mm\}$	Hoop size d_{bw} $d_{bw}>\max\{0.4\, d_{bL}^{max}(f_{ydL}/f_{ydw})^{1/2}; 6mm\}$ Spacing of hoops s_{cr} $s_{cr}<\{b_0/3; 6d_{bL}^{min}; 125mm\}$
Outside critical regions	Hoop size d_{bw} $d_{bw}>\max\{d_{bL}^{max}/4; 6mm\}$ Spacing of hoops s $s<\{h_c; 20d_{bL}^{min}; 400mm\}$	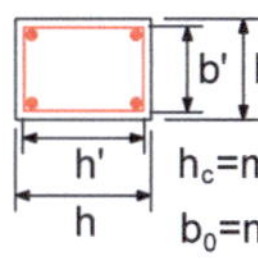$h_c=\min\{h;b\}$ $b_0=\min\{b';h'\}$ $d_{bL}^{max}=\max\Phi_L$ $d_{bL}^{min}=\min\Phi_L$
Critical region (L_{cr}, bottom)	Hoop size d_{bw} $d_{bw}>\max\{d_{bL}^{max}/4; 6mm\}$ Spacing of hoops s_{cr} $s_{cr}<\{b_0/2; 8d_{bL}^{min}; 175mm\}$	Hoop size d_{bw} $d_{bw}>\max\{0.4\, d_{bL}^{max}(f_{ydL}/f_{ydw})^{1/2}; 6mm\}$ Spacing of hoops s_{cr} $s_{cr}<\{b_0/3; 6d_{bL}^{min}; 125mm\}$
		Hoops are provided along length $(L_{cr}+L_{cr}/2)$ as within the critical length in the two lower storeys [5.5.3.2.2(13)P]
Mechanical volumetric ratio of confining hoops ω_{wd}	a) In the critical length of primary seismic column at the base	
	$\alpha\omega_{wd}\geq 30\mu_{\varphi}v_d\varepsilon_{sy,d}(b_c/b_o)-0.035$	
	$\min\omega_{wd}=0.08$ [Clause 5.4.3.2.2(8), (9)]	$\min\omega_{wd}=0.12$ [Clause 5.5.3.2.2(9), (10)]
	b) In the critical length of primary seismic column above the base	
	——	$\alpha\omega_{wd}\geq 30\mu_{\varphi}^{*}v_d\varepsilon_{sy,d}(b_c/b_o)-0.035$ $\min\omega_{wd}=0.08$ [Clause 5.5.3.2.2(9), (10)]

Notes:

(1) Minimum requirements for non-critical regions are provided in paragraph 9.5.3 of EN1992-1-1.

(2) The maximum permissible distance s between ties within non-critical regions shall be reduced to 60 % of the value provided in the figure above:

  (i) in column subsections of length equal to the greatest dimension of the section above or below a beam or slab.

  (ii) along the lap splice length, provided that the maximum diameter of the longitudinal reinforcement is greater than 14 mm. A minimum of 3 ties shall be provided in equal distances within the lap splice length.

Curvature ductility μ_{φ} shall be calculated on the basis of clause 5.2.3.4(3), while μ_{φ}* shall be derived according to 5.2.3.4(3) and 5.5.3.2.2(7).

C.4 Ductile Walls

C.4.1 Sectional Dimensions

EN1998-1 requirements for the sectional dimensions of ductile walls are presented in the following figure.

Definitions	Section squat ratio (length/width): $l_w / b_{w0} \geq 4.0$ [Clause 5.1.2(1)] 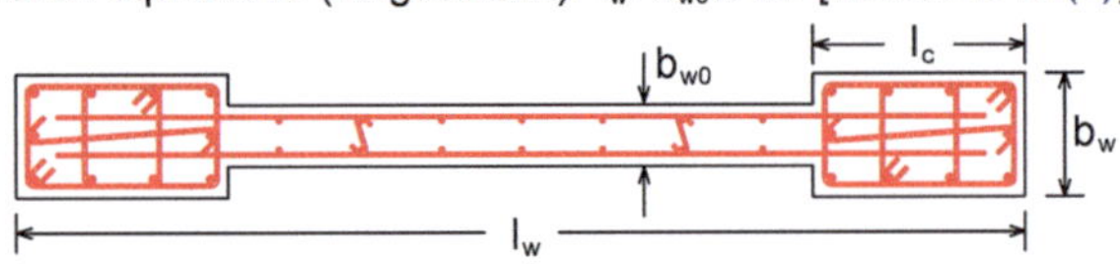	
	DCM	**DCH**
Minimum area of the cross-section	$v_d = \frac{N_{Ed}}{A_c \cdot f_{cd}} \leq 0.4 \Rightarrow A_c \geq \frac{N_{Ed}}{0.4 \cdot f_{cd}}$ Clause 5.4.3.4.1(2)	$v_d = \frac{N_{Ed}}{A_c \cdot f_{cd}} \leq 0.35 \Rightarrow A_c \geq \frac{N_{Ed}}{0.35 \cdot f_{cd}}$ Clause 5.5.3.4.1(2)
Minimum thickness of the web	$b_{w0} \geq \max\{0.15; (h_s/20)\}$ Clause 5.4.3.4.1(2)	Clause 5.5.3.4.1(2)
Minimum length and width of confined boundary elements	$\min l_c = \max\{(0.15 \cdot l_w); (1.50 \cdot b_w)\}$ $\min b_w = \max\{0.2; (h_s/15)\}$ if $l_c \leq \max\{(2 \cdot b_w); (0.2 \cdot l_w)\}$ $\min b_w = \max\{0.2; (h_s/10)\}$ if $l_c > \max\{(2 \cdot b_w); (0.2 \cdot l_w)\}$ Clause 5.4.3.4.2(6) ⟶ $\min l_c$ Clause 5.4.3.4.2(10) ⟶ $\min b_w$	Clause 5.5.3.4.5(6) ⟶ $\min l_c$ Clause 5.5.3.4.5(8) ⟶ $\min b_w$
Length of the critical region	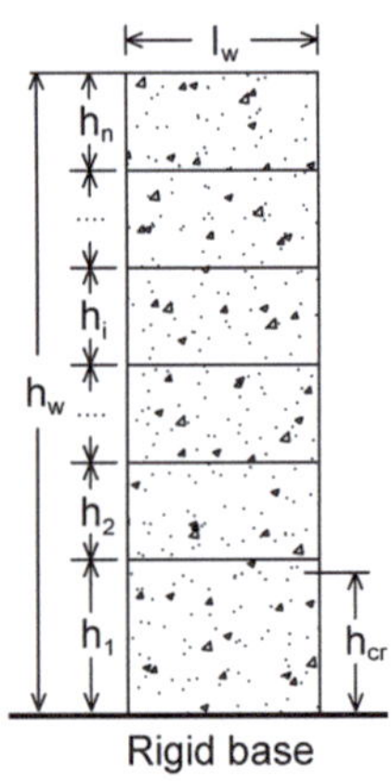 $h_{cr} \geq \max\{l_w; (h_w/6)\}$ and $h_{cr} \leq \begin{cases} \min\{2 L_w; h_s\} & \text{if } n \leq 6 \\ \min\{2 L_w; 2 \cdot h_s\} & \text{if } n \geq 7 \end{cases}$ NOTATIONS ▪ h_s=Clear storey height ▪ In walls which continue (with the same cross section) to a box-type basement, the critical region should be taken to extend below the basement roof level up to a depth of h_{cr} [Clause 5.8.1(5)] Clause 5.4.3.4.2(1)	Clause 5.5.3.4.5(1)

C.4.2 Web Reinforcement

EN1998-1 requirements concerning the web reinforcement of ductile walls are presented in the following table.

		DCM	DCH
Critical region	Horizontal reinforcement	$\rho_{h,min} = \max\{1‰; 0.25\rho_v\}$ Clause 9.6.3(1) / EN1992-1-1	$\rho_{h,min} = 2‰$ Clause 5.5.3.4.5(13)
Critical region	Horizontal reinforcement	$s_h \leq 400mm$ Clause 9.6.3(2) / EN1992-1-1	$8mm \leq d_{bh} \leq (b_{w0}/8)$ $s_h \leq \min\{250mm; 25 \cdot d_{bh}\}$ Clause 5.5.3.4.5(15)
Critical region	Vertical reinforcement	$\rho_{v,min} = 2‰$ Clause 9.6.2(1) / EN1992-1-1	$\rho_{v,min} = 2‰$ Clause 5.5.3.4.5(13)
Critical region	Vertical reinforcement	$\rho_{v,max} = 4\%$ Clause 9.6.2(1) / EN1992-1-1	
Critical region	Vertical reinforcement	$s_v \leq \min\{400mm; 3 \cdot b_{w0}\}$ Clause 9.6.2(3) / EN1992-1-1	$8mm \leq d_{bv} \leq (b_{w0}/8)$ $s_v \leq \min\{250mm; 25 \cdot d_{bv}\}$ Clause 5.5.3.4.5(15)
All storeys above the critical region	Horizontal reinforcement	As in the critical region Clause 9.6.3(1) / EN1992-1-1 Clause 9.6.3(2) / EN1992-1-1	As in the critical region Clause 5.5.3.4.5(13) Clause 5.5.3.4.5(15)
All storeys above the critical region	Vertical reinforcement	As in the critical region Clause 9.6.2(1) / EN1992-1-1 Clause 9.6.2(3) / EN1992-1-1	As in the critical region Clause 5.5.3.4.5(13) Clause 5.5.3.4.5(15)

C.4.3 Reinforcement of the Confined Boundary Elements

EN1998-1 requirements concerning the reinforcement of the confined boundary elements of ductile walls are illustrated in the following table.

<table>
<tr><th></th><th></th><th>DCM</th><th>DCH</th></tr>
<tr><td rowspan="4">Critical region</td><td rowspan="2">Longitudinal reinforcement</td><td colspan="2">$\rho_{v,min} = A_{sv}/(l_c \cdot b_w) = 0.5\%$ (1)</td></tr>
<tr><td>Clause 5.4.3.4.2(8)</td><td>Clause 5.5.3.4.5(7)</td></tr>
<tr><td></td><td colspan="2">$\rho_{v,max} = A_{sv}/(l_c \cdot b_w) = 4.0\%$
Clause 9.6.2(1) / EN1992-1-1</td></tr>
<tr><td>Confining reinforcement</td><td>$\omega_{wd}=0.08$ [Clause 5.4.3.4.2(9)]
$s_{cr} \leq \{b_w/2; (8 \cdot d_{bL}^{min}); 175mm\}^{(2)}$
Clause 5.4.3.4.2(9)
$\alpha \cdot \omega_{wd} \geq 30 \cdot \mu_\varphi \cdot (v_d + \omega_v) \cdot \varepsilon_{sy,d} \cdot (b_w/b_0) - 0.035$ (4)
[Clause 5.4.3.4.2(4)]</td><td>$\omega_{wd}=0.12$ [Clause 5.4.3.4.2(9)]
$d_{bw} > \max\{0.4 \cdot d_{bL}^{max} \cdot (f_{ydL}/f_{ydw})^{1/2}; 6mm\}^{(2)}$
$s_{cr} < \{b_{0w}/3; (6 \cdot d_{bL}^{min}); 125mm\}^{(2),(3)}$
Clause 5.5.3.4.5(10)
$\alpha \cdot \omega_{wd} \geq 30 \cdot \mu_\varphi \cdot (v_d + \omega_v) \cdot \varepsilon_{sy,d} \cdot (b_w/b_0) - 0.035$ (4)
[Clause 5.5.3.4.5(4)]</td></tr>
<tr><td rowspan="2">First storey above the critical region</td><td>Longitudinal reinforcement</td><td>As in the <u>non</u>-critical region</td><td>As in the critical region
[Clause 5.5.3.4.5(11)]</td></tr>
<tr><td>Confining reinforcement</td><td>As in the <u>non</u>-critical region</td><td>As in the critical region
[Clause 5.5.3.4.5(11)]
<u>Exception</u>: Reduction of $\alpha \cdot \omega_{wd}$ by 50% is permitted</td></tr>
<tr><td rowspan="4">Non-critical region</td><td rowspan="2">Longitudinal reinforcement</td><td colspan="2">$\rho_{v,min} = 0.2\% \; [\text{If } \varepsilon_c > 2‰ \rightarrow \rho_{v,min} = 0.5\%]$</td></tr>
<tr><td>[Clause 5.4.3.4.2(11)]</td><td>[Clause 5.5.3.4.5(12)]</td></tr>
<tr><td rowspan="2">Confining reinforcement</td><td colspan="2">Reinforcement according to paragraph 9.6.4 / EN1992-1-1</td></tr>
<tr><td>[Clause 5.4.3.4.2(11)]</td><td>[Clause 5.5.3.4.5(12)]</td></tr>
</table>

(1) According to clauses 5.4.3.4.2(9) and 5.4.3.2.2(11)b: The distance between successive longitudinal rebars restrained by ties shall not exceed 200mm.

(2) Where d_{bL} is the diameter (maximum/minimum) of the longitudinal bars of the edge

(3) Where b_{0w} is the minimum dimension of the concrete core at the edge (inside of the hoops)

(4) Curvature ductility μ_φ can be derived according to clause 5.4.3.4.2(2).

C.5 Beam-Column Joints

EN1998-1 prescribes specific measures for ensuring the seismic capacity of beam-column joints. This is achieved through the provided confining reinforcement, as shown in the following figure. Particularly, for joints of DCM buildings, the relevant guidelines are provided in paragraph 5.4.3.3, while DCH building joints are covered in paragraphs 5.5.2.3 and 5.5.3.3.

DCM Paragraph 5.4.3.3	DCH Paragraphs 5.5.2.3 and 5.5.3.3
• No calculation is required for horizontal confining reinforcement. • Confining reinforcement shall not be less than that derived according to clause 5.4.3.2.2(8)-(11) within the critical regions of the columns. 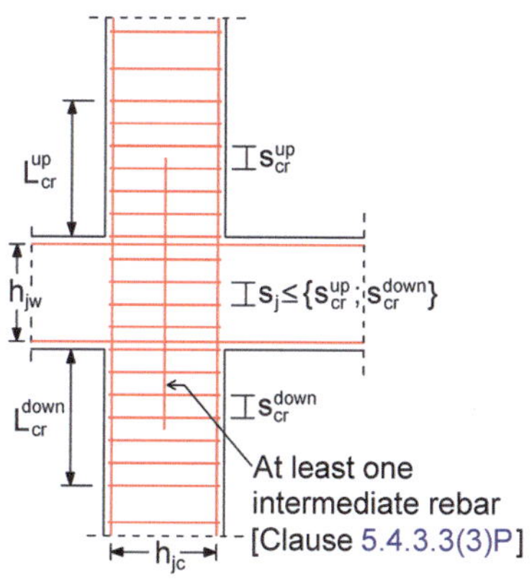 EXCEPTION [Clause 5.4.3.3(2)]: If: 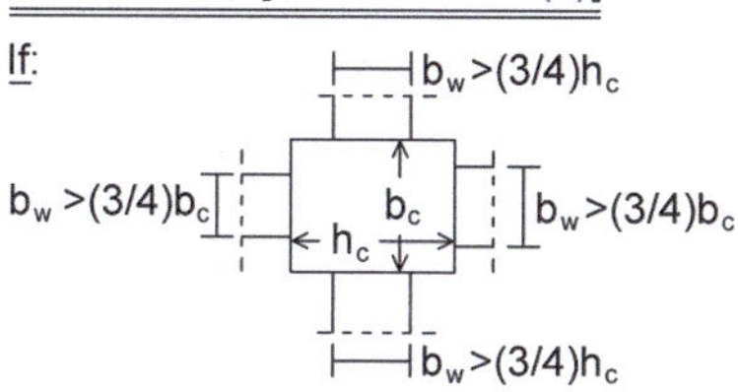 Then: $s_j < 2\{s_{cr}^{up}; s_{cr}^{down}\}$ $s_j < 150\text{mm}$	• Calculation for horizontal confining reinforcement is required. • Total required area of closed horizontal stirrups A_{sh} shall be calculated according to clause 5.5.3.3(3): $A_{sh} \geq \left\{ \frac{\left[V_{jhd}/(b_j \cdot h_{jc})\right]^2}{f_{ctd} + \nu_d \cdot f_{cd}} - f_{ctd} \right\} \cdot \left(\frac{b_j \cdot h_{jw}}{f_{ywd}} \right)$ V_{jhd} the horizontal shear force acting on the concrete core of the joint [Clause 5.5.2.3(2)] h_{jw} the distance between beam top and bottom reinforcement. h_{jc} the distance between extreme layers of column reinforcement in a beam-column. b_j the effective joint width: $b_j=\min\{b_c;(b_w+0.5h_c)\}$ IF $b_c>b_w$ $b_j=\min\{b_w;(b_c+0.5h_c)\}$ IF $b_c<b_w$ • Horizontal hoops with a diameter of not less than 6 mm within the joint [5.5.3.3(3)] • The horizontal hoops should be uniformly distributed within the depth h_{jw} [5.5.3.3(5)]. • Adequate vertical reinforcement of the column passing through the joint should be provided, so that [Clause 5.5.3.3(6)]: $A_{sv,i} \geq (2/3) \cdot A_{sh} \cdot (h_{jc}/h_{jw})$ A_{sh} the required total area of the horizontal hoops in accordance with clause 5.5.3.3(3) • All requirements presented in 5.4.3.3 (see side column) regarding beam-column joints in DCM buildings shall be also satisfied [Clause 5.5.3.3(7)-(9)P].

C.6 Foundation Elements

EN1998-1 and EN1992-1-1 requirements concerning the sectional dimensions and the reinforcement of foundation elements are illustrated in the following figure.

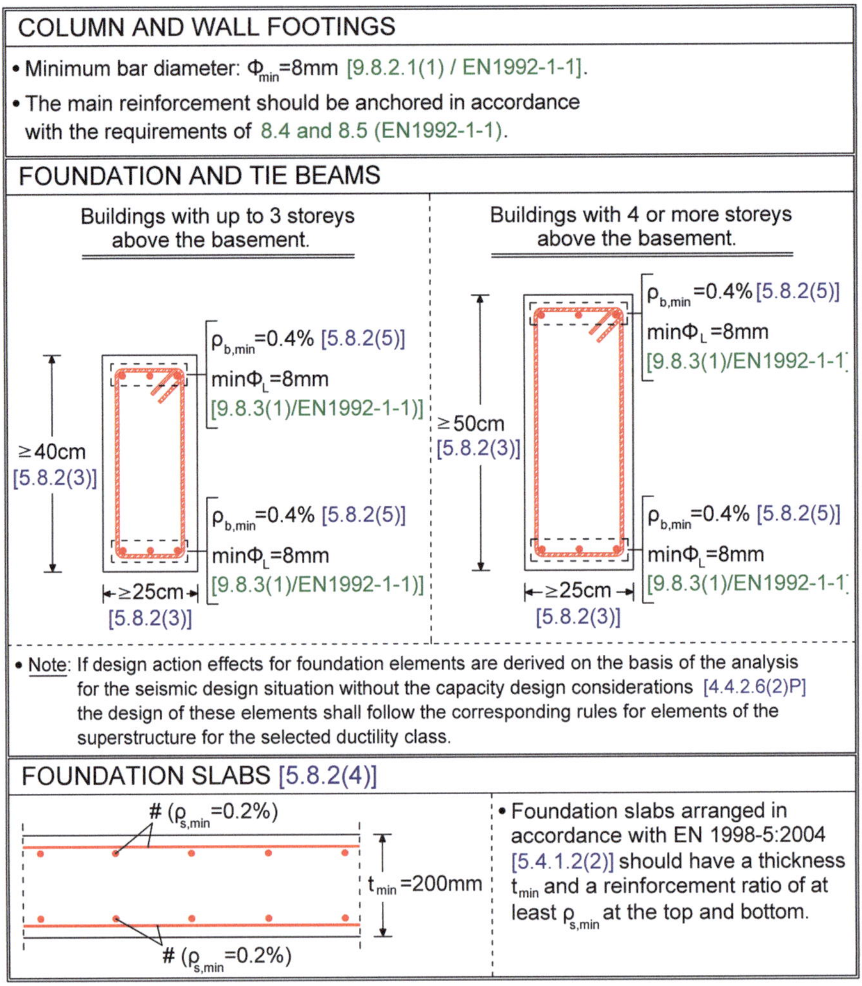

Index

I. Avramidis et al., *Eurocode-Compliant Seismic Analysis and Design of R/C Buildings*, Geotechnical, Geological and Earthquake Engineering 38,
DOI 10.1007/978-3-319-25270-4

MIX
Papier aus verantwortungsvollen Quellen
Paper from responsible sources
FSC® C105338
FSC
www.fsc.org

If you have any concerns about our products,
you can contact us on
ProductSafety@springernature.com

In case Publisher is established outside the EU,
the EU authorized representative is:
Springer Nature Customer Service Center GmbH
Europaplatz 3, 69115 Heidelberg, Germany

Printed by Libri Plureos GmbH
in Hamburg, Germany